Introduction to Geomicrobiology

In memory of my mother

Introduction to Geomicrobiology

Kurt Konhauser

Blackwell Publishing

© 2007 Blackwell Science Ltd
a Blackwell Publishing company

BLACKWELL PUBLISHING
350 Main Street, Malden, MA 02148-5020, USA
9600 Garsington Road, Oxford OX4 2DQ, UK
550 Swanston Street, Carlton, Victoria 3053, Australia

The right of Kurt Konhauser to be identified as the Author of this Work has been asserted in accordance with the
UK Copyright, Designs, and Patents Act 1988.

First published 2007 by Blackwell Science Ltd

1   2007

Library of Congress Cataloging-in-Publication Data

Konhauser, Kurt.
    Introduction to geomicrobiology / Kurt Konhauser.
      p. cm.
    Includes bibliographical references and index.
    ISBN-13: 978-0-632-05454-1 (pbk. : alk. paper)
    ISBN-10: 0-632-05454-9 (pbk. : alk. paper)
  1. Geomicrobiology.   I. Title.

QR103.K66 2006
579—dc22

                2005015459

A catalogue record for this title is available from the British Library.

Set in 10.5/12.5pt Goudy Old Style
by Graphicraft Limited, Hong Kong

The publisher's policy is to use permanent paper from mills that operate a sustainable forestry policy,
and which has been manufactured from pulp processed using acid-free and elementary chlorine-free
practices. Furthermore, the publisher ensures that the text paper and cover board used have met
acceptable environmental accreditation standards.

For further information on
Blackwell Publishing, visit our website:
www.blackwellpublishing.com

Contents

Preface

The past three decades have seen enormous advances in our understanding of how microbial life impacts the Earth. Microorganisms inhabit almost every conceivable environment on the planet's surface, and extend the biosphere to depths of several kilometers into the crust. Significantly, the chemical reactivity and metabolic diversity displayed by those communities makes them integral components of global elemental cycles, from mineral dissolution and precipitation reactions, to aqueous redox processes. In this regard, microorganisms have helped shape our planet over the past 4 billion years and made it habitable for higher forms of life.

Introduction to Geomicrobiology was written in response to the need for a single, comprehensive book that describes our current knowledge of how microbial communities have influenced biogeochemical and mineralogical processes through time. Though other books on geomicrobiology exist, they are either geared more towards the physiological rather than the geological aspects, or they are edited compilations where the individual papers are highly focused on select topics. This has limited the usefulness of those resources in teaching geomicrobiology to my students. Consequently, *Introduction to Geomicrobiology* has taken a different approach. It is process-oriented, with one pervasive theme throughout, that being the geological consequences of microbial activity. I take an interdisciplinary and "global" view on the issues, and although the chapters can stand alone, they are linked through a natural progression of microorganism–environment interactions in modern ecosystems, concluding with a synopsis of how the biosphere evolved during the Precambrian. Designed primarily as a core text for upper undergraduate/graduate students in earth and biological sciences, *Introduction to Geomicrobiology* is also rich in references, with the aim being to provide the reader with a useful starting point for further study. In doing this, particular attention has been paid to making this book a useful resource for researchers from a number of other scientific backgrounds.

Introduction to Geomicrobiology covers the following topics:

- How microorganisms are classified, the physical constraints governing their growth, molecular approaches to studying microbial diversity, and life in extreme environments.

- Bioenergetics, microbial metabolic capabilities, and major biogeochemical pathways.

- Chemical reactivity of the cell surface, metal sorption, and the microbial role in contaminant mobility and bioremediation/biorecovery.

- Microbiological mineral formation and fossilization.

- The function of microorganisms in mineral dissolution and oxidation, and the industrial and environmental ramifications of these processes.

- Elemental cycling in biofilms, formation of microbialites, and sediment diagenesis.

- The events that led to the emergence of life, evolution of metabolic processes, and the diversification of the biosphere.

One feature this book does not have is a methods section. There are many techniques employed in such an interdisciplinary science, and it is my feeling that since they are constantly evolving it is more appropriate for the reader to consult the most recent articles where the methodologies are fully explained.

The preparation of this book was greatly aided by discussions and reviews with a number of colleagues. They include Jill Banfield, Roger Buick, Jeremy Fein, Bill Inskeep, Andreas Kappler, Jon Lloyd, Anna-Louise Reysenbach, Sam Smith, Gordon Southam, and Nathan Yee. Special thanks go out to two of my students, Stefan Lalonde and Larry Amskold, for their help with figures and proofreading. I am also indebted to a number of other colleagues and publishers for making available original photographs or allowing reproduction of previously published material for illustration. Lastly, but most importantly, I owe my wife and son many hours of undivided attention for all the time missed during the writing of this book.

I am grateful to all the individuals and publishers who have given permission to reproduce material in this book.

Kurt O. Konhauser
University of Alberta, Edmonton

For instructors, if you did not receive an artwork CD-ROM with your comp copy, please contact this email address: artworkcd@bos.blackwellpublishing.com

1

Microbial properties and diversity

Since their origin, perhaps some 4 billion years ago, microorganisms have had a profound influence on shaping our planet. From localized niches, that occur on the order of micrometers, to ecosystems as immense as the oceans, microorganisms are intimately involved in transforming inorganic and organic compounds to meet their nutritional and energetic needs. Because the metabolic waste from one type of species nearly always provides substrates for another, there is a continuous recycling of elements throughout the biosphere. This interdependence can exist between species growing in close proximity to one another, where any number of sorption, precipitation, and redox reactions inevitably create unique community-specific biogeochemical and mineralogical signatures. Alternatively, the communities can be spatially separated, and elemental cycling may take on more complex and convoluted pathways, such as the transfer of metabolites across the sediment–water interface or from the ocean water column to the atmosphere. The latter examples are particularly important for global-scale cycling of carbon, nitrogen, sulfur, and oxygen. Given sufficient time, the collective metabolic activities of countless microbial communities can even modify the dynamics of the entire Earth, controlling the composition of the air we breathe, the water we drink, and the soils in which we grow plants. So intimate is this relationship that in order to discover a part of the planet not fundamentally affected by microorganisms, it would be necessary to penetrate deep into the crust where temperatures are outside physiochemical limits

for life. From the early recognition that microbial activity affects the environment, and vice versa, was born the field of Geomicrobiology. As this book is primarily concerned with the study of microbially influenced biogeochemical and mineralogical reactions at present and through time, we begin by briefly examining some of the characteristic properties of microorganisms and their overall diversity in the biosphere.

1.1 Classification of life

Since the early eighteenth century, and the first formal taxonomic classification of living organisms by Carolus Linnaeus, a number of attempts have been made to categorize the different life forms according to morphological and nutritional similarities. Traditionally, all organisms were grouped into the kingdoms Animalia or Plantae. However, with the recognition of the immense diversity of microscopic organisms, Whittaker (1969) proposed a five-kingdom classification system that subsequently added Fungi, Protista, and Monera at the kingdom level. The Monera contained the bacteria that, based on their simplistic prokaryotic cell structure, were distinguished from the more complex eukaryotic organisms. The Protista contained all eukaryotes that were neither animals, plants, or fungi – included algae, protozoa, slime molds, and various other primitive microorganisms. According to Whittaker, the Monera were the first organisms on Earth, and the Protista evolved directly from them. Fungi, Plantae, and Animalia evolved from the Protista via three separate directions of evolution based on differences in how the organisms met their nutritional needs. Fungi evolved as the most complex multicellular organisms that still obtained their nutrients by absorption; animals evolved based on their ability to ingest other organisms; while plants evolved a photosynthetic pathway to synthesize their own organic compounds.

At around the same time as the Whittaker classification scheme was being implemented, others recognized that comparative analyses of specific genes could be used to infer the evolutionary pathways of organisms (e.g., Zuckerkandl and Pauling, 1965). This approach, termed molecular phylogeny, is based on the number and location of differences in the nucleotide sequences of homologous genes from different organisms. This, in turn, provides an indication of the evolutionary relatedness between those organisms because the number of sequence differences identified is proportional to the stable mutational changes that accumulate in a sequence with time. Thus, two closely related species will have accumulated few differences in a sequence shared in common, whereas two more distantly related organisms will have accumulated many more differences in the same sequence.

In order to determine evolutionary relationships it is essential that the gene chosen for sequencing studies: (i) be universally distributed between organisms; (ii) must be properly aligned so regions of homology can be identified; and (iii) have a sequence that changes at a rate commensurate with the evolutionary distance being measured (Olsen et al., 1986). There are several evolutionary conserved genes that are present in most organisms, one of which, however, is particularly important because it encodes a specific component of the protein-synthesizing ribosomes, the ribosomal RNA (rRNA) molecule. In prokaryotic organisms this is known as the 16S ribosomal RNA. In eukaryotes, where the equivalent molecule is slightly larger, it is called 18S ribosomal RNA. Due to the fundamental role they play, rRNA molecules have remained

structurally and functionally conserved. However, some portions of the molecule are more important to its function than others, and, as a result, these regions of the molecule are almost identical in sequence from the smallest bacterium to the largest animal. By contrast, regions that have a lesser role to play in the functionality of the molecule have very different sequences in different organisms. This has produced a molecule that consists of a mosaic of sequences that are highly conserved in the rRNAs of all organisms and sequences that can be extremely variable.

When rRNA sequences from different organisms are compared with each other, the number of changes in the sequence can be used to create a phylogenetic tree, with closely spaced "branches" symbolizing species that are genetically similar. There are a number of procedures used to construct phylogenetic trees, but the first stage in these analyses is always the careful alignment of the rRNA nucleotide sequences. This is a relatively straightforward task for regions that have a highly conserved sequence, yet it is considerably more problematic in regions of greater sequence variability. The mutations leading to sequence variability include additions, deletions, and reversions, further complicating the matter. Once the rRNA sequences have been aligned, phylogenetic trees of relatedness can be constructed using a number of different approaches, including distance, parsimony, maximum likelihood, and Bayesian algorithms/methods (see Schleifer and Ludwig, 1994 for details).

In their pioneering work, Woese and Fox (1977) and Fox et al. (1980) used single gene 16S rRNA sequences to determine the phylogenetic relatedness between many different organisms. What they showed was that all life could be grouped into three major domains (a term that supplanted the use of kingdom), the *Bacteria*, *Archaea*, and *Eukarya*, and that these domains diverged from a "root," a graphical representation of a point in evolutionary time when all extant life on Earth had a common ancestor, also known as the Last Universal Common Ancestor, or LUCA (Fig. 1.1). The comparative sequence analyses further allowed the definition of other major lineages (phyla, class, and order) within the three domains (Woese, 1987). Significantly, the new taxonomic system showed that life is dominated by microbial forms – *Bacteria* and *Archaea* are entirely prokaryotic, while the majority of the phyla in *Eukarya* are also microbial. The "higher" organisms, the plants and animals, are relegated to the peripheral branches of the tree. The degrees of variation in microbial diversity are similarly astounding, with the evolutionary distances between the three domains being much more profound than those that distinguish the traditional kingdoms from each other (Woese et al., 1990). So, in terms of evolutionary distance, humans are only slightly removed from plants or fungi, while the distances between seemingly similar bacteria are considerably greater – in this book the term *bacteria* written with a lowercase "b" is synonymous with the term prokaryote, not the domain.

Although it is presently impossible to calibrate sequence changes against geological time because different lines of descent have evolved at different rates (i.e., sequence change is nonlinear), the universal phylogenetic tree does give a relative timing for major evolutionary change. For example, the *Eukarya* are more closely related to *Archaea* than *Bacteria* (Iwabe et al., 1989). This suggests that *Archaea* and *Eukarya* had a common history that excluded the descendants of the bacterial line. What is more, the major organelles of eukaryotes – the mitochondria and chloroplasts – are derived from bacterial partners that had undergone specialization through co-evolution within a host archaeal cell (see section 7.5.2 for details). Because *Eukarya* contain both bacterial and archaeal parts, their origins would appear to be very ancient (for an opposing view see Cavalier-Smith, 2002). This hypothesis, however, creates problems in trying to root the universal phylogenetic tree because we cannot be sure whether the first divergences in the tree separate: (i) *Bacteria* from a line that was to produce *Archaea* and *Eukarya* (as shown in Fig. 1.1); or (ii) a proto-eukaryotic lineage from

fully developed bacterial and archaeal lineages (Doolittle and Brown, 1994).

Despite the success of single gene taxonomy, phylogenetic reconstructions based on this technique can lead to ambiguous evolutionary relationships because single gene sequences lack sufficient information to resolve much of the divergence pattern of their major lineages. Additional complexity is also introduced by lateral gene transfer, a process whereby genes from one species are passed onto another, leading to the acquisition of physiological properties or metabolic traits that are not concordant with their 16S rRNA phylogenies (Doolittle, 1999). An example we are all familiar with from the news is how certain bacteria have suddenly developed resistance to antibiotics that used to destroy them. What we now suspect has happened is that they have acquired resistance by a donation of genes from those cells immune to the drug. Lateral gene transfer often results in a confusing picture, where different phylogenies for the same organism are obtained when using different gene sequences because of an unusually high degree of similarity between the donor and recipient genes of otherwise distantly related species. As a result of these incongruities, the entire premise of equating single gene phylogeny with organismal phylogeny has been called into question. These problems can, in part, be addressed by multi-gene- and whole-genome-based systematics (i.e., the study of genomics), with phylogenetic trees being built by either analyzing a large number of conserved genes or by analyzing for the presence or absence of genes in each genome, respectively (e.g., Fitz-Gibbon and House, 1999). Thus far, complete genomic sequences of more than 200 microorganisms have been made available. And, one of the most exciting outcomes of these studies is that regardless of domain, all of the microorganisms share a number of core genes, the so-called "housekeeping" genes that are responsible for critical functions (Koonin, 2003). This suggests that life may indeed have had a common ancestor, with perhaps as few as 500–600 genes.

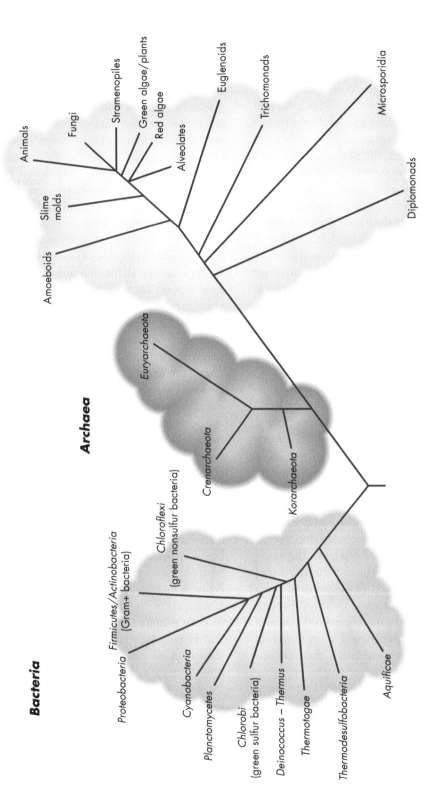

Figure 1.1 The "universal" phylogenetic tree, based on 16S/18S rRNA sequence comparisons, showing the three domains of life: *Bacteria*, *Archaea*, and *Eukarya*. The *Bacteria* and *Archaea* are subdivided by representative phyla, and the *Eukarya* are outlined by familiar grouping for lack of general consensus among taxonomists. Note: all the branch lengths should be considered approximate.

1.2 Physical properties of microorganisms

1.2.1 Prokaryotes

Most *Bacteria* and *Archaea* are between 500 nanometers (nm) and 2 micrometers (μm) in diameter, with a volume between 1 and 3 μm^3 and a wet weight approximating 10^{-12} grams. The only way to observe such small objects is by microscope, hence the term microorganisms or microbes. Individual cells most frequently occur as rods (known as bacilli), spheres (known as cocci), or helical shapes (known as either spirilla or vibrio) (Fig. 1.2). These different shapes in turn affect the relationship between volume and surface area: rods have a surface area to volume ratio of 10:1; spheres, 5.8:1; and helical-shaped, 16:1 (Beveridge, 1988). Individual cells also commonly form together in groups or clusters after division. When they grow end on end, they form what are known as filaments.

The small size of *Bacteria* and *Archaea* has a number of ramifications. First and foremost, they are typically not large enough to differentiate the many cellular processes into compartmentalized organelles. The lack of complex parts, however, benefits them because it means that there is little to fail, perhaps explaining why they are often found in "extreme" environments that preclude the growth of more complex organisms (Nealson and Stahl, 1997). Given the lack of intracellular organelles, many of the important metabolic and biosynthesis reactions take place instead at the cell's periphery where the external sources of nutrition and energy are derived. Hence, establishing an appropriate shape with a high surface area to volume ratio is of considerable importance for optimizing the diffusional properties of the cell (Fig. 1.3). If a cell grows too large in volume, the surface area becomes small in comparison, and the diffusion time becomes excessively long. On the other hand, cells too small would suffer from having insufficient surface area to accommodate the numerous proteins required to transport nutrients into the cell, and not enough volume to house all of the macromolecules and solutes necessary to support normal cell growth (Beveridge, 1989a).

The physical makeup of *Bacteria* and *Archaea* is remarkably simple (see Madigan et al., 2003 for details). All possess a plasma membrane that completely envelopes the cell and separates the inside of the microorganism from the external environment. Its primary function is as a selective and semipermeable barrier that regulates the flow of material into and out of the cell. The plasma membrane is 7–8 nm in thickness (Fig. 1.4). In *Bacteria*, it consists of a lipid bilayer with a fatty acid portion sandwiched in the middle of two glycerol layers, while in *Archaea* the fatty acid is replaced by repeating units of the hydrocarbon molecule isoprene. The placement of ionized phosphate compounds at the outer surface of the glycerol imparts a negative charge that makes the phospholipid hydrophilic, that is, attracted to water (see Box 1.1). Therefore, the phosphate portions of one layer interact with the water outside the cell and those of the other layer interact with the fluids within the cell. The fatty acids, on the other hand, are made up of nonpolar hydrocarbon chains (with H–C bonds) that are hydrophobic. Accordingly, they cluster together and point inwards towards each other in an attempt to minimize their exposure to the aqueous environment. Despite the exclusion of water from their interior, most plasma membranes are actually quite fluid, with a viscosity similar to light grade oil. As a result, the phospholipids have the capacity to move about with a certain degree of flexibility.

The plasma membrane also houses a number of different proteins, making up to 60% of the membrane weight. Some of these proteins are found on the outer surfaces where they bind and process substrates for transport into the cell. Within the plasma membrane, various proteins

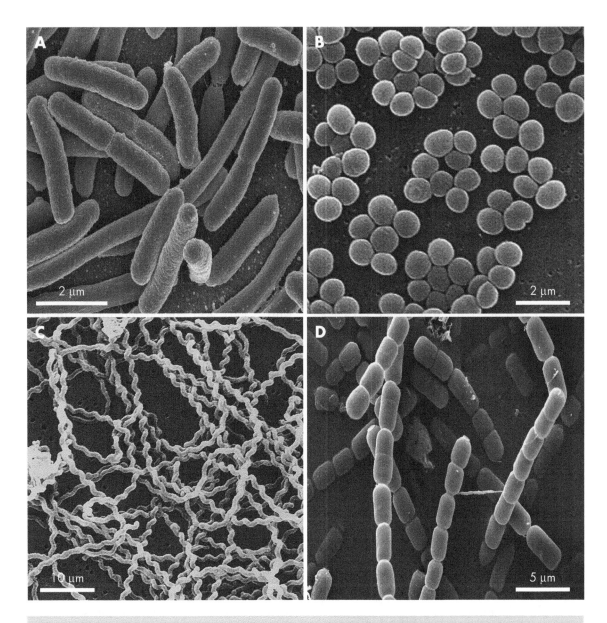

Figure 1.2 Basic shapes of bacteria as seen under the scanning electron microscope (SEM). (A) Rods, *Escherichia coli* (courtesy of Rocky Mountain Laboratories, NIAID and NIH). (B) Cocci, *Staphylococcus aureus* (courtesy of Janice Carr and CDC). (C) Spirilla, *Leptospira* sp. (courtesy of Janice Carr and CDC). (D) Filaments, *Anabaena* sp. (courtesy of James Ehrman).

facilitate energy production in respiration and photosynthesis (see Chapter 2). There are also proteins that completely span the membrane, having portions exposed to both the interior of the cell and the external environment. Some of these proteins, called membrane transport proteins or permeases, provide passageways through which specific molecules can pass.

Enclosed within the plasma membrane is the cell's interior or cytoplasm, which is over 70%

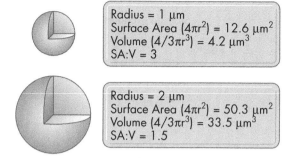

Radius = 1 μm
Surface Area $(4\pi r^2)$ = 12.6 μm^2
Volume $(4/3\pi r^3)$ = 4.2 μm^3
SA:V = 3

Radius = 2 μm
Surface Area $(4\pi r^2)$ = 50.3 μm^2
Volume $(4/3\pi r^3)$ = 33.5 μm^3
SA:V = 1.5

Figure 1.3 The relationship between cell surface area and volume.

water and has a pH maintained near neutrality. Within the cytoplasm, raw materials from the external environment are enzymatically degraded and new organic macromolecules are biosynthesized. It contains a particulate fraction, consisting of: (i) its genome, which is in the form of a large double-stranded molecule (the bacterial chromosome) that aggregates to form a visible mass called the nucleoid, as well as some extrachromosomal DNA called plasmids that confer special properties on the cell and are amenable to lateral gene transfer; (ii) ribosomes, the machinery needed in the manufacture of proteins; (iii) carboxysomes, polyhydroxybutyrate bodies, and various other inclusions and granules, that serve as specific storage sites; (iv) gas vacuoles, that confer buoyancy on the cells; and (v) magnetosomes, the magnetic particles found in some cells. The soluble portion (the cytosol) contains a variety of small organic molecules and dissolved inorganic ions (Fig. 1.5).

Most cells have internal solute concentrations that greatly exceed their external environment. As a result, there is a constant tendency throughout the life of the cell for external water to enter into the cytoplasm to dilute the salt content. This creates an internal hydrostatic pressure that may reach up to 50 atmospheres. External hydrostatic pressures are usually significantly lower. This pressure differential causes the delicate and deformable plasma membrane to stretch near the breaking point, and were it the sole structural support, it would likely rupture and the cell would die (known as lysis). *Bacteria* and most *Archaea* have, however, evolved an additional layer, the cell wall, which superimposes the plasma membrane, providing extra rigidity and support for the cell, as well as governing cell morphogenesis (Beveridge, 1981). It also functions as an exterior armor that protects the cell from physical

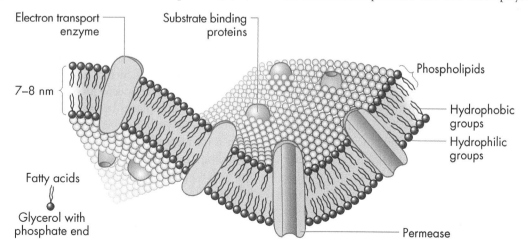

Figure 1.4 Structure of a bacterial plasma membrane, showing the phospholipid molecules with their hydrophilic groups pointing outwards and their hydrophobic groups pointing inwards. Embedded within the membrane are various proteins, each of which serves a specific cellular function.

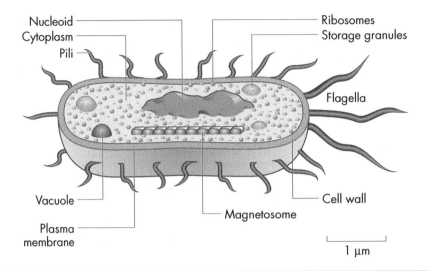

Nucleoid
Cytoplasm
Pili
Ribosomes
Storage granules
Flagella
Vacuole
Magnetosome
Cell wall
Plasma membrane
1 μm

Figure 1.5 Idealized drawing of a prokaryotic cell, showing some of its internal structure.

stresses, such as collision with other particles, and chemical perturbants, such as pH changes, dissolved inorganic ions, and organic solvents (Koch, 1983). Furthermore, the cell wall acts as a filter controlling the passage of dissolved molecules into the cell (see Chapter 3). Without going into detail here, it suffices to mention that there is variation in wall types based on morphology and composition. *Bacteria* are typically classified according to how they react with what is known as the Gram stain. Gram-positive cells stain purple, and this reflects their thick, single-layer walls made up of a material called peptidoglycan. Gram-negative cells stain pink, a reflection of their more complex structure, with a thin peptidoglycan layer, overlain by gel-like space known as the periplasm, on top of which lies the outer membrane. *Archaea* are much more diverse, and have several wall types.

1.2.2 Eukaryotes

Eukaryotes include microorganisms, such as protists and fungi, as well as multicellular life, such as the plants and animals. Of the various protists, two will be covered in the following chapters, the algae and protozoans. Algae are a diverse group of photosynthetic organisms that produce O_2 as a byproduct of their metabolism.

They are typically found in fresh and sea water, but they can also grow on rocks and trees, or in soils when water is available to them. Algae are broadly classified into green, brown, golden, and red based on the biochemical characteristics of their chlorophyll molecules and accessory pigments. Collectively they display a wide variety of morphologies, ranging from unicellular forms to macroscopic seaweeds that can grow tens of meters in length (see van den Hoek et al., 1996 for details). Others can grow symbiotically with animals, such as the dinoflagellates associated with coral reefs. Despite being relatively primitive, green algae are classified in the same clade as the plants because of their shared use of photosynthetic pigments. Some algae even form mineralized shells (or exoskeletons), including the silica-precipitating diatoms and the calcite-precipitating coccolithophores (see Chapter 4). By contrast, protozoans are colorless and unicellular animal-like organisms that obtain food by ingesting other organisms or particulate organic matter (see Sleigh, 1992 for details). They are classified according to their ability to move in a concerted manner (known as motility); some use flagella, which are whip-like appendages that arise from the cell surface, others have shorter appendages called cilia that move in a wave-like manner, while some use extensions of their cytoplasm called pseudopods. Many protozoans live as parasites, absorbing or ingesting organic compounds from their host cell; one well known example is *Plasmodium* sp., a parasite that causes malaria. Some protozoans form mineralized shells as well, such as the calcite-forming foraminifera and the silica-forming radiolarians.

Fungi also have a range of morphologies, from unicellular yeasts, to filamentous molds that clump together to form what is known as mycelia, to macroscopic forms such as mushrooms. Their most important roles are: (i) as decomposers of organic material, leading to nutrient cycling and spoilage of natural and synthetic materials, such as wood or food; (ii) as symbionts of algae and cyanobacteria (e.g., as lichens); and (iii) as producers of economically important substances, such as ethanol, citric acid, antibiotics, and vitamins. They are often dominant in acidic conditions and, in soil, represent the largest fraction of biomass (see Moore-Landecker, 1996 for details).

Eukarya are larger and much more complex in structure than either *Bacteria* or *Archaea*, typically 5–100 μm in diameter. However, this demarcation has been clouded by the discovery of marine bacteria on the seafloor that can reach sizes as large as 750 μm (Schulz et al., 1999) and the recent recognition that a diversity of 0.5–5 μm sized "picoeukaryotes," with compact organization and limited organelles, exist throughout the ocean waters (Moreira and López-García, 2002). Most eukaryotes have a rigid cytoskeleton that varies in type amongst the different phyla. They also have a plasma membrane that contains complex lipids known as sterols. These molecules stabilize the cell structure, making them less flexible than their prokaryotic counterparts, and, depending on the type of cell, sterols can comprise up to 25% of the total lipids in the membrane (Madigan et al., 2003). Eukaryotes also contain a membrane-enclosed nucleus that accommodates its genome in the form of DNA-containing chromosomes, as well as a number of membrane-bound organelles that are used to segregate the various cellular functions from one another. These include, amongst many, the mitochondria for energy production and the chloroplasts, the chlorophyll-containing sites used by algae (and plants) for photosynthesis (Fig. 1.6). This compartmentalization is necessary for localizing the various metabolic processes,

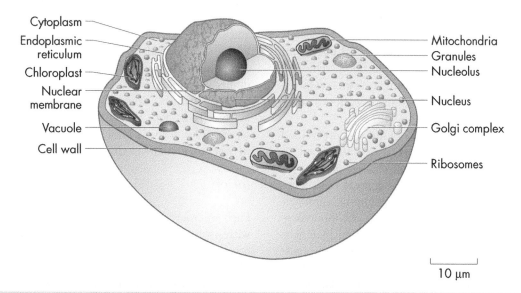

Figure 1.6 Idealized drawing of an eukaryotic cell, showing some of its internal structure.

and thereby minimizing the inherent problems that their large size would face in obtaining nutrients and excreting waste products simply by diffusional processes (Beveridge, 1988).

1.3 Requirements for growth

Although microorganisms are ubiquitous in the surface environment, each particular species is subject to a number of variables that affect their rates of growth. We can divide these requirements into two categories, physical and chemical. Physical aspects include temperature, pH, and osmotic pressure, while the chemical aspects include sources of nutrition and energy.

1.3.1 Physical requirements

(a) Temperature

Each microorganism has a temperature range over which it can grow. Psychrophiles are capable of growth up to 20°C. In nature, they are commonly found in deep ocean waters or in polar regions, but they also include those that grow within our refrigerators. Mesophiles grow between 15 and 45°C. These are the most common types of microorganisms in nature, and include most of the pathogenic species. Thermophiles grow between 45 and 80°C and hyperthermophiles grow above 80°C. The last two groups of microorganisms are associated with terrestrial geothermal regions or seafloor hydrothermal vent systems.

The temperature limits of life are constrained in one of two ways. At high temperatures, subtle changes occur in the configuration of the proteins and nucleic acids causing them to become irreversibly altered. As temperatures approach the surface boiling point of water, the monomers that make up the cells also begin to hydrolyse (bonds break by reacting with water); the half-life of adenosine triphosphate (ATP) is less than 30 minutes at 100°C (Miller and Bada, 1988). Also, as reactions generally proceed faster at higher temperatures, solute transfer across the plasma membrane may become excessively quick, thereby inhibiting the cells' ability to generate energy through what is known as the proton motive force (see section 2.1.4(b)). To date, the uppermost survival temperature measured is 121°C for an archaeal species most similar to *Pyrodictium occultum* (Kashefi and Lovley, 2003). Hyperthermophiles compensate for the high temperatures in a number of ways, including altering some of their protein structures and by possessing plasma membranes composed of hydrocarbons with repeating units of isoprene (instead of fatty acids) that are nearly impermeable to ions and protons (van de Vossenberg et al., 1998).

One principal factor that governs a microorganism's minimum temperatures is whether the cell membrane retains its fluid state, so that its capability for nutrient transport and energy generation still exists. Experiments with psychrophiles have shown that as temperature is decreased, lipids in their plasma membrane change composition by adding an increasing proportion of unsaturated fatty acids that help maintain an optimal degree of fluidity (Gounot, 1986). Other adaptations include producing cold-acclimation enzymes that allow the cells to metabolize at rates comparable to mesophiles (Feller and Gerday, 2003) and formation of extracellular layers (i.e., outside the cell wall) containing compounds that increase the viscosity of the immediate fluid phase, in essence acting as a natural antifreeze agent (Raymond and Fritsen, 2001). Thus far, the minimum temperature recorded for actively growing bacteria is −20°C in Siberian permafrost (Rivkina et al., 2000).

(b) pH

Each microorganism has an external pH range within which growth is possible. For most microorganisms (e.g., the neutrophiles), this is within pH 5–9, although the intracellular – within the cell – pH must remain neutral in order to prevent destruction of its cytoplasmic macromolecules.

Only a few species are tolerant of pH values below 2, these are the so called acidophiles. They are generally restricted to mine drainage and geothermal environments, where acid is generated from the oxidation of reduced sulfur-containing compounds. Other microorganisms, the alkaliphiles, can grow at pH 10–11. Such extremely alkaline environments are often associated with soda lakes and carbonate-rich soils.

(c) Water availability

The primary impediment for microbial life is the availability of water. Water is crucial because it facilitates biochemical reactions by serving as the transport agent for reactants and products. When a substance cannot move across the plasma membrane in response to a chemical gradient, water will move across instead. This occurs because the presence of solutes changes the concentration of water. Such movement of water is called osmosis, and the osmotic pressure is the force with which water moves from low to high solute concentrations when the solutions are separated by a semipermeable barrier. In a hypotonic medium, the solute concentration is higher in the cell. As a consequence, water will move into the cell to attain positive water balance. If the flow into the cell was unrestricted, the plasma membrane would burst. The reverse occurs in a hypertonic medium, where the solute concentration is higher outside the cell than inside. In brines, a cell would inadvertently shrink through plasmolysis if it did not naturally possess the means to deal with osmotic stresses. Thus, microorganisms living in evaporitic environments generally have some specific requirements for sodium ions (Na^+) to maintain osmotic equilibrium. Such organisms are called halophiles.

1.3.2 Chemical requirements

Life is made up of a few basic organic ingredients. Besides water, carbon is the primary requirement for growth, making up 50% of an organism's dry weight. It serves as the structural backbone of living matter, to which carbon and a number of other elements, such as hydrogen, oxygen, nitrogen, sulfur, and phosphorous, can bind. Covalently linked carbon atoms can form linear chains, branched chains (aliphatics), benzene rings (aromatics), or ring structures containing one or more noncarbon atoms (heterocyclic). The other elements usually take the form of anions or neutral species that have a definite geometrical arrangement around the central carbon atom. The resultant molecules form what are known as functional groups, each of which possesses characteristic chemical and physical properties that can be observed in a range of organic compounds. Many organic macromolecules are polyfunctional, containing two or more different kinds of functional groups. Figure 1.7 shows the common functional groups associated with microbial cell surfaces, some of which play an important role in metal sorption processes (see Chapter 3).

The various functional groups are bonded together to form low molecular weight compounds called monomers, which in turn combine to form much larger macromolecules (Box 1.1). Monomers are covalently bonded together, often through reaction of a hydrogen atom from one monomer with the hydroxyl group from another, liberating a molecule of water. These are referred to as dehydration or condensation reactions. By contrast, macromolecules can be broken down into their precursor monomers through the addition of water. These chemical reactions are known as digestion or hydrolysis. When several of the same monomers are repetitively bonded together, the macromolecule is called a polymer. The important macromolecules in living systems, in order of their dry cell percentage, are proteins (55%), nucleic acids (24%), lipids (9%), and carbohydrates (10%).

(a) Nutrition

Organisms obtain their carbon in one of two ways. Those that convert CO_2 to organic carbon, such as glucose ($C_6H_{12}O_6$), function autotrophically,

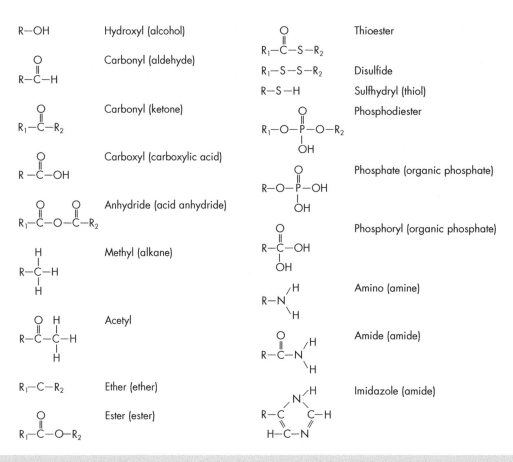

Figure 1.7 Some organic functional groups, shown in their protonated state. The symbol "R" is used to represent any substituents, but typically it is a carbon-containing moiety. When two substituents are shown in a molecule, they are designated "R_1" and "R_2".

whereas those that consume pre-existing organic materials act heterotrophically.

A living cell further requires a number of trace metals to fulfill essential biochemical roles in cell metabolism. For example, iron, nickel, copper, manganese, cobalt, molybdenum, and zinc are all important components of different metalloenzymes. Other trace metals, however, are toxic to the cell (e.g., arsenic, cadmium, copper, and mercury), and in order for a cell to survive when they are present in excess, they must actively prevent their intracellular transport. As such, microorganisms possess a suite of genes that serve to activate specific proteins designed either to facilitate metal transport into the cell or, alternatively, help rid the cells of them through their efflux, biomethylation, volatilization, or immobilization (e.g., Nies, 2000).

Because of the unique physicochemical properties of each individual ion or compound, a cell requires more than one transport mechanisms to obtain the full spectrum of solutes and compounds required. There are a wide variety of mechanisms utilized by microorganisms, of which four will be covered here. These include: (i) passive diffusion; (ii) facilitated diffusion; (iii) active transport; and (iv) cytosis.

Box 1.1 Organic macromolecules

Proteins

The importance of proteins to living organisms stems from their roles as enzymes, as well as their involvement in solute transport and cell structure. A single cell of *Escherichia coli*, a typical bacterium, contains about 1900 different kinds of proteins and 2.4 million total protein molecules (Madigan et al., 2003). The basic building blocks of proteins are amino acids, which in themselves comprise both amino and carboxyl functional groups, as well as an additional side group that distinguishes the various amino acids (there are 20 different kinds of amino acids commonly found in proteins). The side group may be a hydrogen atom, or a more complex organic molecule, such as a sulfhydryl

(–SH) group or an aromatic ringed structure. The side chains also impact the surface characteristics of the amino acid because insertion of a carboxyl group makes it anionic, while insertion of methyl groups (CH_3) makes them more hydrophobic. Dipeptides result from the covalent bonding of the carboxyl group of one amino acid to the amino group of another, releasing water via dehydration. Further addition of amino acids would result in the formation of a long chain-like molecule called a polypeptide. Polypeptides can take on any number of more complex structures resulting from the way they fold in accordance with the positioning of the side groups and the way different polypeptides interact with one another in the same protein. This gives rise to the incredible diversity in protein structures.

Nucleic acids

Nucleic acids consist of basic structural monomers called nucleotides, where each nucleotide contains a nitrogen-containing base, a pentose sugar, and an organic phosphate group. Two principal types of polynucleotides are deoxyribonucleic acid (DNA) and ribonucleic acid (RNA). DNA contains the genetic information of the cell. RNA plays three crucial roles: (i) messenger RNA carries select genetic information from DNA in a single-stranded molecule; (ii) transfer RNA are adaptor molecules that translate mRNA during protein synthesis; and (iii) ribosomal RNA, of which several types exist, are catalytic components of the ribosome that carry out the synthesis of

proteins. The nitrogen-containing bases are all ring structures, with either a single ring (e.g., the pyrimidines: thyamine, cytosine, and uracil) or a double ring (e.g., the purines: adenine and guanine). The pentose in DNA is deoxyribose, while the pentose in RNA is ribose. In a nucleotide, a nitrogenous base is attached to a pentose sugar by a glycosidic bond, while the sugars are held together by phosphodiester bonds. Polynucleotides, such as DNA and RNA, then form by covalent bonding between the phosphate of one nucleotide with the pentose of another. An extremely important nucleotide, aside from being a constituent in nucleic acids, is adenosine triphosphate (ATP). It functions as a carrier of chemical energy.

continued

Box 1.1 *continued*

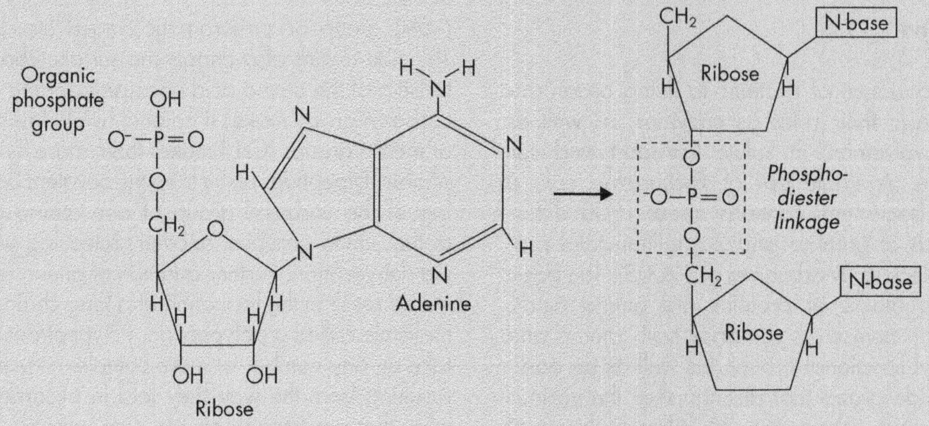

Organic phosphate group

Adenine

Ribose

N-base

Ribose

Phospho-diester linkage

N-base

Lipids

Lipids are a diverse group of organic compounds that are insoluble in water, but dissolve readily in nonpolar solvents such as alcohol. They are essential to the structure and function of cell membranes, as well as serving as fuel reserves. Simple lipids contain two monomers; an alcohol called glycerol and a group of hydrocarbons known as fatty acids. A fat molecule is formed when a molecule of glycerol combines with one, two, or three fatty acid molecules to form a monoglyceride, diglyceride, or triglyceride, respectively. The chemical bond that forms between a fatty acid and a glycerol molecule is called an ester linkage. Complex lipids form when phosphate, carboxylate (COO⁻), or amino groups replace a fatty acid molecule – recall the

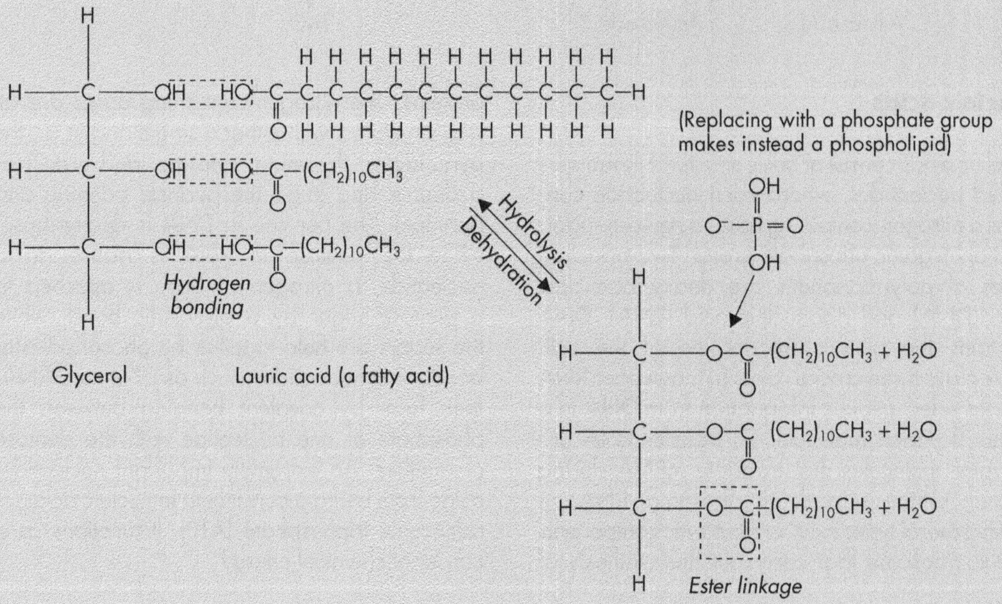

Hydrogen bonding

Glycerol　　　　Lauric acid (a fatty acid)

Hydrolysis

Dehydration

(Replacing with a phosphate group makes instead a phospholipid)

$+ H_2O$

$+ H_2O$

$+ H_2O$

Ester linkage

continued

Box 1.1 *continued*

phospholipids in cell membranes. This can make the macromolecule amphipathic in nature, with a hydrophilic "head" that interacts with water and a hydrophobic "tail" that avoids water.

Carbohydrates

The basic building blocks of carbohydrates consist of simple sugars called monosaccharides, with each molecule containing three to seven carbon atoms. Pentoses (five-carbon sugars) and hexoses (six-carbon sugars) are extremely important to life because they include ribose (found in RNA and ATP), deoxyribose (found in DNA), and glucose (the main constituent of cell walls and an important energy reserve). Derivatives of monosaccharides can be formed by replacing one or more of the hydroxyl groups by another

chemical species, such as through N acetylation (N replacing O in sugar and adding $-CH_3CO$ group) to form one of the sugar derivatives in peptidoglycan. Polysaccharides consist of eight or more monosaccharides covalently bonded to one another via glycosidic bonds. A high proportion of polysaccharides have simple repetitive structures of the same monosaccharide (e.g., starch, cellulose, chitin) or disaccharides (e.g., peptidoglycan). More complicated structures (e.g., heteropolysaccharides) may involve three or more sugars with perhaps a side-branch, or the repeating sequences can be interrupted (i.e., as in alginates). Polysaccharides can even combine with other macromolecules, such as protein or lipids, to form glycoproteins and glycolipids, respectively.

α-glucose   α-glucose   Starch

1 *Passive* – As discussed above, prokaryotes have evolved cell shapes that maximize their diffusional properties. The process of passive diffusion involves the movement of solutes across the plasma membrane due to a concentration gradient, i.e., from an area of high concentration to one that is lower, in order to achieve a state of equilibrium. During the process there is no consumption of energy. The rate of diffusion, in turn, is proportional to the difference in concentration between the external and internal environments, the permeability, and the total surface area. Cells commonly rely on this process to transport nonpolar, fatty-acid-soluble molecules, such as alcohols, or those molecules sufficiently uncharged, such as gases and water.

2 *Facilitated* – The hydrophobic nature of the inside membrane prevents the passive movement of soluble ions and simple organic molecules. Even a substance as small as a proton (H^+) is restricted because it is always hydrated as the charged hydronium ion (H_3O^+). If the ions or organic molecules have external concentrations in excess of the cytosol, then they can diffuse into the cell along a concentration gradient with the aid of specialized membrane transport proteins called permeases. Three kinds of permeases exist (Booth, 1988). Uniporters transfer single ions from one side of the plasma membrane to the other. Symporters move ions or organic molecules, with a coupling ion required for transport of the first, in the same direction. Antiporters move the ions one way, and

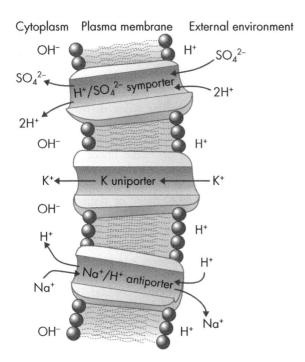

Cytoplasm   Plasma membrane   External environment

Figure 1.8 The three types of permeases; uniporters, symporters, and antiporters. Energy to drive the transport of ions (or simple organic molecules) towards the cytoplasm comes from the pH gradient between the outside and inside of the cell.

a secondary ion the other. Each permease is highly specific to the uptake of certain ions or simple organic molecules, and the cell has the means to regulate which permeases are active depending on its nutritional needs (Fig. 1.8).

3 *Active* – The concentration of most chemicals within the cytosol is generally higher than those outside the cell. Therefore, neither passive or facilitated diffusion allows the cells to acquire their necessary concentrations. As an alternative, active transport is an energy-dependent process that moves solutes or simple organic molecules against the concentration gradient. This process similarly uses permeases, with the energy derived from either ATP or an electrochemical gradient (see section 2.1.4(b)).

4 *Cytosis* – As a consequence of their more rigid cytoskeleton, protozoans use an additional mechanism called cytosis, in which organic molecules or solutes are moved in and out of a cell without passing through the plasma membrane. It essentially involves the plasma membrane wrapping around a substance, engulfing it to form a membrane-bound sphere called a vesicle, which can then open inside the cell and release the substance. Alternatively, it can separate from the plasma membrane and remain intact as a vesicle.

(b) Energy

All life forms require a source of energy to drive cellular biosynthesis and function. The ways in which organisms metabolize are covered in detail in Chapter 2, but at this point it is sufficient to point out that the energy can come either from the conversion of radiant energy into chemical energy via the process of photosynthesis or from oxidation-reduction reactions using organic or inorganic molecules. The former behave phototrophically, while the latter behave chemotrophically. The nutritional patterns among all organisms can thus be distinguished on the basis of both the carbon and energy sources (Table 1.1). Those that use light as an energy source and carbon dioxide as their chief source of carbon are called photoautotrophs. Those phototrophs that cannot convert CO_2 to organic carbon, but instead use organic compounds, are known as photoheterotrophs. Organisms that use inorganic compounds for energy and CO_2 as their carbon source are referred to as chemolithoautotrophs (or simply chemoautotrophs), while

Table 1.1 Nutritional classification of organisms.

Nutritional pattern	Carbon source	Energy source
Photoautotroph	CO_2	Sunlight
Photoheterotroph	Organic compounds	Sunlight
Chemolithoautotroph	CO_2	Inorganic compounds
Chemoheterotroph	Organic compounds	Organic compounds

those that use organic compounds for their carbon source are called chemolithoheterotrophs (or mixotrophs). When organic compounds are used for both energy and carbon the organism is a chemoorganoheterotroph (or simply chemoheterotroph). As we will see in later chapters, chemoheterotrophy is very important for element cycling because it couples the oxidation of organic carbon to the reduction of an oxidized compound, referred to as the terminal electron acceptor (TEA). The major TEAs are O_2, nitrate (NO_3^-), Mn(IV) oxides, Fe(III) oxyhydroxides, sulfate (SO_4^{2-}), or CO_2.

1.3.3 Growth rates

As a microbial population colonizes a new environment (whether in a laboratory experiment or in a natural setting) the microorganisms will collectively undergo a growth cycle that consists of the lag, exponential (or log), stationary, and death phases (Fig. 1.9). The lag phase marks the initial period of time when the cells adjust to their new surroundings, perhaps requiring the synthesis of new enzymes. Once acclimatized, the cells begin to reproduce very rapidly during the exponential phase, with, for instance, the bacteria *Escherichia coli* dividing on the order of every 20–30 minutes under typical laboratory batch culture conditions. Not only is this phase marked by significant increases in biomass, but it is also at this stage that the cells exhibit their most visible characteristics: the shape, color, density, and the way their colonies aggregate (Madigan et al., 2003). If exponential phase were

to continue unchecked, an astronomical number of cells would result. In the example above, *E. coli* would produce a population of 2^{144} cells in just 48 hours, a population weighing about 4000 times the weight of the Earth! This of course does not happen because the cells would soon run out of required nutrients, their wastes would build up to toxic levels, and there may even be significant changes in the localized aqueous composition that would impact negatively on the cells. In reality, growth rates in nature are considerably slower than estimated in the lab, often less than 1% of the maximal experimental rates, because physiochemical conditions in natural environments are rarely ideal and indigenous populations must contend with competition for a limited suite of nutrients.

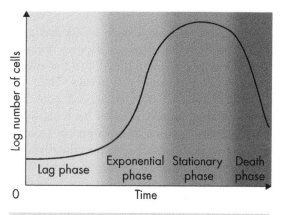

Figure 1.9 Typical growth curve for a bacterial population based on batch culture. Time (0) represents inoculation.

In the stationary phase, there is no net increase or decrease in overall cell number – some cells in the group actively grow while others die. But, if conditions continue to deteriorate for the entire population, a large number of cells enter the final death phase. Microorganisms are, however, extremely adaptable to adverse conditions, and if the supply of nutrients becomes diminished, they will employ a variety of compensatory strategies (Sunda, 2000). One typical response is for the cells to become smaller, thereby increasing their surface area to volume ratios to maximize their diffusional capabilities. This is observed in the surface waters of the oceans where there is a shift in species dominance from large cells to those that are extremely small (<2 µm), such as the so-called picoplankton. Cells also respond by growing at a slower rate, which increases cellular concentrations at a given uptake rate and decreases the metabolic demand for metal-loenzymes. Some species also decrease their requirements for limiting metals by altering metabolic pathways or by changing the type of metal-containing enzyme in key pathways. For instance, under iron-limiting conditions, many marine species are able to replace ferrodoxin, an iron-containing protein, with flavodoxin, a nonmetalloprotein (e.g., La Roche et al., 1996).

If conditions worsen still, and the micro-organism is potentially faced with starvation, it may respond by forming protective structures. Certain bacterial genera, such as *Bacillus* and *Clostridium*, form endospores (spores formed intracellularly) that are released into the environment once the parent cell decays. Then, as conditions improve, the endospore converts back to a vegetative cell. Other bacteria, such as *Methylosinus* species, form exospores (spores formed outside the cell – extracellularly) by growing or budding out from one end of the cell. Both types of spores are extremely durable to heat and chemicals, so much so that they can even survive conditions in space, leading to the hypothesis that life may have been transported to Earth from elsewhere in the solar system (see section 1.5.7). What is truly amazing is that some

Bacillus spores apparently have been resuscitated after preservation in amber for 25–40 million years (Cano and Borucki, 1995) and in brine inclusions within salt crystals that are some 250 million years old (Vreeland et al., 2000), although these claims are controversial. Species of *Azotobacter*, *Bdellovibrio*, *Myxococcus* and some cyanobacteria instead form protective structures called cysts. These are thick-walled structures that, like spores, protect the microorganism from harm, but they are somewhat less durable than spores.

1.4 Microbial diversity

Microorganisms are the most ubiquitous life forms. Unlike multicellular life that is restricted to the Earth's surface, the adaptable nature of microorganisms has allowed them to inhabit the most diverse environments imaginable, often representing the only life forms. In most aquatic systems, microbial cell densities are remarkably similar, ranging from 10^5 to 10^6 cells ml^{-1}. Such consistency reflects less the overall microbial productivity than it does control of numbers by grazing (Fenchel et al., 2000). Much higher cell densities can, however, exist, wherever abundant nutrients and energy are available, and often where predation is minimized. One such setting is in the waters immediately surrounding warm, deep-sea hydrothermal vents, where cell densities up to 10^9 cells ml^{-1} have been reported (e.g., Corliss et al., 1979). The high microbial densities, in general, have important implications for global dispersal because microorganisms are unlikely to be restricted by any geographical barriers. Yet ubiquitous dispersal also means relatively low global species richness when compared with more complex organisms that have geographically restricted ranges (Finlay, 2002). In terms of global biomass, the total amount of microbial carbon is 60–100% of the estimated total carbon found in plants, with their total biomass estimated to be roughly $4–6 \times 10^{30}$ cells or 350–550 Pg of C, where 1 Pg $= 10^{15}$ g

(Whitman et al., 1998). In addition, micro-organisms contain 85–130 Pg of N and 9–14 Pg of P, which amounts to roughly ten times more nutrients than those stored in plants.

The vast majority of microorganisms in aquatic systems grow in microcolonies attached to submerged surfaces in the form of biofilms. The biofilms consist mainly of highly hydrated extracellular polymers (EPS) that are secreted by the microorganisms embedded within it. Benthic strategies can include growing on suspended sediment particles, plants and mineral surfaces (epilithic mode), the latter often leads to mineral weathering and metal corrosion (see Chapter 5). Biofilms are remarkably resilient communities where the cells live and are retained within protected adherent microcolonies, while taking advantage of the inorganic and organic compounds that preferentially accumulate at interfaces. As microbial populations grow, they eventually enshroud available surfaces, while at the same time dispatching mobile "swarmer" cells to reconnoitre neighboring niches and to establish new microcolonies in the most favorable of them (Costerton et al., 1994). Under ideal conditions, the biofilms thicken into what is commonly referred to as a microbial mat (see Chapter 6). These natural ecosystems contain many types of microbial species, and in any given part of the mat there exists a highly organized community where nutrients and metabolites are continuously recycled between cells in close proximity.

One of the primary goals of microbial ecologists has been the isolation and culture of microorganisms of interest. During enrichment culture techniques, specific media and growth conditions are chosen to duplicate as closely as possible the natural conditions of the desired microorganism. Unfortunately, the enrichment cultures favor some species over others, and a frequent outcome is that the microorganisms that thrive and dominate the culture conditions may only have comprised a small component of the natural population. Indeed, estimates suggest that some 99% of the microorganisms visualized microscopically in environmental samples are not cultivated by routine techniques (Amann et al., 1995). This has led to one of the greatest obstacles to understanding the diversity of natural microbial communities – our inability to make sound ecological inferences based on the metabolic properties of a few cultivatable species.

In order to circumvent the inherent culture-based biases, it was suggested that, by extraction of nucleic acids directly from environmental samples, genes that were present in all micro-organisms (e.g., rRNA) could be isolated, sequenced, and compared with pre-existing cultures (e.g., Pace et al., 1986). Such a phylogenetic approach is now routinely used to study microbial diversity, and is often published in the form of a tree that strictly highlights the evolutionary relationship between the microorganisms, without actually inferring an evolutionary path beyond the species of interest (e.g., Fig. 1.10). Consequently, rRNA studies have provided a much more comprehensive view of the microbial world, particularly in the sense that novel species are continuously being reported to exist in environments where we previously thought we had a firm understanding of the microbial ecology (e.g., Dojka et al., 2000).

Ribosomal RNA genes are obtained from DNA isolated directly from natural samples. However, these genes need to be separated from all the other genes in the genomic DNA, and the quickest way to this is to specifically amplify rRNA using the polymerase chain reaction or PCR, a method that multiplies DNA by up to a billion-fold in a test tube – only very small amounts of DNA are initially required to obtain workable amounts of rRNA. Once the different 16S rRNA genes have been amplified from environmental DNA, they have to be sorted or separated so their sequences can be identified. There are numerous approaches to separate this mixture of 16S rRNA genes, including cloning, which separates individual genes into cloning vectors that are then assimilated by a large number of laboratory cultures (also known as a clone library), or more rapid approaches

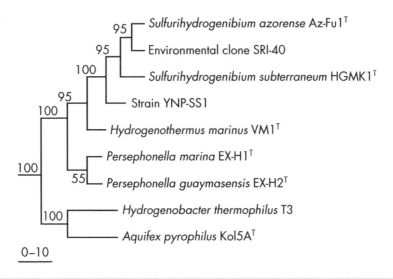

Figure 1.10 Maximum-likelihood phylogenetic tree showing the position of various members of the *Aquificales*, the purported deepest branching lineage within the domain *Bacteria*. The length of the scale bar represents evolutionary distance by the number of fixed mutations per nucleotide position. The numbers associated with the branches are a statistical measure of the confidence of divergence between two lineages (expressed in percentages) based on an algorithmic re-sampling known as bootstrapping. (From Aguiar et al., 2004. Reproduced with permission from the Society for General Microbiology.)

such as denaturing gradient gel electrophoresis (DGGE), which separates different genes by sequence in the form of individual bands on an acrylamide gradient gel. In both cases, the environmental sequences can then be compared with sequences from known cultured isolates found in reference databases such as GenBank (http://www.ncbi.nlm.nih.gov) and the Ribosomal Database Project (RDP, http://rdp.cme.msu.edu). Molecular phylogenetic trees are then constructed by placing the unknown environmental sequences within a phylogenetic framework with their closest cultured relatives (see Theron and Cloete, 2000 for details).

At this stage, we still have very little idea of what the relative abundance of some of these environmental sequences are. To determine this, one can use the sequence obtained from the environmental analysis to design a probe that will identify the cell from which the sequence was originally isolated (e.g., Devereux et al., 1992). These probes are short fluorescently tagged oligonucleotides that bind specifically to the ribosomal RNAs of the probe-target population. This method, known as fluorescence in situ hybridization (FISH), provides a way to visualize the community spatial distribution, and the ability to design these probes to detect groups of related microorganisms means that they can be used to identify and enumerate similar types of microorganisms even if they cannot be grown in culture (e.g., Fig. 1.11). However, the amount of rRNA present in a microorganism is proportional to its metabolic activity. Inactive cells have too few rRNA molecules within their cells to be detected using these techniques and thus only active members of a microbial community contain enough ribosomes to yield a detectable signal (e.g., DeLong et al., 1989). Natural samples can also be subjected to multiple FISH probing, where a suite of probes, each deigned to react with a specific microorganism or group, can lead to phylogenetically characterizing the diversity of an entire population.

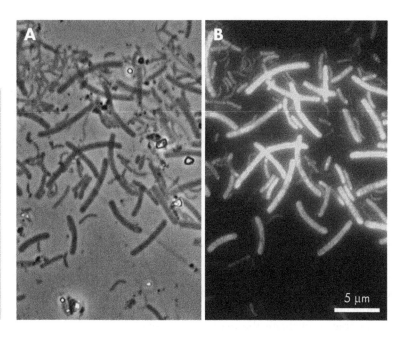

Figure 1.11 (A) Phase contrast micrograph of a microbial community from thermal springs in the Azores. (B) The same cells hybridized with an *Aquificales*-specific rRNA labeled probe. The FISH analyses clearly show the fluorescence of only a fraction of the microbial community, indicating the presence of many non-*Aquificales* species in the same sample. (Courtesy of Paula Aguiar and Anna-Louise Reysenbach).

If, however, the aim is instead to simply identify cell biomass and attain a reasonable estimate of the cell density present, then other stains, such as DAPI (4',6-diamidino-2-phenylindole) can be used as a DNA counterstain for multicolor fluorescence techniques. The blue fluorescence of DAPI stands out in vivid contrast to the green, yellow, or red fluorescence of the other reagents. Cell viability is also important in experimental procedures, and there are a number of assays available to distinguish between live and dead cells in culture. For example, the BacLight kit uses a green fluorescent dye to penetrate the plasma membrane of living cells and stain all nucleic acids present, while a highly-charged red dye only stains damaged membranes. When both stains are used in conjunction, they differentiate between cells with intact membranes and those without.

In addition to determining what species exist in a natural microbial community, new methods have been developed to ascertain which ones are metabolically active. Through the process of stable isotope probing (SIP), nucleic acids (DNA or RNA) from actively metabolizing cells are labeled with ^{13}C from a given substrate, such as $^{13}CH_3OH$, $^{13}CH_3COO^-$, $^{13}CH_4$, or $^{13}CO_2$ (e.g.,

Radajewski et al., 2000). In a wider context, SIP is not exclusively nucleic acid-based, but can also be used in labeling other biomarkers, such as phospholipid-derived fatty acids (e.g., Boschker et al., 1998). Once the labeled substrate is incorporated into the cell, density gradient ultra-centrifugation is used to separate the "heavy" ^{13}C-labeled fraction from the unlabeled "light" fraction of the microbial community that did not incorporate the provided substrate. PCR is then used to amplify the 16S rRNA genes from the labeled ^{13}C-enriched nucleic acids, and these are cloned and sequenced in order to identify the microorganisms actively involved in the specific metabolic process.

Although much of the preceding discussion has been on identifying what's there and what's active, techniques emerging in the field of functional genomics are now being used to ascertain which genes are being expressed, providing some indication of a cell's major activities under different conditions. Such methods tend to focus on the concentration and identity of messenger RNA (mRNA), which are the information carriers between the DNA and the ribosomes. As mRNA has a high turnover and reflects the sequence of the gene from which it was transcribed, the identification and quantification of mRNA provides a snapshot of microbial activity from a genetic standpoint. The reverse transcription PCR method (RT-PCR) generates readily quantified DNA oligonucleotides from the unstable mRNA, and improvements in the efficiency of RNA isolation and the RT-PCR process has recently made this method applicable to environmental samples. DNA microarrays, sometimes referred to as a "genome on a chip" because they consist of thousands of genes individually adhered to different locations on a substrate glass slide chip, are also emerging as an alternative for mRNA assessment. They serve to highlight changes in mRNA levels on a gene-by-gene basis by the use of fluorescent dyes that shift color upon the hybridization of sample mRNA to the DNA on the chip. Besides being able to evaluate the expression of many genes at once, both DNA microarray and RT-PCR technologies have advanced to the stage of revealing the variation in gene expression during growth under different environmental conditions (Lockhart and Winzeler, 2000). This may make it possible to determine how (and how rapidly) microorganisms respond to geochemical and physical changes in their habitat.

A problem inherent to techniques focusing on mRNA is the variability associated with the next step, the production of proteins from mRNA. Proteomics picks up where genomics leaves off by assessing what the microorganisms are doing through the study of the expressional end product, the protein. For example, a mRNA nucleotide might be generated at a continuous rate, while the production of its corresponding protein is dependent upon external/internal factors. In this case, proteomics yield a more robust understanding of microbial activity than genomics. In addition to qualifying the regulation of proteins, proteomics also serve to identify the functions of unknown proteins in cell cultures and environmental samples (Pandey and Mann, 2000). Polyacrylamide gel electrophoresis is used to separate, identify, and sometimes quantify all of the proteins extracted from a sample. These gels can be run in more then one direction (i.e., 2-D gel electrophoresis), allowing for separation according to more than one protein characteristic, such as molecular weight and point of zero charge. Variations in separation techniques include the purification of proteins of interest by high-performance liquid chromatography (HPLC), as well as the use of fluorescent tags to differentially label individual protein extracts so that they may be run and quantitatively compared on the same gel (fluorescence 2-D difference gel electrophoresis – DIGE). Increasingly robust analytical tools, such as matrix-assisted laser desorption/ionization–time of flight mass spectrometry (MALDI–TOF MS) allow isolated proteins to be more accurately characterized according to molecular weight, peptide sequence, and post-translational modifications.

1.5 Life in extreme environments

Microorganisms define the limits of biological tolerance to physiochemical extremes. This includes environments subject to acute variations in temperature (hydrothermal vents to polar ice), pressure (interstellar space to the deep subsurface and ocean trenches); aqueous chemistry (acid rock drainage to alkaline brines), and extreme dessication (Rothschild and Mancinelli, 2001). And, not only have they adapted to thrive in most environmental settings, but remarkably, some bacteria (e.g., *Deinococcus radiodurans*) can even survive sterilizing doses of ionizing radiation and grow within the water core of nuclear reactors (Battista, 1997). Such remarkable feats of survival have led to the genetic engineering of *D. radiodurans* cells to degrade organic solvents in radioactive mixed waste disposal sites (Lange et al., 1998). In each of these environments, the indigenous microorganisms have made unique adaptations in terms of variations in cell size and morphology, metabolic strategies, motility, and many other structural, compositional, and functional aspects of the cell. In fact, microorganisms inhabiting those environments require those conditions to survive, hence their frequent description as extremophiles.

1.5.1 Hydrothermal systems

In 1979 when John Corliss and co-workers published their discovery of animal and bacterial communities living within active hydrothermal vent systems on the Galápagos Rift (Corliss et al., 1979) they initiated a flurry of research into cataloging the life forms that survive under such seemingly inhospitable conditions. Growing at pressures equivalent to water depths in excess of

2500 meters, and around vents emitting fluids as hot as 380°C, are a community of *Bacteria* and *Archaea* feeding off the toxic emissions from hydrothermal vents (Fig. 1.12). Although the hyperthermophiles do not live in the super-heated fluid, they locate themselves along a gradient where the hot water mixes with cold bottom seawater, for instance, as free-living cells in the buoyant metal-laden plumes, and even within the walls and adjacent sediments of the so-called "black smoker" chimneys (see Plate 1). Microorganisms also grow profusely as mats around warm water vents (<40°C), where the effluent is sourced from a subsurface mixing zone of

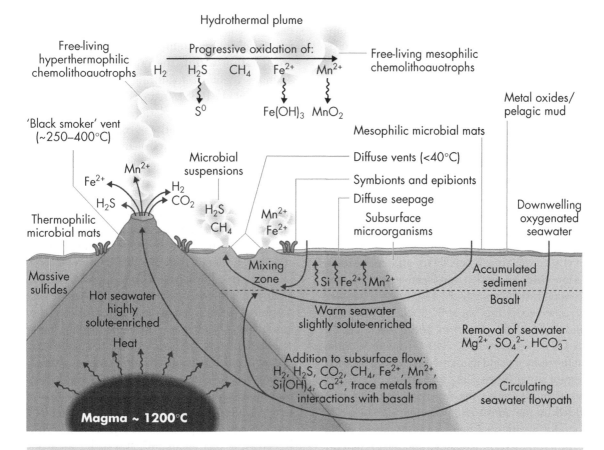

Figure 1.12 A hypothetical hydrothermal deep-sea vent system, showing the emission of a number of reduced gases and solutes from high temperature (black smoker-type) chimneys, warm water vents, and diffuse seepages from fissures underlying sediment. These reductants serve as primary sources of energy for a variety of free-living, benthic and/or sub-seafloor hyperthermophilic to mesophilic bacterial and archaeal communities.

upwelling hot hydrothermal fluids and down-welling cold, oxygenated seawater. Observations of dense, white-colored (sulfur-rich) suspensions of cells being dispersed as plumes from the warm vents further suggest that a substantial microbial community exists in the porous lava and sedimentary layers below the seafloor (Jannasch and Mottl, 1985).

The primary producers in that community obtain their energy chemolithoautotrophically by exploiting the steep chemical gradients and localized redox disequilibrium that result from the discharge of reduced, solute-rich hydrothermal effluent into an oxygenated environment (Reysenbach et al., 2002; Amend et al., 2003). Hydrogen sulfide (H_2S or HS^-, depending on fluid pH) is quantitatively the most important electron donor for chemosynthesis, and it is not uncommon to observe centimeter-thick mats of sulfur-oxidizing bacteria, such as *Beggiatoa* sp. growing at the vent periphery (e.g., Nelson et al., 1989). Other reductants, such as hydrogen gas (H_2), elemental sulfur (S^0), thiosulfate ($S_2O_3^{2-}$) and methane (CH_4) support localized chemolithoautotrophic communities (see Chapter 2 for details on these forms of metabolism). Some vents instead emit high concentrations of dissolved ferrous iron (Fe^{2+}) and manganous manganese (Mn^{2+}), but low H_2S. Such sites are typically characterized by thick accumulations of ferric oxyhydroxide (e.g., $Fe(OH)_3$) and manganese oxide (MnO_2), and it has been proposed that mineralization is partly the result of the activity of metal-oxidizing bacteria (e.g., Mandernack and Tebo, 1993; Emerson and Moyer, 2002). In O_2-depleted niches, chemolithoautotrophic sulfur-reducing and methane-generating bacteria respire using emitted H_2, and either S^0 or CO_2, respectively, while the organic substrates provided by the autotrophic communities serve as food for a variety of heterotrophs (Baross et al., 1982). Significant chemical energy to support biomass production is further available in the form of mineral substrates. This energy can be harnessed from the oxidation of seafloor hydrothermal sulfide deposits (Edwards et al., 2003) or

from particles of elemental sulfur, metal sulfides, Fe^{2+}, and Mn^{2+} entrained within, or settling out of hydrothermal plumes (McCollom, 2000). In fact, the latter study estimated that the total primary productivity potential in the plume may represent a significant fraction of the organic matter present in the deep sea.

In addition to the microbial communities, there are a number of benthic animals that exist within a few meters of the vents. They consist of tube worms and bivalves that often contain endosymbiotic chemolithoautotrophic bacteria that reside within their tissues. At sulfide-rich vents, the host tube worm provides shelter and gases to sulfur-oxidizing bacteria (e.g., O_2 and H_2S via an unusual hemoglobin), while the bacteria supply organic carbon. This activity leads to elemental sulfur and iron sulfide mineral precipitation within the organic tissues of the animals, in effect fossilizing part of the tube (Paradis et al., 1988). In areas where methane is abundant, a symbiotic relationship exists between giant mussels and methane-oxidizing bacteria. Still other bacteria act as epibionts, growing attached to the outer surface of the host (Nelson and Fisher, 1995).

Despite the current excitement garnered by new discoveries at deep-sea vents, much of our actual understanding regarding the phylogenetic diversity of hydrothermal communities comes from detailed studies at the more accessible terrestrial hot springs (Fig. 1.13; see Plate 2). Nearly one hundred thermophilic to hyperthermophilic species have been described in the past decade, and many of the isolated genera were completely unknown until a few years ago. A case in point is the discovery of two uncultured organisms at Yellowstone National Park, USA, that branch so deeply in the archaeal tree that a new phylum-level designation, "Korarchaeota," was subsequently proposed (Barns et al., 1996). Recently, a very small (400 nm wide) member of the *Archaea*, "*Nanoarchaeum equitans*," was identified and also proposed as forming a new phylum in the *Archaea*, the *Nanoarchaeota* (Huber et al., 2002). The purported deepest-

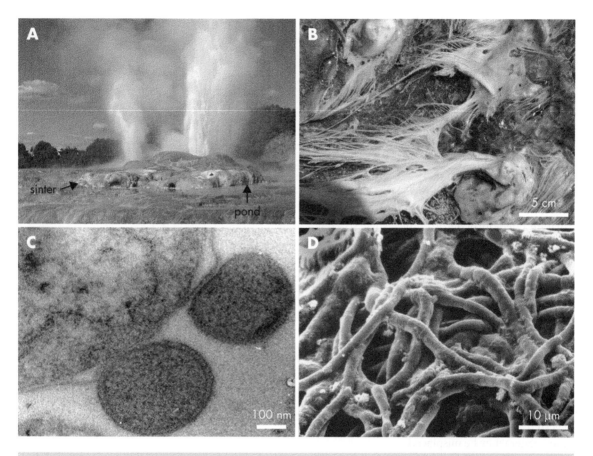

Figure 1.13 (A) The Whakarewarewa Geyser in New Zealand, showing extensive silica sinter deposits flanking the vent. Growing in the geothermal fluids are a diverse microbial community, ranging from hyperthermophiles near the vent to mesophilic cyanobacteria, algae, and fungi downstream. (B) White bacterial streamers of the deeply-branching S^0-oxidizing *Aquificales* predominate, the high-temperature microbial community at Angel Terrace Hot Spring, Yellowstone National Park (courtesy of Anna-Louise Reysenbach). (C) Transmission electron micrograph (TEM) of two newly described *Archaea*, "*Nanoarchaeum equitans*" attached to the surface of a larger archaeal cell (courtesy of Reinhard Rachel, Harald Huber and Karl Stetter). (D) SEM image of silica-encrusted cyanobacteria *Calothrix* growing in 30°C geothermal waters from Krisuvik, Iceland (courtesy of Vernon Phoenix).

branching bacteria, the *Aquificales*, and the highest temperature cyanobacteria have similarly been isolated from a variety of hot spring sites (Reysenbach and Shock, 2002). Viruses showing a wide range of morphology have additionally been found in nearly boiling waters at Yellowstone, hosted by hyperthermophilic *Archaea*, further highlighting our very limited understanding of microbial diversity (Rice et al., 2001). Aside from the ecological discoveries, hot springs offer us the unique opportunity to determine the processes of fossilization because many vents emit high concentrations of dissolved silica that encrust living organisms in a manner reminiscent of early Earth (see section 4.3.1). And, from a biotechnology standpoint, a hot spring near Great Fountain Geyser in Yellowstone, was the source of the culture of *Thermus aquaticus* that is used to make Taq polymerase, a key constituent of the polymerase chain reaction.

One of the most significant outcomes from studies of microbial ecology at hydrothermal systems is the hypothesis that life originated in a hot aqueous setting. The most persuasive argument supporting this notion (aside from the biogeochemical grounds discussed in section 7.1.2(b)), is the recognition that the deeply-rooted lineages within the universal phylogenetic tree are all represented by *Archaea* and *Bacteria* that grow in hot water (Stetter, 1994). Furthermore, the rather short phylogenetic distances between contemporary hyperthermophilic *Archaea*, with disparate chemolithoautotrophies, suggests that major evolutionary adaptations have not been required, perhaps because the *Archaea* evolved in a hot setting and they remain there today. By contrast, the moderately thermophilic and mesophilic *Bacteria* (i.e., Gram positives, *Proteobacteria*, cyanobacteria, etc.) represent long lineages that reflect higher rates of rRNA sequence evolution (Kandler, 1994). However, it must be pointed out that the hot origin hypothesis is controversial, and the precise branching is still likely to be modified because a number of recent studies have suggested that the branching of these deeply rooted genera may be misplaced.

1.5.2 Polar environments

It is estimated that 75–80% of the Earth's surface is permanently cold (less that 5°C). Life, however, has adapted to these conditions and can be found in sea ice (see Plate 3), permafrost, beneath glaciers, within Arctic and Antarctic rocks, and even several kilometers deep in ice cores.

On an annual basis, seawater in Arctic regions is transformed into ice, yet a small fraction of the water remains liquid even at temperatures as low as −35°C due to the presence of trapped salts. The resulting brine channels represent a refuge for a variety psychrophilic microorganisms, with cell activity reported at −15°C (Fig. 1.14). Survival of the microbial communities during winter is extremely important for the polar water ecosystem dynamics because it ensures that a

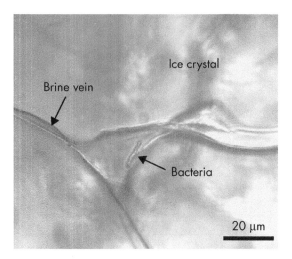

Figure 1.14 Photomicrograph of a bacterium (arrow), in a state of cell division, observed by transmitted light at −15°C within a brine inclusion at a triple point juncture between three grains of ice. The sample is a thin section of natural sea ice sliced from a core obtained in the Chukchi Sea just north of Barrow, Alaska. (From Junge et al., 2001. Reproduced with permission from the International Glaciology Society.)

viable population is available to seed the global oceans with psychrophiles on an annual basis (Deming, 2002).

Permafrost underlies about 20% of the Earth's land surface. About 93–98% of water is present as ice, with the remainder in an unfrozen state as nanometer-thin films on ice crystals and organic/mineral particles. These films facilitate cell survival because they protect against freezing–thawing stresses and they serve as the source of dissolved nutrients and metabolites (Price, 2000). Accordingly, the average number of viable cells in permafrost is 10^4–10^5 cells g^{-1}, while in pure ice they are not more than a few dozen cells in 1 litre of thawed water (Gilichinsky et al., 1993). Despite being essentially an oligotrophic (low nutrient) environment, one of the advantages of living in permafrost is that growth is usually so slow that large amounts of nutrients are not actually necessary. Not surprisingly, cell doubling times at −20°C can be as low as 160 days, with stationary phase being brought on by the formation of diffusion barriers in the thin layers of unfrozen water (Rivkina et al., 2000). At even lower temperatures, growth might stop altogether and the cells enter a state of suspended animation, for perhaps millions of years, until conditions improve (Gilichinsky et al., 1995).

In contrast to living in permafrost, microbial communities beneath alpine glaciers have an easier life. As mean subglacial temperatures reach the pressure-controlled melting point, water becomes present in the underlying sediments, in the grain–boundary network of basal ice and in larger subglacial and englacial water pockets. The microorganisms derive organic carbon and nutrients from permafrozen soils that were overridden by the advancing glacier, then finely ground by subglacial abrasion processes, and eventually released into basally derived meltwaters, while surface meltwaters provide a source of nitrate (NO_3^-) and ammonium (NH_4^+). The debris-rich basal ice communities are dominated by various chemoheterotrophs, including nitrate-reducing bacteria (NRB), sulfate-reducing bacteria (SRB), and methanogens (e.g., Skidmore et al., 2000), with total cell densities ranging from nearly 10^8 cells ml^{-1} in the very fine-grained glacial debris to 10^6 cells ml^{-1} in the meltwaters (Sharp et al., 1999).

The climate in the Antarctic continental interior is dry, creating desert-like conditions where extensive areas of rock and soil are without snow or ice cover. Yet in certain rock types, a comparatively mild and insulated microclimate exists for microorganisms that are able to penetrate into the rock subsurface (Friedmann, 1982). Such microorganisms are called endoliths, and three types exist: chasmoendoliths living in pre-existing rock fissures and cracks (see Plate 4), cryptoendoliths inhabiting structural cavities in porous rocks, and euendoliths that penetrate soluble or easily friable substrata (e.g., Fig. 1.15). Many endolithic communities are dominated by lichens, organisms that are a symbiosis between a photosynthetic primary producer, also known

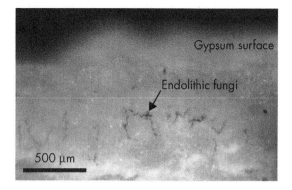

Figure 1.15 Photomicrograph of a black-pigmented community of endolithic fungi living within a translucent crust of salt that formed on the outer surfaces of sandstone boulders at Two Steps Cliffs, Alexander Island, Antarctic Peninsula. (From Hughes and Lawley, 2003. Reproduced with permission from Blackwell Publishing Ltd.)

as the phycobiont (e.g., mostly green algae, but some cyanobacteria), and heterotrophic filamentous fungi (the mycobiont) that deliver water and minerals to the phototroph. There are many kinds of lichens in nature, with the diversity largely driven by the number of different fungi that can form lichen associations. Because of the requirements for light, only translucent rocks (e.g., sandstone, granite) or mineral crusts (e.g., gypsum, calcite, silica) are suitable for colonization. Depending on the optical properties of the rock, phototrophs can be found at depths between 1 and 10 mm beneath the surface, where photosynthetically active radiation (between the wavelengths 400 and 700 nm) is still sufficient for photosynthesis, yet harmful UV radiation is filtered out (e.g. Vestal 1988). Water is generated via solar heating of occasional snowfall, while nutrients are obtained either from atmospheric fallout or fungal organic acid-generated rock weathering. The latter leads to the formation of silica and ferric hydroxide crusts, and inevitably the characteristic exfoliative weathering patterns that occur when microbial biomass reaches the carrying capacity of the endolithic habitat. In Ross Desert, Antarctica this occurs on a timescale of 10^3–10^4 years (Sun and Friedmann, 1999).

One of the most exciting revelations in polar microbiology has been the possible presence of life in subglacial lakes in Antarctica, such as the highly publicized Lake Vostok. The water in Lake Vostok is kept liquid by the pressure of the ice overburden, and possibly by geothermal heating. In 1998, a team of Russian, US, and French scientists drilled a hole 3623 m deep into the overlying ice, reaching to within 120 m of the ice–lake water interface. From that core, ice samples extracted at a depth of 3603 m showed the presence of a bacterial community on the order of 300 cells ml^{-1} (Karl et al., 1999). These measurements complement another ice core analysis from 3590 m at Vostok that showed cell densities as high as 3.6×10^4 cells ml^{-1} (Priscu et al., 1999). Where those bacteria came from is unknown; some may have been atmospherically deposited and then covered by accumulating snow, while others may well have existed in the lake itself during preglacial times. In either case, it is unlikely that those cells actively grow in this environment, but more likely they represent the resistant remnants of a community that suffered a similar fate. Significantly, a portion of those cells could be revived when warmed to 3°C. It has since been shown that bacteria can metabolize *in situ* in snow at ambient subzero temperatures as low as −12 to −17°C (Carpenter et al., 2000). Some of those species are members of the genus *Deinococcus*, the radiation- and dessication-resistant bacteria, that may be one of the few organisms that can survive the extremely low levels of free liquid water and the high UV doses characteristic of the ozone-depleted Antarctic.

1.5.3 Acid environments

When metal sulfide ore is exposed to moisture and air during mining activities, some of the constituent minerals undergo gradual oxidation that leads to the generation of metal-rich and highly acidic waters (see section 5.2 for details).

Within these waters exist a number of acidophilic chemolithoautotrophic Fe(II)- and S-oxidizing bacteria, such as *Acidithiobacillus ferrooxidans* and *Acidithiobacillus thiooxidans* that thrive at pH values below 2. One archaeal species, *Ferroplasma acidarmanus*, was isolated from biofilms attached to the surface of the mineral pyrite (FeS_2), in waters with a pH as low as 0 (Edwards et al., 2000a). Several eukaryotes have also been recovered from waters with a pH lower than 1, including the red alga *Cyanidium caldarium*, the green alga *Dunaliella acidophila*, and several fungi. Recently, molecular analyses have actually demonstrated that eukaryotes can comprise a significant proportion of acid rock drainage environments (up to 60% of the biomass in the Rio Tinto, Spain). Interestingly, some of the species were closely related to neutrophilic species (Zettler et al., 2002). Such observations imply that adaptations from neutral to acid environments might occur relatively rapidly when measured on evolutionary timescales. Meanwhile, other phylogenetic surveys are showing that aside from the few readily cultured species, the extreme pH and temperature conditions at acid-generating sites are dominated by other unknown or little studied microorganisms (e.g., Bond et al., 2000). One intriguing example is the detection of bacteria growing as endosymbionts within acidophilic protists (Baker et al., 2003). If the endosymbionts reside directly in the host's cytoplasm, they are likely neutrophilic, but the question then is how are they transferred between hosts to avoid exposure to the acidic waters?

An even more adverse environment than mine drainage is the acidic geothermally heated waters in solfataras (i.e., fumaroles), where pH can drop to values as low as 0, and temperatures can be in excess of 100°C (see Plate 5). The high concentrations of sulfuric acid (H_2SO_4), coupled with elevated temperatures, limits growth to just a few types of *Archaea*, including species from the orders *Sulfolobales* and *Thermoplasmatales* (e.g., Schleper et al., 1995). These *Archaea* grow chemolithoautotrophically by the oxidation of

S^0 and H_2S with O_2, but they can also grow as aerobic chemoheterotrophs on the remains of their lysed neighbors. They also possess unusual cell envelopes; they either have an extracellular polymer directly above the plasma membrane or they completely lack a cell wall (see section 3.1.3). Considering that these species cannot survive in water of circumneutral pH, then not only is the plasma membrane's composition and construction vital for extreme acid tolerance, but the high acidity may be essential to the membrane's maintenance.

For all organisms, high intracellular acidity levels would destroy essential molecules, such as DNA. The acidophiles have thus had to evolve a mechanism to maintain pH homeostasis within the cell. One way this can be accomplished is with an electroneutral transport system that exchanges protons for certain cations, particularly K^+ (recall Fig. 1.8). For example, any decrease in the cytoplasmic pH could trigger a K^+/H^+ antiport system that would bring K^+ into the cell while extruding protons (Hill et al., 1995). This helps moderate cytoplasmic pH, keeping it at a circumneutral value compared with the acidic external milieu.

1.5.4 Hypersaline and alkaline environments

Hypersaline environments have a great diversity in ionic composition, total salt concentrations, and pH due to variations in regional geology and climate. Great Salt Lake in the USA and the Dead Sea have salt contents over 20%, with circumneutral pH. By contrast, many brines are depleted in divalent cations (e.g., Ca^{2+} and Mg^{2+}). This leads to excess carbonate (CO_3^{2-}) anions, and ultimately very alkaline waters. Some of the most extensively studied hypersaline and alkaline ecosystems are the soda lakes of the Kenyan Rift Valley, where salinities can reach in excess of 30% and pH exceeds 12. Despite these conditions, the high temperatures and daily light intensities make these lakes amongst some of the most productive naturally occurring aquatic environments in the world (Jones et al., 1998). Massive blooms of cyanobacteria (e.g., species of *Spirulina* and *Anabaenopsis* sp.), photosynthetic bacteria (e.g., *Chromatium* sp.), and sometimes diatoms, represent the primary producers in the less saline lakes, while red anoxygenic, photo-autotrophic halophilic (salt-loving) and alkaliphilic *Bacteria* (e.g., *Ectothiorhodospira* sp.) and *Archaea* (e.g., *Natronococcus* and *Natronobacterium* species; see Plate 6) thrive in the concentrated brines (Grant and Tindall, 1986). Associated with the death of the primary producers is a thick accumulation of biomass that subsequently supports a thriving community of aerobic heterotrophs in the uppermost layers, with SRB and methanogens growing underneath. Recent phylogenetic analyses have further highlighted the diversity of fermentative bacteria and the important role they play in nutrient recycling within soda lakes (Rees et al., 2004).

Marine salterns are manmade systems where seawater is sequentially pumped through a succession of shallow ponds, leading to increasingly concentrated brines and evaporite precipitation. Such environments host a large variety of halophilic and alkaliphilic bacteria that develop throughout the entire gradient of salt concentrations, but tend to diminish in diversity the more concentrated the brine. In the most dilute ponds the bacteria are only slightly halophilic, whereas in the intermediate ponds, where the seawater is concentrated to a salinity of 10–20% NaCl, most of the bacteria are moderately halophilic. The final ponds are inhabited by extremely halophilic organisms that have been recovered from underneath crusts of gypsum in ponds with salinities >20%, including several archaeal (e.g., *Halobacterium*, *Natronobacterium*) and bacterial (e.g., *Ectothiorhodospira*) genera, as well as the alga, *Dunaliella* sp. (Ollivier et al., 1994). One of the bacterial species, *Plectonema nostocorum*, has the highest recorded pH values for growth, up to pH 13 (Edwards, 1990).

Similar to the acidiphiles, the central challenge for extremely alkaliphilic bacteria is the need to establish and sustain a neutral cytoplasmic pH. This is done through a complex cycle that begins with the primary extrusion of protons via the electron transport chain and their reintroduction to the cell via ATPase (see section 2.1.4(b) for details). This process then energizes a Na^+/H^+ (or K^+/H^+) antiport system that catalyzes the uptake of protons from the external environment coupled to the release of cytoplasmic Na^+. The cell manages this by circulating Na^+ across the plasma membrane, with the aid of internal pH-sensitive Na^+/solute symporters or Na^+-specific channels (Krulwich et al., 1997). This requirement for Na^+ means that some species, such as those from the genera *Halobacterium* or *Dunaliella*, can actually grow directly on (Fig. 1.16), or within, salt crystals. Amazingly, some species trapped within fluid inclusions of growing NaCl crystals may remain viable for up to 6 months (Norton and Grant, 1988).

Extreme halophiles also require a mechanism to actively maintain a cytoplasmic solute concentration higher than the surrounding brine to prevent plasmolysis (Beales, 2003). Several types of solutes are noninhibitory to biochemical reactions within the cell, including organic compounds such as sugars that are generated within

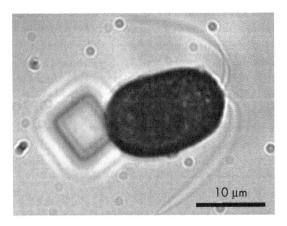

Figure 1.16 Photomicrograph of the green alga, *Dunaliella salina*, attached to a crystal of halite in a salt lake in Australia. (Courtesy of Mike Dyall-Smith.)

the cell, or ions such as K^+ that are extracted from the brine and pumped into the cell until its cytoplasmic concentration is higher than the external concentration of Na^+ – typically the dominant cation in brines but more toxic to the cell. During episodes of extreme dessication, some species even have the capacity to enter a state of anhydrobiosis, when there is little intracellular water and the cells display no metabolic activity (Potts, 1994). Other general adaptations made by the alkaliphiles include modifying the composition of the cell wall to give it a more negative surface charge to inhibit the unwanted entry of hydroxyl anions.

1.5.5 Deep-subsurface environments

Microorganisms are the only life forms that can inhabit deep subsurface environments where the pore spaces are only micrometers in scale. The metabolic processes employed by subsurface species are very similar to those utilized by surface species, with two major exceptions. First, light is not available at depth, so photosynthesis is not possible. Therefore, microbial life is dependent upon energy sources that have been buried in the sediment or enter as dissolved components in recharge waters. Second, in sediments, O_2 is rapidly depleted in the top few meters, hence anaerobic respiratory processes dominate at depth (Lovley and Chapelle, 1995).

In a well-cited study, core samples of marine sediment, collected from > 500 m depth, contained viable populations of sulfate-reducing bacteria that were shown to slowly metabolize using supplies of pore water sulfate (Parkes et al., 1994). Based on extrapolation of direct cell counts, the authors estimated that similarly metabolizing populations of buried SRB, growing globally in sediment to depths of 500 m, could constitute as much as one-tenth of the Earth's living biomass. Moreover, it suggests that sedimentary organic matter will undergo continued bacterial modification long after burial (see section 6.2 for details on sediment diagenesis). Studies of oil-producing formations in deep sediments of the

North Sea and off Alaska similarly indicate that thermophilic archaeal (e.g., *Archaeoglobus* sp.) and bacterial (e.g., *Thermodesulforhabdus* sp.) sulfate reducers can grow at depths over a kilometer, presumably utilizing organic acids and hydrocarbons as their primary source of energy (L'Haridon et al., 1995; Nilsen et al., 1996). Since the isolates grew at the same temperature as that measured in the reservoirs, it was proposed that they are representatives of the indigenous microbial community (versus being mesophilic surface contaminants), and that they may have been deposited with the original sediment, and have survived over the millions of years.

Microorganisms additionally appear to reside in submarine basalts, up to 500 m within the Earth's crust, below which temperatures reach upwards of 100°C and the permeability becomes prohibitively low (e.g., Furnes et al., 2001). Although they have never been cultured, and their metabolic processes are largely unknown, several points of evidence suggest that these microorganisms promoted dissolution of the outermost layers of the glassy substratum, and then grew within the micro-pitted surfaces:

1 Cell-sized, granular and tubular etch marks, and bacteria-like structures infilling the pores, have been identified in cores of fresh basaltic glass. They are invariably connected to, or are directly rooted in, fractures in the glass (e.g., Fig. 1.17).

2 Some of the altered glass is enriched in bioessential elements, such as C, N, P, and K, that suggests the presence of biological material.

3 Carbon isotopes (δ^{13}C) of the disseminated carbonate phases in the glassy margins show fractionations characteristic of microbial activity.

4 Bacterial and archaeal DNA is present in altered basalts.

Interestingly, the basalt-dwelling microorganisms appear unique from those inhabiting the overlying sediment and seawater, once again suggesting that the microbial community is indigenous to the rock (Thorseth et al., 2001). However, it is presently unresolved whether the

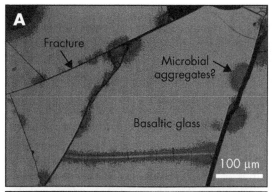

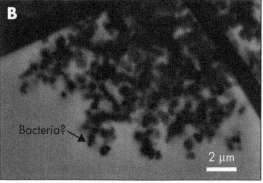

Figure 1.17 Putative biogenerated structures found within fractures of basaltic glass. (A) Fractures of different generations cutting the basaltic glass. Deep Sea Drilling Project (DSDP) core 46–396B. (B) Alteration at the junction between two fractures showing small patches of isolated and coalesced grains that have similar size and morphology to bacteria. DSDP core 69–504B. (From Furnes et al., 2001. Reproduced with permission of American Geophysical Union.)

microorganisms actively inhabit basaltic crust to depths of 500 m, or if they are merely remnants of microorganisms that once colonized seafloor basalt. Microbial surface communities that were buried by younger volcanic flows (and pelagic sediments) may either die during burial, so that only textural and geochemical traces of life are carried into the deeper part of the crust, or the survivors may gradually adapt to the increasing temperatures and pressures associated with burial, with concomitant changes in microbial community structure.

There is now significant evidence that microorganisms also live within deep terrestrial rock fractures and groundwaters. A community of bacteria, slowly migrating vertically with groundwater flow, would need between 1000 and 10,000 years to reach a depth of 1 km (Pedersen, 1993). The extent to which they can migrate is set by the temperature, assuming that adequate energy supplies are available for microbial life. Therefore, even at 6 km deep, as in some gold mines of South Africa, fissure waters still support a deep biosphere of hyperthermophilic *Archaea* (Takai et al., 2001). Remarkably, life is not limited by pressure. Experiments have documented that bacteria survive pressures equivalent to a depth of 50 km below the Earth's crust (Sharma et al., 2002), while bacteria have even been recovered from the world's deepest sediment in the Mariana Trench at 10,898 m, where optimal pressure conditions for growth exceed 700 atmospheres (Kato et al., 1998).

Two distinct biogeochemical end members characterize deep basaltic aquifers: (i) groundwaters with methanogenic signatures; and (ii) those with high sulfate and sulfide concentrations indicative of active sulfate reduction. The discovery of a methanogenic community 1.5 km deep within a crystalline rock aquifer of the Columbia River flood basalts has led to significant controversy regarding what is the primary energy source supporting the microbial community (Stevens and McKinley, 1995). The original authors suggested that the methanogens were supported by H_2 generated via water reacting with reduced ferromagnesian silicates in basalt. However, this view has not been supported by experiments designed to replicate conditions in the aquifer because they yielded insufficient amounts of H_2 to sustain a microbial ecosystem beyond a few hundred years (Anderson et al., 1998). Moreover, molecular analyses of the groundwater from the actual aquifer indicated that less than 3% of the microbial community is composed of methanogens, the bulk of the population being anaerobic respiring bacteria, i.e., the source of H_2 is organic carbon (Fry

et al., 1997). Nonetheless, noncarbon-sourced H_2 can be the primary reductant for methanogen-dominated aquifers, but only when it is of geothermal origin (Chapelle et al., 2002) or from the radiolysis of water (Lin et al., 2005). In some sulfate-rich sedimentary aquifers, where the rocks have been isolated from contact with the surface for at least 10^4 years, *in situ* sulfate reduction activity has been detected, supported by organic matter that was deposited along with the formation sediments (Krumholz et al., 1997).

1.5.6 Life on other planets

If any one particular field in geomicrobiology has attracted an enormous amount of recent interest, it is surely the study of the potential existence of life on other planets (known as astrobiology or exobiology). This is highlighted by the worldwide media attention drawn to the reported findings of life in the ALH84001 meteorite. This meteorite was originally found on the Antarctic ice sheet in 1984. It has a rock crystallization age of 4.5 billion years and is interpreted as originating from Mars. Four lines of evidence for life in this meteorite were initially put forward by McKay et al. (1996): (i) the presence of carbonate globules; (ii) the presence of magnetite (Fe_3O_4) characteristic of microbial formation within the carbonate; (iii) the presence of indigenous, reduced carbon; and (iv) bacteria-like structures within the carbonate globules (Fig. 1.18).

Despite heavy debate, it has finally been accepted that the carbonate globules were probably formed at low temperatures (as originally suggested by Romanek et al., 1994), and that their age of formation is consistent with a period in time when there was relatively abundant water on the surface of Mars; dried remnants of outflow channels, crater lakes, and possibly even an ocean testify to the former presence of water. Based on an exhaustive study of the meteorite's magnetite, Thomas-Keprta et al. (2000) support a biogenic origin for about 25% of the magnetites enclosed within the carbonate globules. These magnetites are single domain, chemically pure

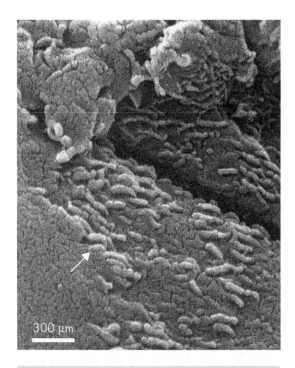

Figure 1.18 SEM image of the 20–100 nm long structures (arrow) associated with the carbonate globules within the Martian meteorite ALH84001. Based on their morphology, they have been interpreted by some researchers as being fossilized remains of nanobacteria. (Courtesy of the NASA Jet Propulsion Laboratory (NASA-JPL).)

Fe_3O_4, and they have prismatic morphologies that at present are only known to be produced by magnetotactic bacteria (see section 4.2.1). This view, however, needs to reconcile the profound implication that early Martian life would have been very similar to extant life on Earth and, crucially, it would have necessitated a complex microbial community comprising primary producers and respiring bacteria, that lived under microaerophilic conditions. Other studies have confirmed the presence of indigenous polycyclic aromatic hydrocarbons (PAHs). Although these might be possible decay products of bacteria in ALH84001 (Clemett et al., 1998), the majority of reduced organic matter appears due to massive terrestrial contamination while the meteorite was in the Antarctic ice (Jull et al., 1998). Finally, the microstructures

interpreted as possibly representing "nanobacteria" (20–100 nm in size) fall below the accepted minimum size range for even the smallest identified cells capable of independent growth, the 200 nm mycoplasma that are phylogenetically related to the Gram-positive bacteria (Madigan et al., 2003). This makes it difficult to conceive how those structures could be of biological origin. Indeed, one very significant problem in assigning a biogenic origin to any structure is that abiological processes commonly form similar morphologies that can easily be misinterpreted as a microorganism. Even preparation of samples for electron microscopy can lead to artefacts that superficially appear biological (e.g., Bradley et al., 1997). The main proponents for the presence of life on Mars have since documented further structures in other, younger, Martian meteorites, that have a markedly more biological appearance (e.g., Gibson et al., 2001). Yet once again, these samples also suffer from contamination problems by terrestrial microorganisms. This means that even if the "fossil-like" structures are biological, they may instead be Earth-based and not extraterrestrial (Steele et al., 2000).

Notwithstanding the critics, many proponents for life on Mars believe that it awaits discovery in a subsurface layer of frozen water, formed during a period over 3.5 billion years ago when the planet still had surface volatiles. Not only does this mean that life may have been present when the carbonate globules were formed in ALH84001, but that it might have persisted until the present day, possibly as an endolithic community in some subsurface cryospheric refuge (i.e., perennially frozen), much as cold-adapted bacteria do on Earth today. One possibility then is that cells frozen in the cryosphere could be brought to the surface of Mars during intermittent melting of the icy layer (due to bolide impact, intermittent volcanism, or solar warming at high obliquities). Once at the surface, the revived cells could metabolize for a few cycles before the surface water sublimated and the cells dried out and/or became fossilized. Thus, there is the possibility of finding fossil traces of Martian life from throughout its history, but the real question is where to look? On Earth life requires a continuous flux of energy, and microorganisms are quite adept at utilizing chemical disequilibrium to support their growth. As will be discussed in Chapter 7, hydrothermal systems are ideal sites for origin of life models, and it would not be unreasonable to presume that if life were to have evolved on Mars evidence for it would be in the vicinity of a thermal spring (Jakosky and Shock, 1998).

As this book is being written, the National Aeronautics and Space Administration (NASA) have two rovers to the planet's surface as part of the Mars Exploration Rover Mission. Their aim is to ascertain the history of climate and water at two sites where conditions might once have been favorable to life. What they have already discovered is that at some nonpolar regions of the planet, there exists mineralogical and geomorphological features suggestive of a watery history. The possible presence of jarosite (a hydrated mineral), putative hematite-rich concretions that typically form in sediment (Fig. 1.19), outcrops with bedding features, and crater-like depressions interconnected to stream-like features are some examples of these findings.

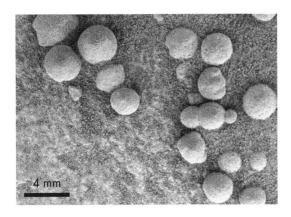

4 mm

Figure 1.19 This microscopic image, taken at the outcrop region dubbed "Berry Bowl" near the Mars Exploration Rover Opportunity's landing site, shows the so-called "blueberries" that appear to be concretions rich in hematite that grew in pre-existing wet sediments. (Courtesy of the NASA Jet Propulsion Laboratory (NASA-JPL).)

Another question receiving significant attention is whether the search for extraterrestrial life can extend beyond Mars to other planetary bodies (Cavicchioli, 2002). One such place is Jupiter's moon, Europa. Its surface is covered with an icy crust some 150 km thick. Given the possibility of abundant subsurface water, a sustained heat source maintained by tidal frictional heating of the moon's interior, and a potential supply of organic molecules generated through chemical disequilibrium between the ice cover and the charged particles in Jupiter's magnetosphere, all the crucial components necessary to support a chemotrophic microbial system may exist on the moon (Chyba and Phillips, 2001). The parent bodies of carbonaceous meteorites may also have harbored life. For instance, the Murchison meteorite contains a diverse assemblage of amino acids that were apparently synthesized during an early transient hydrothermal phase (Cronin, 1989). These are by no means the only examples of potential sites for extraterrestrial life, and as more planned missions to space come to fruition, so too will be our drive to sample other planetary bodies for evidence of life.

1.5.7 Panspermia

Taking the concept of extraterrestrial life one step further, a number of researchers have revived Svante Arrhenius' original theory called Panspermia (Arrhenius, 1908), that if life had existed near the surface of Mars, or any terrestrial planet in a continuously habitable zone, it could feasibly have been ejected as part of a meteorite and delivered intact as a living organism to Earth (e.g., Melosh, 1988). Not only that, but given the continuous supply of meteoric material from Mars to Earth, billions of microbial hitchhikers may have traveled between these planets many times during the history of the solar system (Mileikowsky et al., 2000).

If an organism were to travel from Mars to Earth it would have to survive three stages of transport, namely: (i) ejection from the planet of origin; (ii) travel through space; and (iii) atmospheric entry. During phase 1, the organism would experience extreme acceleration, a change in acceleration (or jerk), and shock heating. Laboratory experiments have been designed whereby spores of *Bacillus subtilis* were accelerated in a centrifuge operating at 100,000 rpm, while others were loaded into lead bullets and fired from a compressed-air pellet rifle into a solid target. Astoundingly, 10% of the centrifuged spores and greater than 40% of ballistic spores survived these conditions (Mastrapa et al., 2001).

Phase 2 includes exposure to extreme vacuum, lethal UV and ionizing radiation, bombardment by high energy particles, desiccation, and freezing temperatures. Experiments aboard the European Space Agency's Long Duration Exposure Facility have shown that *B. subtilis* spores can survive in deep space for nearly 6 years (Horneck et al., 1994). Although solar UV radiation reduces survival of the spores greatly, some spores that were sheltered by layers of dead cells, polysaccharides, or salts remained viable. Furthermore, exposure to the desiccating effects of space did not have an effect on the survival of *B. subtilis* spores (Rettberg et al., 2002). These findings suggest that an organism within a meteorite may survive transfer.

Phase 3 would involve heating upon atmospheric entry and deceleration upon landing. A meteoritic fall through the atmosphere is so fast (taking only a few tens of seconds) that the heat may not have time to penetrate the interior, thereby shielding the inhabiting organism from an extreme rise in temperature. For example, recent analysis of the ALH84001 meteorite has shown that the core was never heated over 40°C since it was ejected from Mars (Weiss et al., 2000). Experiments have also shown that a certain fraction of *B. subtilis* spores survive pressures in excess of 400 kilobar in simulated meteorite impacts (Horneck and Brack, 1992).

The whole premise behind delivering life from another source, however, merely transfers the problem of life's origin to another locality. Yet, it does lead one to speculate whether life could have evolved elsewhere in the solar system. As pointed out by Kirschvink and Weiss (2002), Mars may have had one of the most essential prerequisites for life some 4 billion years ago, that being a large electrochemical gradient between the highly reduced mantle (due to a lack of plate tectonics) and the more oxidized crust. Other terrestrial planets with a chemistry not too dissimilar from Earth may also exist somewhere in the universe. For that matter, models predict that for terrestrial planets like Earth that are older then 1 billion years, the probability of biogenesis is greater than 13% (Lineweaver and Davis, 2002).

1.6 Summary

Microorganisms are found everywhere on the Earth's surface where water is at least temporarily available. Their ubiquity, coupled with their high cell densities and fast growth rates, means that microorganisms are extremely important geochemical agents. As we will examine in the ensuing chapters, by facilitating aqueous redox processes, sorbing and concentrating metal cations and anions, mediating mineral precipitation and dissolution, and driving Earth's biogeochemical cycles, microorganisms have left, and continue to leave, their mark on our planet. Nevertheless, our understanding of microbe–environment interactions is still in its infancy, and imminent discoveries and advances in geomicrobiology are likely to come from all areas of this expanding discipline.

2

Microbial metabolism

Over the past 4 billion years the evolutionary forces acting on microorganisms have molded them in such a way that they are now able to inhabit every conceivable niche where liquid water is at least periodically available. Each species employs a biogeochemical lifestyle that is optimally suited to its particular environmental conditions, and in this regard, microorganisms show tremendous diversity in the ways in which they obtain the energy needed for growth. This leads us to the concept of metabolism, that refers to all of the biochemical processes occurring within a cell. It involves two basic kinds of transformations, building up or biosynthetic processes, called anabolism, and breaking down or degradative processes, called catabolism. In the process of anabolism, cells use chemical energy to convert nutrients and simple compounds into more complex structural and functional macromolecules. This includes the formation of proteins from amino acids, nucleic acids from nucleotides, polysaccharides from simple sugars, and lipids from glycerol and fatty acids. The opposing process, catabolism, oxidizes organic and/or inorganic compounds, accompanied by the release of chemical energy and the excretion of waste products into the surrounding environment. Some of the chemical energy is captured and utilized by the cells for cell movement, the transport of nutrients into the cell, and anabolic reactions, while the remaining energy is lost to the environment in the form of heat. In this chapter we will briefly examine the general principles underpinning energy production in microorganisms and look at how cells couple energy–

electrons–carbon through phototrophy, chemolithotrophy, and chemoheterotrophy. Crucially, the various metabolic pathways employed by microorganisms directly influence the chemistry and distribution of nearly all elements in the periodic table. Here, we will focus on the major biogeochemical cycles.

2.1 Bioenergetics

2.1.1 Enzymes

In order for catabolic reactions to take place, a certain amount of initial energy needs to be invested to align reacting groups and break the pre-existing chemical bonds. This is referred to as the activation energy barrier, and unless the reactants can be raised to a high enough energy level to overcome this barrier, the reactions cannot proceed spontaneously. Cells, however, are equipped with enzymes, those proteins capable of increasing the rates of biological reactions up to 10^{17} times the abiological rate (Nelson and Cox, 2005). Enzymes do not affect the free energy change of a reaction, but only the speed with which it proceeds by lowering the activation energy. In other words, they serve as catalysts, and without them most reactions in living organisms would simply not occur at the required rates. Enzymes are also highly specialized in the reactions they catalyze. For example, a particular enzyme may facilitate the hydrolysis of a peptide bond only between two specific amino acids, while another may help form a particular sugar. Not surprisingly, a cell possesses thousands of different enzymes.

The catalytic event that converts reactant (also known as substrate) to product occurs at a site on the enzyme (the active site) that has been evolutionarily structured, in terms of stereo-chemistry, polarity, and charge, to provide high-affinity binding for that particular substrate. During the subsequent enzyme-catalyzed reaction the substrate can be altered in a number of ways; through oxidation-reduction, its chemical structure being simplified, or by combining with another substrate molecule. After any of those reactions, the product is released and the enzyme is then returned to its original state without having undergone any physical alteration in the process. In any typical enzyme-catalyzed reaction, substrate and product concentrations are usually hundreds or thousands of times greater than the enzyme concentration. Consequently, each enzyme molecule catalyzes the conversion of many substrate molecules to product. However, the enzymes have the propensity of becoming "saturated" with substrate, causing the reaction rates to become independent of substrate concentration. The simplest scheme for an enzyme-catalyzed reaction is given in the following equation:

$$E + S \underset{k2}{\overset{k1}{\longleftrightarrow}} ES \overset{k3}{\rightarrow} E + P \qquad (2.1)$$

where E is the enzyme, S is the substrate, and P is the product; $k1$ (forward reaction), $k2$ (reverse reaction), and $k3$ are the rate constants associated to each step of the equation. There are two ways to determine the relationship between the observed rate and those of the above scheme: (i) by making the rapid-equilibrium approximation and assuming that the equilibration between E, S, and ES is fast compared to the subsequent reaction of ES to $E + P$, thus product formation is the rate limiting step; or (ii) by making what is known as the steady-state assumption, where substrate is converted into product at a constant rate, while the rate of change of ES is essentially zero. Both models obey what is known as the Michaelis–Menten equation:

$$V_0 = V_{max}[S]/[S] + K_m \qquad (2.2)$$

where V_0 is the initial rate of reaction, K_m (the Michaelis constant) is an indicator of enzyme–substrate affinity calculated as $(k2 + k3)/k1$, and V_{max} is the rate of reaction at enzyme saturation.

Some enzymes consist entirely of proteins, but others also contain nonprotein molecules, called cofactors, that participate in catalysis. These cofactors can be divided into two broad categories based on the nature of their association with the enzyme. Prosthetic groups are very tightly bound to their enzyme, usually permanently. By contrast, coenzymes are bound rather loosely to enzymes and they can associate with a number of different enzymes at different times. Coenzymes are particularly important because they increase the diversity of oxidation-reduction reactions by bridging chemically dissimilar molecules during reaction and serving as intermediate carriers of small molecules from enzyme to enzyme (e.g., Fig. 2.1). Some coenzymes are derived from vitamins, while others have metal cations. The metalloenzymes comprise greater than one-third of all characterized enzymes, with iron, magnesium, molybdenum, manganese, copper, zinc, cobalt, and vanadium being the most important metals (see da Silva and Williams, 2001 for details).

2.1.2 Oxidation-reduction

The utilization of chemical energy in living systems involves the transfer of electrons from one reactant to another. These are known as oxidation-reduction reactions, or redox reactions (Box 2.1). The energy source, known as the primary electron donor (PED), gives up its electrons because it has the most negative electrode potential in a given system. Those electrons are then transferred, either one or two at a time, via a series of intermediate carrier enzymes to a terminal electron acceptor (TEA), the molecule with the most positive electrode potential (see Jones, 1988 for details).

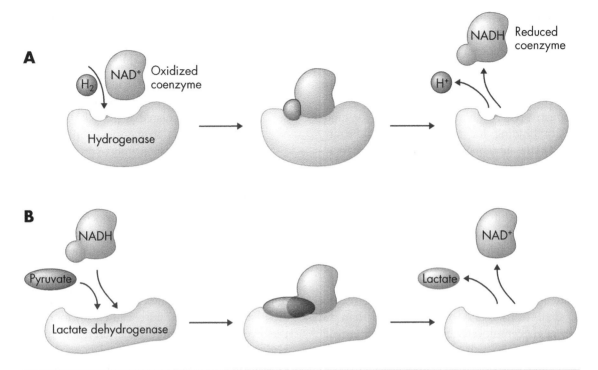

Figure 2.1 (A) Representation of how an enzyme (hydrogenase)–coenzyme (NAD⁺) partnership can lead to the formation of a reduced compound, such as NADH. (B) Then as a diffusible carrier, NADH can reduce pyruvate to lactate via lactate dehydrogenase. After that reaction, NAD⁺ can help facilitate another oxidation-reduction reaction.

Box 2.1 Oxidation and reduction reactions

Oxidation is defined as the removal of an electron(s) or hydrogen atom(s) from a compound, while reduction is defined as the gain of an electron(s) or hydrogen atom(s). When hydrogen atoms are involved, the loss of an electron causes the hydrogen atom to become a proton (H^+):

$$H_2 \rightarrow 2e^- + 2H^+$$

Because electrons cannot exist independently in aqueous solution, they must associate with another molecule. For example, the oxidation of H_2 (written as electrons and protons) can be coupled to the reduction of oxygen gas:

$$0.5O_2 + 2e^- + 2H^+ \rightarrow H_2O$$

In this case H_2 is the electron donor and O_2 is the electron acceptor. In other words, H_2 (with a valence of 0) has lost its two electrons to become $2H^+$ (each atom now has a charge of +1), while O_2 (with a valence of 0) becomes reduced to water with a valence of -2.

The tendency of a compound to accept or donate electrons is expressed as its electrode potential (E) which is referenced to a standard substance, H_2, of 0.00 volts. Standard electrode potentials ($E°$) are given for 25°C, 1 atmosphere, and unit activity of substances (in dilute solutions,

continued

Box 2.1 *continued*

activity is equal to molar concentration). In biology the electrode potentials are usually given at neutrality (pH 7) because the cytoplasm of the cell is near neutral, thus we add a superscript ($E^{\circ\prime}$) when dealing with electrode potentials at pH values of 7. Using these conventions, $E^{\circ\prime}$ of the half reactions: $H_2 \rightarrow 2e^- + 2H^+$ is -0.41 V, and that of: $0.5O_2 + 2e^- + 2H^+ \rightarrow H_2O$ is $+0.82$ V. When half reactions are written with the oxidized form on the left, such potentials indicate the direction in which electrons will be transferred. For example, in the above half reactions, the -0.41 V electrode potential of $2H^+/H_2$ suggests that H_2 is a very good electron donor (i.e., a strong reducing agent), whereas the high positive potential of the $0.5O_2/H_2O$ couple suggests that oxygen has a very high tendency to accept electrons (i.e., a strong oxidizing agent).

Commonly, redox reactions are presented in a comparative way with the most negative electrode potentials written at the top and the most positive written at the bottom. This forms the basis of what is commonly referred to as the electromotive series. Those reduced substances at the top right have the greatest tendency to donate electrons, whereas those on the bottom left have the greatest tendency to accept electrons. The alkali metals and alkaline earths are all strong reducing agents because they are good electron donors; some members of the transition groups are moderate to weak reducing agents compared to H_2 (e.g., Zn^0, Cr^0, Ni^0, Pb^0), while others (Fe^{3+}, Mn^{4+}) are strong oxidizing agents.

The difference in electrode potential between two half reactions is expressed as ΔE°; the greater the ΔE° value, the greater the energy associated with those reactions. For example, coupling the half reactions of $0.5O_2/H_2O$ with CO_2/acetate (CH_3COO^-) at pH 7 yields an $\Delta E^{\circ\prime}$ of 1.11 V, whereas the couple of NO_3^-/NO_2^- with CO_2/acetate yields only an $\Delta E^{\circ\prime}$ of 0.72 V at the same pH.

Microbiologically important electrode potentials at pH 7. (Values from Thauer et al., (1977) and Nelson and Cox (2005).)

Redox pair	$E^{\circ\prime}$ (V)
$2H^+/H_2$	-0.41
Ferredoxin ox/red	-0.39
$NAD^+/NADH$	-0.32
Cytochrome c_3 ox/red	-0.29
CO_2/acetate	-0.29
S^0/H_2S	-0.27
CO_2/CH_4	-0.24
$FAD/FADH_2$	-0.22
SO_4^{2-}/H_2S	-0.22
Pyruvate/lactate	-0.19
$FMN/FMNH_2$	-0.19
Ubiquinone ox/red	$+0.05$
Cytochrome b ox/red	$+0.08$
AsO_4^{3-}/AsO_3^{3-}	$+0.14$
$Fe(OH)_3 + HCO_3^-/FeCO_3$	$+0.20$
Cytochrome c_1 ox/red	$+0.23$
Cytochrome c ox/red	$+0.25$
Cytochrome a ox/red	$+0.29$
NO_2^-/NH_3	$+0.34$
Cytochrome c_2 ox/red	$+0.36$
Cytochrome a_3 ox/red	$+0.39$
NO_3^-/NO_2^-	$+0.43$
Fe^{3+}/Fe^{2+}	$+0.77$
Mn^{4+}/Mn^{2+}	$+0.80$
$0.5O_2/H_2O$	$+0.82$
NO/N_2O	$+1.18$
N_2O/N_2	$+1.36$

Although the transfer of electrons through the intermediate carrier enzymes involves a series of independent redox reactions, the net free energy change, known as the standard Gibbs free energy ($\Delta G^{\circ\prime}$) at pH 7 of the complete reaction sequence is determined by the difference in standard electrode potentials ($\Delta E^{\circ\prime}$) at pH 7 between the PED and the TEA (Box 2.2). Even if cells could transfer electrons directly from PED to TEA, this reaction would end up liberating most of the energy as heat. In fact, there is no reaction sufficiently endergonic as to consume all that energy in one step. Therefore, the systematic transfer of electrons in discrete and controlled steps down the redox gradient is the only means by which the cell can reduce the free energy, yet maximize energy gains. As will be discussed below, cells display various approaches to managing electron flow, but given the link between energy and $\Delta E^{\circ\prime}$, it is not surprising that they prefer to couple PEDs and TEAs with the greatest differences in electrode potential.

Some electron carriers are fixed in the membrane, whereas others are freely diffusible in the cytoplasm or periplasm, transferring electrons from one place to another. Two of the most common freely diffusible electron carriers are the coenzymes NAD$^+$ (nicotinamide adenine dinucleotide), that is involved in catabolic reactions, and NADP$^+$ (NAD with an additional phosphate), that is involved in anabolic reactions. Both compounds contain derivatives of

---

Box 2.2 Gibbs free energy

The change in free energy during a reaction is expressed as ΔG°, the superscript implying a free energy value determined under standard conditions of 25°C, with all products and reactants initially at 1 molar concentration – of course, in cells, reactants are not found in molar concentrations, so ΔG° values must be considered reasonable estimates. Moreover, since many of the reactions discussed in this book are biologically mediated, we adopt the symbol $\Delta G^{\circ\prime}$ to represent the standard free energy change at pH 7. When the reactants possess substantially more free energy than the products, the $\Delta G^{\circ\prime}$ is negative, free energy is released, and the reaction should spontaneously run from reactants to products (in terms of the way reactions are written, from left to right). Such reactions are termed exergonic. If the $\Delta G^{\circ\prime}$ for the reaction has a large negative value, then it is likely that the reaction will run to completion, i.e., very little reactant remains when equilibrium is achieved. A $\Delta G^{\circ\prime}$ value near zero is characteristic of a readily reversible reaction; reactants and products have almost the same free energies. When $\Delta G^{\circ\prime}$ is positive, energy needs to be consumed

to drive the reaction to the right. These reactions are called endergonic. In general, the magnitude of $\Delta G^{\circ\prime}$ is a measure of the driving force of a chemical reaction. However, these values give no real indication of how fast equilibrium will be achieved, i.e., the kinetics.

Another useful term of be aware of is the Gibbs free energy of formation of individual compounds (known as G_f°). Compounds that form spontaneously from the elements have negative G_f° values; those that do not form spontaneously have positive G_f° values. Elements (H$_2$, O$_2$, N$_2$, etc.) have G_f° equal to zero. Most simple compounds have negative values, reflecting the fact that they tend to form spontaneously from the elements.

It is possible to calculate the change in free energy of a reaction by knowing the individual G_f° of the relevant compounds, such as those listed below. As an example, take the oxidation of acetate by oxygen:

$$CH_3COO^- + 2O_2 \rightarrow H_2O + 2CO_2 + OH^-$$

We can calculate the $\Delta G^{\circ\prime}$ of the reaction (all values in kJ mol^{-1}) by simply subtracting the

continued

Box 2.2 *continued*

Free energies of formation (G_f°) for some compounds (in kJ mol^{-1}).
(Values from Thauer et al. (1977) and Faure (1998).)

Carbon compounds	G_f° (in kJ mol^{-1})	Metals/ metalloids	G_f° (in kJ mol^{-1})	Nonmetals	G_f° (in kJ mol^{-1})
CO_2	−394.40	Fe^{2+}	−78.87	H^+	−5.69/pH unit
CO	−137.15	Fe^{3+}	−4.60	OH^-	−197.20 (pH 7)
H_2CO_3	−623.16	$Fe(OH)_3$	−655.21	H_2O	−237.17
HCO_3^-	−586.85	$FeCO_3$	−673.23	H_2O_2	−134.10
CO_3^{2-}	−527.90	FeS_2	−166.90	SO_3^{2-}	−486.60
CH_4	−50.75	Fe_3O_4	−1015.41	SO_4^{2-}	−744.63
Acetate	−369.41	Mn^{2+}	−227.93	$S_2O_3^{2-}$	−513.40
Glucose	−917.22	MnO_2	−456.71	H_2S_{aq}	−27.87
Lactate	−517.81	$HAsO_4^{2-}$	−714.71	HS^-	12.05
Methanol	−175.39	$H_2AsO_3^-$	−587.22	NO_2^-	−37.20
Pyruvate	−474.63	UO_2	−962.32	NO_3^-	−111.34
		UO_2^{2+}	−949.77	NH_3	−26.57
				NH_4^+	−79.37

sum of the G_f° of the reactants from that of the products. In this case this leads to:

$$\Delta G^{\circ\prime} = [G_f^\circ \text{ products}] - [G_f^\circ \text{ reactants}]$$

$$[(-237.17: \textbf{H}_2\textbf{O}) + 2(-394.4: \textbf{CO}_2)$$
$$+ (-197.20: \textbf{OH}^- \text{ at pH 7})]$$
$$- [(-369.41: \textbf{CH}_3\textbf{COO}^-)]$$

$$\Delta G^{\circ\prime} = [-1223.17] - [-369.41]$$

$$\Delta G^{\circ\prime} = -853.76 \text{ kJ mol}^{-1}$$

The difference in $\Delta G^{\circ\prime}$ can also be described in terms of $\Delta E^{\circ\prime}$, using the equation:

$$\Delta G^{\circ\prime} = -nF\Delta E^{\circ\prime} \quad \text{or} \quad \Delta E^{\circ\prime} = \frac{-\Delta G^{\circ\prime}}{nF}$$

where n is the number of electrons transferred and F is the Faraday constant (96.48 kJ V^{-1}). Using the acetate oxidation reaction above, we determine that $\Delta E^{\circ\prime} = +1.11$ V simply by looking at the paired reactions. Now, if we substitute 8 for n in the $\Delta G^{\circ\prime}$ equation (since 8 electrons have moved from left to right (i.e., $2O_2$ gains 8 electrons), then $\Delta G^{\circ\prime}$ solves as −856.74 kJ mol^{-1}. This value is virtually identical to that of the −853.76 kJ mol^{-1} calculated above.

the B vitamin nicotinic acid (niacin) and function with their respective enzymes as hydrogen atom carriers (recall Fig. 2.1). Strictly, however, when NAD$^+$ and NADP$^+$ are reduced, they carry two electrons and one proton, the second H$^+$ is released into solution. Consider, for instance, the oxidation of lactate ($C_3H_5O_3^-$) to pyruvate ($C_3H_3O_3^-$) (reaction (2.3)). When lactate is oxidized, two hydrogen atoms are removed and

both electrons are transferred to reduce NAD$^+$. One of the hydrogen atoms combines with NAD$^+$, while the other remains as H$^+$ in the cell.

$$\text{COO}^- + \text{NAD}^+ \rightarrow \text{COO}^- + \text{NADH} + \textbf{H}^+ \quad (2.3)$$

CHOH	C=O
CH$_3$	CH$_3$
lactate	pyruvate

Note, at the circumneutral pH values normally encountered in cells, organic acids usually deprotonate, yielding anions and H^+. Thus, on the grounds of chemical accuracy, the names of anions (e.g., lactate, pyruvate), not the names of their acids (e.g., lactic acid, pyruvic acid), are used herein.

2.1.3 ATP generation

Some of the chemical energy released in redox reactions is conserved by the formation of biochemical compounds that contain high-energy bonds. The most important in the cell is the nucleotide, adenosine triphosphate (ATP). ATP consists of an adenine base, a ribose sugar, and three phosphate groups (Fig. 2.2), with the phosphorous and oxygen atoms attached together by either anhydride or ester bonds. In ATP, breaking the anhydride bonds yields $31.8 \, kJ \, mol^{-1}$, whereas the ester bond yields only $14 \, kJ \, mol^{-1}$ (Thauer et al., 1977). Yet, given that the cellular concentrations of ATP, ADP, and inorganic phosphate (P_i) are much lower than the $1.0 \, mol \, L^{-1}$ concentrations at standard conditions, the actual free energy (ΔG) may be fundamentally different, and in the case of ATP hydrolysis, the energy released can range from -50 to $-65 \, kJ \, mol^{-1}$ (Nelson and Cox, 2005). Consequently, during the hydrolysis of one of its phosphate anhydride bonds (from ATP to ADP, adenosine diphosphate) large amounts of useable free energy are released to drive biosynthetic reactions (reaction

(2.4)). Conversely, that also means that significant energy is required to form ATP. It should also be pointed out that during ATP hydrolysis, a molecule of inorganic phosphate is liberated into the cytoplasm: P_i is pH dependent, and can take the form of dissolved $H_2PO_4^-$ or HPO_4^{2-}, both of which are abundant in the cytoplasm because the dissociation reaction: $H_2PO_4^- \rightarrow HPO_4^{2-} + H^+$ occurs at pH 7.

$$ATP + H_2O \longleftrightarrow ADP + P_i \qquad (2.4)$$

An active cell requires more than two million molecules of ATP per second to drive its biochemical machinery (Purves and Orians, 1983). Because the supply of ATP at any particular time is limited, a mechanism must exist to replenish it. This is done by the addition of a phosphoryl group to ADP, using either chemical or light energy to drive the reaction from right to left. The process by which high energy phosphate bonds are synthesized is called phosphorylation, and three different variations exist: (1) photophosphorylation; (2) substrate-level phosphorylation; and (3) oxidative phosphorylation:

1 *Photophosphorylation* – during photosynthesis light is absorbed by specific chlorophyll molecules so that its electrons are energized to more negative electrode potentials. These electrons are then passed along a series of intermediate electron carriers back to the original chlorophyll, with a net release of energy used to convert ADP into ATP.

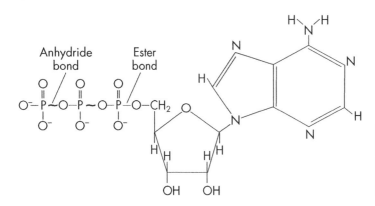

Figure 2.2 Structure of ATP, showing the three phosphate groups, two of which are joined by high-energy anhydride bonds.

2 *Substrate-level phosphorylation* – energy from the enzymatic oxidation of a phosphorylated intermediate compound is transferred directly into forming ATP. No terminal electron acceptors are required.

3 *Oxidative phosphorylation* – electrons are stripped from a reduced molecule and then passed through a series of intermediate electron carriers to a terminal electron acceptor. The stepwise oxidation-reduction reactions, similar to those in photophosphorylation, release sufficient energy to generate ATP. This process forms the basis of the chemiosmosis model.

2.1.4 Chemiosmosis

(a) Membrane carriers

Before moving onto the specifics by which different cells generate energy, let us first examine the concept of chemiosmosis. The chemiosmosis model, initially published by Mitchell (1961), describes how all living cells utilize an electrode potential gradient between a primary electron donor and a terminal electron acceptor to ultimately generate energy for phosphorylation reactions. Because $\Delta E^{\circ\prime}$ may be quite large, electrons (usually as pairs, but not always, i.e., $Fe^{2+} \rightarrow Fe^{3+}$) must sequentially be moved through an electron transport chain consisting of a series of carrier enzymes associated with the plasma membrane of bacteria, or mitochondria and chloroplasts in eukaryotes. Some carrier enzymes are found in a range of metabolic pathways (see Saier, 1987 for details). Those covered in our ensuing discussions include:

1 *Hydrogenases* – these proteins catalyze the reduction of a substrate by adding H_2. Soluble hydrogenases are found within the cytoplasm and periplasm, where they transfer electrons from H_2 to NAD^+ (forming $NADH + H^+$) or some low-redox enzymes (e.g., cytochrome c_3), respectively. The intracellular H_2 comes from various respiratory processes, while the extracellular H_2 comes from aqueous solution. Other hydrogenases catalyze the formation of H_2 from H^+ and electrons (reaction (2.5)):

$$2H^+ + 2e^- \longleftrightarrow H_2 \qquad (2.5)$$

2 *Flavoproteins* – some of these proteins initiate the electron transport chain by catalyzing the removal of H_2 from a reduced coenzyme (e.g., NADH) or molecule (e.g., lactate), and then transferring either H_2 or the electrons to the next carrier enzyme. Flavin adenine dinucleotide (FAD) or flavin mononucleotide (FMN) are two prosthetic groups that form the backbone of a number of dehydrogenases (e.g., NADH and lactate dehydrogenases).

3 *Iron-sulfur proteins* – these electron carriers range from simple molecules containing one Fe-S center to complexes containing multiple types of Fe-S clusters. Fe_2S_2 (ferredoxin) and Fe_4S_4 are the most common. Each Fe-S center has at least two redox states, a reduced form, Fe(II), and an oxidized form, Fe(III), and each center carries only one electron at a time.

4 *Ubiquinones* – these are lipid-soluble, ring-shaped (aromatic), nonprotein carriers that can freely diffuse through the membrane matrix because they are hydrophobic and not charged. They have great versatility in being able to collect electrons from many different donor carriers and subsequently passing them off to various electron accepting enzymes. By possessing three different oxidation states, they further can cycle electrons in what is known as the "Q cycle." The fully oxidized form, called quinone or Q, has both oxygen atoms connected to the ring by double bonds; attaching a hydrogen atom to one of the oxygen atoms creates the semiquinone form or QH$^{\bullet}$, while the fully reduced form, called hydroquinone or QH_2, has both oxygen atoms with hydrogen atoms attached.

5 *Cytochromes* – these proteins have an iron-containing porphyrin ring (a prosthetic group known as heme) that is capable of alternating between Fe(II) and Fe(III). There are a number of different cytochromes based on differences in the side groups of the porphyrin ring (heme *a*, *b*, *c*, *d*, and *o*), each with a different electrode potential, and hence different locations in the electron transport chain. Some serve as reductases, passing electrons onto the next carrier enzyme in the electron chain (e.g., cytochrome *c*), while others serve as the terminal oxidase, passing instead the electrons onto a TEA (e.g., cytochrome aa_3). Other cytochromes specifically facilitate the transfer of electrons from the external environment into the transport chain (e.g., those that oxidize $S_2O_3^{2-}$), and vice versa (e.g., those that reduce external mineral phases, such as ferric oxide minerals).

6 *Copper proteins* – these electron carriers undergo Cu(I) to Cu(II) transitions, and include plastocyanin and rusticyanin. Some also form prosthetic groups in complex catalytic enzymes such as cytochrome aa_3 or nitrate reductase.

Although the carriers described above are common to a wide range of microorganisms, different species employ different combinations depending on what redox pair is used, i.e., where the electrons are initially fed into the chain and at what level they are consumed. The greater the $\Delta E^{\circ\prime}$ the more carriers enzymes are required for electron transfer. When $\Delta E^{\circ\prime}$ is small, due to either a poor electron donor (e.g., Fe^{2+} at low pH) or a poor terminal electron acceptor (e.g., SO_4^{2-}), then some carriers are either simply not present in a particular cell, or they become repressed by signals in response to changing environmental conditions. Take for example *Escherichia coli*. This versatile bacterium contains the genetic information to synthesize at least two terminal oxidases, and several other carrier enzymes, but not all of them are expressed at the same time (Jones 1988).

As shown in Box 2.2, differences in $\Delta E^{\circ\prime}$ relate directly to the amount of free energy liberated. A potential difference of around 165 mV (for a two-electron transfer between donor and acceptor) is required to ensure the generation of one mole of ATP. Thus, in terms of energy, a large $\Delta E^{\circ\prime}$ (e.g., NADH to O_2) means that less substrate needs to be oxidized compared to when the $\Delta E^{\circ\prime}$ of the PED–TEA couple are more similar (e.g., Fe^{2+} to O_2).

(b) Proton motive force

The carrier enzymes are arranged in the plasma membrane in order of their electrode potentials, from most negative to most positive. This results in a stepwise release of energy as the electrons are progressively passed down the chain until they are used to reduce the TEA. Simultaneously, there is a transfer of protons from one side of the plasma membrane to the other, i.e., from the cytoplasm to the periplasm in Gram-negative cells or the cell wall region in Gram-positive cells (Hinkle and McCarty, 1978). This occurs because there is an asymmetrical arrangement of the carrier enzymes. Some carriers, such as flavoproteins and ubiquinones, acquire or transfer H_2, while iron-sulfur complexes and cytochromes carry only electrons. When a carrier can only accept electrons, protons are released as ions external to the plasma membrane, resulting in the net accumulation of protons outside the cell (Fig. 2.3). This process of proton translocation can occur up to a maximum of three times during aerobic respiration (see section 2.4.1), and fewer times with all other metabolic pathways.

The intermediate carrier enzymes release protons in two ways. First, flavoproteins are oriented so that they span the entire membrane, thus taking up H_2 on one side and releasing two protons outside into the periplasm or cell wall equivalent. Second, ubiquinones can diffuse through the membrane from side to side. After, the protons are extruded, the next carrier that then requires H_2 must obtain other protons from the dissociation of cytoplasmic water into H^+ and OH^-: both the extrusion of H^+ from the cell's cytoplasm and their consumption during H_2 formation causes a net accumulation of OH^- inside the membrane. Additional H^+ can also be released from the action of oxidase enzymes within the periplasm (e.g., the oxidation of $S_2O_3^{2-}$, reaction (2.51)).

The plasma membrane is normally impermeable to protons (and OH^-), so proton translocation and oxidative reactions establish a proton, or pH gradient, across the membrane that cannot be spontaneously restored. Accordingly, the pH immediately outside the plasma membrane drops relative to the cytoplasmic pH, which is maintained at values between 6.5 and 7.5 by various homeostatic mechanisms (Padan et al., 1981). For metabolizing Gram-positive cells, this may lead to an acidification front that extends throughout much of the cell wall, leaving just the outermost regions deprotonated and electronegative (e.g.,

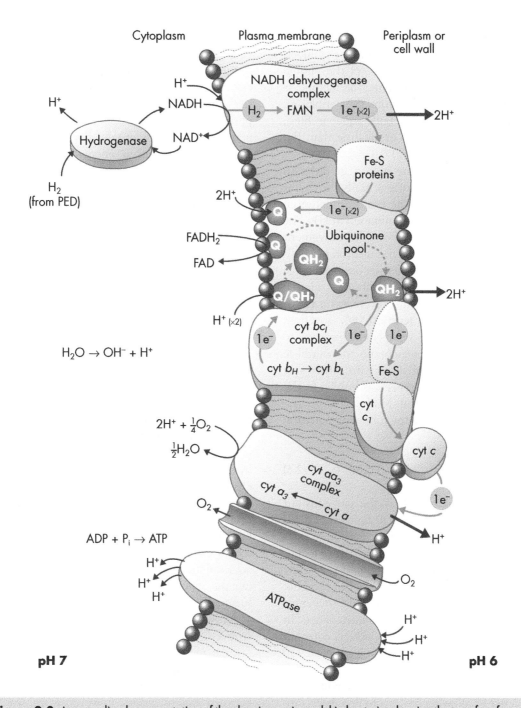

Figure 2.3 A generalized representation of the chemiosmosis model in bacteria, showing the transfer of electrons from a primary electron donor (NADH or $FADH_2$), through various intermediate carrier enzymes associated with the plasma membrane, to a terminal electron acceptor, in this case O_2. During electron transport, protons are extruded to the exterior of the periplasm (Gram-negatives) or cell wall (Gram-positives), causing an electrochemical imbalance that is neutralized when protons are driven back into the cytoplasm via the enzyme, ATPase. Q, ubiquinones; cyt, cytochromes; FMN, flavin mononucleotide. (Modified from Madigan et al., 2003.)

Calamita et al., 2001). By contrast, in Gram-negative bacteria, the pH gradient may be much more localized to the outer face of the plasma membrane given that the outer membrane is permeable to ions, and therefore, the periplasm's pH takes on that of the bulk aqueous fluid, which can be more acidic or alkaline than the cytoplasm (Beveridge, 1981).

Energy is stored in the proton gradient in two forms. One component is actually the chemical potential caused by the proton differential on opposite sides of the membrane (ΔpH). The second component arises from the movement of electrical charges across the membrane, with extruded protons leading to greater positive charge immediately outside the membrane relative to the cytoplasm. In turn, this creates a trans-membrane electrical potential ($\Delta\psi$). Together, these two components compel the protons to drive back across the membrane to neutralize the electrochemical imbalances on either side of the membrane. This process is known as the protonmotive force or pmf. Mathematically the pmf can be described by the following equation (Mitchell, 1966):

$$pmf = \Delta\psi - (2.3RT/F)(\Delta pH) \qquad (2.6)$$

In the above, equation R is the gas constant ($8.314 \, J \, mol^{-1} \, K^{-1}$), T absolute temperature (in K), and F is Faraday constant ($96.48 \, kJ \, V^{-1}$). It should be noted, however, that the relative contributions of $\Delta\psi$ and ΔpH to the pmf are highly dependent on the $\Delta E^{\circ\prime}$ and the pH of the bulk fluid/periplasm, respectively, and together must provide sufficient pmf for cell maintenance. In most species the pmf remains relatively constant over the growth pH range, varying between 150 and 200 mV (Kashket, 1985).

The free energy associated with the pmf can be recovered by allowing protons to flow back into the cell. The actual diffusion of protons back into the cytoplasm is facilitated with the aid of an enzyme called adenosine triphosphatase (ATPase). This enzyme is oriented perpendicular to the membrane and it acts as a specific channel through which protons are transported inwards (Fig. 2.3). When this controlled movement occurs, the passage of these protons back into the cell provides the release of chemical energy for the synthesis of ATP from ADP and inorganic phosphate. So, just as the formation of a proton gradient was energy-driven, the controlled dissipation of the pmf is energy releasing, with some of this energy conserved in the synthesis of ATP. This is oxidative phosphorylation. Those protons that re-entered the cell are subsequently consumed by combining with free OH^- groups (from the previous dissociation of water) and in the reduction of some TEAs (see reaction (2.25) as an example of H^+ consumption during aerobic respiration).

The pmf has benefits that extend beyond ATP generation. As mentioned in Chapter 1, the pmf facilitates pH homeostasis within the cell, with some electroneutral transport systems exchanging protons for certain cations, particularly Na^+ and K^+ (Krulwich et al., 1997). Most microorganisms also have the means to use the pmf to facilitate the transport of nutrients into the cell (Booth, 1988). For example, certain cations may actively be transported into the cytoplasm by uniporters in response to the transmembrane electrical potential, since the interior of the cell is negative when the cell is energized. Uptake of anions occurs together with that of protons by symporters, and so it is effectively the undissociated acid that enters the cell. The pmf can additionally be used to drive the transport of organic macromolecules into the cell, such as sugars, etc.

As will become evident in the following sections, chemiosmosis is a unifying model for cellular metabolism in the sense that all living organisms employ some form of electron transport for the generation of ATP. What differs between various species, however, is the way in which the electron donor is generated. It can either be from light energy (as in phototrophy) or the oxidation of either reduced inorganic compounds (chemolithotrophy) or organic compounds (chemoheterotrophy).

2.2 Photosynthesis

Photosynthesis is the utilization of solar energy by plants, algae, and certain bacteria for the synthesis of organic molecules. Two different processes of photosynthesis exist. The first, known as anoxygenic photosynthesis, is used by green bacteria, purple bacteria, and heliobacteria. These organisms use one photosystem (each with its own distinct reaction center) and generate ATP via cyclic photophosphorylation. The second, oxygenic photosynthesis, is employed by plants, algae, and cyanobacteria. They use two different photosystems and generate energy through noncyclic photophosphorylation.

2.2.1 Pigments

The basis for photosynthesis is the transformation of light energy into chemical energy using light-sensitive pigments such as chlorophyll. Chlorophyll consists of a porphyrin ring structure, similar to cytochromes, but with a central magnesium atom (Fig. 2.4). Chlorophylls also contain specific functional groups bonded to the porphyrin ring, as well as a hydrocarbon side chain that enables them to associate with lipid and hydrophobic proteins of the cell membrane. There are a number of chemically unique chlorophylls that are distinguished on the basis of their absorption spectra (see Ke, 2001 for details). Chlorophyll *a*, the principal chlorophyll in cyanobacteria, most algae, and plants, is green in color because it absorbs red light (680 nm) and blue light (430 nm), thus transmitting green light (Fig. 2.5a). Chlorophyll *b* absorbs maximally at 660 nm. Purple bacteria, green bacteria, and heliobacteria have chlorophyll of a different structure, called bacteriochlorophyll (bchl), that absorbs light of longer wavelengths (Fig. 2.5b). A number of bacteriochlorophylls exist due to variations in the composition of the ring structure. Those modifications change the absorption spectra, so for example, bchl *g* absorbs light maximally at 670 and 790 nm; bchl *c*, bchl

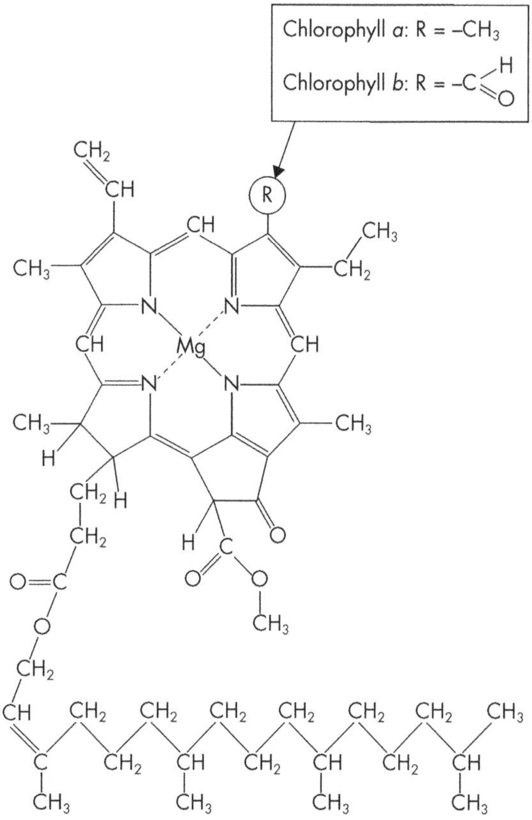

Figure 2.4 The chlorophyll molecule.

d, and bchl *e* absorb light between the range of 705 and 755 nm, bchl *a* from 805 to 890 nm, and bchl *b* absorbs at wavelengths as high as 1040 nm. Radiation above 1040 nm is too low in energy to support photosynthesis.

The chlorophyll pigments are associated with special photosynthetic membranes, the location of which differs between eukaryotes and prokaryotes (see Madigan et al., 2003 for details). Within the membrane they are organized into clusters of 200–300 molecules, of which only a few participate directly in the conversion of light energy into chemical energy. These special chlorophyll molecules, referred to as reaction center chlorophylls, receive their energy by transfer from the more numerous light-harvesting chlorophyll molecules. The latter make it possible to dramatically increase the rates of

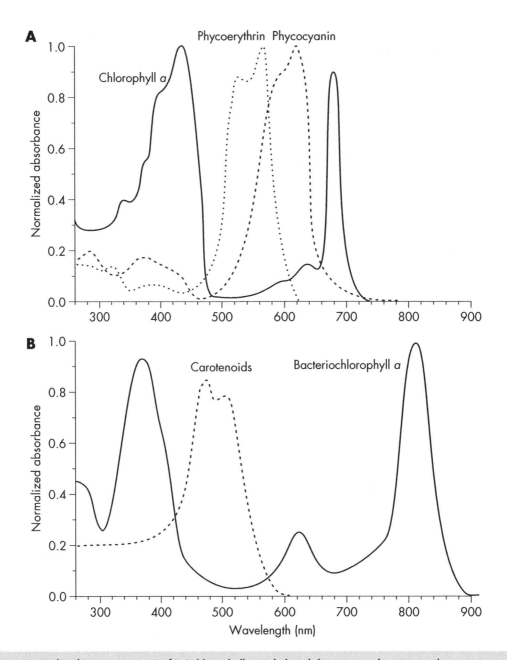

Figure 2.5 The absorption spectra of (A) chlorophyll *a* and phycobilins in cyanobacteria and (B) bacteriochlorophyll *a* and carotenoids in purple bacteria.

photosynthesis because they increase the effective cross-sectional area for light absorption.

In purple and green nonsulfur bacteria, the reaction centers contain a pair of bacteriochlorophyll *a* or *c* molecules, respectively, two molecules of

bacteriopheophytin (bacteriochlorophyll *a* without its Mg atom), as well as two molecules of both ubiquinone and carotenoid pigments. This combination has given these reaction centers the designation "type II" reaction centers. In green

sulfur bacteria, the reaction center bacteriochlorophyll *a* does not contain pheophytins or
ubiquinones, but instead it has Fe-S proteins.
These types of reactions centers are known as
"type 1." Heliobacteria have the same reaction
centers as green sulfur bacteria, but differ by
possessing bacteriochlorophyll *g*, a pigment that
closely mimics chlorophyll *a*. Interestingly, when
heliobacteria are exposed to oxygen, the original
brownish pigmentation turns green as a result
of the conversion of bacteriochlorophyll *g* to
chlorophyll *a*. Cyanobacteria use a combination
of the above (see section 2.2.4).

If chlorophylls were the only pigments
involved in photosynthesis, much of the visible
spectrum would go unused. Therefore, all photosynthetic cells contain accessory pigments that
absorb light over a range of different wavelengths (recall Fig. 2.5). Since only light that
is absorbed can be harnessed for energy, more
light absorbency means more energy is available
for cell growth. Carotenoids, such as β-carotene,
absorb in the blue and green wavelengths (Frank
and Cogdell, 1996). They also have a protecting
role by quenching toxic oxygen species generated during photooxidation reactions under UV
light. The phycobilins, such as the blue phyco-
cyanin and red phycoerythrin, absorb within the
yellow and orange wavelengths (Glazer, 1985).
Indeed, the appearance of red-colored cyanobacteria is an example of how microorganisms
utilize accessory photosynthetic pigments under
low light conditions.

2.2.2 The light reactions – anoxygenic photosynthesis

Anoxygenic photosynthesis begins when
photons of light are transferred from the light-
harvesting bacteriochlorophylls to a pair of
reaction center bacteriochlorophyll molecules.
The absorption of energy converts the latter
into stronger reductants with a lower electrode
potential (Fig. 2.6a). This reaction is the step
where light energy is transformed into chemical
redox energy. In purple bacteria, before excita-

tion their reaction center bchls (P870, the
number referring to the wavelength of radiation most strongly adsorbed) have an $E^{o'}$ of
+0.5 V; after excitation they are converted to
good electron donors with a potential of about
−1.0 V; this energy difference is exactly equal
to the energy of the absorbed photon (Brune,
1989). Following the primary photochemical
event, each excited electron is subsequently
passed to an enzyme called bacteriopheophytin,
thereby leaving behind a positive charge on
the bacteriochlorophyll molecules in the reaction center. The speed with which this reaction
occurs is of the order of billionths of a second,
ensuring that the energized electrons are not
simply returned to the donor, with the energy
being dissipated as heat. From bacteriopheophytin, the electrons are transferred to an intermediate ubiquinone (Q_A) and then exported
from the reaction center to the photosynthetic
membrane, where they are cycled between a pool
of ubiquinone molecules and a transmembrane
cytochrome bc_1 complex, and eventually transferred to a soluble periplasmic cytochrome c_2.
The latter returns one electron at a time back
to the reaction center bacteriochlorophyll *a*
molecules so that the entire light reactions can
be repeated. Green nonsulfur bacteria have a
similar reaction center and electron transport
system, except they use bchl *c*.

Energy generation during photosynthesis arises
during what is known as the Q cycle (Fig. 2.7).
As the electrons are passed from the reaction
center they reduce a molecule of quinone (Q_B) to
hydroquinone, with two protons extracted from
the cytoplasm. In turn, the hydroquinone diffuses
to the periplasmic side of cytochrome bc_1, a complex consisting of two cytochromes (*b* and c_1) and
a Fe-S protein. From there, it is oxidized back to
quinone, releasing both protons to the periplasm.
Meanwhile, one of the electrons is systematically
transferred to cytochrome c_2, while the other is
temporarily retained within the cytochrome bc_1
complex (QH•), until a second turn of the cycle,
when another electron is made available to reduce
QH• to QH_2. Also, after the second turn of the

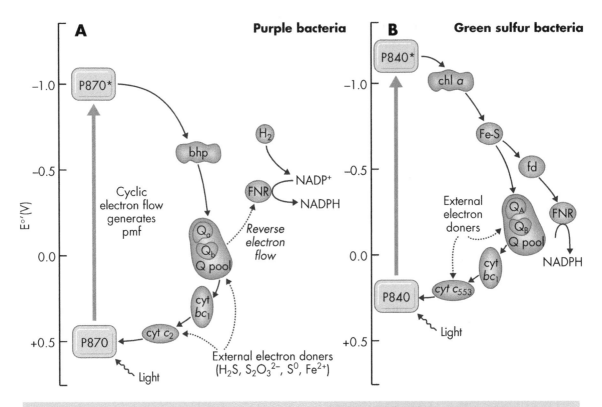

Figure 2.6 Photophosphorylation in (A) purple bacteria and (B) green sulfur bacteria. bph, bacteriopheophytin; chl, chlorophyll; cyt, cytochrome; fd, ferredoxin; FNR, ferrodoxin NADP+ reductase; Fe-S, iron-sulfur protein; Q, ubiquinones.

cycle, the other bchl a molecule is reduced (via cytochrome c_2) to repeat the photochemical reaction (Jackson, 1988). Thus, the net effect of the Q cycle is to increase the number of protons pumped across the membrane by recycling electrons. The synthesis of ATP then results from the activity of ATPase in coupling the dissipation of the protonmotive force. Because electrons are repeatedly moved around in a closed circle, and there is no actual net input or consumption of electrons, this method of making ATP is called cyclic photophosphorylation.

In green sulfur bacteria, photophosphorylation is noncyclic (Brune, 1989). The input of light energy causes P840 to become energized to a sufficiently electronegative state that its electrons can be transferred, via a series of electron carriers, directly to an enzyme called ferrodoxin NADP+

reductase. As the name suggests, it facilitates the reduction of NADP+ to NADPH (Fig. 2.6b). So that the reaction center bacteriochlorophyll does not remain in an oxidized state, electrons are consequently stripped from external sulfur compounds (H_2S, S^0, or $S_2O_3^{2-}$) or H_2, fed into the membrane electron transport chain at the ubiquinone or cytochrome c level, and then transferred to the reaction center. The latter once again becomes reduced and ready for another repeat of the cycle. The oxidized substrates are waste products and excreted from the cell, while ATP is generated by protons extruded during electron transfer in the ubiquinone pool. Heliobacteria have a very similar reaction center (but with different bchl molecules) and electron transport system to green sulfur bacteria (Nitschke et al., 1990).

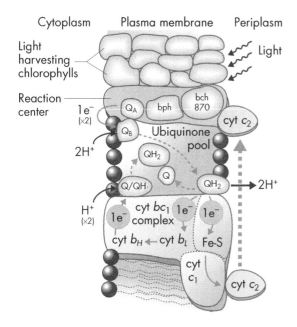

Cytoplasm   Plasma membrane   Periplasm

Figure 2.7 The arrangement of the reaction center, the quinone pool, and the cytochrome bc_1 complex in the photosynthetic membrane of a purple bacterium. While electrons are shuttled in the Q cycle, protons are systematically extruded to the periplasm setting up a proton motive force. This occurs because the cytochrome bc_1 complex has both hydroquinone-oxidizing and ubiquinone-reducing sites arranged so that protons liberated during the oxidative reaction are released into the periplasm, while those taken up during the reductive reaction are removed from the cytoplasm. The net effect is that $3H^+$ (on average) are extruded from the cytoplasm to the periplasm per electron pair that goes around the cycle. bchl, bacteriochlorophyll; bph, bacteriopheophytin; cyt, cytochrome; Fe-S, iron-sulfur protein; Q, ubiquinones. (Modified from Trüper, 1989.)

In addition to generating ATP, purple bacteria and green nonsulfur bacteria must also produce reductants, such as NADPH, to reduce CO_2 to glucose (see section 2.2.5 for details). $NADP^+$ is reduced by electrons originating from various electron donors. In anoxic environments (when O_2 is lacking) the source of electrons is from H_2, H_2S, S^0, $S_2O_3^{2-}$, and Fe^{2+}. The electrode potential of the $NADP^+/NADPH$ pair is -0.32 V, so using H_2 as an electron donor ($E^{\circ\prime}$ for $2H^+/H_2$ pair is -0.41 V) is clearly preferable as long as the microorganism has the hydrogenase to accept H_2. However, what happens when only sulfur compounds or ferrous iron are available to the cells as electron donors? Under such prevailing conditions, electrons from those donors are fed either to a c-type cytochrome in the electron transport system with a similar electrode potential or directly into the ubiquinone pool. From there the electrons are then forced backwards against the electropotential gradient (using the pmf) until the electrons can reduce $NADP^+$ (Brune, 1989). This energy-consuming process is called reversed electron flow because electrons are forced to move in a thermodynamically unfavorable direction.

2.2.3 Classification of anoxygenic photosynthetic bacteria

Based on their pigmentation, as well as the type of reductant they use, photosynthetic bacteria can be classified into: (i) green sulfur bacteria; (ii) green nonsulfur bacteria; (iii) purple sulfur bacteria; (iv) purple nonsulfur bacteria; and (v) heliobacteria (Ormerod, 1992).

The green sulfur bacteria, such as the *Chlorobium* genus, are strictly anaerobic phototrophs that require sulfide-rich environments to not only support CO_2 assimilation, but also to assimilate H_2S as their predominant source of sulfur for biosynthesis (Imhoff, 1992). Commonly they form the lowermost layers of phototrophic microorganisms in microbial mats (see section 6.1.2(b)), shallow marine sediment, and the hypolimnion of stratified lakes, where they are overlain by purple bacteria and algae. *Chlorobium* contains Bchl a, c, d, and e, and oxidizes H_2S to sulfate:

$$3H_2S + 6CO_2 + 6H_2O \rightarrow C_6H_{12}O_6 + 3SO_4^{2-} + 6H^+ \tag{2.7}$$

During the oxidation of H_2S to SO_4^{2-}, eight electrons are transferred, and as is often the case, this can only occur via a series of intermediate two-electron steps, written hereon as $2e^-$. Elemental sulfur is formed as the first oxidation product and is frequently deposited epicellularly (i.e., on the outer side of the cell wall) in the form of membrane-bound vesicles. Although elemental sulfur is chemically stable and very insoluble in oxygenated environments, it is used by *Chlorobium* as an additional energy source when H_2S is limiting. Under these conditions they oxidize it completely to sulfate:

$$12S^0 + 18CO_2 + 30H_2O \rightarrow$$
$$3C_6H_{12}O_6 + 12SO_4^{2-} + 24H^+ \qquad (2.8)$$

The elemental sulfur is, however, not available to other cells because it is physically attached to the *Chlorobium* cell surface (van Gemerden, 1986). *Chlorobium* has also been shown capable of growing on thiosulfate (reaction (2.9)) (e.g., Khanna and Nicholas, 1982) and ferrous iron (reaction (2.10)) (e.g., Heising et al., 1999). Additionally, green sulfur bacteria can assimilate some simple organic compounds (e.g., acetate) for photoheterotrophic growth, provided that a reduced sulfur compound is available as a source for assimilation (Kelly, 1974).

$$3S_2O_3^{2-} + 6CO_2 + 9H_2O \rightarrow$$
$$C_6H_{12}O_6 + 6SO_4^{2-} + 6H^+ \qquad (2.9)$$

$$24Fe^{2+} + 6CO_2 + 66H_2O \rightarrow$$
$$C_6H_{12}O_6 + 24Fe(OH)_3 + 48H^+ \qquad (2.10)$$

Green nonsulfur bacteria, such as the thermophilic genus *Chloroflexus*, are ubiquitous in hot springs systems around the world, where they grow in waters as hot as 60°C (Ward et al., 1989). Being filamentous, *Chloroflexus* species have the propensity to grow as mats in association with cyanobacteria, where they take advantage of the organic compounds produced from the latter. They are also a major constituent of marine intertidal and hypersaline environments, where they are often associated with purple sulfur bacteria and the chemolithoautotrophic sulfur-oxidizing

bacterial genus *Beggiatoa* (Mack and Pierson, 1988). They use Bchl *c* with an absorption maxima at 740 nm. Their metabolism is much more versatile than the green sulfur bacteria, growing preferentially as aerobic photoheterotrophs on a wide variety of sugars, amino and organic acids, but also photoautotrophically under anoxic conditions with H_2S, H_2 (reaction (2.11)), and possibly even Fe(II), as electron donors (e.g., Pierson and Parenteau, 2000). Under fully aerobic conditions, bacteriochlorophyll synthesis is repressed, the color of the culture changes from dull green to orange, and the organisms begin to grow chemoheterotrophically.

$$12H_2 + 6CO_2 \rightarrow C_6H_{12}O_6 + 6H_2O \qquad (2.11)$$

The purple sulfur bacterial genera, such as *Chromatium*, *Thiospirillum*, *Thiocapsa*, and *Ectothiorhodospira*, utilize bacteriochlorophylls with absorption maxima at long wavelengths, between 800 and 1040 nm wavelengths (e.g., Bchl *a* and Bchl *b*), as well as a number of red and purple carotenoid pigments. The latter are particularly important because it enables the bacteria to live in deeper waters where infrared radiation is filtered out, but light between 450 and 550 nm still penetrates (Imhoff, 1992). The cells often form massive blooms in stratified lake and marine waters, where underlying sulfate-reducing bacteria (SRB) produce large amounts of H_2S (e.g., Camacho et al., 2000). Most purple bacteria are flagellated and exhibit photo- and chemotactic responses to changing environmental conditions, that is, they move towards or away from light or chemicals, respectively. They are also usually larger than green bacteria, and the oxidation of H_2S can lead to the deposition of numerous intracellular elemental sulfur granules (Fig. 2.8): the exceptions are species of the halophilic genus, *Ectothiorhodospira* that form S^0 epicellularly. The sulfur is a storage product, formed when sulfide concentrations in the surrounding environment are high. If sulfide becomes temporarily limiting, or if S^0 sufficiently accumulates, the intracellular S^0 can alternatively be used as an electron donor,

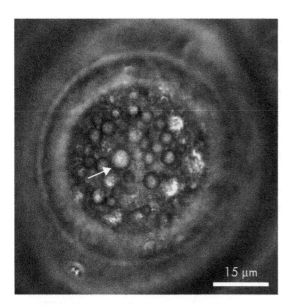

Figure 2.8 Transmission electron micrograph of *Chromatium* sp. with elemental sulfur granules found within the cytoplasm (arrow). (Courtesy of David Patterson and provided by micro*scope.)

leading to sulfate formation (van Gemerden, 1986). Purple sulfur bacteria can also grow with H_2, $S_2O_3^{2-}$, or Fe^{2+} as electron donors, while in the case of *Thiocapsa roseopericina*, the organic constituents in sewage induce them to grow heterotrophically in the dark, forming distinctive pink to reddish blooms (Pfennig, 1978). Another important feature of *Thiocapsa roseopersicina* is that it can grow chemolithoautotrophically with H_2S or $S_2O_3^{2-}$ (De Wit and van Gemerden, 1987), and it can even grow in the presence of O_2, although bchl synthesis is inhibited by its presence (De Wit and van Gemerden, 1990).

Despite their nomenclature, the purple nonsulfur bacteria are very similar to the purple sulfur bacteria except that they cannot tolerate high concentrations of H_2S, and thus they tend to be precluded from the same microenvironments that favor the green and purple sulfur species (e.g., Hansen and van Gemerden, 1972). When grown on sulfide, however, some species from the genera *Rhodospirillum* and

Rhodopseudomonas precipitate elemental sulfur epicellularly. Other genera, such as *Rhodomicrobium* and *Rhodobacter*, have the ability to use Fe^{2+} or H_2 as electron donors (e.g., Widdel et al., 1993). Unlike other photoferrotrophs, *Rhodomicrobium vannielli* produces epicellular ferric hydroxide ($Fe(OH)_3$) crusts that impede further Fe(II) oxidation after two or three reactions, indicating that this metabolic mode is only a side activity of the bacterium (Heising and Schink, 1998). The purple nonsulfur bacteria can additionally utilize a wide variety of organic compounds (acetate and pyruvate are the most universally used) and thus metabolize as photoheterotrophs. Some species grow well on fatty acids, ethanol (C_2H_6O), or methanol (reaction (2.12)), with minor amounts of CO_2 necessary to serve as an electron acceptor:

$$12CH_3OH \text{ (methanol)} + 6CO_2 \rightarrow 3C_6H_{12}O_6 + 6H_2O \qquad (2.12)$$

They are not, however, capable of breaking down organic macromolecules such as cellulose, lipids, or proteins. Therefore, in natural habitats they maintain a syntrophic relationship with fermentative/chemoheterotrophic bacteria, whereby the two organisms combine their metabolic capabilities for each others' mutual benefit (Pfennig, 1978).

Heliobacteria are a phylogenetically separate group of anoxygenic phototrophic bacteria. Three genera, *Heliobacterium*, *Heliophilum*, and *Heliobacillus*, are physiologically most similar to purple nonsulfur bacteria, but they are strict anaerobic photoheterotrophs. Photoautotrophic growth using H_2 or H_2S has not been observed, although if sulfide is added to the culture media, it is frequently oxidized to S^0 (Madigan and Ormerod, 1996). Heliobacteria are found most abundantly in rice paddies, suggesting that a syntrophic relationship exists between the rice plant and the bacterium, with the bacterium receiving organic compounds excreted by the rice, and the plant benefiting from the nitrogen-fixing capabilities of the heliobacteria.

2.2.4 The light reactions – oxygenic photosynthesis

In oxygenic photosynthesis, light is still used to generate ATP, but the electrons for CO_2 reduction come from the splitting of water into O_2 and electrons:

$$6H_2O + 6CO_2 \rightarrow C_6H_{12}O_6 + 6O_2 \qquad (2.13)$$

The process uniquely involves two coupled photochemical reactions, known as Photosystem I and Photosystem II (see Nelson and Ben-Shem, 2004 for details). Both are necessary because no single reaction center is capable of being a strong enough oxidant to remove electrons from H_2O but sufficiently reducing to generate NADPH. Oxygenic photosynthesis also relies on a unique water oxidizing complex (WOC) that contains a tetramanganese–calcium cluster (Mn_3CaO_4-Mn) which binds H_2O and extracts the electrons needed to replace those lost during photooxidation (Ferreira et al., 2004).

Photosynthesis begins when the reaction center chlorophyll *a*, called P680 (type II), receives four electrons (one at a time) from water, forming O_2 via the WOC, and thus initiating Photosystem II (Fig. 2.9). This process is feasible because the electrode potential of P680 is about +1.1 V, sufficiently higher than that of the O_2/H_2O couple that is +0.82 V (at pH 7). Light energy then transforms P680 into a moderately strong reductant (with an $E^{\circ\prime}$ of about −0.7 V), capable of reducing a pheophytin *a* molecule (chlorophyll *a* without the magnesium atom). From the reaction center, the electrons are then shuttled through the Q cycle, cytochrome *bf*, and eventually ending up at the copper-containing protein, plastocyanin (Merchant and Sawaya, 2005). It is at this stage that the overall process diverges from that discussed earlier for anoxygenic photosynthesis. Instead of plastocyanin returning the electrons to the reaction center chlorophyll *a* of Photosystem II, the protein donates the electrons to the reaction center chlorophyll *a* of Photosystem I, called P700 (type 1). Simultaneously, P700 absorbs light and the donated electrons are energized to a very electronegative potential ($E^{\circ\prime}$ of about −1.3 V), that are then shuttled down through a sequence of chlorophyll *a*, a quinone, Fe-S proteins, and ferrodoxins, to eventually reduce $NADP^+$ to NADPH. The net effect of this two-photosystem set-up is that oxygenic photosynthetic microorganisms have an unlimited supply of electrons (from water) and they do not have the burden of reverse electron transport.

During the transfer of electrons from P680 to P700 a proton gradient is generated akin to the anoxygenic bacteria, resulting in a pmf from which ATP is produced. This form of ATP synthesis is known as noncyclic photophosphorylation because the electrons are not returned to the original reaction center, but alternatively are used to reduce $NADP^+$. When sufficient reducing power already exists within the cell, ATP can also be produced by cyclic photophosphorylation involving only Photosystem I. In this case, the electrons from ferrodoxin are transferred to cytochrome *bf*, that then passes the electrons to P700 via plastocyanin.

Photosystems I and II normally function together in oxygenic photosynthesis. However, certain cyanobacteria (e.g., some *Oscillatoria* species) are also able to carry out cyclic photophosphorylation using only Photosystem I, growing similar to the purple and green bacteria. In doing so, they are forced to obtain reducing power from sources other than water, such as reduced sulfur compounds (Cohen et al., 1975). When sulfide is used, the cyanobacteria form elemental sulfur which is excreted from the cell. Then during the night, when O_2 is depleted, the cyanobacteria are further able to reassimilate S^0 and use it as an electron acceptor during anaerobic respiration of endogenously produced polysaccharides formed earlier in the day. *Oscillatoria limnetica* has even been observed to generate energy anaerobically by fermenting the polysaccharides to lactic acid. This versatility is greatly advantageous to them because they are not restricted to specific layers in mat systems, like those of other phototrophic bacteria (Oren and Shilo, 1979).

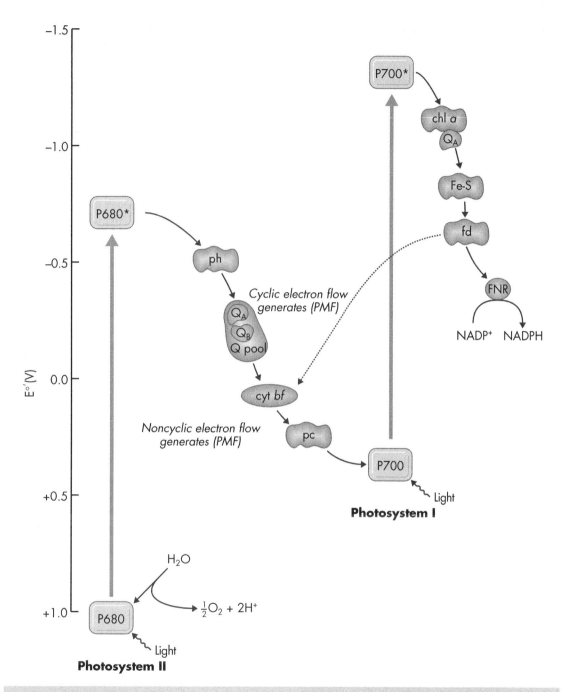

Figure 2.9 Noncyclic photophosphorylation in cyanobacteria, algae, and plants. cyt, cytochromes; fd, ferredoxin; Fe-S, iron-sulfur protein; FNR, ferrodoxin NADP+ reductase; pc, plastocyanin; ph, pheophytin; Q-ubiquinones.

2.2.5 The dark reactions

Irrespective of how the microorganisms create NADPH and ATP, photoautotrophic cells must also reduce CO_2 to organic carbon. This forms the basis of what is known as the dark reactions – the latter nomenclature aptly named because these reactions do not directly involve the trapping of light energy, and hence can occur in the dark. The dark reactions do, however, require the products of the light reactions, ATP and reducing power. There are three mechanisms of carbon fixation exhibited by photoautotrophic organisms (Ormerod, 2003). The one most widely used by plants, algae, cyanobacteria, and purple photosynthetic bacteria (as well as most chemolithoautotrophs) is the Calvin cycle, also known as the Calvin–Benson cycle or reductive pentose cycle. In green sulfur bacteria, CO_2 fixation occurs by a reversal of steps in the citric acid cycle, such that CO_2 is assimilated at the same points that it is released during the oxidative process (see section 2.3.2(b)). This pathway is referred to as the reductive citric acid cycle, and it is also used in a number of deeply branching thermophilic chemolithoautotrophs (Hügler et al., 2003). Meanwhile, *Chloroflexus* uses what is known as the hydroxypropionate pathway. There are also three other CO_2 fixation pathways exclusively employed by chemolithoautotrophs: (i) the reductive acetyl-CoA pathway; (ii) the ribulose monophosphate pathway; and (iii) the serine pathway. These will be discussed in more detail in sections 2.5.2 and 2.5.3.

Being the CO_2 fixation mechanism preferred by most microorganisms, we shall limit our discussion to the Calvin cycle (see Shively et al., 1998 for details). It consists of three main stages (Fig. 2.10):

1 To begin the cycle, 6 molecules of CO_2 are reacted with 6 molecules of the five-carbon sugar ribulose bisphosphate to form 12 molecules of the three-carbon phosphoglycerate. This reaction is catalyzed by the Fe-S-containing enzyme, ribulose bisphosphate carboxylase/oxygenase (rubisco), and is the first step in which atmospheric CO_2 becomes fixed into organic molecules.

2 The carbon atoms in 3-phosphoglycerate are still at the same oxidation state as CO_2, thus in the next steps the compound is rearranged and reduced to the oxidation state of glucose by 12 molecules each of ATP and NADPH (from the light reactions) to form 1 molecule of fructose-6-phosphate that goes to biosynthesis of a six-carbon sugar (e.g., glucose) or other organic compounds (e.g., amino acids and lipids).

3 The remaining 10 molecules of glyceraldehyde-3-phosphate are recycled through a series of reactions to form 6 molecules of five-carbon ribulose phosphate, that ultimately, through the consumption of 6 additional molecules of ATP, are converted back to 6 molecules of ribulose bisphosphate to repeat the cycle again.

Because CO_2 diffuses slowly in water, it can commonly be the limiting step in photosynthesis. As a result, many aquatic microorganisms and plants have evolved mechanisms to actively transport and increase the dissolved inorganic carbon (DIC) concentrations entering the photosynthesizing cell (Lucas, 1983). The DIC in mildly acidic solutions is mostly in the form of solvated CO_2 (by convention written as H_2CO_3, carbonic acid), but in neutral to mildly alkaline waters DIC is mostly in the form of HCO_3^- (bicarbonate). Therefore, photoautotrophs must also have the means to convert the HCO_3^- transported intracellularly into the CO_2 used by rubisco. This reaction is catalyzed by the enzyme carbonic anhydrase.

As alluded to above, some photosynthetic cells do not require CO_2 as their sole carbon source, but instead are able to attain their carbon from existing organic compounds, thus behaving as photoheterotrophs. They are at a competitive advantage relative to the photoautotrophs because they do not have to burden themselves with CO_2 fixation, but they are instead reliant upon extraneous sources of organic carbon. Whether or not CO_2 is still reduced when an organic compound is added depends, in part, on the oxidation state of that compound. Glucose and acetate are the same oxidation state of cell material and, therefore, can be assimilated directly as carbon sources with no requirement for any oxidation or reduction reactions. Conversely, fatty acids longer than acetate are more reduced than cell

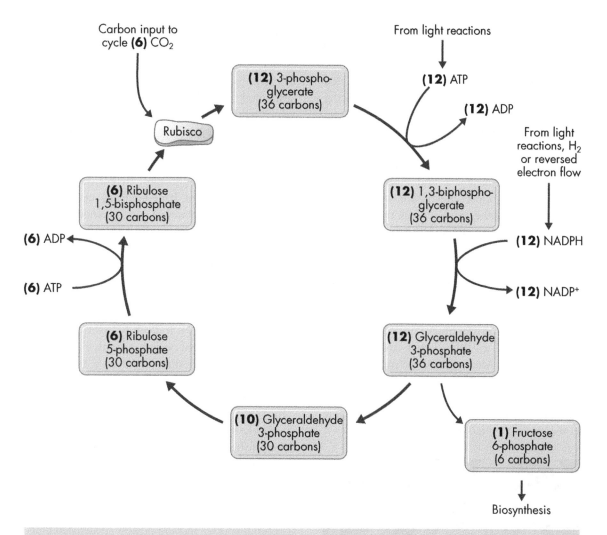

Figure 2.10 The Calvin cycle. For each 6 molecules of CO_2 incorporated, 1 molecule of fructose-6-phosphate is produced. 12 molecules each of NADPH and ATP are required to form 1 molecule of hexose sugar.

material, and some means of disposing of excess electrons, such as CO_2 reduction, or N_2 fixation, is thus required.

2.2.6 Nitrogen fixation

Nitrogen is an element essential to all life. It is abundant in the atmosphere, mostly as dinitrogen gas (N_2), but owing to the stability of the N≡N triple bond, it is extremely inert and its activation energy is very high. This translates into a very energy-intensive process for those species that use it as their nitrogen source, requiring approx-

imately 16 ATP molecules per N_2 fixed (Newton and Burgess, 1983). There are a number of N_2-fixers, the majority of which are phototrophic, which is why the process is covered here.

Six electrons must be transferred to reduce N_2 to $2NH_3$ (reaction (2.14)), and it is believed that three successive reduction steps occur within the nitrogenase complex, an oxygen-intolerant enzyme containing iron, and either molybdenum or vanadium. The reducing power needed for N_2 reduction comes in the form of ferredoxin, with an electrode potential low enough (-0.38) to directly accept electrons from H_2 (-0.41 at pH 7).

$$N_2 + 3H_2 \rightarrow 2NH_3 + 2R\text{-}OH \rightarrow 2R\text{-}NH_2 + 2H_2O$$
$$(2.14)$$

Nitrogenase also provides cells with the added benefit of being able to dissipate excess reducing power when they grew under photosynthetic conditions in the presence of organic carbon, as alluded to above (Joshi and Tabita, 1996).

Many nitrogen-fixing bacteria (e.g., members of the genera *Rhizobium* and *Anabena*) grow symbiotically in close association with specific plants (e.g., legumes and several nonleguminous angiosperms) or fungi. In some plants, the nitrogen-fixer may be localized on the root tissue in the form of nodules (e.g., Fig. 2.11). In effect, the N_2-fixer infects the plant, with the latter providing the nitrogen-fixer with a source of energy, in the form of simple sugars, and a sheltered environment in which access by oxygen is controlled so that nitrogenase is not inactivated. Mean-while, the nitrogen-fixer supplies the plant with a source of nitrogen. Symbiotic nitrogen-fixers are very important agriculturally because they provide a renewable source of nitrogen to soils.

Figure 2.11 Nodules of *Rhizobium* bacteria (arrow) provide fixed nitrogen to soybean roots in an infectious, symbiotic relationship. The nodules are approximately 2 mm in diameter. (Courtesy of Stephen Temple.)

By contrast, most N_2-fixing species, including chemoheterotrophs (e.g., some species of the genera *Azotobacter*, *Clostridium*), chemolithoautotrophs (e.g., some *Acidithiobacillus* species), and most cyanobacteria, live freely in soil or water where they fix nitrogen when extraneous nitrogenous supplies are limiting. In soils, these microorganisms are considerably less significant in terms of nitrogen input to their environment than the symbiotic species. By contrast, in the oceans, planktonic (free-floating) cyanobacterial genera, *Trichodesmium* and *Synechococcus*, are the primary N_2-fixers, fuelling up to half of the new production, despite making up only 1% of the marine biomass (Karl et al., 1997). Their importance is such that during periods when fixed nitrogen is scarce (e.g., nitrate or ammonium), they compensate by increasing nitrogen fixation rates (Tyrrell, 1999). Indeed, N_2 fixation in subtropical and tropical oceans not only has a major role on the global marine N budget, but it impacts the coupling of N–C–P–Fe cycles, in particular net oceanic sequestration of atmospheric CO_2 (Karl et al., 2002).

Unlike symbiotic nitrogen-fixers, the nonsymbiotic forms require a means of protecting their nitrogenases from O_2. Some aerobic respirers accomplish this by maintaining a very high rate of oxygen utilization, thereby minimizing the diffusion of oxygen into the cell. Many cyanobacterial species instead possess specialized cells within their filaments, called heterocysts, that serve as O_2-free microenvironments, while other cyanobacteria photosynthesize strictly during daylight and fix nitrogen at night (e.g., Bergman et al., 1997).

2.3 Catabolic processes

In most heterotrophic microorganisms the oxidation of carbohydrates provides the bulk of the cell's energy. To degrade sugars, cells must first secrete specific extracellular hydrolytic enzymes that break down the glycosidic bonds in polysaccharides into mono- and disaccharides that can then be transported through the plasma membrane (using permeases and pmf) into the cytoplasm. To be most effective, it is advantageous for close contact to exist between the microorganism and substratum, thereby creating as short a diffusional path from the products of hydrolysis to the cell surface. From there, the breakdown continues in two different ways (see Nelson and Cox, 2005, for details). The partial breakdown of glucose begins with glycolysis and is immediately followed by fermentation. The complete breakdown via respiration comprises four discrete, but interrelated steps beginning with glycolysis, followed by a transition reaction that forms acetyl-CoA, the citric acid cycle, and eventually the electron transport chain.

2.3.1 Glycolysis and fermentation

The most common pathway of glycolysis is the Embden–Meyerhof pathway. Through a number of complex steps, each catalyzed by their own unique enzyme, a six-carbon glucose molecule is partially broken down into 2 molecules of pyruvate (Fig. 2.12). In addition, glycolysis yields a net release of 2 molecules of ATP through substrate-level phosphorylation and 2 molecules of NADH (reaction (2.15)). Glycolysis does not require oxygen, and thus can occur under both oxic (O_2 present) and anoxic (O_2 absent) conditions.

$$C_6H_{12}O_6 + 2NAD^+ + 2ADP + 2P_i \rightarrow$$
$$2C_3H_3O_3^- + 2NADH + 4H^+ + 2ATP \quad (2.15)$$

In addition to glycolysis, many other bacteria have the means to break down five-carbon sugars via the pentose phosphate pathway. Its importance lies in its ability to produce important intermediate molecules required in the synthesis of nucleic acids and some amino acids, as well as being a key producer of NADPH from $NADP^+$. Unlike glycolysis, however, the pentose phosphate pathway only produces one

ATP molecule from each glucose molecule. Bacteria, such as *Escherichia coli*, *Bacillus subtilis*, and *Enterococcus faecalis*, use this pathway.

Once formed, the three-carbon pyruvate can then either undergo fermentation or respiration. In fermentation, there is no electron transport chain as in respiration (see next section) and there is an absence of externally supplied electron acceptors. Rather, pyruvate is reduced, via the transfer of electrons from the NADH generated during glycolysis, into various end products depending on the type of microorganism (Fig. 2.12). Some of these include lactate by *Lactobacillus* species (reaction (2.16)); acetate and H_2 by *Escherichia coli* (reaction (2.17)); and ethanol by species of the yeast *Saccharomyces* (reaction (2.18)):

$$C_3H_3O_3^- + NADH + H^+ \rightarrow C_3H_5O_3^- + NAD^+$$
$$(2.16)$$

$$C_3H_3O_3^- + H_2O + NADH + H^+ \rightarrow$$
$$CH_3COO^- + CO_2 + 2H_2 + NAD^+ \quad (2.17)$$

$$C_3H_3O_3^- + NADH + 2H^+ \rightarrow$$
$$C_2H_6O + CO_2 + NAD^+ \quad (2.18)$$

In reactions (2.17) and (2.18), some of the carbon atoms end up incorporated into CO_2, a more oxidized form of carbon than in pyruvate, while other carbon atoms in acetate and ethanol are more reduced. Fermentation of sugars can lead to many other end products, including propionate, butyrate, succinate, formate, citrate, and methanol, depending on the dehydrogenase present in the cell. Proteins are also important substrates for fermentation, and a large quantity of amino acids ingested by fermentative microorganisms are deaminated and enter some of the same pathways used in carbohydrate metabolism.

Fermentation occurs for two reasons.

1 In the absence of external electron acceptors, glycolysis/fermentation provides the means by which cells can generate ATP for biosynthesis. In essence, it allows them to perform internally balanced redox reactions regardless of the external environment.

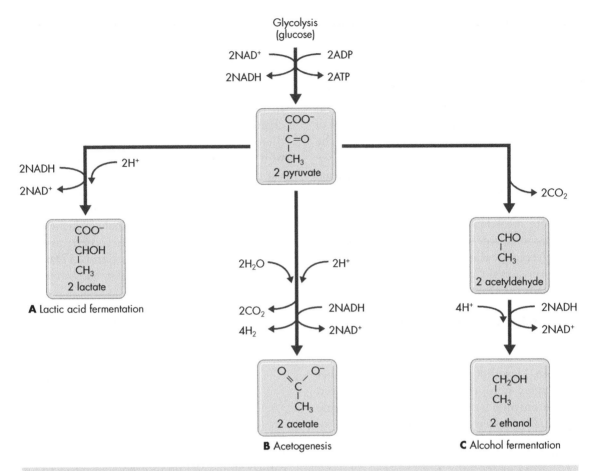

Figure 2.12 A number of different fermentation products can be formed from pyruvate generated during glycolysis.

2 During one of the steps in glycolysis, NAD⁺ is reduced to NADH. This step could potentially inhibit further growth because cells have a finite amount of NAD⁺ available, and if it were all reduced to NADH, then the oxidation of glucose would stop. This limitation is resolved within the cell by oxidizing NADH back into NAD⁺ through reactions involving the reduction of pyruvate to any of a variety of fermentation byproducts.

Although essential to the cell, the partial breakdown of glucose during fermentation is not particularly efficient in terms of utilizing the potential chemical energy stored in the bonds of organic matter. This largely stems from the fact that the end products (lactate, acetate, ethanol) are still rather complex organic molecules themselves. This means that only 2 molecules of ATP are generated during the partial oxidations in glycolysis, while the rest of the energy stored in the C–H bonds remains unused. The inefficiency is compounded by the small difference in electrode potential between the glucose and the fermentation products. As will be covered below, this is not the case for respiration, a process that completely oxidizes organic carbon to CO_2 or CH_4 and utilizes a TEA with a more positive electrode potential.

2.3.2 Respiration

(a) Oxidative decarboxylation

In the presence of an electron acceptor, all of the reducing equivalents associated with the organic substrate can be utilized for energy generation through the process of respiration. The first step in respiration involves the transformation of pyruvate into 2 molecules of the two-carbon acetyl-CoA through a process called oxidative decarboxylation. Acetyl is the radical of acetic acid, containing a methyl group single-bonded to a carbonyl to yield CH_3CO^- (recall acetylation of glucose to form peptidoglycan). Acetyl-CoA is a derivative of the coenzyme A, that functions to conserve the energy released in exergonic reactions by forming an energy-rich bond between the acetyl and sulfur atom of CoA.

Initially, the carboxyl groups of each pyruvate are removed as CO_2, and the remaining molecule quickly combines with coenzyme A, forming acetyl-CoA. For each molecule of pyruvate decarboxylated, 1 molecule of CO_2 is formed (and released to the atmosphere) and 1 molecule of NAD^+ is reduced to NADH:

$$2C_3H_3O_3^- + 2NAD^+ + H_2 + 2CoA \rightarrow$$
$$2acetyl\text{-}CoA + 2NADH + 2CO_2 \quad (2.19)$$

Only some of the energy from pyruvate oxidation is stored in NADH, the rest is temporarily stored in the acetyl-CoA molecules.

(b) Citric acid cycle

Once the pyruvate has been primed, it is ready for entry into the citric acid cycle, also known as the tricarboxylic acid cycle or Krebs' cycle. It consists of a series of chemical reactions in which the chemical energy stored in acetyl-CoA

is released on a step-by-step basis and transferred, in the form of electrons, to a number of intermediate compounds.

There are eight steps to the citric acid cycle (each driven by its own enzyme), and it takes two turns of the cycle to metabolize both acetyl-CoA molecules from above (Fig. 2.13). The cycle begins when acetyl-CoA initially reacts with the four-carbon oxaloacetate to form a six-carbon citrate molecule. When this happens, coenzyme A is released and freed to await another pyruvate molecule or citric acid intermediate. The citrate molecule then undergoes a number of transformations, including oxidation steps that transfer electrons to NAD^+ to form NADH. This coenzyme now contains some of the stored energy originally present in glucose. During some of the oxidative steps, CO_2 molecules are liberated, ultimately converting the six-carbon organic compound back to the four-carbon oxaloacetate. At about mid-way through the cycle, when succinyl-CoA is oxidized to succinate, the energy released is used to make guanosine triphosphate (GTP) from guanosine diphosphate (GDP) and free inorganic phosphate by substrate-level phosphorylation. GTP is later converted to ATP. The succinate formed is then oxidized to fumarate. This transformation does not release enough energy to make NADH outright, so instead the energy is captured by FAD, which reduces to $FADH_2$.

The production of reduced coenzymes in stages is an example of how heterotrophic microorganisms control the movement of electrons. As discussed in the next section, those electrons are then systematically fed into the electron transport chain in pairs. This ensures that all the energy that is stored in the reduced organic compounds is not transferred in one step, so energy is not lost as heat. In terms of energy, 1 ATP molecule is generated when succinyl-CoA is converted to succinic acid. Thus, at the end of both turns of the citric acid cycle, both acetyl-CoA molecules are completely broken down into $4 CO_2$ molecules, 6 NADH, 2 $FADH_2$, and 2 GTP molecules (reaction (2.20)):

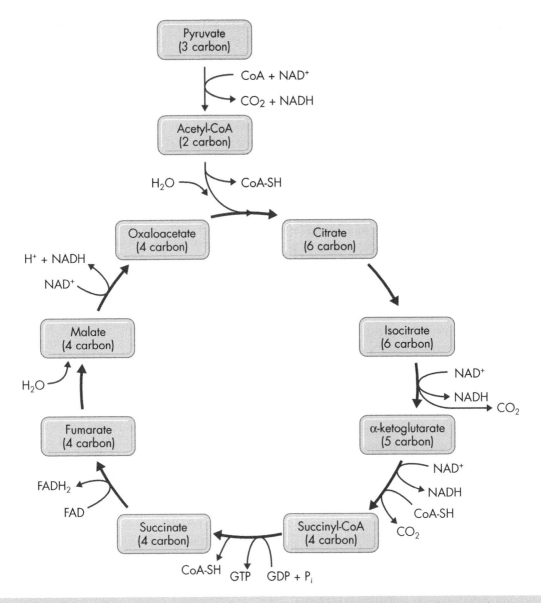

Figure 2.13 The citric acid cycle. The effect of this cycle is to transfer electrons originally from acetyl-CoA to NAD^+ and FAD, thereby creating strong reductants for the electron transport chain. Some of the organic carbon is transformed to CO_2, and lost from the cell, while the remainder is recycled.

$$2\text{acetyl-CoA} + 6NAD^+ + 2FAD + 2GDP + 2P_i$$
$$+ 4H_2O + 4e^- \rightarrow 6NADH + 2FADH_2$$
$$+ 2H^+ + 4CO_2 + 2CoA\text{-}SH + 2GTP \quad (2.20)$$

The citric acid cycle is relevant beyond simple energy production. During many of its intermedi-ate stages, several key compounds for biosynthesis are generated (see Madigan et al., 2003 for details). For example, oxaloacetate is a precursor to a number of amino acids, while succinyl-CoA can be used in the synthesis of the porphyrin ring of cytochromes and chlorophyll. When any of the carbon compounds are removed, the acid cycle stops unless those intermediates are replenished. Several of the four-, five- and six-carbon anions of the cycle, such as malate, fumarate, succinate, and citrate, can directly be incorporated as both electron donors and carbon sources. Utilization of two- and three-carbon acids, however, cannot occur by means of the citric acid cycle alone. So, when an organic substrate such as acetate is available, it needs to be converted into the four-carbon oxaloacetate to be used for biosynthesis in the normal citric acid cycle reactions. The process that does this is called the glyoxylate cycle.

In addition to the oxidation of glucose, bacteria are also able to generate energy from a number of other organic compounds (see Nelson and Cox, 2005 for details). For instance, they degrade proteins into their constituent amino acids by secreting different types of hydrolytic enzymes, the proteases. Some proteases specifically hydrolyze the peptide linkage between particular amino acids, whereas others are nonspecific and hydrolyze any peptide bond. The amino acids are then taken into the cell where they are degraded by different pathways depending on their particular chemistry. The breakdown of simple lipids into glycerol and fatty acids occurs through the activity of lipid hydrolyzing enzymes, known as lipases. Glycerol is then metabolized as part of the Embden–Meyerhof pathway to pyruvate, while the fatty acids are degraded to acetyl-CoA and then oxidized in the citric acid cycle.

(c) Electron transport chain

The electron transport chain contains a number of redox active complexes that shuttle electrons to a TEA, whilst pumping protons across the plasma membrane to generate a pmf. For simplicity, the chain could be viewed as a hill, with the PED at the peak and the TEA at the bottom, and hence electrons travel down the gradient and free energy is released. The first step in the electron transport chain occurs when electrons from either glycolysis ($4e^-$), formation of acetyl-CoA ($4e^-$), or the citric acid cycle ($16e^-$) are transferred as part of NADH or $FADH_2$ to NADH

dehydrogenase or the ubiquinone pool, respectively. This causes NADH to become oxidized to NAD^+, making it once again available for the citric acid cycle, while $FADH_2$ becomes reduced to FAD (recall Fig. 2.3).

The first complex in the chain is the flavoprotein, NADH dehydrogenase. It consists of a number of subunits, including FMN and Fe-S proteins. FMN initially splits H_2 into $2H^+$ and $2e^-$, with the two electrons *only* being transferred (one at a time) to the Fe-S proteins. Unlike the flavin component, the latter only carry electrons. Meanwhile, the two protons (1st and 2nd) are extruded outside the plasma membrane envelope. In the following step, the Fe-S proteins return the electrons to the cytoplasmic side of the membrane by donating the electrons to a quinone molecule, making the fully reduced form, hydroquinone, QH_2. The hydrogen component comes from the acquisition of two protons from the cytoplasmic dissociation of water. The hydroquinone then crosses from the inner surface of the membrane to the periplasmic side of cytochrome bc_1, a complex that also contains several subunits, including cytochromes with different hemes (b and c_1) and a Fe-S protein (called the Rieske protein).

At this stage electron flow strongly parallels that of the Q cycle in the photosynthetic membrane, with the oxidation of QH_2 causing the translocation of protons to the periplasm or cell wall (3rd and 4th), while electrons are split in two directions (Brandt, 1996). One electron is transferred to cytochrome b, which has two hemes, one with low potential (b_L, 0.02 V) and the other with relatively higher potential (b_H, 0.05 V). From the latter the electron is used to reduce an ubiquinone, forming a semiquinone (using a proton extracted from the cytoplasm) that stays tightly bound to the inner surface of the plasma membrane. In a second reaction, the semiquinone is fully reduced to hydroquinone, thus permitting two more protons to subsequently be translocated, as before. Meanwhile, the other electron is shuttled through the Rieske protein and cytochrome c_1 to a periplasmic cytochrome c. From there, the electron is passed from outside the plasma membrane back to the terminal oxidase. In aerobic bacteria the terminal oxidase comprises an iron- and copper-containing complex that transects the entire membrane. It too has many subunits, two copper centers and cytochrome aa_3, the latter possessing two hemes (a and a_3). The role of the terminal oxidase is to facilitate the transfer of the electrons to the TEA (in this case O_2), along with the release of a proton each time an electron is transferred (5th). As O_2 requires 4 electrons to reduce it to water, so 4 cytochrome c proteins have to bind sequentially to one docking site on the periplasmic side of the cytochrome aa_3 complex.

It has been estimated that on average 1 molecule of ATP is generated for every 3 protons transported into the cell by ATPase (Schink, 1997). Therefore, in terms of ATP generation, the longer the electron transport chain, the more protons are extruded from the cell, the larger the pmf, and ultimately, the more ATP formed. The length of the electron transport chain, in turn, is governed by the difference in electrode potential ($\Delta E^{\circ\prime}$) between the primary electron donor and the terminal electron acceptor: more electron carriers are required to shuttle the electrons down the redox gradient from a highly reducing PED to a highly oxidizing TEA. For any given reductant, the amount of energy ($\Delta G^{\circ\prime}$) generated is highest when the TEA is the most electropositive (Fig. 2.14). In nature that TEA is typically O_2, and when the reductant is organic carbon (e.g., NADH), 6 protons (on average when 2NADH are oxidized) are translocated during electron flow, although it has been proposed that up to 10 protons may be translocated per NADH (Nelson and Cox, 2005). Sequentially less electropositive TEAs, such as NO_3^-, Mn(IV) oxides, Fe(III) oxides, SO_4^{2-}, and CO_2, correspond to decreasing $\Delta G^{\circ\prime}$ when using the same organic electron donor. Since many of the C_{org}-TEA reaction combinations will have similar number of proton translocations, differences in energy yield is based on the magnitude of the electrical current in the membrane ($\Delta\Psi$), which, predictably, is based on differences in electrode potential between the PED and TEA, i.e., NADH $\rightarrow$ Mn(IV) versus NADH $\rightarrow$ Fe(III).

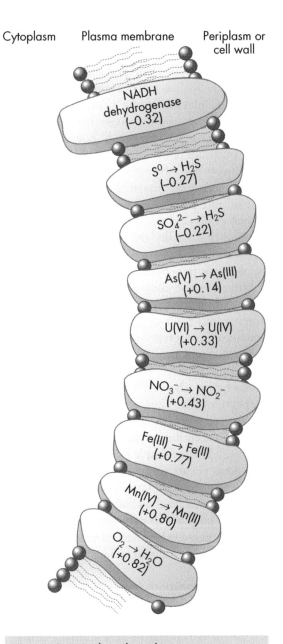

Figure 2.14 A hypothetical representation showing approximate electrode potentials of major environmental TEAs and the primary electron donor in chemoheterotrophic metabolism, NADH. The amount of free energy in any given metabolic coupling is based on the difference between the electrode potential of the primary electron donor (NADH) and the terminal electron acceptor. Hence, the NADH-O_2 couple yields more energy, for example, than the NADH-S^0 couple.

2.4 Chemoheterotrophic pathways

In any given environment, microorganisms are in continuous competition with each other for available substrates. Those species that can maximize the potential energy yield by using TEAPs (terminal electron accepting processes) that promote a higher yield of biomass per mole of substrate oxidized may then be in a position to monopolize that substrate to the detriment of species using less energetic reaction pathways. The form of microbial respiration that yields the most energy to support microbial growth is oxidation of organic carbon coupled to the reduction of O_2. This process is called aerobic respiration, and the environment in which the microorganisms reside is referred to as the oxic zone. Metabolism in the absence of O_2 is less energetically favorable, and is called anaerobic respiration. Facultative anaerobes are unique in that they are capable of aerobic respiration but switch to anaerobic processes (i.e., fermentation or using an alternative TEA) when O_2 is unavailable. They grow in what is referred to as the suboxic zone, where O_2 is depleted, yet redox potential is kept positive due to the availability of NO_3^-, NO_2^-, Mn(IV), and Fe(III). The lack of anaerobic respiration in the presence of O_2 is due to the fact that microorganisms that can respire both aerobically and anaerobically regulate their metabolism to use only O_2 when it is available (e.g., Tiedje, 1988). Anaerobic microorganisms that cannot use O_2 as an electron acceptor are generally inhibited in the presence of O_2. These microorganisms are called obligate anaerobes, and the most used TEAs in the anoxic zone are SO_4^{2-} and CO_2.

2.4.1 Aerobic respiration

Prokaryotic and eukaryotic aerobic respirers are ubiquitous in environments where dissolved oxygen is available. One of their distinguishing features is their ability to completely oxidize a wide variety of natural and synthetic organic compounds to carbon dioxide and water. This capability arises because aerobes possess a suite of hydrolytic enzymes that can be released into the environment to selectively degrade individual classes of compounds. For example, the degradation of crystalline cellulose by fungal species requires the synergistic action of three types of cellulases (Senior et al., 1990). Even the breakdown of refractory compounds, such as lignin, is made possible by the use of highly reactive oxygen-containing radicals, such as the superoxide anion (O_2^-) and hydrogen peroxide (H_2O_2); both generated as byproducts of the biochemical reduction of O_2. The oxygen derivatives, however, can invariably lead to the oxidation of a number of useful organic compounds in the cell if left unchecked. For that reason, the cells have additionally evolved enzymes that eliminate excess radicals. The enzyme superoxide dismutase combines 2 molecules of superoxide to form 1 molecule of hydrogen peroxide and 1 molecule of O_2:

$$2O_2^- + 2H^+ \rightarrow H_2O_2 + O_2 \qquad (2.21)$$

Catalase and peroxidase both convert H_2O_2 into water (reactions (2.22) and (2.23), respectively) without the production of free radicals, but peroxidase requires NADH as a reductant. Collectively, these enzymes facilitate the conversion of superoxide back to O_2 (see Chelikani et al., 2004 for details).

$$2H_2O_2 \rightarrow 2H_2O + O_2 \qquad (2.22)$$

$$H_2O_2 + NADH + H^+ \rightarrow 2H_2O + NAD^+ \qquad (2.23)$$

The electron transport chain in aerobic bacteria is the most extensive of all the heterotrophs, using the complete suite of electron carriers described earlier. For each molecule of glucose oxidized, 6 molecules of O_2 are required (reaction (2.24)). Consequently, as the cell metabolizes O_2, the gas must continuously be replenished from the external environment. From the cell's perspective, this is actually quite easy because the gas freely diffuses through the cell membrane due to passive diffusion.

$$C_6H_{12}O_6 + 6O_2 \rightarrow 6CO_2 + 6H_2O \qquad (2.24)$$

$$\Delta G^{\circ\prime} = -2872 \text{ kJ mol}^{-1} \text{ glucose or } -240 \text{ kJ/2e}^-$$

Although not readily apparent from the overall reaction above, during the reduction of O_2 to H_2O, protons are also required (reaction (2.25)). They are supplied by those re-entering the cell through the activity of ATPase.

$$0.5O_2 + 2H^+ + 2e^- \rightarrow H_2O \qquad (2.25)$$

In terms of the overall energy balance there is a net yield of at least 26 molecules of ATP generated from the catabolism of each molecule of glucose (based on a conservative stoichiometry of 2ATP per 1NADH). The breakdown is as follows:

2 ATP (from substrate-level phosphorylation in glycolysis)

4 ATP (2 NADH in glycolysis)

4 ATP (2 NADH in acetyl-CoA formation reaction)

2 ATP (substrate-level phosphorylation in citric acid cycle)

12 ATP (6 NADH in citric acid cycle)

2 ATP (2 FADH$_2$ in citric acid cycle)

If hydrolysis of the high energy anhydride bonds in ATP generates 31.8 kJ mol^{-1}, then 827 kJ of energy is available to the cell for each molecule of glucose oxidized. However, free-energy calculations show that the total amount of energy available from the oxidation of glucose is 2872 kJ mol^{-1}. This indicates that aerobic respiration is approximately 30% efficient based on standard free energy changes; but when corrected for actual free energy changes required to form ATP within cells using the 50 kJ mol^{-1} discussed earlier, then the calculated efficiency increases to 49%. Subsequently, growth yields for microorganisms, based on theoretical reaction stoichiometry, are generally less than predicted. This holds true for the aerobes, as well as the anaerobes discussed below.

2.4.2 Dissimilatory nitrate reduction

Anaerobic respiratory pathways normally employ similar carriers to aerobic respiration, but they replace the cytochrome aa_3 complex as the terminal oxidase because their TEAs are not sufficiently electropositive to make use of it. In environments where O_2 concentrations are very low, the form of metabolism most similar to aerobic respiration is dissimilatory nitrate reduction. The term dissimilatory refers to the use of nitrate molecules as electron acceptors in energy metabolism, compared to assimilatory nitrate reduction, that refers to the use of nitrate as a nitrogen source for the cell. Assimilatory nitrate reduction occurs in all plants, most fungi, as well as in many bacteria, whereas dissimilatory nitrate reduction is restricted to bacteria.

Dissimilatory nitrate reduction plays an important role in soils and freshwaters subject to agricultural or sewage pollution. In those environments the nitrate reducers are capable of completely degrading complex organic matter to carbon dioxide, with the attendant depletion of pore water nitrate. The overall process of dissimilatory nitrate reduction actually comprises two separate pathways, namely nitrate ammonification (end product ammonia – NH_3) and, more globally important, denitrification (end product dinitrogen gas – N_2). The first step in either pathway involves the reduction of NO_3^- to nitrite (NO_2^-), using the iron-molybdenum containing nitrate reductase enzyme whose synthesis is repressed in the presence of O_2. Certain bacteria, such as *Escherichia coli* or *Staphylococcus carnosus*, either excrete the nitrite or reduce it further to ammonia (Neubauer and Götz, 1996). Conversely, other nitrate reducers, such as various species from the *Pseudomonas* genera, take advantage of the entire 8e$^-$ reduction through a series of intermediate stages from nitrate to nitrite, then nitric oxide (NO), nitrous oxide (N_2O), and eventually dinitrogen gas (reaction (2.26)). Each step in the reaction is triggered by a specific reductase enzyme, and the denitrification reactions can stop at any stage in the pathway, with the gaseous products released (Fig. 2.15). Those intermediate products can subsequently be used as TEAs by other N-respiring species. In other words, denitrifiers can use exogenous sources of nitrite through to nitrous oxides as their TEAs (Carlson and Ingraham, 1983).

$$2.5C_6H_{12}O_6 + 12NO_3^- \rightarrow$$
$$6N_2 + 15CO_2 + 12OH^- + 9H_2O \qquad (2.26)$$
$$\Delta G^{\circ\prime} = -2715 \text{ kJ mol}^{-1} \text{ glucose or } -226 \text{ kJ/2e}^-$$

The species that can completely oxidize glucose to CO_2, while reducing NO_3^- to N_2, have a distinct advantage over *E. coli* because they maximize every dissolved nitrate molecule they transport into the cell. For example, in the complete reduction, only 4.8 NO_3^- molecules are required to oxidize the 24e$^-$ provided by 1 mol of glucose. However, if a cell, such as *E. coli*, only reduces nitrate to nitrite, then it would instead require 12 NO_3^- molecules. In addition to organic carbon serving as the electron donor, other denitrifiers (e.g., *Thiobacillus denitrificans*) can behave chemolithoautotrophically by using H_2S or S^0 as their reductants (e.g., Aminuddin and Nicholas, 1973). The use of H_2 also yields the denitrifiers considerable free energy but its concentration seldom reaches significant levels in the same environment where nitrate is abundant.

Denitrifiers employ an electron transport system in which electrons are originally fed into the system at the NADH dehydrogenase stage, then transferred progressively down through various carrier enzymes to nitrate reductase. Many species have at least three different nitrate reductases depending on whether reduction is assimilatory, as a means to dissipate excess reducing power, or for energy, with nitrate as the TEA (Moreno-Vivián et al., 1999). In the case of the latter, reduction takes place in the periplasm. Interestingly, when O_2 is present, these same bacteria use it instead, but activate the cytochrome aa_3 complex (Fig. 2.15). Then, within a few hours after O_2 is exhausted, nitrate reductase is regenerated (Knowles, 1982). The O_2 preference arises because in the anaerobic process, where the NO_3^-/NO_2^-

is +0.43 V, only two proton-translocating processes occur, the first at the flavoprotein stage and the second at the ubiquinone stage. By contrast, aerobic respiration has an additional proton-translocation at the terminal oxidase stage, thus generating more ATP. The ability to respire both O_2 and NO_3^- may be an advantage for the facultative anaerobes because they tend to inhabit environments with fluctuating oxic–suboxic conditions (Robertson and Kuenen, 1984). Some nitrate reducers can even utilize ferric iron as their electron acceptor if supplies of nitrate (and nitrite) become exhausted (Sørensen, 1982).

The pathways for organic matter oxidation by most other anaerobic processes are much different from those of aerobic respiration and denitrification. Whereas aerobes and denitrifiers can by themselves completely oxidize a wide variety of organic compounds to carbon dioxide, most anaerobes are limited to a few types of simple organic substrates. Therefore, organic matter released from hydrolysis of complex polymers must first be metabolized by fermentative microorganisms. These microorganisms convert the larger polymers into smaller degradation products, the most common being H_2, pyruvate, lactate, acetate, and formate. In turn, they are utilized by various acetogenic bacteria and anaerobic respirers that, growing syntrophically, bring about the complete oxidation of the organic matter to CO_2 or CH_4 in suboxic to anoxic sediments.

2.4.3 Dissimilatory manganese reduction

There are a number of Mn(IV) oxides in soils and sediment, principally those with an average composition of MnO_2 (along with additional cation contributions from Ca, Na, K, or Mg), including todorokite, vernadite, pyrolusite, and birnessite. There are also mixed valence and trivalent forms, such as hausmannite (Mn_3O_4) and manganite (MnOOH). Despite the similar formula for the Mn(IV) oxides, they are not equal with regard to their propensity to be reduced. For

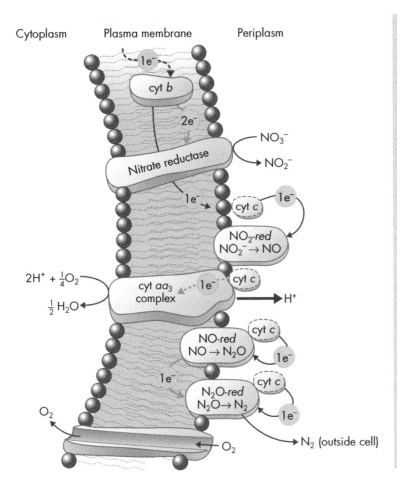

Figure 2.15 Possible electron pathways in some denitrifying bacteria such as *Pseudomonas species*. Pathway 1 (solid grey line) has electrons (one at a time) passed from cytochrome *b* directly to nitrate reductase. Two electrons are then needed to reduce nitrate to nitrite. Pathway 2 (solid black lines) have the electrons passed from cytochrome *b* to a periplasmic cytochrome *c*, where they are then shuttled to either nitrite reductase, nitric oxide reductase, or nitrous oxide reductase, depending on what intermediate nitrogen species is present. Upon completion of pathway 2, nitrogen is lost from the cell as N_2. In pathway 3 (stippled grey line), O_2 is used as the TEA, and as such, electrons are passed from cytochrome *c* to cytochrome aa_3. Note, not all enzymes are depicted, and location of enzymes not to scale. Cyt, cytochrome; NO_2-*red*, nitrite reductase; NO-*red*, nitric oxide reductase; N_2O-*red*, nitrous oxide reductase. (Adapted from Knowles, 1982.)

instance, amorphous phases with high surface areas are considerably more reactive than highly crystalline oxides with low surface area (Burdige et al., 1992). Also, thermodynamic calculations show that Mn(III) oxyhydroxides are more reactive than Mn(IV) oxides.

Many bacteria are capable of carrying out Mn(IV) reduction. The most prominent genera include *Pseudomonas*, *Bacillus*, *Geobacter*, and *Shewanella*. Some of them are facultative anaerobes, and in the presence of O_2 may use it as a TEA instead, while some grow preferentially as Fe(III) reducers, only switching to the reduction of MnO_2 when it becomes available (Nealson and Saffarini, 1994). They are also quite versatile with regards to the electron donors that they

utilize. For example, *Geobacter metallireducens* can use butyrate, propionate, lactate, succinate, and acetate, which are all completely oxidized to CO_2 (Lovley and Phillips, 1988a). By contrast, *Shewanella putrefaciens* can utilize lactate and pyruvate but oxidizes these only to acetate (Myers and Nealson, 1988a). It has also been recognized that *S. putrefaciens* can reduce Mn(III) hydroxides to Mn^{2+} (Larsen et al., 1998).

For any of the dissimilatory Mn(IV) reducers (as well as the Fe(III) and sulfate reducers discussed below), electron transport begins with the transfer of H_2 to a hydrogenase. The H_2 can come from the external environment (reaction (2.27)), whereby the bacteria behave either chemolithoautotrophically as hydrogen-oxidizing bacteria (see section 2.5.1) or mixotrophically, whereby they assimilate organic compounds as their carbon source:

$$H_2 + MnO_2 \rightarrow Mn^{2+} + 2OH^- \qquad (2.27)$$

$$\Delta G^{o\prime} = -166 \text{ kJ mol}^{-1} \text{ H}_2 \text{ (or 2e}^-)$$

More commonly, H_2 is removed from organic electron donors such as acetate (reaction (2.28)). Under these conditions, the microorganisms behave chemoheterotrophically.

$$CH_3COO^- + 4MnO_2 + 3H_2O \rightarrow$$
$$4Mn^{2+} + 2HCO_3^- + 7OH^- \qquad (2.28)$$

$$\Delta G^{o\prime} = -558 \text{ kJ mol}^{-1} \text{ acetate or } -139 \text{ kJ/2e}^-$$

The reduction of MnO_2 (or ferric oxyhydroxides, as discussed in the next section) presents the bacteria that reduce it with a rather unique situation. Unlike aerobes that rely on cross-membrane diffusion of O_2, or anaerobes that transport NO_3^- and SO_4^{2-} into the cell through special permeases, mineral oxides cannot traverse the plasma membrane to become incorporated into the electron transport chain. Instead, metal-reducing bacteria need to reside on the mineral surface (e.g., Fig. 2.16) to transfer the electrons derived from central metabolism to the cells' outer membrane where c-type cytochromes function as reductases (Lovley et al., 2004). Aside from direct contact, reduction can also be facilitated by the use of organic intermediates that shuttle electrons between the cell and the oxide surface (Stone, 1987), similar to the processes described for Fe(III) reduction (see below).

In addition to direct enzymatic reductive processes, many bacteria excrete certain metabolic products that may indirectly mediate Mn(IV) reduction. For instance, *S. putrefaciens* has been shown to reduce MnO_2 with Fe^{2+} generated during Fe(III) reduction (Myers and Nealson, 1988b). Along similar lines, any sulfate-reducing bacterium that produces H_2S or $S_2O_3^{2-}$ can potentially cause the reduction of MnO_2 (Nealson et al., 1989). A number of dissolved

Figure 2.16 Image of *Shewanella putrefaciens* CN32 on a crystal of hematite. (Courtesy of Yuri Gorby and Alice Dohnalkova.)

organic compounds can similarly reduce MnO_2 abiologically, with high-molecular-weight compounds (e.g., siderophores) reducing them at very fast rates, while low-molecular-weight compounds (e.g., pyruvate, oxalate, formate, and lactate) generally show poor reactivity (Stone and Morgan, 1984).

2.4.4 Dissimilatory iron reduction

Dissimilatory Fe(III) reduction is broadly distributed amongst several known bacterial genera. *G. metallireducens* and *S. putrefaciens* were among the first bacteria studied in pure culture that could gain energy from coupling Fe(III) reduction to the oxidation of H_2 and/or simple fermentation products, including short- and long-chain fatty acids, alcohols and various monoaromatic compounds:

$$0.5H_2 + Fe(OH)_3 \rightarrow Fe^{2+} + 2OH^- + H_2O \quad (2.29)$$

$$\Delta G^{o\prime} = -110 \text{ kJ/2e}^-$$

$$CH_3COO^- + 8Fe(OH)_3 \rightarrow$$
$$8Fe^{2+} + 2HCO_3^- + 15OH^- + 5H_2O \quad (2.30)$$

$$\Delta G^{o\prime} = -337 \text{ kJ mol}^{-1} \text{ acetate or } -84 \text{ kJ/2e}^-$$

Since then, many more species, including a number of hyperthermophilic *Archaea*, some sulfate and nitrate reducers, and a few methanogens, have shown the capacity for reducing ferric iron minerals (Lovley et al., 2004). The importance of bacterial Fe(III) reduction will be discussed in more detail in section 6.2.4(c), but it suffices to mention here that this process is accountable for a significant fraction of the organic matter oxidized and Fe(III) minerals reduced in deeply buried sediments. It may even be responsible for such late post-depositional phenomena as the formation of variegated red beds and the release of high concentrations of dissolved Fe^{2+} in anoxic ground waters (e.g., Lovley et al., 1990).

Ferric iron minerals occur in soils and sediment in a wide variety of forms, ranging from amorphous to crystalline phases. The amorphous to poorly ordered iron oxyhydroxides, such as ferric hydroxide or goethite (FeOOH), are the preferred sources of solid-phase ferric iron for Fe(III)-reducing bacteria (Lovely and Phillips, 1987a). More crystalline Fe(III) oxides, such as hematite (Fe_2O_3) and magnetite (Fe_3O_4) are also microbially reducible, and some experimental observations suggest that these minerals may provide energy for cellular growth comparable to that derived from the poorly crystalline phases (e.g., Roden and Zachara, 1996). This view is, however, controversial and it has been argued instead that those studies were conducted under artificially high nutrient levels that would not otherwise support the reduction of Fe(III) oxides in nature (Lovley et al., 2004). Other studies have even shown that bacteria can generate sufficient energy for growth by reducing the ferric iron component in smectite clays (e.g., Kostka et al., 2002). The variations in reductive rates are related to a number of factors, including the amount of surface area exposure, crystal morphology, particle aggregation, the composition of the aqueous solution in which the microorganisms grow, and the amount of Fe^{2+} sorbed to the oxide surface (e.g., Urrutia et al., 1998). Importantly, with such great heterogeneity in reactivity towards microbial reduction, it is not surprising that Fe(III) minerals can represent long-term electron acceptors for organic matter oxidation, even at depths where other anaerobic respiratory processes are thermodynamically predicted to dominate (Roden, 2003).

Until recently, it was believed that the reduction of Fe(III)-containing minerals (dissolved Fe^{3+} only dominates at pH < 4) necessitated direct contact of the microorganism with the mineral surface, and that only those Fe(III) reducers attached to the iron mineral were viable (e.g., Tugel et al., 1986). Furthermore, once in contact with the surface, the Fe(III) reducers are still faced with the problem of how to effectively access an electron acceptor that cannot diffuse into the cell. These criteria thus require that Fe(III)-reducing bacteria must not only be able to recognize an iron mineral surface and attach to it, but that they must also be capable of activating or producing proteins that specifically interact with that mineral surface. Recent studies now show how this is done. Some species, such as G. *metallireducens*, are chemotactic towards Fe(II) and Mn(II), in that they are able to sense, and move towards, the gradient of reduced metals emanating from the dissolution of their respective oxide phases. Once in their proximity, the bacteria specifically express surface appendages, such as pili, that help them adhere to the oxide surfaces (Childers et al., 2002). Alternatively, *Shewanella algae* relies on the production of hydrophobic surface proteins that facilitate cell adhesion (Caccavo et al., 1997). Fe(III)-reducing bacteria can also accumulate nanometer-scale iron oxide particles onto their surfaces for subsequent reduction (Fig. 2.17).

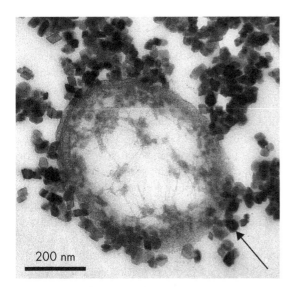

Figure 2.17 TEM image of *Shewanella putrefaciens* with a cluster of very fine-grained hematite crystals (arrow) attached to the outer membrane and penetrating into the peptidoglycan layers. (Courtesy of Susan Glasauer and Sean Langley.)

Once the bacteria attach to the mineral surface they begin shuttling electrons from a reduced source within the cytoplasm, across the plasma membrane and periplasm, to the outer membrane. Located there are iron reductase enzymes (primarily *c*-type cytochromes) that transfer those electrons directly to the Fe(III) mineral surface, causing a weakening in the Fe–O bonds and invariably its reductive dissolution (Lower et al., 2001). The expression of the reductases is growth dependent, with higher nutrient availability leading to more rapidly metabolizing populations, increased enzyme production, more favorable redox conditions and, correspondingly, the reduc-

tion of a wide range of crystalline ferric iron minerals (Glasauer et al., 2003). Most recently, it has also been demonstrated that the pili of certain *Geobacter* species could serve as biological 'nonowires', transferring electrons from the cell to the surface of Fe(III) oxides (Reguera et al., 2005).

Some *Shewanella* species also overcome the insolubility problem by utilizing organic compounds as electron shuttles between the cell surface and the Fe(III) oxides, which may be located at some distance away from the cell. One example is the quinone moieties in exogenous humic compounds that bacteria can reduce to the semiquinone and hydroquinone oxidation state via the oxidation of acetate or lactate (Lovley et al., 1996a). The reduced humics subsequently transfer electrons abiotically to the Fe(III) minerals, producing Fe^{2+}, and in doing so, regenerate the oxidized form of the humic compound for another cycle (Fig. 2.18): in fact, any number of oxidants can fulfill a similar role. Some *Shewanella* species (*S. oneidensis*) have also been shown to produce and excrete their own quinone compounds that function in a similar manner to natural humics (Newman and Kolter, 2000), while the closely related *S. algae* produces soluble melanins which might serve as another type of electron conduit for ferric mineral reduction (Turick et al., 2002). Significantly, Fe(III) reduction rates are faster in the presence of organic electron shuttles than in their absence because they are likely to be more accessible for microbial reduction than the insoluble mineral phases (Nevin and Lovley, 2000). Electron shuttles may be particularly beneficial when Fe(III)-bearing minerals are occluded in pore spaces too small for bacteria to enter.

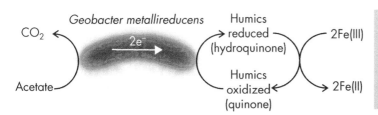

Figure 2.18 The mechanism by which some bacteria, such as *Geobacter metallireducens*, reduce ferric oxyhydroxides through the use of humic compounds as electron shuttles.

Fe(III)-reducing bacteria may also have an important role in the bioremediation of refractory organic contaminants in oxygen-depleted aquifers where sufficient ferric iron is present. A number of studies have shown that the bacteria directly oxidize a wide range of synthetic aromatic hydrocarbons, such as toluene, phenol, and p-cresol (e.g., Lovley and Lonergan, 1990). Benzene tends to be more persistent in the Fe(III) reduction zone of contaminated aquifers, and to bring about its degradation requires either the addition of soluble chelated iron (e.g., Fe^{3+} chelated to ethylenediaminetetraacetic acid, known as EDTA, or humic acids) or by using chelators on their own (e.g., nitrilotriacetic acid) that are capable of solubilizing solidphase Fe(III) minerals in the aquifer sediments (Lovley et al., 1996b). For other contaminants, such as the chlorinated organic compounds (e.g., chlorophenol and chlorobenzoate), degradation proceeds via a syntrophic community of bacteria, whereby one or more species are responsible for dehalogenating the substrate, while others (e.g., G. *metallireducens*) utilize the aromatic ring and its degradation products (Kazumi et al., 1995). Fe(III) reducers also contribute indirectly to organic oxidation reactions when the Fe^{2+} formed at circumneutral pH binds to mineral surfaces, and in doing so becomes a powerful reductant capable of inorganically reducing some chlorinated compounds (e.g., Amonette et al., 2000).

2.4.5 Trace metal and metalloid reductions

Microorganisms can obtain energy to support growth from the dissimilatory reduction of any number of multivalent trace metals and metalloids. Of particular interest has been the recognition that those chemoheterotrophic pathways may facilitate the amelioration of a number of radioactive and toxic contaminants in the environment, including As(V), Se(IV), Cr(VII), U(VI), Tc(VII), and even Pu(V) (see Lovley, 1995 for review). We will focus on two contaminants that have received significant recent attention, namely uranium and arsenic.

(a) Uranium

Uranium-contaminated groundwater is of considerable concern in areas of past uranium extraction and milling operations. Reduced uranium, U(IV), is highly insoluble and is in the oxidation state associated with most uranium-containing ores. Conversely, oxidized uranium, U(VI), in the form of a uranyl ion (UO_2^{2+}) or U(VI)–carbonate complexes, is much more mobile in oxygenated waters. Accordingly, there is interest in understanding the role microorganisms might play in its reduction, particularly in light of the fact that the abiological reductive processes are by comparison kinetically sluggish.

A variety of bacteria are known to enzymatically reduce U(VI), but at present only the Fe(III)-reducing species G. *metallireducens* and S. *putrefaciens* are known to obtain energy for growth from the process (Lovley et al., 1991). These bacteria use either acetate (reaction (2.31)) or H_2 (reaction (2.32)) as their electron donors, forming the insoluble mineral uraninite (UO_2) extracellularly as byproduct of their metabolism (Gorby and Lovley, 1992):

$$CH_3COO^- + 4UO_2^{2+} + 4H_2O \rightarrow$$
$$4UO_2 + 2HCO_3^- + 9H^+ \qquad (2.31)$$
$$\Delta G^{\circ\prime} = -264 \text{ kJ mol}^{-1} \text{ } CH_3COO^- \text{ or } -66 \text{ kJ/2e}^-$$

$$H_2 + UO_2^{2+} \rightarrow 2H^+ + UO_2 \qquad (2.32)$$
$$\Delta G^{\circ\prime} = -92 \text{ kJ/2e}^-$$

The free energy yield per mole of acetate, coupled to U(VI) reduction, is -264 kJ mol^{-1}, a value that lies between the yields for the reduction of Fe(III) and SO_4^{2-}. Not surprisingly, high levels of nitrate, Mn(IV), or Fe(III) diminish the rates of enzymatic U(VI) reduction, whereas high sulfate levels do not (Anderson and Lovley, 2002). Despite their inability to compete with microorganisms that use TEAs with more positive electrode potentials, sulfate-reducing bacteria nonetheless increase U(VI) reduction rates by: (i) directly catalyzing the reduction of U(VI) by coupling it to the oxidation of H_2 or lactate; or

(ii) promoting U(VI) reduction indirectly through reaction with H_2S, leading to the subsequent precipitation of sulfidic phases (Mohagheghi et al., 1985). The presence of organic compounds are also important because they contain functional groups (e.g., carboxyl, phosphate, hydroxyl) that form coordinate-covalent bonds with U(VI), that may, or may not, decrease the rates of reduction (Ganesh et al., 1997).

Given the ability of bacteria to reduce U(VI), a novel strategy for the treatment of uranium-contaminated aquifers is through the creation of reducing conditions that foster uranium immobilization. This concept is similar to the use of permeable reactive barriers in that the addition of organic substrates into the groundwater flow path consumes dissolved oxygen and subsequently stimulates the growth of indigenous Fe(III)-reducing bacteria to reduce and precipitate uranium after the Fe(III) is eliminated. For instance, in experiments that injected acetate into the subsurface, it was observed that after a few days the majority of the microbial community consisted of *Geobacter* species, which coincided with the most active periods of Fe(III) and U(VI) reduction (Anderson et al., 2003). In terms of bioremediation, on a per cell basis, enzymatic reduction of U(VI) has a greater potential to remove uranium from solution than does the biosorptive process described in the next chapter (see section 3.7.1). It can also be potentially coupled to the oxidation of organic contaminants, and thus be employed in treating multicomponent wastes (Lovley and Phillips, 1992).

(b) Arsenic

The contamination of groundwaters by sediment-derived arsenic threatens the health of tens of millions of people worldwide, most notably in Bangladesh and West Bengal. Arsenic has two main oxidation states, As(V) as $H_2AsO_4^-$ or $HAsO_4^{2-}$, and As(III) as $H_2AsO_3^-$. The oxidized form, arsenate, has the propensity to adsorb onto iron and aluminum oxyhydroxides, which can limit its hydrological mobility. By contrast, arsenite adsorbs less readily to mineral surfaces, making it the more mobile oxyanion.

There are a number of microorganisms that directly reduce As(V) to As(III), using fermentation products as their electron donors (Oremland and Stolz, 2003):

$$CH_3COO^- + 4HAsO_4^{2-} + 3H^+ \rightarrow$$
$$2HCO_3^- + 4H_2AsO_3^- \qquad (2.33)$$

$$\Delta G^{o\prime} = -175 \text{ kJ mol}^{-1} \text{ CH}_3\text{COO}^- \text{ or} -44 \text{ kJ/2e}^-$$

Some strains have even been found to degrade more complex aromatic compounds like benzoate and toluene. None of those species, however, are obligate As(V) reducers. Instead, all naturally respire using other TEAs (e.g., Fe(III), NO_3^-), but are capable of switching to arsenate reduction when their preferred substrates are unavailable (Laverman et al., 1995). A number of sulfate-reducing bacteria have also been observed to switch to As(V) reduction as it offers them more free energy than sulfate as their TEA (Ahmann et al., 1994). Notably, recent studies have uncovered the presence of a highly conserved pair of proteins (ArrA and ArrB) that form a dedicated complex for As(V) respiratory reduction. These genes are present in virtually every As(V)-respiring bacterium isolated to date, regardless of the phylogenetic placement of the bacterium (Malasarn et al., 2004). In addition, many microorganisms (e.g., *E. coli*) reduce As(V) as a means of detoxification. Such species possess a different suite of genes that specifically encode for As(V) reduction via an As(V) reductase, followed by As(III) removal from the cell via an efflux pump (Cervantes et al., 1994).

Fe(III)-reducing bacteria further promote the increased mobility of arsenic when they reduce ferric oxyhydroxides, thereby leading to the indirect release and solubilization of arsenate. This, in turn, can serve as an electron acceptor once the Fe(III) source is depleted (Dowdle et al., 1996). That As(V) reduction occurs after ferric iron is predicted on the basis of the relevant electrode potentials, $Fe(OH)_3/Fe^{2+}$ at pH 7 being 0.20 V, while As(V)/As(III) is 0.14 V at similar pH. Rates of As(V) reduction from ferric oxy-hydroxide surfaces is dependent on mineralogy, mineral surface area, and As surface coverage (Jones et al., 2000). Another factor affecting the capacity for arsenic release is the availability of fermentation products in sediment. Therefore, human activities such as irrigation pumping, that have the effect of delivering organic carbon into subsurface communities, may dramatically increase arsenic mobility in shallow ground waters (e.g., Islam et al., 2004).

Despite its greater mobility under reducing conditions, in the presence of hydrogen sulfide, arsenite can be removed from solution through sulfide mineral precipitation. In most cases this process is indirectly related to the activity of sulfate-reducing bacteria, but in one study it was shown how the SRB, *Desulfotomaculum auripigmentum*, could form orpiment (As_2S_3) via the simultaneous dissimilatory reduction of As(V) to As(III) and S(VI) to S(-II) (Newman et al., 1997). Biomineralization begins intracellularly on the inside of the plasma membrane in association with reductive enzymes associated with the respiratory electron transport chain. Then, as the concentration of As(III) and H_2S increases sufficiently, diffusion out of the cell leads to mineral formation in the bulk medium (Fig. 2.19).

2.4.6 Dissimilatory sulfate reduction

As will have become obvious in earlier sections, a diverse and heterogeneous range of bacteria are able to use sulfate as a terminal electron acceptor. Within the domain *Bacteria*, many SRB belong to the δ-group of the phylum *Proteobacteria*. These species are typically mesophiles. Three genera, *Desulfotomaculum*, *Desulfosporosinus*, and *Thermoacetogenium*, are also mesophilic but belong to the Gram-positive phylum. The mesophiles are prevalent in anoxic marine sediments, where sulfate is the major pore water anion, but they are usually not found in the water column unless the oxygen concentrations are limited by stratification. With that said, it has been shown that species of SRB differ significantly with respect to oxygen sensitivity, and some can even survive hours of exposure to oxygen and regain activity when anoxic conditions return.

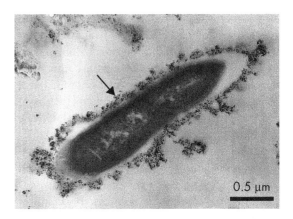

0.5 µm

Figure 2.19 TEM image of epicellular orpiment crystals (arrow) precipitated by *Desulfotomaculum auripigmentum* as a direct result of its respiratory growth on arsenate and sulfate. (Courtesy of Dianne Newman.)

If a habitat becomes homogenously oxic over a long period of time, vegetative cells of SRB die off and only spores may survive (Perry, 1995). SRB have also been isolated from waterlogged soils, brackish waters, sewage, and mine waste, basically wherever anoxic conditions prevail and sufficient concentrations of organic material and sulfate exist. Three bacterial genera, *Thermodesulfobacterium*, *Thermodesulfovibrio*, and *Thermodesulfobium*, as well as two archaeal genera, *Archaeoglobus* and *Caldivirga*, constitute deeply branching lineages that are thermophilic isolates from hot springs, oil and natural gas wells, and deep-sea hydrothermal vents.

Two major metabolic groups of sulfate-reducing bacteria are known, depending on whether or not they can oxidize acetate (Widdel, 1988). The first group utilize lactate, formate, propionate, butyrate, pyruvate, and aromatic compounds, which they typically oxidize to acetate, while the second group oxidize acetate all the way to CO_2. Many SRB from both groups are additionally able to grow chemolithoautotrophically, some using H_2 as the electron donor, while others show incredible metabolic plasticity in that they can oxidize hydrogen sulfide, sulfite (SO_3^-), thiosulfate, and elemental sulfur with oxygen, nitrate, or nitrite, as electron acceptors (e.g., Dannenberg et al., 1992). Still other SRB have proven capable of switching to chemoheterotrophic NO_3^- or Fe(III) reduction when those oxidants are available (e.g., Seitz and Cypionka, 1986; Coleman et al., 1993).

In terms of their most widely used metabolic pathway, that being the complete reduction of sulfate to hydrogen sulfide, the overall dissimilatory reduction process is an $8e^-$ reduction (reactions (2.34) and (2.35)), requiring several $2e^-$ reductions at a time:

$$4H_2 + SO_4^{2-} \rightarrow H_2S + 2OH^- + 2H_2O \qquad (2.34)$$

$$\Delta G^{\circ\prime} = -152 \text{ kJ mol}^{-1} \text{ or } -38 \text{ kJ } 2e^-$$

$$CH_3COO^- + SO_4^{2-} + H_2O \rightarrow$$
$$H_2S + 2HCO_3^- + OH^- \qquad (2.35)$$

$$\Delta G^{\circ\prime} = -48 \text{ kJ mol}^{-1} \text{ acetate or } -12 \text{ kJ } 2e^-$$

SRB use a unique system for reducing sulfate. First, H_2, either as a gas from the environment or extracted from simple organic substrates, donate their electrons to a hydrogenase, which is situated in the periplasm (Fig. 2.20). The protons from the former oxidation reaction remain outside the membrane (setting up the pmf), while the electrons are transferred to a very electronegative periplasmic cytochrome called c_3. From there the electrons cross the plasma membrane via a cytochrome complex (called HMC) to a cytoplasmic Fe-S protein. If the H_2 is sourced from the intracellular oxidation of lactate, then H_2 is transported across the membrane to the same hydrogenase. When acetate is the electron donor, it must first be cycled through a modified citric acid cycle to generate the H_2.

Second, because of sulfate's stability, it must first be activated by means of ATP via the enzyme

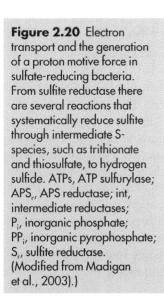

Figure 2.20 Electron transport and the generation of a proton motive force in sulfate-reducing bacteria. From sulfite reductase there are several reactions that systematically reduce sulfite through intermediate S-species, such as trithionate and thiosulfate, to hydrogen sulfide. ATPs, ATP sulfurylase; APS_r, APS reductase; int, intermediate reductases; P_i, inorganic phosphate; PP_i, inorganic pyrophosphate; S_r, sulfite reductase. (Modified from Madigan et al., 2003).)

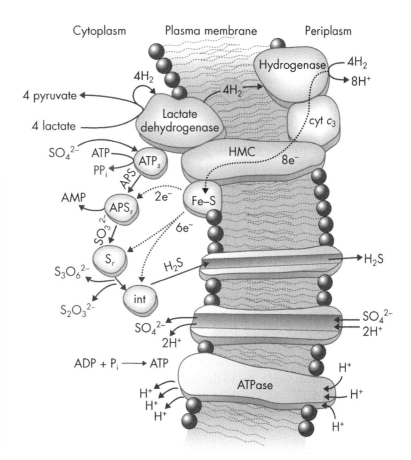

ATP-sulfurylase. The reaction involves the attachment of sulfate to a phosphate group of ATP after the anion is transported into the cytoplasm via a symporter. This leads to the formation of adenosine phosphosulfate (APS), with the concomitant release of inorganic pyrophosphate, $P_2O_5(OH)_2^{2-}$. APS is then reduced to sulfite by a pair of electrons derived from FeS by the enzyme APS reductase, with the release of AMP (adenosine monophosphate). Once formed, sulfite is reduced to H_2S through a series of intermediate reactions that include trithionate ($S_3O_6^{2-}$) and thiosulfate ($S_2O_3^{2-}$) and perhaps some organic sulfur compounds (Peck, 1993). The S-intermediates can also be excreted from the cell, thus providing the surrounding aqueous environment with an assortment of dissolved S-species that can either serve as TEAs for other chemoheterotrophs or as electron donors for photo- or chemolithotrophic populations.

Several heterotrophic species unable to carry out dissimilatory sulfate reduction are instead able to reduce elemental sulfur (reaction (2.36)) or sulfite (reaction (2.37)) to hydrogen sulfide (e.g., Moser and Nealson, 1996):

$$H_2 + S^0 \rightarrow H_2S \qquad (2.36)$$

$$\Delta G^{o\prime} = -28 \text{ kJ/2e}^-$$

$$3CH_3COO^- + 4SO_3^{2-} + 5H^+ \rightarrow 4H_2S + 6HCO_3^-$$
$$(2.37)$$

$$\Delta G^{o\prime} = -126 \text{ kJ mol}^{-1} \text{ acetate or } -32 \text{ kJ/2e}^-$$

For S^0 reduction, the cells grow attached to the mineral surface, whereas sulfite is transported into the cytoplasm. Coupling the reaction of S^0 reduction with acetate oxidation ($CH_3COO^- + 4S^0 + 4H_2O \rightarrow 4H_2S + 2HCO_3^- + H^+$) only yields -7 kJ mol^{-1} of acetate, a value below the theoretical limits of -20 kJ mol^{-1} of substrate for cells to exploit the free energy from the reaction (Schink, 1997). However, a recent study with syntrophic fermentative bacteria has lowered this limit to values very close to thermodynamic equilibrium ($\Delta G^{o\prime} = 0 \text{ kJ mol}^{-1}$) in situations where the degradation product by one species is rapidly removed by another species (Jackson

and McInerney, 2002). This likely explains early observations that SRB grown on elemental sulfur and various organic compounds did not thrive in pure culture (e.g., Biebl and Pfennig, 1977). By contrast, growing chemolithoautotrophically using H_2 gives the cells -28 kJ mol^{-1}. Similarly, sulfur-reducing bacteria (e.g., *Desulfuromonas acetoxidans*) commonly grow well in syntrophic communities with green-sulfur bacteria because the latter excrete copious amounts of organic compounds into the growth medium, as well as extracellular sulfur granules as a byproduct of their oxidative metabolism (Biebl and Pfennig, 1978).

Certain sulfate-reducing bacteria are additionally capable of generating energy by disproportionation, where one of the compounds ends up more oxidized, while the other becomes more reduced. For example, *Desulfovibrio sulfodismutans* can disproportionate sulfite (reaction (2.38)), thiosulfate (reaction (2.39)), and sulfur (reaction (2.40)) into both sulfide and sulfate (Bak and Cypionka, 1987; Lovley and Phillips, 1994):

$$4SO_3^{2-} + 2H^+ \rightarrow 3SO_4^{2-} + H_2S \qquad (2.38)$$

$$\Delta G^{o\prime} = -236 \text{ kJ/reaction}$$

$$S_2O_3^{2-} + H_2O \rightarrow SO_4^{2-} + H_2S \qquad (2.39)$$

$$\Delta G^{o\prime} = -22 \text{ kJ/reaction}$$

$$4S^0 + 4H_2O \rightarrow SO_4^{2-} + 3H_2S + 2H^+ \qquad (2.40)$$

$$\Delta G^{o\prime} = +41 \text{ kJ/reaction}$$

Reaction (2.40) is written as an energetically unfavorable reaction. However, if the H_2S formed is oxidized by an available electron acceptor or incorporated into a solid-phase sulfide mineral, then the decrease in sulfide concentration will drive the disproportionation reaction to the right (Thamdrup et al., 1993). Sulfur disproportionation can be viewed as a type of fermentation because it does not require an external electron donor or acceptor, using instead inorganic energy sources when organic compounds are in short supply.

2.4.7 Methanogenesis and homoacetogenesis

In many instances, the terminal step in the anaerobic degradation of organic material is methanogenesis. In fact, it has been estimated that nearly half of the organic carbon degraded globally by anaerobically respiring bacteria is eventually converted to methane (Higgins et al., 1981). The methanogens are strict anaerobic *Archaea*, and many are killed quickly by exposure to oxygen. They commonly occupy an important niche in anoxic, sulfate-deficient environments such as swamps, water-logged soils, tundra, marine sediment, hydrothermal vents, landfill sites, and sewage. Some can even be found in aerated environments that contain anoxic microenvironments, such as the interior of soil particles. Methanogens currently define the largest temperature range of any microorganisms, from as high as 110°C (*Methanogenium kandleri*) to below 0°C (*Methanogenium frigidum*).

Three classes of methanogens are known, two of which are heterotrophic, and will be discusses here. One of those classes disproportionate methyl-containing compounds to methane. Some genera, such as *Methanococcus*, grow on methanol (reaction (2.41)) or methylamine (CH_3NH_2), while others, such as *Methanospirillum*, oxidize propanol, ethanol, and butanol (Lovley and Klug, 1983). Importantly, the methylated compounds do not stimulate sulfate-reducing bacteria, which means the two types of species can coexist in sediments where the compounds are abundant (Oremland et al., 1982).

$$4CH_3OH \rightarrow 3CH_4 + CO_2 + 2H_2O \quad (2.41)$$
$$\Delta G^{o\prime} = -319 \text{ kJ/reaction}$$

Another class disproportionate acetate to form methane and bicarbonate (reaction (2.42)). Only two genera of methanogens, *Methanosarcina* and *Methanothrix*, have species that are acetotrophic, but they have been found to be a significant pathway for methane production in sewage sludge (Mah et al., 1976).

$$CH_3COO^- + H_2O \rightarrow CH_4 + HCO_3^- \quad (2.42)$$
$$\Delta G^{o\prime} = -31 \text{ kJ/reaction}$$

Methanogens cannot use long-chain fatty acids (with more than two carbons) or aromatic compounds, so they depend on the activity of other microorganisms to break down complex organic compounds into more suitable and simple substrates. In reality, the simple conversion of cellulose into methane may require as many as five major physiological groups of bacteria (Fig. 2.21). One such group are the homoacetogenic bacteria, such as *Acetobacterium woodii* and *Clostridium aceticum* (Drake, 1994). These bacteria ferment sugars, fatty acids, and aromatics to acetate and H_2 via a pyruvate intermediate (reaction (2.43)):

$$2C_3H_3O_3^- \text{ (pyruvate)} + 2H_2O \rightarrow$$
$$2CH_3COO^- + 2CO_2 + 2H_2 \quad (2.43)$$
$$\Delta G^{o\prime} = -104 \text{ kJ/reaction}$$

These substrates are then consumed by the methanogens, sometimes so readily that the metabolites are not even measurable. This is another example of a syntrophic relationship, and when H_2 is the metabolite, it is referred to as interspecies H_2 transfer. For the homoacetogens, they require a sink for their metabolic waste because the degradation of the above compounds becomes thermodynamically unfavorable unless the concentration of H_2 is kept low. Meanwhile the methanogens use that waste as a substrate, and in doing so, remove the final electrons from the ecosystem (Thiele and Zeikus, 1988). Such mutual dependence with respect to energy limitation can go so far that neither partner bacteria can operate without the other, and together they exhibit a metabolic activity that neither one could accomplish on its own. Such syntrophy has been applied in anaerobic sludge digestion, with fermentative bacteria hydrolyzing large polymers into H_2, CO_2, and volatile fatty acids. Those acids are then converted to H_2, CO_2, and acetate by the homoacetogens, while the H_2 and acetate serve as substrates for the

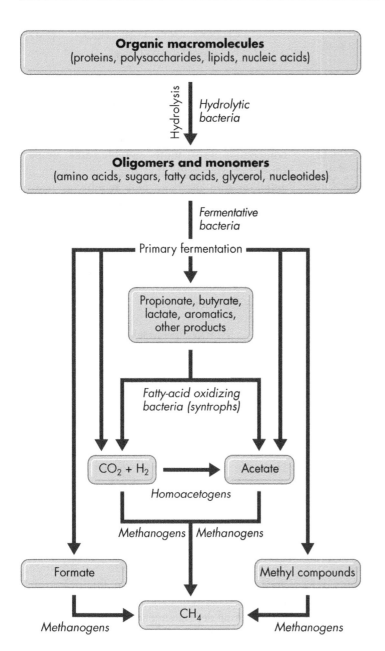

Figure 2.21 The overall process of organic matter degradation to CO_2 and CH_4. The initial input of complex organic macromolecules is progressively broken down into simpler substrates, and eventually to gas phase, by the cooperative interaction of a number of species growing as part of a syntrophic community. (Modified from Madigan et al., 2003.)

methanogens. It has been estimated that 70% of the methane produced in sludge digestion arises from the methyl group of acetate, while the remainder is produced from CO_2 reduction by H_2 (see section 2.5.2) (Frea, 1984).

Without going into detail, unlike all the chemoheterotrophs discussed above, the conventional electron transport system involving cytochromes and ubiquinones are absent in methanogens. Even the common coenzymes such as NAD^+ and FMN are of diminished importance. Instead, the methanogens possess a number of enzymes unique to methane production (see Ferry, 1993 for details).

2.5 Chemolithoautotrophic pathways

Chemolithoautotrophic bacteria employ various aspects from both photoautotrophs and chemoheterotrophs. Similar to the photoautotrophs, these microorganisms use CO_2, generally processed via the Calvin cycle, as their source of carbon. However, sunlight is not their source of energy. Instead, they have evolved the metabolic capacity to oxidize reduced inorganic substrates, such as H_2, CH_4, H_2S, S^0, $S_2O_3^{2-}$, Fe(II), Mn(II), NH_4^+, and NO_2^-, coupled to a terminal electron acceptor, for the production of ATP in a manner reminiscent of oxidative phosphorylation (Wood, 1988). There is, however, one very important distinction between chemolithoautotrophs and chemoheterotrophs, namely the latter utilize preformed organic compounds, and hence are able to generate abundant biomass per mole of substrate oxidized. By contrast, the chemolithoautotrophs have the added burden of autotrophy. This means that the reductants must be used for both CO_2 fixation and ATP production, with the net result being that most chemolithoautotrophs synthesize only small amounts of biomass, while oxidizing large amounts of substrate.

In most of the examples below, O_2 serves as the TEA, but different microorganisms can substitute it for less electropositive TEAs, i.e., NO_3^-, Mn(IV), Fe(III), SO_4^{2-}, and S^0 when oxic conditions become diminished (e.g., Table 2.1). The reducing power for generating NADPH can be obtained either directly from the inorganic substrate, if it has a sufficiently low electrode potential (e.g., H_2), or the cell must employ reverse electron flow reactions, feeding electrons from the reductants directly into the transport chain at a point slightly more electronegative than those of the intermediate carriers. In most cases that carrier is either a c-type cytochrome or an enzyme that transfers electrons directly into the cytochrome.

Characteristic of chemolithoautotrophs is their need to oxidize external reductants. Thus, they tend to possess oxidase enzymes situated either on the outer membrane or within the periplasm. These locations provide the cells with a number of benefits, including: (i) it contributes to the proton electrochemical gradient, while preventing cytoplasm acidification; (ii) it negates energy-consuming transport systems for the substrate, particularly when it is charged (e.g., Fe^{2+}) or insoluble (e.g., S^0); (iii) it excludes potentially toxic oxidation products (e.g., NO_2^-) or mineral precipitates (e.g., $Fe(OH)_3$) from the cytoplasm; and (iv) they avoid overcrowding in the plasma membrane, where a number of other enzymes are already embedded (Ferguson, 1988). Indeed, almost all chemolithoautotrophs are Gram-negative bacteria, and it has been suggested that this restriction may be a reflection of the need for a volume (such as the periplasm) sufficient to accommodate the essential oxidase enzymes (e.g., DiSpirito et al., 1985).

The chemolithoautotrophic examples discussed below are ordered according to their theoretical energy yields, which, similar to the chemoheterotrophs, are proportional to the difference in electrode potentials between the PED and TEA. However, such ordering disguises the reality that $\Delta G°$ calculations do not take into account the kinetics of the redox reaction, the actual concentration of substrate (i.e., not standard 1 mol L^{-1}), or factors such as the cell's physiology, in that a given cell may perhaps favor one reductant over another, e.g., *Leptospirillum ferrooxidans* uses Fe(II) not H_2S, even if hydrogen sulfide is more available. The implications of the latter are that some lower energy redox reactions may occur preferentially, and this makes metabolic predictions based solely on thermodynamics sometimes misleading.

2.5.1 Hydrogen oxidizers

Chemolithoautotrophic bacteria are classified on the basis of the type of inorganic substrate oxidized. Hydrogen is a common fermentation product, and a number of chemolithoautotrophs, called H_2-oxidizing bacteria, are able to use it to

Table 2.1 Some documented examples of anaerobic chemolithoautotrophic redox couplings in *Archaea* (top species) and *Bacteria* (bottom species). Question marks represent potential metabolic reactions, but for which no species have yet been described.

Donors electron	Electron acceptors				
	S^0	SO_4^{2-}	Fe(III)	Mn(IV)	NO_3^-
H_2	(Sulfur reducers, H_2 oxidizers) *Thermoproteus tenax* *Persephonella marina*	(Sulfate reducers, H_2 oxidizers) *Archaeoglobus fulgidus* *Desulfovibrio desulfuricans*	(Iron reducers, H_2 oxidizers) *Pyrobaculum islandicum* *Geobacter metallireducens*	(Manganese reducers, H_2 oxidizers) *Pyrobaculum islandicum* *Shewanella putrefaciens*	(Nitrate reducers, H_2 oxidizers) *Pyrolobus fumarii* *Paracoccus denitrificans*
CH_4	(Methanotrophs) ?? ??	(Methanotrophs) ?? *Desulfovibrio desulfuricans*	(Methanotrophs) ?? ??	(Methanotrophs) ?? ??	(Methanotrophs) ?? ??
H_2S			(Sulfur oxidizers) ?? *Acidithiobacillus ferrooxidans*	(Sulfur oxidizers) ?? ??	(Sulfur oxidizers) *Ferroglobus placidus* *Thioploca araucae*
S^0			(Sulfur oxidizers) ?? *Acidithiobacillus thiooxidans*	(Sulfur oxidizers) ?? ??	(Sulfur oxidizers) ?? *Thiobacillus denitrificans*
Fe(II)				(Iron oxidizers) ?? ??	(Iron oxidizers) *Ferroglobus placidus* *Thiobacillus denitrificans*
Mn(II)					(Manganese oxidizers) ?? ??

generate ATP. The H_2-oxidizing bacteria are the most widespread chemolithoautotrophs, ranging from aerobes, to facultative anaerobes that can grow on NO_3^-, Mn(IV) or Fe(III), to obligate anaerobes that use SO_4^{2-}, and finally to methanogens and homoacetogens. An interesting aspect of H_2-oxidizing bacteria is that they will incorporate organic compounds if a limited amount is added to the medium, thus behaving as mixotrophs. Most even grow as facultative chemolithoautotrophs. In other words, they can grow chemoheterotrophically when sufficient levels of organic compounds are available, apparently suppressing the synthesis of Calvin cycle enzymes and hydrogenases. This is the major distinction between them and the obligate chemolithoautotrophs; most representatives of the latter cannot grow in the absence of an inorganic energy source.

In aerobic H_2-oxidizing bacteria, such as *Ralstonia eutropha*, H_2 can be taken up by two hydrogenases, one membrane-bound and the other freely soluble in the cytoplasm (Moat et al., 2002). The membrane-bound hydrogenase is associated with energy production, and following H_2 binding to the enzyme, electrons from it are transferred to a ubiquinone, and from there through a series of cytochrome reductases to the terminal oxidase, cytochrome aa_3, which facilitates the reduction of O_2. As the electrons are shuttled down the transport chain this leads to the generation of the pmf, and ATP. The second hydrogenase is involved in reducing power for CO_2 fixation. This soluble hydrogenase

takes up H_2, and because the $E^{o\prime}$ of H_2 is -0.41 V, it can directly reduce NAD^+ to NADH, the latter then converted to NADPH (by enzymes called transhydrogenases) for use in the Calvin cycle. The ability to directly reduce CO_2 without the need for reverse electron flow allows for high yields of H_2-oxidizing bacteria compared to other chemolithoautotrophs.

Reaction (2.44), and many of the others below, are written as the sum of both the chemolithotrophy (energy generation) and autotrophy (CO_2 reduction) "half reactions." The decoupled reactions are written directly underneath to highlight the partitioning of substrate molecules for both individual reactions.

2.5.2 Homoacetogens and methanogens

Homoacetogens have already been discussed in the previous section, but they are mentioned here as well because many of them (e.g., *Clostridium acetium*) actually use CO_2 (or more accurately HCO_3^- at circumneutral pH) as both their terminal electron acceptor and carbon source, while H_2 serves as their electron donor (reaction (2.45)). Under these conditions, the homoacetogens metabolize as chemolithoautotrophs, using the reductive acetyl-CoA pathway, not the Calvin cycle, for autotrophic carbon fixation (Hamilton, 1988).

The reductive acetyl-CoA pathway, also known as the Ljungdahl–Wood pathway, involves the direct reduction of CO_2 to acetyl-CoA. It has

$$13H_2 + 0.5O_2 + 6CO_2 \rightarrow C_6H_{12}O_6 + 7H_2O$$
$$12H_2 + 6CO_2 \rightarrow C_6H_{12}O_6 + 6H_2O$$
$$H_2 + 0.5O_2 \rightarrow H_2O$$

theoretical overall reaction (2.44)
autotrophy $^1/_2$ reaction
chemolithotrophy $^1/_2$ reaction
$(\Delta G^{o\prime} = -237$ kJ/2e$^-)$

$$16H_2 + 8HCO_3^- + 7H^+ \rightarrow C_6H_{12}O_6 + CH_3COO^- + 16H_2O$$
$$12H_2 + 6HCO_3^- + 6H^+ \rightarrow C_6H_{12}O_6 + 12H_2O$$
$$4H_2 + 2HCO_3^- + H^+ \rightarrow CH_3COO^- + 4H_2O$$

theoretical overall reaction (2.45)
autotrophy $^1/_2$ reaction
chemolithotrophy $^1/_2$ reaction
$(\Delta G^{o\prime} = -26$ kJ/2e$^-)$

thus far only been found in certain obligate anaerobes. In the pathway, one molecule of CO_2 is reduced to the CH_3 group of acetate by a series of enzymatic reactions involving the coenzyme tetrahydrofolate (see Ragsdale, 1991 for details). Hydrogen is also required as an electron donor. The other CO_2 molecule is reduced to the carbonyl group ($-COO^-$) by carbon monoxide dehydrogenase. In the final synthesis of acetate, the methyl group is combined with the carbonyl group.

In direct competition with the homoacetogens for available H_2 are the methanogens. They produce methane rather than acetate from HCO_3^- and H_2 (reaction (2.46)). Quantitatively, this reaction is the most important process involving H_2 in nonsulfate environments, and it accounts for some 40% of the methane produced in sediments (Wolfe, 1972). Like the homoacetogens, the methanogens are obligate anaerobes that employ the reductive acetyl-CoA pathway for autotrophic growth.

2.5.3 Methylotrophs

A variety of bacteria, the methylotrophs, make a living by oxidizing one-carbon compounds, ranging from those highly reduced, such as methane (C^{-4}), to progressively more oxidized, like methanol (C^{-2}) and methylamine (C^{-2}), through to formaldehyde (C^0) and formate (C^{+2}) (Ribbons et al., 1970). Those that oxidize methane are known as methanotrophs. These bacteria are widespread in aquatic and terrestrial systems where a stable source of methane is available. One typical environment is in marine sediment, where methane from underlying anoxic zones diffuses upwards into the overlying oxic zone. Some methanotrophs can even switch to ammonia oxidation if the supply of methane is temporarily diminished (e.g., O'Neill and Wilkinson, 1977).

The pathway of methane oxidation to CO_2 (reaction (2.47)) involves a series of $2e^-$ oxidations, with the initial step involving diffusion of methane into a unique enzyme complex called methane monooxygenase. This enzyme catalyzes the incorporation of an oxygen atom into the methane molecule as a hydroxyl group, leading to the formation of methanol. The methanol can be further oxidized to completion via formaldehyde (CH_2O) and formate ($HCOO^-$) intermediates; it can be disproportionated into CH_4 and CO_2; or it can be excreted, later serving as a substrate for other methylotrophs and methanogens. The electrons from the oxidative steps are then fed into the electron transport system (Anthony, 1986).

Several methanotrophs assimilate methane as their carbon source via the serine pathway. In it acetyl-CoA is synthesized from one molecule of CH_2O and one molecule of CO_2. Other methylotrophs utilize the ribulose monophosphate pathway for carbon fixation. It is a more efficient process than the serine pathway because all the carbon is derived from formaldehyde, which is

$$
\begin{aligned}
&16H_2 + 7HCO_3^- + 7H^+ \rightarrow C_6H_{12}O_6 + CH_4 + 15H_2O \\
&12H_2 + 6HCO_3^- + 6H^+ \rightarrow C_6H_{12}O_6 + 12H_2O \\
&4H_2 + HCO_3^- + H^+ \rightarrow CH_4 + 3H_2O
\end{aligned}
$$

theoretical overall reaction (2.46)
autotrophy $^1/_2$ reaction
chemolithotrophy $^1/_2$ reaction
($\Delta G^{o'} = -34$ kJ/$2e^-$)

$$
\begin{aligned}
&7CH_4 + 8O_2 \rightarrow C_6H_{12}O_6 + CO_2 + 8H_2O \\
&6CH_4 + 6O_2 \rightarrow C_6H_{12}O_6 + 6H_2O \\
&CH_4 + 2O_2 \rightarrow CO_2 + 2H_2O
\end{aligned}
$$

theoretical overall reaction (2.47)
autotrophy $^1/_2$ reaction
chemolithotrophy $^1/_2$ reaction
($\Delta G^{o'} = -204$ kJ/$2e^-$)

at the same oxidation state as cell material (see Hou, 1984 for details).

Two methane-rich environments are currently receiving considerable attention, namely polar regions and along continental margins. The amount of methane that is present as gas hydrates in permafrost and seafloor sediment is perhaps 3000 times the amount present in the atmosphere (Kvenvolden, 1999). From an economic perspective, if these resources can be harnessed, they would complement traditional fossil fuel reserves. Yet environmentally, if high methane fluxes were allowed to freely mix into the atmosphere, they could significantly alter the global climate (e.g., Dickens, 2003). In marine sediments, much of that methane is biologically oxidized, some in the water column by obligate aerobes, but most on the seafloor by a mixed community of chemolithoautotrophic bacteria growing anaerobically with nitrate or sulfate as electron acceptors. Collectively, those anaerobes transform the reactive hydrocarbons into more innocuous authigenic carbonate phases, making those chemosynthetic communities important moderators of the methane carbon cycle (see section 6.2.5(c)).

Indications for such anaerobic methane oxidation were for many years presumed based on the observations that methane diffusing upwards from deep sediments often disappeared long before any contact with oxygen was possible. Moreover, the presence of ^{13}C-enriched methane and ^{13}C-depleted CO_2 in anoxic sediments is consistent with kinetic fractionation that would result from bacterial conversion of a proportion of that methane into carbon dioxide. Subsequently, the site of maximum anaerobic methane oxidation (AOM) was shown to exist at the methane–sulfate reduction interface, where the supplies of both oxidant and reductant are concurrently available (Reeburgh, 1980). The overall reaction is given below:

$$CH_4 + SO_4^{2-} \rightarrow HCO_3^- + HS^- + H_2O \quad (2.48)$$

$$\Delta G^{\circ\prime} = -17 \text{ kJ/reaction}$$

Many years ago it was suggested that some sulfate reducers were capable of directly coupling the oxidation of methane to sulfate reduction, albeit slowly, as long as they had access to another carbon source (e.g., Davis and Yarborough, 1966). Yet, it is now recognized that the more efficient process seems to involve a syntrophic relationship between methanogens and SRB. It was recently proposed that some methanogens from the *Methanosarcinales* order, have the limited capacity to reverse their normal metabolism, thereby oxidizing methane under anoxic conditions, rather than producing it from H_2 and HCO_3^- (reaction (2.46) in reverse). Such "reverse methanogenesis" is energetically favorable only so long as the H_2 end product is maintained at a low concentration by another group of microorganisms, e.g., the H_2-oxidizing sulfate reducers (recall reaction (2.34)), that use it as a substrate for their growth (Hoehler et al., 1994). Evidence for such syntrophy comes from methane-rich marine sediments, where lipid biomarkers and fluorescently labeled rRNA-targeted probes have identified methanotrophs growing in tight clusters, closely surrounded by their sulfate-reducing bacterial partners, e.g., the *Desulfosarcina* and *Desulfococcus* genera of the δ-*Proteobacteria* (Fig. 2.22) (e.g., Hinrichs et al., 1999; Boetius et al., 2000).

Reaction (2.48) has since been experimentally documented, but instead of H_2 serving as the reductant in AOM, it has been suggested that the transfer of reducing equivalents occurs via an electron shuttle operating in direct physical contact between the *Archaea* and the SRB, much the same as that postulated for Fe(III)-reducing bacteria (Nauhaus et al., 2002). Moreover, the archaeal partners may be dedicated methane oxidizers, and not just the usual methanogens that reverse methanogenesis. It should also be pointed out that in theory the role of a H_2-consumer may not be limited to SRB, but any microorganism capable of maintaining low enough H_2 concentrations (e.g., nitrate-, Mn(IV), or Fe(III)-reducers) might enable methanogens to bring about net methane oxidation.

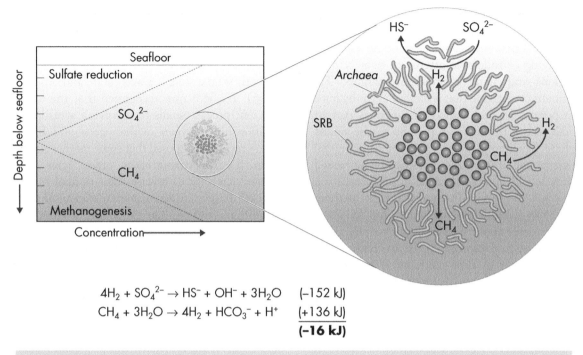

$$4H_2 + SO_4^{2-} \rightarrow HS^- + OH^- + 3H_2O \quad (-152 \text{ kJ})$$
$$CH_4 + 3H_2O \rightarrow 4H_2 + HCO_3^- + H^+ \quad \underline{(+136 \text{ kJ})}$$
$$\mathbf{(-16\ kJ)}$$

Figure 2.22 The spatial arrangement between methane-oxidizing *Archaea* and sulfate-reducing bacteria (SRB) growing at the sulfate–methane interface in marine sediments. The SRB grow on the outer periphery of the archaeal community, utilizing the H_2 excreted by the latter as part of their "reversed" metabolism. (From DeLong, 2000. Reproduced with permission from the Nature Publishing Group.)

2.5.4 Sulfur oxidizers

The colorless sulfur-oxidizing bacteria are a diverse group of microorganisms capable of growing chemolithoautotrophically on reduced sulfur compounds. Two broad ecological classes exist. The first class, including the genera *Beggiatoa*, *Thiothrix*, *Thiovulum*, *Thiomicrospira*, and *Thioploca*, are considered "gradient organisms" that grow at neutral pH under microaerophilic conditions with O_2 concentrations around 5% present atmospheric levels (PAL). Such environments exist at sulfur springs, within microbial mats, low-oxygenated marine waters, and freshwaters polluted in sewage. Similar to Fe(II) oxidation (see next section) the rates of biological sulfide oxidation are considerably enhanced in comparison to chemical oxidation rates when oxygen concentrations are low. In other words, the bacteria act as catalysts, increasing the reaction

kinetics (e.g., Nelson et al., 1986). The second class grow in acid environments, characteristic of metal and coal mine drainage. The main genera are *Acidithiobacillus*, *Sulfolobus*, and *Acidianus*.

Hydrogen sulfide is the most common electron donor, sulfate is the final oxidation product and oxygen is typically the oxidant. When O_2 is absent, some species can instead substitute NO_3^-, NO_2^-, or Fe(III) as their TEA (e.g., Fossing et al., 1995). While many of the sulfide oxidizers are obligate chemolithoautotrophs, some species (e.g., *Acidithiobacillus ferrooxidans*) can also assimilate a mixture of organic substrates and CO_2 for cell carbon synthesis (Gottschal and Kuenen, 1980).

The overall oxidation of H_2S to SO_4^{2-} is an $8e^-$ step (reaction (2.49)), necessitating a series of $2e^-$ transfer reactions that successively lead to the intracellular formation of several different sulfur species and the release of protons into the surrounding medium (Brune, 1989).

In the initial oxidation step, hydrogen sulfide either diffuses into the cell as H_2S or it is actively transported as HS^- (recall the speciation is pH dependent, with the latter more abundant at pH>8). Furthermore, depending on dissolved sulfide availability, the cells may, or may not, require direct attachment to sulfide minerals for growth. Irrespectively, the hydrogen sulfide is oxidized to elemental sulfur, and the pair of electrons are fed into the electron chain at the flavoprotein stage: $E^{o\prime}$ for S^0/H_2S is -0.27, making it a stronger reductant than FMN. The flavoprotein then transfers the electrons via a suite of electron carriers to cytochrome aa_3, where they reduce O_2 to H_2O. The elemental sulfur formed is then oxidized to sulfate via an intermediate sulfite stage. The sulfate that forms is transported out of the cell by a proton symporter mechanism that additionally adds to the pmf. When sufficient ATP has formed, electrons from further sulfide oxidation reactions can be used in reverse electron flow to reduce $NADP^+$ to NADPH, yielding elemental sulfur as a byproduct.

In many species, sulfide oxidation stops with the intermediate formation of elemental sulfur. In *Beggiatoa* species, membrane-bound elemental sulfur granules form inside the cell, with up to 30% of its biomass as S^0 (Fig. 2.23). Similar to the sulfur phototrophic bacteria, this can then serve as an additional energy reserve (being oxidized to sulfate) if the supply of H_2S becomes depleted (Strohl and Schmidt, 1984). *Beggiatoa* can even use the accumulated sulfur in the cells as an electron acceptor and reduce

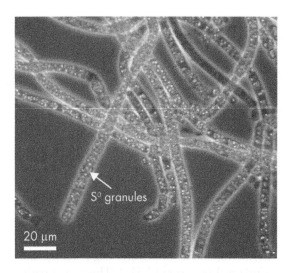

S^a granules

20 μm

Figure 2.23 Phase contrast micrograph of *Beggiatoa* sp. filaments displaying intracellular granules of elemental sulfur accumulated as the result of metabolism by sulfide oxidation. (Courtesy of Rolf Schauder.)

it back to H_2S if conditions become anoxic (Nelson and Castenholz, 1981). It is noteworthy that *Beggiatoa* species do not produce thiosulfate, because if they did the thiosulfate would be difficult to retain within the microniche that they inhabit, and as such the energy associated with its oxidation would be lost to the bacteria.

When the supply of elemental sulfur is epicellular, other species (e.g., *Acidithiobacillus thiooxidans*) must grow attached to the sulfur grains to use it as a substrate for energy (reaction (2.50)). Under these conditions, elemental

$$4H_2S + 2O_2 + 6CO_2 + 6H_2O \rightarrow C_6H_{12}O_6 + 4SO_4^{2-} + 8H^+$$ theoretical overall reaction (2.49)
$$3H_2S + 6CO_2 + 6H_2O \rightarrow C_6H_{12}O_6 + 3SO_4^{2-} + 6H^+$$ autotrophy $^1/_2$ reaction
$$H_2S + 2O_2 \rightarrow SO_4^{2-} + 2H^+$$ chemolithotrophy $^1/_2$ reaction
($\Delta G^{o\prime} = -199\,kJ/2e^-$)

$$5S^0 + 1.5O_2 + 6CO_2 + 11H_2O \rightarrow C_6H_{12}O_6 + 5SO_4^{2-} + 10H^+$$ theoretical overall reaction (2.50)
$$4S^0 + 6CO_2 + 10H_2O \rightarrow C_6H_{12}O_6 + 4SO_4^{2-} + 8H^+$$ autotrophy $^1/_2$ reaction
$$S^0 + 1.5O_2 + H_2O \rightarrow SO_4^{2-} + 2H^+$$ chemolithotrophy $^1/_2$ reaction
($\Delta G^{o\prime} = -196\,kJ/2e^-$)

sulfur is oxidized directly by O_2 to SO_3^{2-} and then SO_4^{2-} (see section 5.2.2(a)).

Upon oxidation, the electrons from S^0 enter the electron transport chain at the level of cytochrome c due to the more positive electron potential of elemental sulfur versus hydrogen sulfide. This results in less free energy available to those cells compared to the sulfide oxidizers simply because ($\Delta E^{\circ\prime}$) is less. Cell suspensions of *A. ferrooxidans* have also been reported capable of coupling the oxidation of S^0 with the reduction of Mo(VI) to Mo(V) (Sugio et al., 1988) and Cu(II) to Cu(I) (Sugio et al., 1990).

Many S-based chemolithoautotrophs can oxidize thiosulfate to sulfate (e.g., *Thiobacillus denitrificans*) (reaction (2.51)). The actual pathway for thiosulfate oxidation involves it being disproportionated into S^0 and SO_3^{2-}. The sulfite is then oxidized to sulfate and excreted from the cell, while the elemental sulfur remains stable until external thiosulfate concentrations become exhausted (Silver and Kelly, 1976). At that point it too is oxidized as above.

A number of thiobacilli further appear capable of oxidizing sulfite by one of two different processes. In the first pathway, *Thiobacillus thioparus* and *T. denitrificans* react AMP with SO_3^{2-} to form APS; the same molecule as that involved in bacterial sulfate reduction. The subsequent reaction of APS with pyrophosphate, via the enzyme ATP sulfurylase, liberates one molecule each of sulfate and ATP (Aminuddin, 1980). The second pathway involves the direct oxidation of sulfite to sulfate by the enzyme,

sulfite oxidase (reaction (2.52)), and the subsequent shuttling of electrons from SO_3^{2-} through cytochrome c to cytochrome aa_3 (Charles and Suzuki, 1966).

2.5.5 Iron oxidizers

The occurrence of bacteria that gain energy from the oxidation of Fe(II) to Fe(III) are generally limited by the availability of dissolved Fe. This is not an insignificant problem because at neutral pH and under fully aerated conditions, Fe(II) rapidly oxidizes to Fe(III), which then hydrolyzes to form ferric hydroxide. The kinetic relationship that describe chemical Fe(II) oxidation at circumneutral pH values is:

$$\frac{-d[\text{Fe(II)}]}{dt} = k[\text{Fe(II)}][\text{OH}^-]^2[\text{O}_2] \qquad (2.53)$$

where $k = 8(\pm 2.5) \times 10^{13}$ min^{-1} atm^{-1} mol^{-2} L^{-2} at 25°C (Singer and Stumm, 1970).

As is evident from the equation, pH and oxygen availability have strong influences on the reaction rate, which explains why at low pH or low oxygen concentrations, Fe^{2+} is quite stable (e.g., Liang et al., 1993). Not surprisingly then, the most efficient way for a microorganism to overcome the stability limitations is to either grow under acidic conditions (as an acidophile) or under low oxygen concentrations at circumneutral pH (as a microaerophile). In both cases the chemical reaction kinetics are sufficiently

---

$3.5S_2O_3^{2-} + O_2 + 6CO_2 + 9.5H_2O \rightarrow C_6H_{12}O_6 + 7SO_4^{2-} + 7H^+$	theoretical overall reaction	(2.51)
$3S_2O_3^{2-} + 6CO_2 + 9H_2O \rightarrow C_6H_{12}O_6 + 6SO_4^{2-} + 6H^+$	autotrophy $^1/_2$ reaction	
$0.5S_2O_3^{2-} + O_2 + 0.5H_2O \rightarrow SO_4^{2-} + H^+$	chemolithotrophy $^1/_2$ reaction $(\Delta G^{\circ\prime} = -205 \text{ kJ}/2e^-)$	

$13SO_3^{2-} + 0.5O_2 + 6CO_2 + 6H_2O \rightarrow C_6H_{12}O_6 + 13SO_4^{2-}$	theoretical overall reaction	(2.52)
$12SO_3^{2-} + 6CO_2 + 6H_2O \rightarrow C_6H_{12}O_6 + 12SO_4^{2-}$	autotrophy $^1/_2$ reaction	
$SO_3^{2-} + 0.5O_2 \rightarrow SO_4^{2-}$	chemolithotrophy $^1/_2$ reaction $(\Delta G^{\circ\prime} = -258 \text{ kJ}/2e^-)$	

slow that the microorganisms can harness Fe(II) oxidation for growth.

(a) Acidophiles

There are a number of acidophilic Fe(II)-oxidizing bacteria that grow autotrophically on reduced iron (or sulfur compounds), using O_2 as its TEA (Blake and Johnson, 2000).

The best characterized acidophiles are the species A. *ferrooxidans* and *Leptospirillum ferrooxidans*. They grow well at mine waste disposal sites where reduced sources of iron (and sulfur) are continuously regenerated during sulfide mineral oxidation (see section 5.2.2(a)). Another iron-oxidizing bacterium is the archaean *Sulfolobus acidocaldarius* that lives in hot, acidic springs at temperatures near boiling (e.g., solfataras).

All of the Fe(II)-oxidizing bacteria use ferrous iron for both ATP generation and as reducing equivalents. For the acidophiles, the energy available from coupling the half reactions of O_2/H_2O with Fe^{3+}/Fe^{2+} yields only 33 kJ mol^{-1} at pH 2 per mol Fe, barely enough to synthesize 1 mol of ATP, which recall requires 31.8 kJ mol^{-1} to form an anhydride bond under standard conditions. Furthermore, A. *ferrooxidans* is not 100% efficient in using energy available from Fe(II) oxidation, and based on a 20% efficiency estimate (Silverman and Lundgren, 1959), and a $24Fe:1C_6H_{12}O_6$ autotrophy ratio, it implies then that a population of A. *ferrooxidans* would require 120 mol of Fe(II) to generate 1 mol of glucose. Consequently, species such as A. *ferrooxidans* must oxidize a large amount of Fe(II) in order to grow, and even a small number of bacteria can be responsible for generating significant concentrations of dissolved Fe(III).

The transfer of electrons from Fe(II) oxidation begins when Fe^{2+} is transported through the outer membrane into the periplasm, where an electron is passed onto a highly electropositive and acid-stable copper-containing protein called rusticyanin. The location of the initial Fe(II) oxidation reaction is thought to be outside the cytoplasm because of iron toxicity and the insolubility of Fe(III) at circumneutral pH (Ingledew et al., 1977). From rusticyanin, the electron is shuttled to cytochrome c, and then to a membrane-bound cytochrome a of the aa_3-type (called a_1), that is oxidized by O_2 on the inner face of the plasma membrane to form water: oxygen gas passively diffuses through the plasma membrane (Fig. 2.24). Meanwhile, the Fe^{3+} formed within the periplasm is then driven out by a diffusional gradient, where it will bind weakly (if at all) to the outer membrane surface because of competition with the abundant protons available in acidic solutions (i.e., the surface is protonated – see section 3.2.1). The electron chain for iron oxidizers at pH 2 is very short ($E^{o'} \sim +0.08$ V because $Fe^{2+}/Fe^{3+} = +0.77$ V at pH 2 and $0.5O_2/H_2O = +0.85$ V at pH 6.5), thus the membrane potential derived from the transfer of electrons from Fe^{2+} to O_2 is virtually negligible. However, the acidophiles get around this predicament due to the extreme acidity of their external environment. The pH difference across the plasma membrane may be as great as 5 pH units, leading to a natural pmf that drives ATP synthesis (Cox and Brand, 1984). However, to retain a circumneutral cytoplasmic pH, protons entering the cell through ATPase must be consumed. This is where the oxidation of Fe^{2+} to Fe^{3+} plays an important role, because reaction (2.54) is a proton-consuming reaction.

Reducing power for all the Fe(II) oxidizers is formed by reverse electron flow, with the electrons from Fe^{2+} transferred via the electron transport chain against the electropotential gradient.

$$26Fe^{2+} + 0.5O_2 + 6CO_2 + 26H^+ \rightarrow C_6H_{12}O_6 + 26Fe^{3+} + 7H_2O \qquad \text{theoretical overall reaction} \qquad (2.54)$$
$$24Fe^{2+} + 6CO_2 + 24H^+ \rightarrow C_6H_{12}O_6 + 24Fe^{3+} + 6H_2O \qquad \text{autotrophy } ^1/_2 \text{ reaction}$$
$$2Fe^{2+} + 0.5O_2 + 2H^+ \rightarrow 2Fe^{3+} + H_2O \qquad \text{chemolithotrophy } ^1/_2 \text{ reaction}$$
$$(\Delta G^{o'} = -66 \text{ kJ/2e}^- \text{ at pH 2})$$

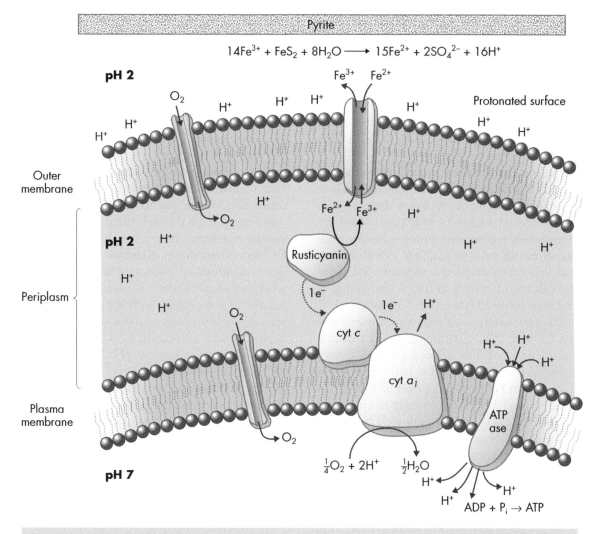

Figure 2.24 Electron transport and the generation of a proton motive force in acidophilic Fe(II)-oxidizing bacteria. (Modified from Ingledew et al., 1977.)

Significant amounts of ATP are consumed in this process because the Fe^{3+}/Fe^{2+} couple has such a positive electrode potential compared to the $NADP^+/NADPH$ couple. NADPH is subsequently used in the Calvin cycle for the reduction of CO_2 into organic matter (Ingledew, 1982).

(b) Neutrophiles

Under neutral pH, but with O_2 levels below 1.0 mg per litre and redox conditions about 200–300 mV lower than typical surface waters (characteristics of some iron springs, stratified bodies of water, and hydrothermal vent systems), microaerophilic bacteria, such as *Gallionella ferruginea*, play an important role in Fe(II) oxidation. They are bean-shaped cells that grow at the terminus of a helical structure called a stalk (Fig. 2.25). The stalk is composed largely of ferric hydroxide, with an uncharacterized organic matrix underneath (Hanert, 1992). *G. ferruginea* survives under low oxygen conditions because

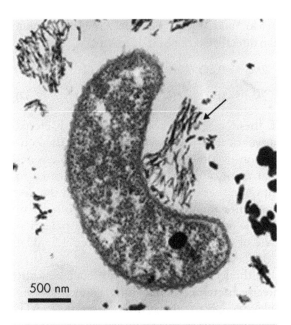

Figure 2.25 TEM image of *Gallionella ferruginea*, showing a portion of its stalk (arrow) attached to the cell. (From Emerson, 2000. Reproduced with permission from the American Society for Microbiology.)

the half-life of Fe^{2+} is much longer, and thus it has a more efficient oxidizing capacity relative to inorganic oxidation processes.

Unlike the acidophiles, the neutrophiles can harness much more energy because at pH 7 the electrode potential of the couple $Fe(OH)_3/Fe^{2+}$ ($E^{\circ\prime} = -0.23$ V) is substantially less compared to the pH 7 redox couple of O_2/H_2O ($\Delta E^{\circ\prime} = 0.81$ V). This indicates that Fe(II) oxidation can generate significant energy under circumneutral to support ATP production. Although G. *ferruginea* grows chemotrophically at a pH just below 7 on a medium with Fe(II) salts and it fixes all of its carbon autotrophically from CO_2 (Hallbeck and Pedersen, 1991), there is at present no conclusive evidence that it actually derives energy from Fe(II)-oxidation. Interestingly, G. *ferruginea* does not form a stalk at a pH < 6 or under very microaerobic conditions, where O_2 is present but the redox potential is -40 mV (Hallbeck and

Pederson, 1990). The authors have suggested that these might be the most favorable growth conditions for the microorganism, and that the stalk represents a survival structure to protect it from the reducing capacity of Fe(II) as it becomes unstable in an oxidizing environment. Perhaps the mineralized stalk gives G. *ferruginea* the capability to colonize and survive in habitats unfeasible to bacteria without a defense mechanism against iron toxicity. Support for this supposition also comes from the observation that when G. *ferruginea* grows on hydrogen sulfide or thiosulfate as electron donors, it does not form a stalk (Lütters-Czekalla, 1990). In a similar manner, it has been suggested that *Leptothrix ochracea* induces ferric hydroxide precipitation on its sheath as a means of detoxifying any free oxygen in its environment (Nealson, 1982). These examples certainly imply that Fe(II) oxidation need not be directly tied to energy production.

Ferrous iron has also been observed to undergo microbial oxidation under anoxygenic conditions. In addition to the photoferrotrophs discussed earlier, biological oxidation can occur in the absence of oxygen, by light-independent, chemotrophic activity with nitrate as the electron acceptor (Straub et al., 1996). In the study of Straub et al., the observation that nitrate reducers, that had never been previously grown in iron media, exhibited the capacity for Fe(II) oxidation implied that this form of microbial oxidation may be commonplace. Indeed, a thermophilic archaean, *Ferroglobus placidus*, that grows by oxidizing ferrous iron and reducing nitrate at neutral pH has been isolated from a shallow marine hydrothermal vent system in Italy (Hafenbrandl et al., 1996).

2.5.6 Manganese oxidizers

At the pH and Eh values characteristic of the Earth's surface the activation energy of Mn(II) to Mn(III) or Mn(IV) is much larger than that of Fe(II) to Fe(III), and under fully oxygenated conditions at neutral pH, the inorganic oxidation reaction proceeds rather slowly (Stumm and

Morgan, 1996). These kinetic constraints, however, afford a number of bacteria the opportunity to catalyze the oxidation of Mn(II) to form Mn(III) oxyhydroxides or MnO_2 (e.g., reaction (2.55)). In fact, it takes only a relatively small population of bacteria capable of oxidizing Mn(II) to sufficiently ensure that the initial production of MnO_2 is dominated by biological catalysis at circumneutral pH (Zhang et al., 2002). Accordingly, it is argued that this process is responsible for the majority of Mn(II) oxidation that occurs in natural waters and sediment. And, once initiated, Mn(II) oxidation is autocatalytic because MnO_2 has a zero point of charge of less than 3 (i.e., it is negatively charged at circumneutral pH). Thus it absorbs more Mn^{2+} from solution, greatly increasing the rates of reaction (Nealson et al., 1988).

Much of our early understanding of manganese oxidation came from inorganic experiments that demonstrated a two-step process for MnO_2 formation. Initially, Mn^{2+} is oxidized to either a metastable Mn(III) oxyhydroxide (e.g., manganite) or a mixed valence mineral (e.g., hausmannite); with the type of product dependent upon the initial Mn^{2+} concentration, pH, ionic strength, temperature, and redox conditions of the medium (Hem and Lind, 1983). Manganite subsequently transforms into MnO_2, while hausmannite disproportionates very slowly to form Mn^{2+} and MnO_2. Those steps then become rate limiting in the abiological formation of MnO_2. However, because bacteria that oxidize Mn^{2+} form MnO_2 much more rapidly, it is believed that they may instead utilize a completely different pathway of Mn(II) oxidation (Ehrlich, 2002). It is possible that the bacteria first enzymatically oxidize Mn^{2+} to Mn^{3+}, and that this intermediate form remains bound to a manganese-oxidizing enzyme (reaction (2.56)). The same enzyme then catalyzes the oxidation of bound Mn^{3+} to MnO_2 (reaction (2.57)), rather than a disproportionation

to Mn^{2+} and MnO_2. Therefore, a single bacterium can directly oxidize Mn(II) to Mn(IV).

$$2Mn^{2+} + 0.5O_2 + 2H^+ \rightarrow 2Mn^{3+} + H_2O \qquad (2.56)$$

$$2Mn^{3+} + 0.5O_2 + 3H_2O \rightarrow 2MnO_2 + 6H^+ \qquad (2.57)$$

If these reactions hold true, then bacterial enzyme catalysis not only accelerates the conversion of Mn^{3+} and MnO_2, but it also ensures that more of the Mn^{3+} formed is converted to MnO_2, rather than just 50% of it as in the disproportionation reaction. Not surprisingly, in aquatic environments, the average oxidation state of manganese oxides is between 3.5 and 4 (Mandernack et al., 1995).

Although a wide variety of metabolically diverse bacteria oxidize Mn(II), only a few are able to couple the energy from the process to ATP synthesis (e.g., select *Pseudomonas* and *Bacillus* species). Because of the high electrode potential for the Mn^{2+}/MnO_2 couple, electrons from Mn^{2+} are transported into the periplasm and then fed into the electron chain at the cytochrome *c* level, from where the pair of electrons are used to reduce O_2 on the inner surface of the plasma membrane. Even though the electrode potential of the *c*-type cytochrome is more negative than that of the Mn^{2+}/MnO_2 couple, the high oxidase activity keeps the cytochrome *c* almost completely oxidized, thus making the cytochrome's electrode potential more positive (Tebo et al., 1997).

Most other Mn(II)-oxidizing microorganisms simply possess suitable ligands that either facilitate the oxidation process or allow for the accumulation of Mn(IV) oxides onto their cell surfaces (e.g., Fig. 2.26). For example, *Leptothrix discophora* grows for a portion of its life as a filament of cells encased in a sheath that possess a unique protein that can oxidize Mn^{2+} to MnO_2 extracellularly (Boogerd and de Vrind, 1987). Sheathless variants of *L. discophora* also secrete Mn(II)-oxidizing

---

$13Mn^{2+} + 6CO_2 + 0.5O_2 + 19H_2O \rightarrow C_6H_{12}O_6 + 13MnO_2 + 26H^+$	theoretical overall reaction (2.55)
$12Mn^{2+} + 6CO_2 + 18H_2O \rightarrow C_6H_{12}O_6 + 12MnO_2 + 24H^+$	autotrophy $^1/_2$ reaction
$Mn^{2+} + 0.5O_2 + H_2O \rightarrow MnO_2 + 2H^+$	chemolithotrophy $^1/_2$ reaction
	($\Delta G^{o\prime} = -71$ kJ/2e$^-$)

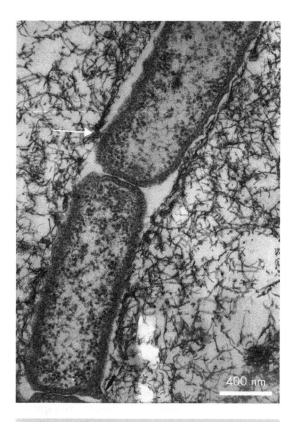

Figure 2.26 TEM image of *Leptothrix discophora*, showing a Fe-Mn oxide encrusted sheath attached to the cell's outer membrane (arrow). (Courtesy of Bill Ghiorse.)

proteins, but in this case the MnO_2 forms as a disorganized mass of amorphous particulates that are free in the medium (Adams and Ghiorse, 1986). It has been suggested that these proteins are generated as a means of protecting the cells against high Mn^{2+} concentrations. Along similar lines, some aerobic respiring bacteria, such as *Arthrobacter siderocapsulatus*, use Mn^{2+} to remove excess intracellular levels of hydrogen peroxide when their catalases cannot break it down fast enough (Ehrlich, 2002).

2.5.7 Nitrogen oxidizers

The most frequent inorganic nitrogen compound used as an electron donor is ammonia (or more accurately, the cation NH_4^+ at neutral pH). It is oxidized to nitrite and nitrate by the nitrifying bacteria, with O_2 serving as the electron acceptor (Knowles, 1985). Nitrifying bacteria are pervasive in habitats where the supply of ammonia/ammonium are high, i.e., streams that receive a supply of sewage or stratified lakes where extensive protein decomposition occurs in the bottom sediment. Most recently, a group of anaerobic ammonia oxidizers, the so-called "anammox bacteria" of the order *Planctomycetales*, have been shown capable of combining ammonium with nitrite directly into N_2 gas (reaction (2.58)); the nitrite originating as an intermediate during bacterial nitrate reduction. These bacteria are found naturally in marine sediments and below the chemocline of the water column in anoxic basins (see section 6.2.4(a)), as well as in laboratory reactors and waste water treatment systems designed to remove ammonium from agricultural waters contaminated by fertilizers (e.g., Strous et al., 1999).

$$NH_4^+ + NO_2^- \rightarrow N_2 + 2H_2O \qquad (2.58)$$
$$\Delta G^{\circ\prime} = -358 \text{ kJ/reaction}$$

In terms of nitrification, two groups of microorganisms are required for the complete $8e^-$ transfer; one group of species (e.g., from the *Nitrosomonas*, *Nitrosospira*, and *Nitrosolobus* genera) oxidize ammonium through to nitrite (reaction (2.59)), while another group (e.g., *Nitrobacter*, *Nitrospina*, and *Nitrococcus* genera) oxidize nitrite to nitrate (reaction (2.60)). All primarily fix CO_2 by the Calvin cycle (Wood, 1988).

During the initial stage of nitrification, ammonia/ammonium is actually oxidized by O_2 within the cytoplasm to produce hydroxylamine

$5NH_4^+ + 1.5O_2 + 6CO_2 + H_2O \rightarrow C_6H_{12}O_6 + 5NO_2^- + 10H^+$	theoretical overall reaction	(2.59)
$4NH_4^+ + 6CO_2 + 2H_2O \rightarrow C_6H_{12}O_6 + 4NO_2^- + 8H^+$	autotrophy $^1/_2$ reaction	
$NH_4^+ + 1.5O_2 \rightarrow NO_2^- + H_2O + 2H^+$	chemolithotrophy $^1/_2$ reaction ($\Delta G^{\circ\prime} = -92$ kJ/$2e^-$)	

(NH$_2$OH) as an intermediate product (reaction (2.61)):

$$NH_4^+ + O_2 + H_2 \rightarrow NH_2OH + H_2O + H^+ \quad (2.61)$$

Because the electrode potential of the reaction is +0.9 V at pH 7, which is more positive than for the reduction of O$_2$ to water, a simple coupling to a cytochrome oxidase is impossible. Instead, ammonium oxidation requires a supply of reductant (Hooper, 1984). The enzyme that catalyzes this reaction is called ammonia monooxygenase, a nonspecific, membrane-bound enzyme that can also be used by the nitrifying bacteria to catalyze the oxygenation of methane to CO$_2$, and as such act as methanotrophs (e.g., Jones and Morita, 1983). The hydroxylamine formed from the oxidation reaction is then transported through the plasma membrane into the periplasm where another enzyme, hydroxylamine oxidoreductase, oxidizes it to form nitrite (reaction (2.62)). This is important because reaction (2.61) is endergonic and accordingly depends on the subsequent enzymatic oxidation of hydroxylamine in order to proceed in a forward direction (see Madigan et al., 2003).

$$NH_2OH + O_2 \rightarrow NO_2^- + H^+ + H_2O \quad (2.62)$$

Nitrite oxidizers use the enzyme nitrite oxidase to oxidize nitrite to nitrate without any detectable intermediates. Because the $E^{o\prime}$ of the NO$_3^-$/NO$_2^-$ couple is very high (+0.43 V), the bacteria must donate electrons into the chain at a late stage in the overall process, i.e., to cytochrome aa_3 (Wood, 1988).

2.6 Summary

The wealth of new information on microbial diversity, metabolic capabilities, and environmental constraints links directly to perhaps the most fundamental characteristic dictating the progression of microbial processes, namely the amount of energy consumed or released in a specific metabolic reaction. All living forms require energy to synthesize the molecules needed for cell maintenance and growth. In turn, energy is derived from chemical reactions that generate ATP via one of four major processes, photoautotrophy, photoheterotrophy, chemoheterotrophy, and chemolithoautotrophy. What ultimately defines these different processes is: (i) how are the initial electron equivalents attained – is it from sunlight, the oxidation of pre-existing organic matter or the oxidation of reduced inorganic molecules; (ii) what electron acceptors are immediately available to the organism; and (iii) how do they obtain carbon? Many microorganisms are also able to switch metabolic pathways in accordance with their prevailing environmental conditions. The implications are that many metabolic processes are utilized by diverse species across the phylogenetic tree (e.g., Fe(III) reduction). From a global perspective, such commonality ensures widespread occurrence of certain major microbially driven geochemical and mineralogical processes throughout the biosphere. Yet, given that individual species tend to grow within microniches, variability in microgeochemical conditions can also lead to community-specific microbial signatures that are manifested into the sedimentary record in the form of organic residues and solid-phase precipitates with unique isotopic signals and/or mineralogical compositions. As we will examine in Chapter 6, the various metabolic processes are intrinsically linked regardlessly of scale, so for example, as autotrophic species generate organic carbon and excrete oxidants, heterotrophic species degrade the biomass and reduce those oxidants, and in doing so, release CO$_2$ and any number of other reduced components, thus closing the chemical loop.

$13NO_2^- + 0.5O_2 + 6CO_2 + 6H_2O \rightarrow C_6H_{12}O_6 + 13NO_3^-$	theoretical overall reaction $\quad$ (2.60)
$12NO_2^- + 6CO_2 + 6H_2O \rightarrow C_6H_{12}O_6 + 12NO_3^-$	autotrophy $^1/_2$ reaction
$NO_2^- + 0.5O_2 \rightarrow NO_3^-$	chemolithotrophy $^1/_2$ reaction
	$(\Delta G^{o\prime} = -74 \text{ kJ/2e}^-)$

3

Cell surface reactivity and metal sorption

One of the consequences of being extremely small is that most microorganisms cannot out swim their surrounding aqueous environment. Instead they are subject to viscous forces that cause them to drag around a thin film of bound water molecules at all times. The implication of having a watery shell is that microorganisms must rely on diffusional processes to extract essential solutes from their local milieu and discard metabolic wastes. As a result, there is a prime necessity for those cells to maintain a reactive hydrophilic interface. To a large extent this is facilitated by having outer surfaces with anionic organic ligands and high surface area:volume ratios that provide a large contact area for chemical exchange. Most microorganisms further enhance their chances for survival by growing attached to submerged solids. There, they may adopt a more hydrophobic nature to take advantage of the inorganic and organic molecules that preferentially accumulate. Accordingly, throughout a microorganism's life, there is a constant interplay with the external environment, in which the surface macromolecules are modified in response to changing fluid compositions and newly available colonizing surfaces. In this chapter we focus on how cellular design can facilitate the accumulation of metals onto microbial surfaces, often in excess of mineral saturation states. We then examine how modeling their chemical reactivity can be applied to the environmental issues of contaminant bioremediation and biorecovery of economically valuable metals.

3.1 The cell envelope

3.1.1 Bacterial cell walls

Bacterial surfaces are highly variable, but one common constituent amongst them is a unique material called peptidoglycan, a polymer consisting of a network of linear polysaccharide (or glycan) strands linked together by proteins (Schleifer and Kandler, 1972). The backbone of the molecule is composed of two amine sugar derivatives, N-acetylglucosamine and N-acetylmuramic acid, that form an alternating, and repeating, strand. Short peptide chains, with four or five amino acids, are covalently bound to some of the N-acetylmuramic acid groups (Fig. 3.1). They serve to enhance the stability of the entire structure by forming direct or interchain cross-links between adjacent glycan strands. The peptide chains are rich in carboxyl (COOH) groups, with lesser amounts of amino (NH$_2$) groups (Beveridge and Murray, 1980).

Despite the enormous variety of bacterial species, most can be classified into two broad categories: Gram-positive and Gram-negative. This terminology has its basis in the cell's response to the differential staining technique developed by Christian Gram in 1884. The Gram stain involves using four chemicals on dried smears of bacteria in the following sequence: crystal violet, iodine, ethanol, and safranin. Bacteria that are able to retain the crystal violet–iodine complex, even after decolorization with ethanol are called Gram-positive. Those that lose their purple

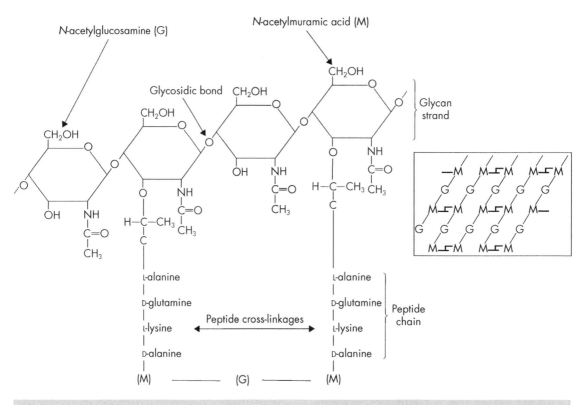

Figure 3.1 Structure of peptidoglycan. It is composed of strands of repeating units of N-acetylglucosamine and N-acetylmuramic acid sugar derivatives. The sugars are connected by glycosidic bonds, but the overall resilience comes from the cross-linking of the glycan strands by peptide chains.

coloration and are counterstained with safranin to become red are Gram-negative. It is now recognized that these staining characteristics highlight some fundamental differences in the chemical and structural organization of the cell wall (Beveridge and Davies, 1983). In both cell types, the crystal violet–iodine complex penetrates the cell wall and stains the cytoplasm. Then during the decolorization step, the ethanol solubilizes some of the membranous material. This is where the inherent differences lie. In Gram-positive cells, their thick peptidoglycan walls become dehydrated by the alcohol, the pores in the wall close, and the crystal violet–iodine complex is prevented from escaping. By contrast, Gram-negative cell walls have thin peptidoglycan walls that cannot retain the stain when the membranes are dissolved. In terms of global

abundances, most bacteria are Gram-negative, while the Gram-positive cells are distinguished on the universal phylogenetic tree as two sister phyla (the *Firmicutes* and the *Actinobacteria*), united by their common cell wall structure.

(a) Gram-positive bacteria

A large proportion of the work conducted on the ultrastructure and metal binding properties of Gram-positive cells has been done using a common soil constituent, *Bacillus subtilis*. Under the transmission electron microscope (TEM), a technique that permits resolution of objects as small as a few nanometers, these species are observed having a single wall layer averaging 25–30 nm thick, which consists of 30–90% peptidoglycan. The remaining materials are secondary polymers that are covalently attached to the peptidoglycan (Fig. 3.2). For instance, when *B. subtilis* is grown in the presence of phosphate, its wall has essentially two chemical components of roughly equal proportion; peptidoglycan and teichoic acid (Beveridge, 1989a). Teichoic acids are either glycerol- or ribitol-based polysaccharides, with a terminal (H_2PO_3) phosphoryl group and glucose or amino acid residues (Ward, 1981). A phosphodiester group links the teichoic acid chain to *N*-acetylmuramic acid of the peptidoglycan. Teichoic acids provide a distinct asymmetry in composition between the wall's inner and outer surfaces because half extends perpendicularly outwards into the external milieu,

while the other half is embedded in the peptidoglycan matrix by penetrating through its interstices (Doyle et al., 1975). Some teichoic acids are also bound to membrane lipids, and they are called lipoteichoic acids.

When growth of *B. subtilis* is limited by the availability of phosphate, teichoic acid synthesis ceases and it is totally replaced by teichuronic acid, a polymer made up of alternating sequences of *N*-acetylgalactosamine and carboxyl-rich glucuronic acid, but lacking phosphate. Variations in the type and quantity of secondary polymer indicate that the wall composition, at least for *B. subtilis*, may be a phenotypic expression of the environment (Ellwood and Tempest, 1972). Stated

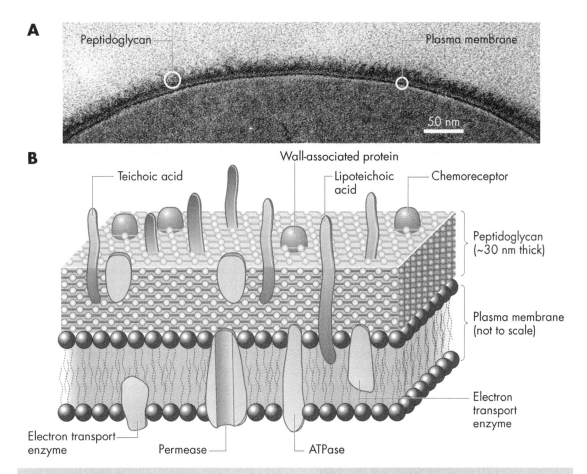

Figure 3.2 (A) A TEM image of a *Bacillus subtilis* cell wall (courtesy of Terry Beveridge). (B) Representation of the overall structure of a Gram-positive bacterium.

simply, the bacterium has the means to adapt its biochemistry to compensate for geochemical changes in its environment, and as will be discussed below, this has implications for it retaining high surface reactivity.

(b) Gram-negative bacteria

Much as *Bacillus subtilis* is the archetypal Gram-positive bacterium, *Escherichia coli* has largely become the model Gram-negative bacterium. The walls of *E. coli* are structurally and chemically complex (Beveridge, 1989a). External to the plasma membrane is a very thin (3 nm thick) peptidoglycan layer that makes up a mere 10% of the cell wall. This, in turn, is overlain by another bilayered structure, the outer membrane, that serves as a barrier to the passage of many unwanted molecules from the external environment into the cell (Fig. 3.3). The narrow region separating the plasma and outer membranes, called the periplasm, contains a hydrated, gel-like form of peptidoglycan. In *E. coli* it is 12–15 nm thick and occupies approximately 10–20% of the total cell volume. Within the periplasm is the peptidoglycan layer itself, a number of dissolved components such as amino acids, sugars, vitamins and ions, and various macromolecules that are attached to the boundary surfaces (Hobot et al., 1984). As discussed in Chapter 2, the periplasm also houses a number of enzymes involved in catabolism, e.g., the hydrolytic enzymes and those employed in electron transport.

The outer membrane has an asymmetric lipid distribution, with phospholipids limited to the inner face. The outer face (exposed to the external environment) contains a uniquely prokaryotic molecule, lipopolysaccharide (LPS). Typically, the outer membrane contains 20–25% phospholipid and 30% LPS. The LPS possesses three distinct chemical regions (Ferris, 1989). The innermost hydrophobic region, called "lipid A," is the segment of the LPS that shows the least chemical variation between different species. It has a disaccharide of glucosamine that is acetylated and attached to short-chain fatty acids. Covalently

bound to the lipid A is the "core", consisting of the unique sugars 3-deoxy-D-mannooctulosonate (also known as KDO) and L-glycero-D-mannoheptose (or heptose), along with N-acetylglucosamine, galactose, and a number of other sugars whose exact combination varies between species. Chemically the core contains carboxyl and cationic amino groups (NH_4^+) that are cross-linked, usually with carboxyl groups present in excess. The outermost region of the LPS is the "O-antigen." It is made up of repeating carbohydrate units that are interspersed with uronic acids and/or organic phosphate groups, the latter comprising 75% of the total phosphorous associated with the outer membrane, while the remainder is in the phospholipid.

The remaining fraction of the outer membrane contains two major types of proteins. Lipoproteins are confined to the inner face of the outer membrane, and they serve to anchor the outer membrane to the peptidoglycan (Di Rienzo et al., 1978). The other proteins are porins. They puncture the bilayer and function as small-diameter (up to a few nanometers), water-filled channels that completely span the outer membrane and regulate the exchange of low-molecular-weight hydrophilic solutes into and out of the periplasm along a concentration gradient (Hancock, 1987). Some porins contain specific binding sites for one or a group of structurally related solutes that they allow in. Other porins are nonspecific, in that the width of their channel largely determines the exclusion limit for dissolved compounds. Therefore, porins with restrictive channel widths can both sieve out potentially harmful enzymes and other large hydrophilic molecules, while preventing internal enzymes that are present in the periplasm from diffusing out of the cell. There are also a number of other proteins located on the exterior surface of the outer membrane, including those that function as: (i) mediators for the cellular adsorption, processing, and transport of essential ions into the cytoplasm; (ii) as receptors for bacteriophages, the viruses that infect bacteria; and (iii) chemoreceptors, helping direct the cell towards or away from specific chemicals (Beveridge, 1981).

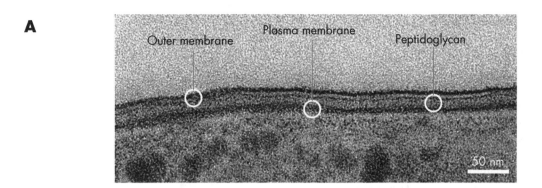

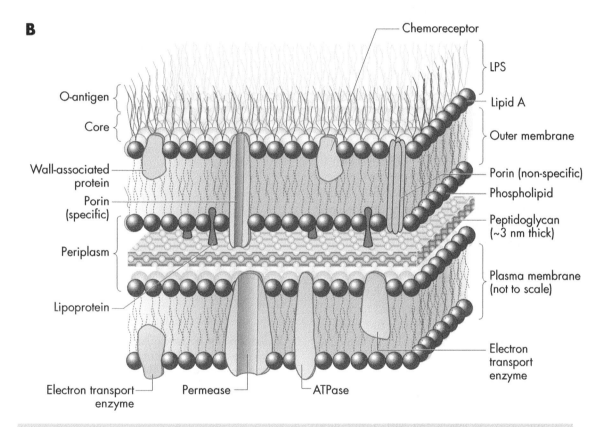

Figure 3.3 (A) A TEM image of a *Synechococcus* PCC7942 cell wall (courtesy of Maria Dittrich and Martin Obst). (B) Representation of the overall structure of a Gram-negative bacterium.

3.1.2 Bacterial surface layers

Direct examination of bacterial cells under the TEM reveals that most possess supplementary layers located external to the cell wall (Beveridge and Graham, 1991). These layers are defined by both their composition and physical characteristics.

(a) Extracellular polymers (EPS)

Extracellular polymers, also known as exopolymers, extracellular polysaccharides, or glycocalyces, are highly hydrated structures (up to 99% water) that are produced inside the cell and excreted to the cell surface. Their consistency is often thixotrophic, that is, they are able to alternate between a gel and a liquid state. The solid material is predominately a heteropolysaccharide, composed of repeating units of several types of sugar monomers, as well as various carboxyl-rich uronic acids that may make up to 25% of the solid capsular material (Sutherland, 1972). Other cells have EPS dominated by proteins (Nielsen et al., 1997). In general, the chemical composition of EPS is extremely diverse, reflecting the different microorganisms that produce them. In fact, even a single strain of bacterium may secrete several types of EPS, each having different physical and chemical properties depending on nutrient availability, their growth stage, and other environmental parameters.

EPS also range in their complexity. Capsules are structured and stable forms firmly attached to the cell (e.g., Fig. 3.4). Their thickness can extend several micrometers from the cell surface, and in many instances, the production of capsular material is so extensive that entire colonies are encapsulated. By contrast, slime layers range from those materials loosely attached to the cell surface to those that are shed into the environment. The latter forms when the bacterium overproduces its capsular material or, for some reason, fails to anchor them securely to their surfaces. Subsequently, the slime layers are sloughed off into their surroundings, where they float freely until they become associated with other solid surfaces (Whitfield, 1988).

The production of EPS involves a significant expenditure in energy and carbon by the microorganism. Accordingly, its formation must have benefits to those cells that produce it (Wolfaardt et al., 1999). Some of those benefits are:

1 They protect cells from periodic desiccation, extreme pH values, elevated temperatures, or freezing.

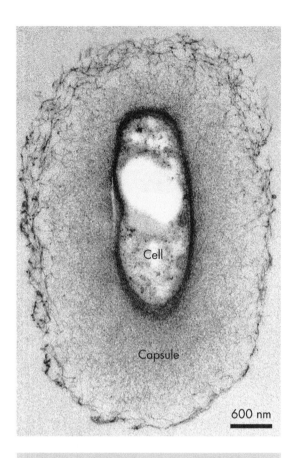

Figure 3.4 TEM image of a capsule surrounding *Rhizobium trifolii*. (Courtesy of Frank Dazzo.)

2 They help the bacteria adhere to surfaces and maintain the overall stability of biofilm/mat communities (see section 6.1.1). This is important because organic and inorganic compounds are preferentially concentrated at interfaces, hence surfaces are desirable locations for growth.

3 They provide microorganisms with a reserve of carbon and energy.

4 They bind metals, form minerals and serve as chemical buffers at the cell's periphery, where essential ions are accumulated and toxic substances immobilized. Consequently, EPS can be considered as an additional design strategy by which bacteria control the concentration of metals actually reaching the vital constituents within the cell.

(b) Sheaths

Several filamentous bacteria are completely encased in a structure that resembles a hollow cylinder when devoid of cells. These are known as sheaths, and for some species they represent the outermost surface layer (e.g., Fig. 3.5). Sheaths come in two varieties: (i) highly ordered and made up of proteins, such as those associated with several species of methanogens (e.g., Patel et al., 1986); and (ii) fibrillar and predominantly made up of neutral sugars, along with variable quantities of uronic and amino acids. They form in association with several iron-and manganese-depositing bacteria, as well as a number of cyanobacteria (e.g., Weckesser et al., 1988). It is interesting to note that in filamentous cyanobacteria mature cells are fixed within the sheath, yet binding of the sheath to the underlying cell wall can be temporarily disrupted to form hormogonia. These are short, motile (i.e., they can move independently) filaments that in response to light can penetrate the viscous extracellular layers to disperse into the environment, and subsequently develop back into mature filaments once they colonize a new substratum (Hoiczyk, 1998).

One common feature seems to exist amongst the broad group of ensheathed microorganisms, that is, their sheaths have minute particle spacing that makes them impervious to large molecules. It is thus likely that they serve as an additional permeability layer or chemical sieve that filters out harmful macromolecules (e.g., Phoenix et al., 1999). Moreover, the sheath material in cyanobacteria has a different surface charge from that of the underlying wall material (see section 3.2.3). This implies that the sheath may mask the charge characteristics of the wall, possibly by exposing a hydrophobic or uncharged surface to the external milieu. This has the effect of mediating physicochemical reactions between the cell and ions/solids in the external environment (Phoenix et al., 2002).

(c) S-layers

Regularly structured layers, also known as S-layers, are more highly organized than both EPS and sheaths (Koval, 1988). They consist of proteinaceous layers, with carbohydrates occasionally present as a minor component. The regularity of their ordering is so great that the S-layers can be considered paracrystalline. S-layers are ubiquitous in nature and are found as part of the cell envelope in virtually every taxonomic group of both *Bacteria* and *Archaea* (e.g., Fig. 3.6).

Although compositionally and structurally different from sheaths, both structures have similar roles. Many S-layers are arranged so as to form aqueous channels 2–3 nm in diameter, just large enough to allow essential nutrients to enter and metabolic wastes to exit, but small enough to exclude some external enzymes (e.g., lyzosymes that degrade peptidoglycan) from passing through to the underlying fabric.

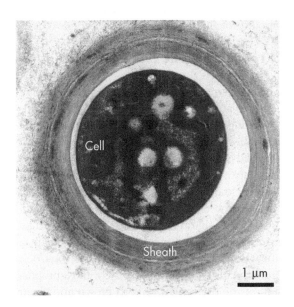

Figure 3.5 TEM image of *Calothrix* sp.. These cyanobacteria produce extremely thick sheaths that can often double to triple the size of the cell. (Courtesy of Vernon Phoenix.)

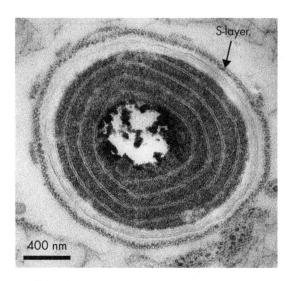

Figure 3.6 TEM image of the cyanobacterium *Synechococcus* strain GL24 showing the S-layer as its outmost surface layer. (Courtesy of Susanne Douglas.)

(Stewart and Beveridge, 1980). S-layers can also have different surface charges than that of the underlying wall. At times this may lead to an uncharged surface that is unreactive to metal cations, while at other times the S-layers may bind considerable amounts of metals, even to the point where they nucleate mineral phases. Once mineralized, S-layers can be shed from the cell surface, allowing the cells to rid themselves of minerals when the burden becomes excessive (e.g., Schultze-Lam et al., 1992).

3.1.3 Archaeal cell walls

The cell envelopes of *Archaea* are much more variable than those in *Bacteria* (König, 1988). In *Crenarchaeota* (an archaeal phylum), the most common cell envelope is represented by a single S-layer that is closely associated with the plasma membrane. There is no external cell wall, and all extreme thermophiles rely entirely on this layer for maintaining cell viability. Other thermophiles, such as *Thermoplasma* sp., are supported just by the plasma membrane.

The *Euryarchaeota* (another archaeal phylum) show a wider range of wall types. Extreme halophiles (e.g., *Halococcus* sp.) have cell walls composed of complex heteropolysaccharides consisting of several sugars, uronic acids, and amino acids. Within the methanogens, there are a number of wall variations, with each genus having invented its own cell envelope. Some methanogenic genera, such as *Methanospirillum*, are surrounded by a proteinaceous sheath, while *Methanococcus* has a S-layer. Others have walls composed of a material similar to peptidoglycan, called pseudomurein, that instead contains *N*-acetylglucosamine and *N*-acetyltalosaminuronic acid (e.g., *Methanobacterium* sp.). Interestingly, of all the *Archaea* subjected to the Gram stain, only the *Methanobacterium* genus stained Gram-positive since its pseudomurein wall remained intact after treatment with ethanol (Beveridge and Schultze-Lam, 1996). Another wall variety, possessed by species of *Methanosarcina*, has a thick layer (up to 200 nm) containing a polymer called methanochondroitin that is made up of uronic acid, *N*-acetylgalactosamine, and minor amounts of glucose and mannose.

3.1.4 Eukaryotic cell walls

The main structural components of all eukaryotic cells are polysaccharides. Most algae have walls consisting of a skeletal layer and an encompassing amorphous matrix (Fig. 3.7). The main skeletal material is cellulose, but it can be modified by the addition of other types of polysaccharides that give an individual species a unique chemical composition (Hunt, 1986). Three such polysaccharides are mannans, pectins, and xylans. The amorphous matrix typically consists of alginate, a linear polymer of repeating units of carboxyl-rich uronic acids that can constitute a large proportion of the dry weight of both brown and green algae. Other amorphous components include sulfated heteropolysaccharides called fucoidan. Accessory amorphous compounds include sulfated galactans (e.g., agar, carrageenan, and porphyran). In some algae, the wall is additionally strengthened

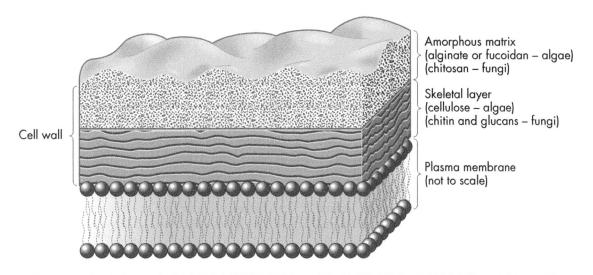

Figure 3.7 The main organic components comprising algal and fungal walls.

by the precipitation of calcium carbonate (as in coralline algae and coccolithophores) or silica (as in diatoms).

The fungal wall is also bilayered, with an inner skeletal layer of chitin (a highly crystalline polymer of N-acetylglucosamine) and glucans, while the outer layer is made up of amorphous compounds such as chitosan (a deacetylated chitin). The secondary components of the wall include proteins, lipids, polyphosphates, phenols (a compound with an –OH group attached to an aromatic ring), and melanin pigments, as well as various inorganic ions that make up part of the wall-cementing matrix (Gadd, 1993).

3.2 Microbial surface charge

3.2.1 Acid–base chemistry of microbial surfaces

One of the characteristic properties of many organic functional groups is that they are amphoteric, that is, they can each either bind or release protons (H^+) into solution depending on the solution pH. To chemically describe the acid–base properties of a microorganism, let us begin our examination with the straightforward release (or dissociation) of a proton from a hypothetical surface functional group on a bacterium's wall. This deprotonation process, which accurately describes the behavior of a number of functional groups associated with cell surfaces (e.g., hydroxyl, carboxyl, sulfhydryl, and phosphate), leads to the formation of an organic anion, or ligand, and the concomitant release of H^+. On the other hand, amino and amide groups are neutral when deprotonated and positively charged when protonated. The combined protonation states of the functional groups on a cell surface largely determines its hydrophilic/hydrophobic characteristics at any given pH.

In its most simplistic form, deprotonation can be expressed by the following equilibrium reaction:

$$R\text{-}AH \longleftrightarrow R\text{-}A^- + H^+ \qquad (3.1)$$

where R denotes the parent organic compound to which each protonated ligand type, A, is attached. The distribution of protonated and deprotonated ligands can be quantified with the corresponding mass action equation:

$$K_{eq} = \frac{[\text{R–A}^-][\text{H}^+]}{[\text{R–AH}]} \qquad (3.2)$$

where K_{eq} is the equilibrium constant for the reversible reaction. Equilibrium constants for ionization reactions are also called dissociation or acidity constants (K_a). [R–A$^-$] and [R–AH] represent the concentration of exposed deprotonated and protonated species on the bacterium, respectively (in mol L^{-1}), and [H$^+$] represents the activity of protons in solution. The term activity reflects the "effective concentration" of the chemical species, and it is calculated by multiplying the molar concentration by an activity coefficient based on ionic concentration (see Langmuir, 1997 for details). In freshwater, the activity coefficient approaches 1, so for simplicity, the two terms, activity and concentration,

are often used interchangeably. The larger the value of K_a, the more dissociation of protons into solution (i.e., the stronger the acid). Importantly, each functional group has its own K_a, and based on equation (3.2), the pH at which [R–A$^-$] and [R–AH] are equivalent is known as the pK_a value, where $pK_a = -\log_{10}K_a$. At pH < pK_a a functional group is protonated and at pH > pK_a it is deprotonated (Fig. 3.8a).

The ionization of functional groups in the cell wall provides an electrical charge at the bacterium's surface that results in the formation of an electric field surrounding the entire cell. In dilute solutions the surface charge is established solely by H$^+$ exchange with the organic ligands, whereas in more concentrated solutions the inherent surface charge can be modified by the adsorption of ions. For any given condition, the mean charge excess of a microbial surface, [L]$_T$ (mol mg^{-1}), can be calculated as a function of pH from the difference between total

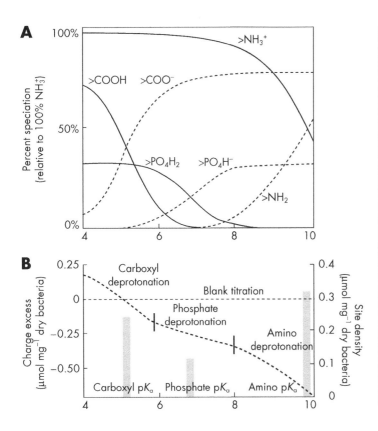

Figure 3.8 (A) Estimated speciation profile of the major functional groups associated with bacterial cell walls e.g., *Shewanella putrefaciens* (modified from Haas et al., 2001). (B) A hypothetical titration profile illustrating charge excess (i.e., net surface charge) resulting from the deprotonation of functional groups (in *S. putrefaciens*) and calculated as the difference in proton concentration between the bacterial and blank titrations. The blank titrations are free of functional groups and correspond to the dissociation of water. The site densities of distinct functional groups (drawn as solid gray bars) are modeled from charge excess and plotted according to the pH at which half are deprotonated, the pK_a value. The various methods of modeling site density and pK_a distribution account for the fact that a single type of functional group deprotonates over a pH range.

base or acid added to a microbial suspension and the equilibrium H^+ and OH^- ion activities:

$$[L]_T = \frac{C_a - C_b + [OH^-] - [H^+]}{B} \qquad (3.3)$$

The components, C_a and C_b, are the concentrations of acid and base added, respectively, $[OH^-]$ and $[H^+]$ are the number of moles of OH^- and H^+ in the solution at the measured pH, respectively, and B is the quantity of bacterial biomass (mg). If the density of cells, ρ (cells mg^{-1}), and the cell wall volume, v (m^3 $cell^{-1}$), are known, then the corresponding cell wall surface charge q (mol m^{-3}) can be calculated:

$$q = \frac{[L]_T F}{\rho v} \qquad (3.4)$$

In this reaction, F is the Faraday constant (the amount of electric charge carried by one mole of electrons). When such calculations are done over an entire pH range, a so-called acid–base titration curve is created that shows the pH range over which some functional groups are chemically active and how the net surface charge of the cell varies with pH (Fig. 3.8b).

The calculations above can also yield important information about the number of moles of reactive surface sites, which reflects the buffering capacity of the cell over a given pH range (e.g., Fein et al., 1997). A large difference in the total base or acid added and the free H^+ ion activity indicates significant pH buffering and a high concentration of surface functional groups. On a plot of charge excess this is shown by a steep slope. A small difference in the total base or acid added and the free H^+ ion activity indicates weak pH buffering and a low concentration of surface functional groups. This translates into a gentle slope on the charge excess plot. Crucially, the proton–bacteria reactions are fully reversible, with the adsorption or desorption of protons reaching the same equilibrium concentration at any give pH value. Consequently, similar acid–

base titration curves are generated regardless of whether the experiments were initiated from acidic or alkaline conditions, and apparently these extremes in pH do not cause changes in the cell wall structure through saponification of lipids or destruction of peptide bonds (e.g., Daughney and Fein, 1998).

The charge difference between the microbial surface and the proximal aqueous solution (at any pH) gives rise to an electrical potential that strongly affects the concentration and spatial distribution of ions at the cell–water interface. The electrical potential can be modeled using either electric double layer (EDL) theory, analogous to the classical representation of mineral surfaces (see Dzombak and Morel, 1990 for details) or Donnan exchange, which has been used to characterize the charge associated with ion-penetrable cell walls (e.g., Yee et al., 2004).

The EDL model describes the distribution of charge on a surface, with an inner electrical layer consisting of the surface proper (the surface potential, ψ_0), and an outer layer of oppositely charged ions, or counter-ions, fixed both directly to the surface (referred to as the Stern layer) or more diffusely (referred to as the Gouy layer). Most of the surface charge is neutralized by the tightly bound (usually covalently) counter-ions in the Stern layer, forming what is known as the inner-sphere complex. The remaining charge is balanced by the Gouy layer whose concentration of counter-ions declines rapidly away from the solid surface. Outer-sphere complexes are those in which the solute ions and surface species are attracted by electrostatic forces alone (Fig. 3.9).

The Donnan model describes the distribution of electrical potential within the wall matrix (the Donnan potential, ψ_D), which it treats as a porous structure with homogeneous cross-linked ionizable functional groups. It further assumes that the transition between cell wall and solution is very thin compared to the thickness of the wall, and that exchange reactions are strictly electrostatic, controlled by differences in the valency of the ions, not their size.

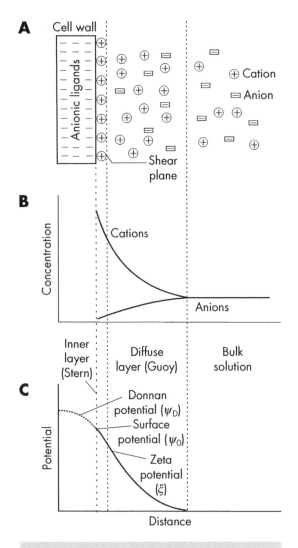

Figure 3.9 (A) Representation of the electrical double layer, with anions and cations surrounding the cell surface. (B) Change in concentration of cations and anions away from the negatively charged bacterial cell wall. (C) The electrical potential across the cell wall. (Modified from Blake et al., 1994.)

3.2.2 Electrophoretic mobility

When subjected to an electric field, microorganisms move in a direction, and at a rate, commensurate with the polarity and density of the overall surface charge. This process is termed electrophoresis, and the electrophoretic mobility of a microbial suspension can be quantified by measuring their velocity in an electric field. This measurement can, in turn, be used to calculate the zeta potential (ξ), using the Smoluchowski equation below (Hunter, 2001):

$$\xi = \frac{\eta\mu}{\varepsilon_0 \varepsilon} \qquad (3.5)$$

The component "μ" is the electrophoretic mobility of a particle, ε_0 and ε are the relative dielectric constants of the vacuum and solution, respectively, and η is the viscosity of the solution. The zeta potential reflects the electrical potential at the interfacial region (viewed as a shear plane) separating the Stern layer, where cations are held tightly in place and move with the bacterium, and the Gouy layer, where ions are mobile (Wilson et al., 2001). A negative ξ potential indicates that the bacterium is negatively charged and migrates towards the positively charged electrode in an electrical field, while a positive ξ potential indicates the opposite. The greater the absolute value of the ξ potential, the greater is the charge density on the surface (Blake et al., 1994). Electrophoretic mobility measurements, however, underestimate the true surface potential because they actually measure the adsorbed cations in the Stern layer as well as the cell's anionic ligands. Unfortunately, techniques are not presently available to measure the surface potential itself, so the closest we can measure is at the shear plane.

The electrophoretic mobility of different species varies significantly with the elemental composition of the cell wall or extracellular layers. For instance, there is a direct correlation between the cell surface N/P atomic concentration ratio and the cell's electrostatic charge. By studying various bacteria and fungi, Mozes et al. (1988) showed that the presence of phosphate groups played a major role in determining the anionic surface charges, while nitrogenous groups were linked to increased positive charge.

Electrophoretic mobility is also pH dependent because the activity of protons in solution controls the ionization reactions of functional groups at the microbial surface (Ahimou et al., 2001). This leads to the concept of isoelectric point, which is defined as the pH value where net surface charge equals zero. The isoelectric point can be estimated from acid–base titrations, but it can also be directly measured with electrophoretic mobility experiments because at the isoelectric point microorganisms do not exhibit motion in an electric field. The isoelectric point of bacterial walls is typically between pH 2 and 4, and no fundamental differences exist in the isoelectric behavior of Gram-positive and Gram-negative cells (Harden and Harris, 1953). This means that at low pH, when the surface functional groups are fully protonated, bacteria are either neutral or positively charged, the latter being the result of a cell possessing abundant amino groups. Meanwhile, at the growth pH of most bacteria, cells inherently display a net negative charge and the magnitude of negativity increases with higher pH values (e.g., reaction (3.1)). Therefore, under low pH conditions, most bacterial surfaces behave hydrophobically, and become increasingly hydrophilic with increasing pH.

3.2.3 Chemical equilibrium models

One of the major challenges facing researchers today is how to interpret acid–base titration data in terms of cell wall biochemistry. Ascribing pK_a values from a titration curve to specific functional groups is not so clear-cut because there can be considerable variation in pK_a values for the same functional group. This occurs because the magnitude of the dissociation constant is controlled by the structure of the molecule to which it is attached (see Martell and Smith, 1977 for details). Consequently, a single carboxyl group in two different organic acids will have different pK_a values, as will an organic acid with multiple carboxylic groups. As might then be expected, the pK_a for a carboxyl group on

one microbial species versus another could yield widely different values simply because of subtle conformational variations within the wall macromolecules. Furthermore, titration experiments are only able to resolve those groups that contribute significant amounts of protons to solution. Minor groups are simply undetectable with the resolution of current techniques. Therefore, it is important to keep in mind that the model-derived binding sites do not directly represent the functional groups of the cell surface; their identity can only be inferred by comparison of the functional group pK_a values with pK_a values of model compounds. Unequivocal identification of the types of functional groups responsible for acid–base buffering can be obtained by spectroscopic techniques, such as Fourier transform infrared spectroscopy (FTIR), or gas/liquid chromatography of cell wall extracts.

Despite the inherent variability, acid–base modeling of bacteria has clearly shown that they have a quantifiable and characteristic geochemical reactivity that reflects a suite of functional groups in their outermost structures. One method of modeling is to constrain the number of pK_a values to fit the titration data. In experiments with intact B. *subtilis* cells, Fein et al. (1997) demonstrated that a three-pK_a model could effectively quantify the buffering effect provided by the cell wall surfaces. For instance, at low pH the deprotonation of carboxyl groups could accurately predict the buffering capacity of the biomass from pH 2 to 6 (reaction (3.6)). A two-pK_a model, including phosphate groups, accurately mimicked the titration curves up to pH 7.5 (reaction (3.7)). At pH values above 7.5, a three-pK_a model yielded an excellent fit to experimental data. Although the authors inferred that the third deprotonation reaction involved the loss of protons by hydroxyl groups (reaction (3.8)), those sites were more likely to be cationic amino groups NH_3^+ because they deprotonate at pH values above 8–11 (reaction (3.9)), while hydroxyl groups tend to deprotonate at pH values above 10 (Hunt, 1986).

Table 3.1 Relative total concentration of reactive ligands amongst various bacteria.

Species	Total ligand concentration (μmol mg^{-1} dry bacteria)	Ionic strength (mol L^{-1})	Reference
Rhodococcus erythropolis	0.93	0.01–1.0	Plette et al., 1995
Bacillus subtilis	0.22	0.3	Fein et al., 1997
Bacillus subtilis	0.50	0.01–0.025	Cox et al., 1999
Bacillus subtilis	1.60	0.001–0.1	Yee et al., 2004
Bacillus cereus	2.29	0.01	He and Tebo, 1998
Bacillus licheniformis	0.29	0.1	Daughney et al., 1998
Shewanella putrefaciens	1.77	0.1	Sokolov et al., 2001
Shewanella putrefaciens	0.08	0.1	Haas et al., 2001
Calothrix sp.	0.80	0.01	Phoenix et al., 2002

$$R\text{-}COOH + OH^- \rightarrow R\text{-}COO^- + H_2O \quad (3.6)$$

$$R\text{-}PO_4H_2 + OH^- \rightarrow R\text{-}PO_4H^- + H_2O \quad (3.7)$$

$$R\text{-}OH + OH^- \rightarrow R\text{-}O^- + H_2O \quad (3.8)$$

$$R\text{-}NH_3^+ + OH^- \rightarrow R\text{-}NH_2 + H_2O \quad (3.9)$$

When the total site densities were calculated over the entire pH range of their experiments, the distribution of proposed functional groups were as follows: 0.12 μmol of carboxyl groups/mg bacteria; 0.04 μmol phosphate groups/mg bacteria; and 0.06 μmol hydroxyl groups/mg bacteria, making a total concentration of 0.22 μmol mg^{-1} of bacteria (dry weight).

Other models fix the acidity constants and determine the minimum number of ligand sites required to achieve a good fit to the titration data. For instance, Cox et al. (1999) used this technique to resolve five proton binding sites on the cell walls of *B. subtilis*: two types of carboxyl sites at low pK_a values, phosphoryl sites at circumneutral pK_a values, and two sites with high pK_a values, which were attributed to either amino or phenol (pK_a 8–12) groups. Very acidic sites, such as some carboxylic acids and phosphodiesters, and very basic sites, such as the hydroxyl groups, could not be observed in the titration range of the experiment. Their results instead yielded a total ligand density of 0.50 μmol mg^{-1} of bacteria.

Titration studies have been performed on a wide range of bacteria. What their collective results demonstrate is that the relative total concentrations of surface functional groups can vary by over an order of magnitude amongst a range of bacteria, and even within the same species (Table 3.1), and that the acid–base behavior of the surface ligands are weakly affected by solution ionic strength. So, the question is, aside from pK_a variations due to coordination of any given functional group, what else might be causing variations in the magnitude of the surface charge? As discussed above, differences in surface charge between different species can be ascribed to the types and densities of exposed functional groups, i.e., the Gram-positive wall of *B. subtilis* versus the Gram-negative wall of *E. coli*. Variations amongst the same species, however, are most likely a function of either subtle changes in: (i) nutrient conditions; or (ii) population growth phase:

1 An individual bacterium has the means to alter its surface chemistry to compensate for the chemical composition of the aqueous environment. Take for example *B. subtilis*, it can change the secondary polymers associated with the peptidoglycan in response to the levels of dissolved inorganic phosphate. When teichoic acids are produced, phosphoryl groups (pK_a between 5.6 and 7.2) and phosphodiester groups (pK_a between 3.2 and 3.5) are most abundant, while a cell loaded with teichuronic acids has instead an abundant supply of carboxyl groups (pK_a between 4 and 6). At other times, when the cells may need to behave more hydrophobically, they can strategically place positively charged functional groups into the wall that markedly reduce the net negative surface charge (Beveridge et al., 1982). Similarly, Gram-negative bacteria, such as *Shewanella putrefaciens*, have variable sugar arrangements in their LPS which not only impact the carboxyl:amino ratios, but also affect the cross-linking of functional groups and limit which remain unoccupied, and hence may be ionizable (Moule and Wilkinson, 1989).

2 Daughney et al. (2001) have observed that exponentially growing cells of *B. subtilis* possess four times more carboxyl sites, twice as many phosphate sites, and 1.5 times as many amino sites (per unit weight) as cells in either the stationary or sporulated phase. It would appear that the higher nutrient availability prompts exponential phase cells to modify their cell wall to be more efficient at metal sequestration (see next section), whereas diminished nutrient availability causes the cells to return to perhaps their "default" setting. During starvation, the effects on cell surface reactivity become even more pronounced. Frequently a large reduction in cell volume and a tendency towards increased hydrophobicity occurs, i.e., a reduction in carboxyl and phosphate groups (Kjelleberg and Hermansson, 1984). Associated with growth phase is the cell's metabolic state. Actively respiring cells pump protons into their wall matrix during respiration. These, in turn, can protonate the surface functional groups, rendering them electrically neutral (Koch, 1986). By contrast, dead cells no longer produce a proton gradient, and for that reason, they are likely to be more anionic (Urrutia et al.,1992).

One other point needs emphasizing – all those studies on cell wall material have not taken into account the fact that most benthic bacteria possess extracellular layers. Microorganisms with EPS, sheaths, or S-layers will have a more complex charge distribution because those layers differ compositionally from that of the underlying cell wall. This was highlighted in a recent study of isolated sheaths and intact filaments of the cyanobacterium, *Calothrix* sp. (Phoenix et al., 2002). Electrophoretic mobility measurements of cell walls showed completely different profiles to those of isolated sheath material (Fig. 3.10). While the wall was characterized by a net negative surface charge, the sheath's charge was found to be near neutral, indicating that the dominant electronegative carboxyl and electropositive amino groups must occur in approximately equal proportions. Significantly, this study confirmed that under normal growth conditions, some species possess a dual-layered surface charge, i.e., a highly electronegative cell wall surrounded by an electroneutral sheath.

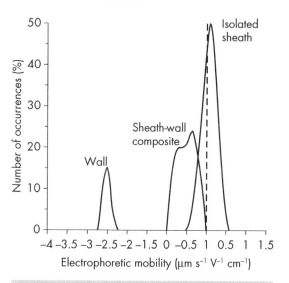

Figure 3.10 Electrophoretic mobility measurements performed on *Calothrix* sp. at pH 5.5. The very electronegative peak at $-2.5 \, \mu m \, s^{-1} \, V^{-1} \, cm^{-1}$ is characteristic of exposed cell wall material with deprotonated carboxyl and phosphate groups. The peak around +0.1 is isolated sheath material, made up predominantly of polysaccharides. The broader peak at around −0.3 is likely a composite of wall and sheath material. (Modified from Phoenix et al., 2002.)

3.3 Passive metal adsorption

As the chemical equilibrium models show, bacteria are not inert objects immune to the physicochemical conditions in the environment surrounding them. On the contrary, all bacteria have low isoelectric points, meaning that they should interact with metal cations and have them intimately associated with their surfaces. Considering their ubiquity in the surface environment, their high population densities wherever suitable sources of nutrition exist, and their characteristically large surface area to volume ratios, it is easy to understand why they are very important agents in metal sequestration.

Some bound metals serve the purpose of stabilizing the negative charges of the anionic functional groups, and thus are relatively "fixed" into place (see section 3.4.1). Others metals are much more exchangeable and merely provide a temporary positive charge to counter the negative charge induced by the deprotonation of the cell's surface functional groups (Carstensen and Marquis, 1968). The strength of the metal–ligand bond is given by the surface complexation/binding constant (K_M), where M refers to the specific metal of interest. The greater its surface complex formation constant, the less likely a metal cation will be desorbed into solution. Metal cation sorption is also directly affected by pH, which dictates metal partitioning (or speciation) within, and between, soluble and solid phases, and hence controls its mobility, reactivity, and toxicity in aquatic environments. Of particular importance here is the hydrolysis constant, which measures the tendency of a metal cation to react with water and form a hydroxide phase, e.g., $Fe(OH)_3$.

There have been literally hundreds of studies discussing the metal sorption properties of microorganisms (see Ledin, 1999 for review). It would be impossible to cover more than a fraction of them here, so instead, the goals of the following section are to highlight a few studies that have addressed where metal cations bind to bacterial cell surfaces and then examine how cell surface composition can influence the partitioning of certain metals from multi-elemental solutions, as encountered in the natural environment.

3.3.1 Metal adsorption to bacteria

In their pioneering work on metal binding to *B. subtilis*, Beveridge and Murray (1976) showed that when the cell walls were chemically separated and suspended in a supersaturated salt-rich solution, so much metal was bound that they formed dense aggregates. Transition metals, in particular, impart such strong electron-scattering power that some of them have subsequently been used as contrasting agents for electron microscopy. Indeed, many of the advances made on bacterial ultrastructure over the past three decades were made possible by metal staining and visualization of biological thin sections under the TEM (e.g., Fig. 3.11). Alkali and alkaline earth metals that were freely soluble in water (e.g., Na^+, Mg^{2+}, etc.) can also be sequestered from solution, but then tend to yield diffusely stained walls. Interestingly, alteration of the charge density within the wall fabric due to the introduction of different metal cations elicits a dimensional response in the peptidoglycan strands. Thus, the cell wall can be made to shrink or swell according to the metal staining agent used to give it contrast (Beveridge and Murray, 1979).

To reveal the functional groups in the wall to which metal cations react, a variety of chemical treatments can be used to modify or remove electronegative and electropositive groups. Anionic carboxyl groups can be neutralized or converted into electropositive sites by treatment with water-soluble carbodiimides or ethylenediamide, respectively; teichoic acids can be removed by dilute base; while amino groups can be made anionic by replacing them with succinyl groups or removed by deamination using nitrous acid (Doyle, 1989). When carboxyl groups in the peptidoglycan of *B. subtilis* are chemically

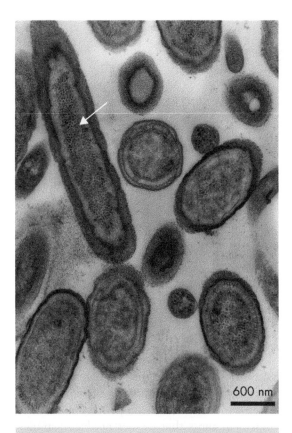

Figure 3.11 The routine use of transition metals to stain cell tissue for transmission electron microscopy works because the cell surface functional groups are anionic and they react electrostatically with multivalent metal cations. This is an unidentified species growing at a hot spring in Kenya. Notice how staining with uranyl acetate and lead citrate has revealed many of the structural details of the cells, including the ribosomes (arrow).

600 nm

neutralized, substantial reduction in the amounts of metals sorbed occurs (e.g., Beveridge and Murray, 1980). Similarly, when teichoic acids are extracted to ascertain how much metal is bound to the peptidoglycan only, metal binding decreases in all instances, yet for most metals not to the same extent as by the loss of carboxyl groups (Matthews et al., 1979). The introduction of positive charges into the B. subtilis wall also results in a marked decrease in the total

number of sites that can bind metals, while decreasing the number of cationic amino groups leads to an increase in metals bound (Doyle et al., 1980). Collectively, these experiments demonstrate that with B. subtilis walls, carboxyl groups are the most electronegative sites, and that the bulk of the binding capacity in Gram-positive bacteria remains associated with the peptidoglycan.

Another important feature governing how much metal is bound is the degree of interstrand cross-linking in peptidoglycan. Cross-linking occurs by means of peptide bond formation, which involves the loss of two charged groups for every bond formed. The more cross-linking, the greater the physical compactness of the cell wall, and hence the fewer cations that are required to neutralize the excess anionic ligands in the wall matrix (Marquis et al., 1976). What's more, metal probes, such as polycationic ferritin (PCF), have shown an inherent heterogeneity in charge distribution, with PCF binding preferentially to the negatively charged peptide chains and teichoic acids on the outer wall surface of B. subtilis (Sonnenfeld et al., 1985a). The positioning of these polymers likely represents a favored orientation that exposes carboxyl and phosphate groups to the external aqueous environment where cations can be scavenged. PCF also tends to label the polar ends of the walls, further implying that the tips of the cells are more electronegative than other sites on the wall (Sonnenfeld et al., 1985b).

Bacillus licheniformis walls are unlike those of B. subtilis, in that they contain up to 26% teichuronic acid and 52% teichoic acid, thus having much less peptidoglycan. In similar metal binding studies as above, the walls bind an order of magnitude less metal (e.g., Beveridge et al., 1982). However, unlike B. subtilis, the phosphate groups in the teichoic acid and carboxyl groups in teichuronic acid play a greater role in metal adsorption. Therefore, the overall metal binding ability of these particular Gram-positive cells is determined by the amount and type of secondary polymers present. The fact that both secondary polymers bind metals implies that the bacterium maintains a tight control on the negative:positive charge ratio, regardless of the aqueous conditions in which they grow.

When the wall material of Gram-negative cells, such as *E. coli*, are subjected to metal-rich solutions, it becomes readily evident that they do not adsorb as much metal from solution as do their Gram-positive counterparts. Typically the quantities are less than 10% (e.g., Beveridge and Koval, 1981). Unlike the Gram-positive bacteria, there is a bilayered distribution of metals associated with the outer membrane. This pattern stems from the higher phosphate:lipid ratio of the LPS compared to the phospholipids, ensuring that the outer face is more electronegative and subsequently can bind more metal (Ferris and Beveridge, 1984). Although carboxyl groups are also present in the LPS core, only one-third of the groups are available for metal binding; the others being cross-linked to the cationic amino groups (Ferris and Beveridge, 1986a). The peptidoglycan in *E. coli* is chemically similar to *B. subtilis*, and even though it is only a monolayer, it reacts more strongly to some metals than the outer membrane. As in the Gram-positive bacteria, this feature is attributed to the availability of carboxyl groups (e.g., Hoyle and Beveridge, 1984).

Many Gram-positive and Gram-negative bacteria also produce EPS. Due to their hydrated nature, dissolved metals can freely diffuse throughout the extracellular layers, binding to the anionic carboxyl groups of uronic acids and the neutrally charged hydroxyl groups of sugars (Geesey and Jang, 1989). Under metal-deficient conditions in the growth media or in naturally dilute solutions, capsules can appear diffuse and extensive, whereas higher metal concentrations can lead to flocculation, and even the precipitation of metal–capsule composites (e.g., Fig. 3.12). By possessing a large and reactive surface area, it is thus not unexpected that a number of studies have also documented that encapsulated bacteria bind more metals than nonencapsulated varieties

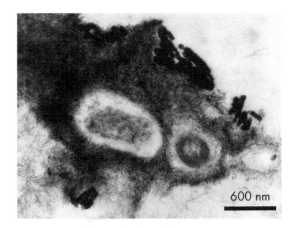

Figure 3.12 TEM image of two encapsulated bacteria that are naturally iron-stained from growing in Fe-rich hydrothermal fluids.

(e.g., Rudd et al., 1983). In fact, species producing capsules can tolerate higher metal concentrations than those that do not, and it has been shown that the proportion of encapsulated bacteria increases in metal-polluted sediment, whereas mutants that cannot produce capsules die off (Aislabie and Loutit, 1986). Interestingly, many isolates from metal-contaminated sediment lose their capsules upon subculture in metal-free media, suggesting that the role of capsular production may be linked to protection against metal toxicity.

Irrespective of the bacteria studied, what has repeatedly been shown in metal binding studies is that there is no apparent stoichiometry between the quantity of metals that bind to cell walls and the amount of anionic ligands. In some instances, large metallic deposits line the wall, while at other times so much metal may be fixed to the cell surface that it forms a distinct mineral phase. This led Beveridge and Murray (1976) to originally propose a two-step mechanism for the metal adsorption process; the first step in time is an electrostatic interaction between the metal cations and the anionic ligands in the cell wall. This interaction then acts as a nucleation site for the deposition of more metal cations from solution, potentially leading to biomineralization (see Chapter 4). The size of the deposit depends on a number of variables, including the concentration of the metals in solution and the amount of time through which the reactions proceed. If sufficient time exists, the metal/mineral product grows in size within the intermolecular spaces of the wall fabric or on the outer surface until it is either physically constrained by the wall polymers or the saturation state diminishes. The end result is a bacterial wall that contains copious amounts of metal, often approaching the mass of the bacterial cell itself (Beveridge, 1984).

3.3.2 Metal adsorption to eukaryotes

Algae possess a number of functional groups that deprotonate under normal growth conditions. As a result, algal populations can sequester a wide range of metals, commonly with uptake values in excess of 100 mg of metal g^{-1} biomass dry weight (Volesky and Holan, 1995). The functional groups of greatest importance are the carboxyls associated with uronic acids in alginate because they are appropriately spaced to cross-link and neutralize a number of multivalent metals (e.g., Majidi et al., 1990). Consequently, the removal of alginate from algal biomass can cause significant decreases in metal binding capacity. Many green algae also contain sulfate esters in cellulose, that, because of their low pK_a values (between 2.5 and 1), can facilitate metal sorption under very acidic conditions (Crist et al., 1992).

For a number of algae, biosorption is clearly a passive process, accumulating available metals irrespective of whether they serve a physiological role or not. This statement is supported by a study comparing biosorption by green benthic algae in two chemically dissimilar river systems, where it was observed that metal sorption by biomass in a solute-rich river was often greater by an order of magnitude relative to biomass

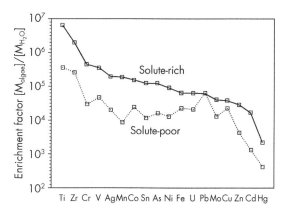

Figure 3.13 Comparison of metal enrichments in filamentous green algae from two contrasting rivers, one solute-rich and the other solute-deficient. (Modified from Konhauser et al., 1993.)

from a solute-deficient river (Konhauser and Fyfe, 1993). Plotting the concentration of metals sorbed to algae versus their dissolved concentrations further showed algal biomass characterized by enrichments of between 10^2 and 10^7 for the metals studied (Fig. 3.13). These patterns reflect the strong complexing ability of the reactive ligands, leading to the natural conclusion that metal concentrations within the algae are a direct reflection of availability. For that reason, algal populations in metal-rich rivers will have correspondingly high metal accumulations, an observation that might have great merit when prospecting for mineral deposits.

The capacity of fungi and yeasts to bind metals has also been extensively explored (see Gadd and Sayer, 2000 for review). The importance of chitin and chitosan in metal binding has been demonstrated by the observation that their removal from biomass results in a significant decrease in metal sorption (Galun et al., 1983). Within these polymers, protonated amino groups are strongly linked to the adsorption of anionic species (e.g., Tsezos, 1986), while phosphate and carboxyl groups are important in the adsorption of cationic species (e.g., Tobin et al., 1990). The carboxyl groups in particular appear to be responsible for metal accumulation in the biomass from *Aspergillus niger*, since 90% of the metal binding capacity was irretrievably lost when these groups were chemically modified (Akthar et al., 1996). The secondary components of fungal cells, such as the phenols and melanins, are also effective at metal sorption (e.g., Saiz-Jimenez and Shafizadeh, 1984; Caesar-Tonthat et al., 1995).

3.3.3 Metal cation partitioning

As the studies above show, binding of cations to a microbial surface is largely an electrostatic phenomenon. However, the structural and compositional variability of the wall or extracellular layers, as well as the unique physicochemical properties of each element, adds a level of complexity to the overall process. In other words, protons and each different cation should be capable of interacting in a distinctive way with the reactive ligands on a cell's surface.

This realization has led to a number of metal binding studies that have compared the relative affinities of protons and various cations for the exposed functional groups of different microorganisms, by techniques that involve displacing one by another. Those studies have highlighted two very important points, the first being that metals and protons compete for the same surface sites. As solution pH decreases, the functional groups become protonated, displacing loosely bound metal cations. Conversely, at circumneutral pH, the functional groups deprotonate and electrostatic interactions with metal cations increasingly takes place (e.g., Crist et al., 1981). The second finding is that some metals preferentially bind to different ligands in the cell wall, but most importantly, they are not equally exchangeable. For instance, trivalent (e.g., La^{3+}, Fe^{3+}) and divalent metal cations (e.g., Ca^{2+}, Mg^{2+}) are strongly bound to the wall of *B. subtilis*, while monovalent cations (e.g., Na^+, K^+) are easily lost in competition with those metals for binding sites (Beveridge and Murray, 1976).

Despite the different ligand-metal affinities, the cell surface can still be viewed as being largely "non-specific", in that adsorbed metals can be desorbed as geochemical conditions change (Ledin et al., 1997). In this regard, metal-ligand interactions are reliant upon thermodynamics, just the same as inorganic systems. A particularly compelling example of nonspecificity comes from Fowle and Fein (1999) who demonstrated that in mixed metal experiments with *B. subtilis* and *B. licheniformis*, the cell walls consistently had a higher affinity for Cd^{2+}, even though Ca^{2+} concentrations were two orders of magnitude higher. Moreover, Ca is an important element for cell structure and Cd is toxic, but the sorption behaviors of these two elements were not manifestations of the different effects they had on the cell. Instead, they reflected the chemical properties of the metal cations (in this context also known as Lewis acids) and those of the oxygen-, nitrogen-, and sulfur-containing ligands that reside within the cell wall (known as Lewis bases). These properties are largely understood, and given sufficient information about the environment in which a microorganism is growing, it may be possible to extrapolate and predict metal binding patterns on a cell surface.

The supply of cations to the cell depends on several external factors, such as their aqueous concentrations or the presence of co-ions and other organic anions that can complex the metals in solution and affect their bioavailability. Once at the cell periphery, competition between those metals for organic ligands will ensue. To fully understand metal adsorption processes, we must consider both the ionic forces, which are the initial electrostatic attractions between a metal cation and the organic ligands, and the subsequent covalent forces that arise from electron sharing across a metal cation–ligand molecular orbital (see Williams, 1981; Hughes and Poole, 1989; Stone, 1997 for details). Some of the most important factors that influence metal binding to cells are briefly examined below:

1 *Ionic potential* – Cations in solution vary in their propensity to coordinate water molecules as a function of their charge density, that being the relationship between the charge of the nucleus (z) and its radius (r). Most monovalent, and many divalent, cations remain unhydrated or form aquoanions. Contrastingly, trivalent metals displace protons from coordinating water molecules. They also displace protons from functional groups with O^- ligands. Thus, the affinity of a metal cation for an organic ligand increases dramatically in going from a +I metal cation (e.g., Na^+) to a +III metal cation (e.g., Fe^{3+}).

2 *Ligand spacing/stereochemistry* – Monovalent cations are generally preferred by isolated or widely spaced sites, where they replace single protons on individual ligand sites to neutralize the negative charge (e.g., phosphates in phospholipids). Under such conditions, a divalent cation may not be able to satisfy the two distant sites of negative charge. By contrast, multivalent cations are preferred by closely opposed sites (e.g., carboxyl groups of LPS), where either the ligands are unable to accommodate two monovalent cations or where greater steric stability is achieved by increased coordination between the metal and two ligand sites. In EPS, the ability of the macromolecules to accommodate intra- and intermolecular cation bridging will dictate which metals are preferentially sorbed. The latter is the principle behind chelation, the binding of a metal cation to two or more coordinating anionic sites in the same biomolecule (known as bidentate or multidentate ligand bonding, respectively).

3 *Ligand type* – Different metals are favored by different ligands. Oxygen atoms have a distinct affinity for the alkali and alkaline earth metals (e.g., K^+, Mg^{2+}, Ca^{2+}) and some transition metals (e.g., Fe^{3+}), while nitrogen and sulfur atoms preferentially bind a number of transition metals (e.g., Ni^{2+}, Co^{2+}, Cu^{2+}, Zn^{2+}, Cd^{2+}, Fe^{2+}).

4 *Covalent bonding* – Once a metal cation is adsorbed, the ability to form covalent bonds with the ligand is important for complex stability. In general, cations increasingly form inner-sphere (and stronger) complexes with a given ligand as the difference in electronegativity between the two decreases (see Faure, 1988). Covalent bonding is

also most effective when either the highest occupied molecular orbital or the lowest unoccupied molecular orbital is a d-orbital. These factors account for why transition metals (e.g., V, Cr, Mn, Fe, Co, Ni, Cu, Zn, and Mo) exhibit greater covalent bonding than lighter metals (e.g., Al) and metal cations to the left on the periodic table (e.g., the alkali and alkaline earth metals). Generally, cations that are bound only weakly through electrostatic attraction, like Na^+, are effective in competing only with other weakly bound ions. This also explains why protons, which are mainly covalently bound, are only displaced during transition metal uptake, and not during light metal uptake (Crist et al., 1981).

When we consider the log K values for divalent metal cations and a particular organic ligand, we note that as we move from left to right in the periodic table, the values increase, reaching a maximum with Cu^{2+} (Fig. 3.14). This trend, called the "Irving–Williams Series," is observed with practically all oxygen- and nitrogen-bearing ligands (Williams, 1953). Because of their greater charge and smaller radii, the trivalent metals form even stronger bonds with organic ligands (Stone, 1997).

Although trivalent cations typically have higher affinities for wall material, in solutions where divalent cations, such as Ca^{2+} and Mg^{2+}, are several orders of magnitude more abundant, the outer face of the cell wall would be predominantly in the Ca-Mg form. The ubiquitous association of Fe with cell walls also stems from its greater concentration in natural waters compared to other trace metals (e.g., Cu). It thus appears that cell wall material can show different metal binding patterns under different geochemical conditions. On the one hand, it displays distinct preferentiality in binding one metal from a range of competing cations. On the other hand, it reacts to soluble ions as if it were an open ion exchange resin. In this regard, microorganisms bind the cations that are in highest concentration and, accordingly, it is not surprising that microorganisms are considered ideal metal scavengers for bioremediation purposes (see section 3.7.1).

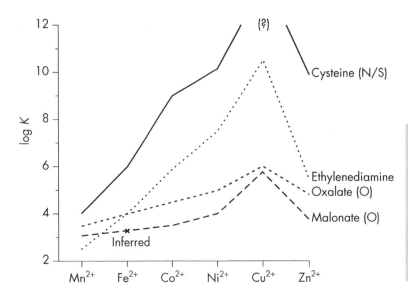

Figure 3.14 Complexation of divalent metals to a number of organic compounds. The adsorption of the metals follows the "Irving–Williams Series." Note the preferential adsorption of Cu^{2+}, irrespective of the donor ligands (in brackets). (Data from Williams, 1953; Stone, 1997.)

3.3.4 Competition with anions

Studies comparing the biosorption of metals onto microbial surfaces have generally been limited to systems involving dissolved cations only. However, in nature dissolved inorganic and organic anions balance the positive charges of metal cations, often leading to dissolved complexes or mineral precipitation. Such reactions impact metal biosorption in two ways. First, the metal–anion complexes typically have lower affinity for the biomass than the free cation does, and subsequently, they bind less strongly (e.g., Shuttleworth and Unz, 1993). Second, some anions bind metals more strongly than biomass, making the metals completely unavailable for biosorption. For example, in studies with dead biomass of *Sargassum natans*, a brown marine alga, the presence of NO_3^- and PO_4^{3-} suppressed the amount of gold uptake capacity at acid pH values (Kuyucak and Volesky, 1989a), while high concentrations of halide ions (e.g., Cl^-, Br^-, and I^-) diminished gold adsorption to the green alga *Chlorella vulgaris* in the order consistent with their reactivity towards Au(III) (Greene et al., 1986).

Organic ligands can have an even more pronounced effect. In the case of gold, commercially produced organic compounds, such as EDTA, have proven to be so effective as gold scavengers that they have been employed in the elution (re-solubilization) of the metal from biomass (see section 3.7.2). Other organic compounds used to bind gold (in order of their complexation capacity) are thiourea > cyanide > mercaptoethanol; each has been shown more effective than the inorganic anions above (Greene et al., 1986). Many microorganisms also produce extracellular organic exudates (see below), with metal complexation constants equal to, or stronger, than the adsorption constants associated with the cell surface functional groups (e.g., Santana-Casiano et al., 1995).

3.4 Active metal adsorption

It would be mistaken to view bacterial metal uptake simply as a passive process in which sorption occurs as a consequence of cells growing in concentrated solutions where metals abound. Instead, microorganisms require a variety of metals to fulfill internal and external cell functions (see Silver, 1996, for details). It is, therefore, not unusual for them to manipulate the type and abundance of their organic functional groups to retain those metals specifically required for structural or operative integrity, and to subsequently regulate their uptake rates to maintain intracellular concentrations at optimal levels for growth and metabolism. At other times, microorganisms may need to immobilize toxic metals away from the cell periphery, and as such, they produce and expel specific metal binding chelates into the bulk fluid phase. In either situation, the metal–ligand complexes formed can often be so tenacious that those bound metals are not easily displaced by other metal ions (e.g., Hoyle and Beveridge, 1983). Quite clearly, the cell surface is a dynamic layer that continuously interacts with those metal cations in its immediate vicinity.

3.4.1 Surface stability requirements

Metal adsorption onto a cell's surface has an important bearing on the its dielectric properties because those metal cations have an effect on the conduction of low frequency electric currents. Their conductivity can actually be measured by the following equation:

$$\sigma_w = c_f^w u^w [1 + 2c^0/c_f^w)^2]^{1/2} \qquad (3.10)$$

where σ_w is the cell wall conductivity, c_f^w is the fixed charge concentration in the wall, u^w is the mobility of ions in the wall, and c^0 is the environmental ion concentration (Carstensen and Marquis, 1968). This equation indicates which metals serve as counter-ions for fixed anionic charges (i.e., those likely to have some requirement), and which metals remain mobile. By measuring the electrical conductance of various wall–cation combinations, it appears, for example, that K^+ is less free to move around in the wall than Na^+, Mg^{2+} appears to be still more tightly bound, while protons are essentially immobile (Marquis et al., 1976). These results correlate well with the affinity series discussed earlier where it was pointed out that H^+ is bound more strongly than alkali and alkaline earth metals at low pH values. The conductivity is also directly related to solution ionic strength. At low

ionic strength the conductivity of the bacterial cell wall is dominated by ions confined to the Stern layer, but at high ionic strength, the walls become saturated with exogenous salts and they obtain a conductivity that is roughly proportional to the surrounding aqueous environment (Carstensen et al., 1965).

The presence of metal cations in the wall matrix influences its stability in a number of ways:

1 *Surface wettability* – In Gram-negative cells, such as *E. coli*, the addition of Mg^{2+} makes the surfaces much more hydrophobic than if they only bound Ca^{2+} or Na^+ (Ferris, 1989). Along similar lines, if the LPS is removed from the surface, there is an overall increase in hydrophobicity since the outermost layer is inherently electronegative. The degree of hydrophobicity is also related to: (i) the quantity of metals bound to the outer surface of the cell; and (ii) the charge density, and hence hydration, of the cations. If the bacterium can control surface wettability, then numerous benefits ensue. For instance, a high degree of hydrophobicity can help the bacterium contact unwettable surfaces and then stick to them. This is particularly useful in micro-colony formation on inert solid surfaces, in the utilization of apolar hydrocarbons for nutrition, and in the exclusion of solvated particles such as bacteriophages from adhering onto the cell (Beveridge, 1989b).

2 *Surface stability* – In Gram-positive bacteria, a single Mg^{2+} cation can cross-link the anionic ligands between two teichoic acid molecules. In this regard, it eliminates the repulsive anionic charges between adjacent molecules, giving rise to more dense but stable structures (Doyle et al., 1974). Under sufficiently alkaline conditions (where the phosphoryl group is completely deprotonated), Mg^{2+} may even stabilize both O^- ligands on the same teichoic acid molecule. The peptidoglycan also binds Mg^{2+} cations, with the peptide chains bending in such a way that the carboxyl group of one peptide is only the diameter of one Mg^{2+} cation away from the carboxyl group of another peptide (Fig. 3.15). Studies have also shown that decreasing the number of cationic amino groups in peptidoglycan, through chemical treatment, leads to an increase in metals bound (Doyle et al., 1980). This suggests that the cationic amino groups must function normally as competitive counter-ions, and

to some extent neutralize the negative charges associated with the carboxyl or phosphate groups of the cell wall (Marquis, 1968). In Gram-negative bacteria, Ca^{2+} functions in a similar manner to neutralize the numerous electronegative charges of the LPS, thereby bridging adjacent molecules of LPS together and anchoring the outer membrane to the underlying peptidoglycan layer. In experiments where calcium is removed from the membrane, an increase in the electrostatic repulsion between the constituent anionic ligands occurs, thereby limiting how close the individual components of the membrane can approach one another (Ferris and Beveridge, 1986b). Subsequently, the LPS is forced to adopt a tighter curvature, causing it to bleb and become sloughed off. Calcium is also required for the proper assembly of S-layers in a number of species, and in calcium-deficient growth media, no surface layers are formed (Smit, 1987).

3 *De-activation of autolysins* – Autolysins are enzymes that break down the cross-links in peptidoglycan so that the cell wall can be restructured during growth and cell division. At the growth pH of most cells autolytic activity is controlled by the active extrusion, and retention, of protons in the cell wall (Doyle and Koch, 1987). Thus, if the proton motive force becomes dissipated by the death of the cell, then the bound protons in the wall would systematically be lost and the autolytic enzymes would become activated. This results in the uncontrolled breakdown of cell wall material, and eventually the exposure of cytoplasmic material to the external aqueous environment (Jollife et al., 1981). Experiments, however, have documented that the addition of metal cations, such as Fe(III), to lysed cells limits cellular degradation. This increases the preservation potential of cellular remains, and might explain why some organic remains are retained in the geological rock record (e.g., Ferris et al., 1988).

3.4.2 Metal binding to microbial exudates

The production of microbial exudates is of global significance in terms of trace metal cycling. Although many of these organic ligands are poorly characterized, recent studies have shown that a number of dissolved metals, such as Cu(II), Fe(III),

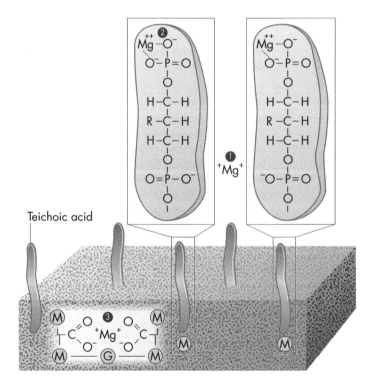

Teichoic acid

Figure 3.15 Representation of a *Bacillus subtilis* cell wall, showing how magnesium possibly functions to cross-link (1) the phosphodiester groups of two teichoic acid molecules, (2) the terminal phosphoryl groups within a single teichoic acid molecule, and (3) between two carboxyl groups associated with the peptide stems in peptidoglycan. M, *N*-acetylmuramic acid; G, *N*-acetylglucosamine.

Zn(II), and Cd(II), exist in nature predominantly as organo-metallic complexes (Sunda, 2000). The exudates can have both detrimental and beneficial properties for microbial communities. Sometimes the exudates are in direct competition with cell surfaces for metals, limiting their availability. At other times, organic exudates can be utilized by microorganisms as either a means of sequestering metals from the external environment to supplement their nutritional needs or to immobilize some metals extracellularly as a method of detoxification. An example of each is given in the next section.

(a) Siderophores

Iron is a key element for all microorganisms, yet the insolubility of Fe(III) at circumneutral pH means that it is often the limiting nutrient for growth. Many bacteria and fungi get around this impasse by excreting low molecular weight, Fe(III)-specific ligands known as siderophores (Neilands, 1989). Siderophores, and their breakdown products, make up a large component of the strong Fe(III)-binding ligands that regulate Fe(III) species in surface ocean water (e.g., Wilhelm and Trick, 1994).

Siderophores have several properties that make them ideal Fe(III) chelators, namely a high solubility, an abundance of oxygen ligands, and a tendency towards bi- and multidentate ligation that forms coordinative positions around the central Fe(III) cation. Significantly, they form especially strong 1:1 surface complexes, and their stability constants for Fe(III) can be as high as 10^{52}. This important property maintains iron in a "soluble" form that minimizes its loss from the aqueous environment by the precipitation of solid-phase ferric hydroxide (Hider, 1984).

So far over 200 siderophores have been identified, with most broadly divided into two classes based on their metal chelating properties, either hydroxamates or catecholates (Winklemann, 1991). The most common siderophores contain the hydroxamic functional group, which forms a five-member ring with Fe^{3+} between the two oxygen atoms, with the hydroxyl proton being displaced. Frequently, three such groups are found on a single siderophore molecule, and hence it requires six oxygen ligands to satisfy the preferred octahedral geometry of ferric iron, each having partial double-bonded characteristics (Fig. 3.16). Hydroxamate siderophores are produced by many types of fungi, and they are the most effective Fe chelators at mildly acidic to neutral pH. Like hydroxamates, catecholates also occur in triplicate so that they can facilitate tridentate bonding with ferric iron. They are produced by all classes of bacteria, and they tend to be the more important Fe chelators at alkaline pH (Hersman, 2000). Irrespective of which siderophore is used, once the iron is bound, the Fe(III)–siderophore complex reacts with a receptor site on the cell's surface and is then transported to the plasma membrane. There it is dismantled, and Fe(III) is released and reduced to Fe(II).

The biosynthesis of siderophores is tightly controlled by iron levels, such that they only become activated when dissolved Fe(III) concentrations are negligible. Interestingly, higher levels of siderophores are produced in response to increasingly insoluble iron sources, such as hematite (e.g., Hersman et al., 2000). Production of siderophores is also related to cell growth phase. Siderophores are produced most abundantly during exponential phase, they then level out during stationary phase, and with time, decrease in concentration as bacteria run out of nutrients and begin to lyse (Kalinowski et al., 2000a). On a much larger scale, recent experiments in the equatorial Pacific have demonstrated that with the addition of iron, a threefold increase in the concentration of Fe-binding organic ligands occurred, leading to a concomitant increase in microbial biomass production (Hutchins and Bruland, 1998). Interestingly, many species produce siderophores in great excess of their requirements (because many are lost via diffusion and advection), yet when levels of iron become sufficiently high, i.e., an order of magnitude above micromolar levels, their production is repressed

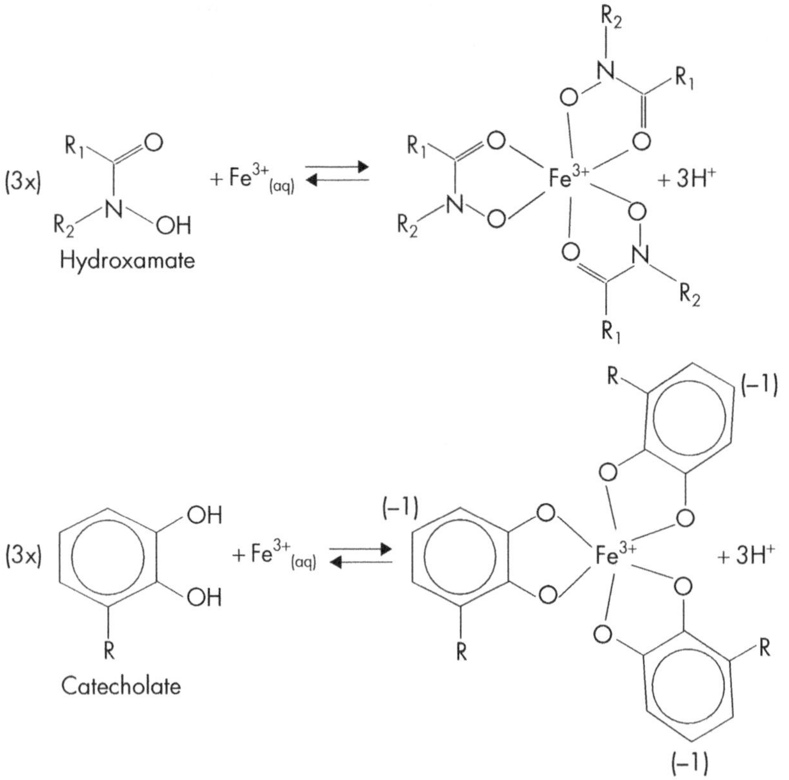

Figure 3.16 Fe(III) complexation reactions with hydroxamate and catecholate siderophores.

and the cells meet their iron needs via low-affinity Fe uptake systems (Page, 1993).

(b) Metal binding ligands

The preference of a given ligand for certain metals provides the cell with an opportunity to specifically sequester individual essential elements. Consider the competition between Mg^{2+}, which is biologically required, with a strong Lewis acid such as Cu^{2+}, which can be toxic to microorganisms when found in high concentrations. A cell surface studded with oxygen ligand-containing functional groups would facilitate the accumulation of the alkaline earth metal because copper has low affinity for such sites. Conversely, if the same metals were competing for a ligand that included nitrogen or sulfur, then copper would prevail (Hughes and Poole, 1989). In some ways possessing sufficiently high amounts of N- and S-ligands on extracellular layers or in microbial exudates can actually benefit the binding of Mg^{2+} because exposing such ligands could effectively cleanse the waters of such transition metals, preventing them from interfering with binding to the O-donor ligands.

As an adaptation to repeated exposure to toxic metal concentrations, cells have evolved inducible detoxification mechanisms, such as the intracellular production of thiol-containing ligands that complex undesirable metals, and in doing so, mask their presence. One such example is the synthesis of metallothionein proteins in response to the presence of high intracellular copper levels (Williams, 1953). Microorganisms can also produce and excrete extracellular ligands that have extremely high surface complexation constants for a number of toxic metals, including copper, cadmium, and lead. This is desirable for the microbial community as a whole because, in most cases, the toxicity of a free hydrated ion is greater than that of metals complexed with other ligands. One prevalent example of this is in lakes and oceans, where greater than 99% of copper is bound to organic ligands (e.g., Coale and Bruland, 1988; Xue et al., 1996). Two lines of evidence suggest that the copper-binding ligands are produced by phytoplankton (the microbial portion of the plankton community, versus the animal component, the zooplankton) in order to regulate Cu^{2+} levels in their environment.

1 Their distribution varies with biological productivity, such that the ligands occur at a maximum concentration in the illuminated euphotic zone during seasonal blooms.

2 Several species of the cyanobacterial genus, *Synechococcus*, produce extracellular ligands with stability constants similar to those ligands identified in seawater. They can reduce the free Cu^{2+} concentration in seawater by 1000-fold, to levels within their tolerance limits (Moffett and Brand, 1996).

3.5 Bacterial metal sorption models

Many of the early studies described above were carried out in conditions supersaturated with respect to the metal of interest. Today, metal sorption experiments are placing greater emphasis on developing geochemical speciation models that describe how microorganisms interact with metals and mineral surfaces under natural, and more realistic, undersaturated geochemical conditions (see Fein, 2000; Warren and Haack, 2001 for reviews). Metal sorption reactions can be quantified using two different approaches: (i) bulk partitioning relationships; or (ii) surface complexation models (SCM). In the first instance, partitioning relationships, such as K_d, Freundlich and Langmuir isotherms, can easily be applied to complex systems because they do not require a detailed understanding of the nature of the surfaces or the adsorption/desorption mechanisms involved. However, they are system-specific, meaning that the results from a set of experiments are not applicable to different

systems. By contrast, the SCM takes into account the effects of changing pH, solution composition and ionic strength, the acid–base properties of surface functional groups, competitive sorption with other solutes, and solid-phase mineralogy. It then draws upon that information to extrapolate to conditions beyond those tested in the laboratory.

3.5.1 K_d coefficients

When a solute adsorbs onto a surface, the surface is termed the sorbent and the solute is termed the sorbate. A plot that quantifies the amount of sorbate sorbed to a solid surface versus the concentration of solute in solution is known as a sorption isotherm. Several models have been proposed to quantify metal adsorption (and conversely desorption) associated with microbial surfaces. The simplest is when there is a linear relationship between the amount of metal adsorbed onto the microorganism and the concentration of metal in solution (Fig. 3.17). Under these conditions a distribution coefficient (K_d), that predicts the quantity of metal sorbed to the biomass, can be used to model the adsorption reaction (see Langmuir, 1997 for details):

$$M_B = K_d M_D \qquad (3.11)$$

In the equation, M_B is the mass of metal adsorbed per dry unit mass of bacteria ($\mu g\ g^{-1}$) and M_D is the concentration of dissolved metals in equilibrium with the bacterial surface ($\mu g\ ml^{-1}$).

Distribution coefficients are simple to apply. They do not require a detailed knowledge of the surface or sorption mechanisms, and as such, they are an uncomplicated means to model the distribution of metals at low concentrations (e.g., Hsieh et al., 1985). However, linear sorption isotherms do not describe sorption in terms of binding sites, and as the concentration of metal increases, the relationship between M_B and M_D eventually becomes nonlinear and K_d coefficients become inapplicable. This occurs because at higher metal concentrations the available reactive ligands

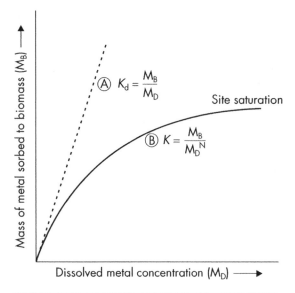

Figure 3.17 Typical K_d (A) and Freundlich (B) isotherms.

become occupied and the affinity between the surface and metal gradually decreases (e.g., Gonçlaves et al., 1987). Since K_d coefficients cannot account for site saturation, they cannot be applied to define an upper adsorption limit. Another limitation with K_d isotherms is that they are specific to each experiment, and as a result, K_d values for the same sorbate–sorbent combination can vary by orders of magnitude depending on aqueous conditions.

3.5.2 Freundlich isotherms

A more flexible sorption model is the Freundlich isotherm. The sorption relationship is expressed as:

$$M_B = KM_D^{N} \tag{3.12}$$

where N is a fitting parameter. The Freundlich isotherm can generally describe sorption over a wider range of metal concentrations (from trace to saturation) than K_d coefficients, and it can account for nonlinear sorption behavior. If N equals 1 then the equation becomes a linear

sorption isotherm (a K_d isotherm). If N is greater than 1, then the extent of sorption increases with increasing metal concentrations, and if N is less than 1, sorption decreases with increasing metal concentrations. For both cases, $N > 1$ and $N < 1$, a curvilinear line is obtained when M_B is plotted versus M_D (Fig. 3.17). The isotherm plot can be linearized by taking the logarithm of the Freundlich equation, such that N becomes essentially the slope of the isotherm (equation (3.13)).

$$\log M_B = \log K + N\log M_D \tag{3.13}$$

The Freundlich equation has been widely applied to quantify metal adsorption onto microbial surfaces. One typical observation is that the highest fraction of metal adsorption occurs at the lowest dissolved concentrations, corresponding to the steepest part of the isotherm plot. This indicates that the propensity for cation binding progressively diminishes in the presence of increasing concentrations as all available sorption sites become occupied. At this stage, the curve plateaus out and no more cations are adsorbed (e.g., Small et al., 1999). Another observation with bacterial biomass is that K values can also decrease at higher cell densities because the production of significant amounts of organic exudates competes directly with wall ligands for available cations (Harvey and Leckie, 1985). This has similarly been reported in fungi where uptake of metals was lower at higher cell densities because of (i) reduced cell surface area due to cell–cell attachments and (ii) diminished mixing of metals with surface ligands (Junghans and Straube, 1991).

The Freundlich isotherm more effectively describes metal distribution in complex systems and surfaces with heterogeneous properties, such as bacterial communities, as long as the conditions can be directly simulated in the lab. However, the Freundlich isotherm is obtained by an empirical fit to experimental data, and similar to K_d coefficients, the sorption constants can vary by many orders of magnitude as a function of solution and system parameters.

3.5.3 Langmuir isotherms

The Langmuir sorption isotherm was developed with the concept that a sorbent contains a finite number of reactive sites, and once all the sites are occupied by a monolayer of cations, the surface will no longer adsorb the solute from solution. It also assumes that all sorbed species interact only with the ligand and not with each other. The metal sorption reaction can be expressed by a site-specific equilibrium reaction:

$$A^- + M^+ \longleftrightarrow AM \qquad (3.14)$$

A^- is the ligand on the surface and M^+ is the dissolved metal. An equilibrium constant can be determined from the law of mass action:

$$K = \frac{[AM]}{[A^-][M^+]} \qquad (3.15)$$

Surface species concentrations can be expressed in terms of moles per liter of solution, per gram of solid or per cubic centimeter of solid surface.

The upper limit of sorption is defined by the concentration of the ligands on the surface. The maximum concentration of surface sites, A_{max}, is given by:

$$[A_{max}] = [A^-] + [AM] \qquad (3.16)$$

From equations (3.15) and (3.16), we can derive the Langmuir equation:

$$[AM] = \frac{K[A_{max}][M^+]}{1 + K[M^+]} \qquad (3.17)$$

If [AM] is plotted versus $[M^+]$, then the line would be a curve that reaches a plateau at the maximum sorption value. The Langmuir isotherm can also be expressed in a linearized form:

$$\frac{1}{[AM]} = \frac{1}{K[A_{max}][M^+]} + \frac{1}{[A_{max}]} \qquad (3.18)$$

If 1/[AM] is plotted versus 1/[M], then the intercept, $[1/A_{max}]$, represents the maximum sorption capacity, while the slope of the line, $1/K[A_{max}]$, can be used to determine the sorption constant.

In natural solutions the cation concentration may exceed the solubility product of a mineral phase before the ligand sites are filled. On the isotherm plot this is shown by a vertical upward line. A relevant example of the continuum between adsorption and precipitation is given with ferric iron (e.g., Warren and Ferris, 1998). The relationship describing the hydrolysis of Fe(III), and its adsorption to cell surface ligands is:

$$R\text{-}AH + Fe^{3+} + 2H_2O \longleftrightarrow R\text{-}AFe(OH)_2^0 + 3H^+ \qquad (3.19)$$

In the above equation R-AH represents a protonated functional group and $R\text{-}AFe(OH)_2^0$ is Fe associated with the ligand. The mass action equation is represented by:

$$K = \frac{[R\text{-}AFe(OH)_2^0][H^+]^3}{[R\text{-}AH][Fe^{3+}]} \qquad (3.20)$$

In double logarithmic plots of experimental equilibrium data, the initial portion of the curve is linear, indicating that adsorption is directly proportional to the number of available organic ligands, i.e., Langmuir-type behavior (Fig. 3.18). In the Warren and Ferris study, the amount of iron adsorbed at this stage approached the micromole per milligram range when normalized to cell dry weight, an amount within an order of magnitude of the total surface ligand concentrations determined independently by the acid–base titrations discussed previously. The second stage of surface site saturation and the onset of supersaturation is evidenced where the curves plateau as $[Fe_D]/[H^+]^3$ values increase. The third stage, that being nucleation, begins when $[Fe_D]/[H^+]^3$ values exceed the recognized equilibrium solubility product of poorly ordered ferric hydroxide. The curve then undergoes a reversal as the dissolved Fe(III) concentration ratio decreases during mineral precipitation, and any additional Fe added to the system goes directly towards mineralization.

An advantage of the Langmuir equation is that it can be expanded to model the sorption

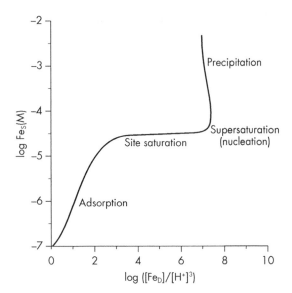

Figure 3.18 Modified Langmuir isotherm describing the amount of Fe(III) bound (Fe$_S$) to *Bacillus subtilis* as a function of the amount of dissolved Fe(III) added to the solution (Fe$_D$). The graph shows a continuum of three stages beginning with adsorption, followed by site saturation, and then nucleation/precipitation. (Adapted from Warren and Ferris, 1998.)

of two competing solutes onto a surface, and/or a surface with two sorption sites. The Langmuir equation can also be employed to evaluate nonideal competitive adsorption (termed the NICA model). The NICA model was originally developed to describe proton and metal complexation with humic substances, but it has been extended to describe competitive metal sorption onto bacterial surfaces (e.g., Plette et al., 1996). This approach can take into account component heterogeneity or nonideality through experimentally calibrated fit parameters that are system specific and must be determined for each system composition of interest.

Similar to the Freundlich isotherms, Langmuir isotherms can accurately quantify metal–bacteria sorption as a function of metal concentration, but their system specificity means that they are only applicable to the conditions at which they were determined. They do not explicitly account for all of the changing parameters in a natural setting, and subsequently cannot be used to estimate the extent of sorption in systems not directly studied in the laboratory (Davis and Kent, 1990).

3.5.4 Surface complexation models (SCM)

In order to better account for all of the environmental variability that underpins the adsorption and desorption reactions of metals onto microbial surfaces, a number of studies have since turned to the surface complexation model. Unlike the bulk partitioning models, the SCM treats surface complexes formed on minerals and microorganisms in a similar manner to aqueous complexes, deriving for them equilibrium constants that describe their thermodynamic stability. This means that the SCM must take into account all the changes in aqueous composition, as well as the acid–base properties of microbial and mineral surface functional groups (Fein, 2000). Such a process has the potential of being extremely useful in extrapolating results from select experiments to conditions beyond those directly studied in the laboratory; clearly it is not possible to conduct experiments using every combination of microbial species and every type of fluid composition. Importantly, the equilibrium constants obtained from such isolated metal–bacteria or metal–mineral laboratory experiments can be combined with others to ultimately model and accurately predict the extent of sorption that occurs in more complex, multicomponent systems.

The pH dependence of metal sorption is depicted in Figure 3.19. This feature is commonly known as a sorption edge. Under acidic conditions, the cell wall functional groups are fully protonated and no adsorption of cationic metal species occurs. Only metals present as oxyanions adsorb at low pH. As pH increases, the functional groups systematically deprotonate, forming discrete anionic metal binding ligands. At low pH those ligands are provided solely by carboxyl groups, at circumneutral pH

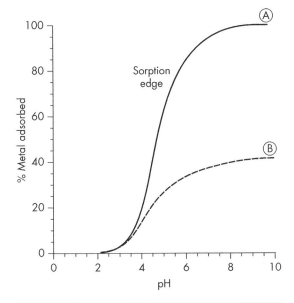

Figure 3.19 Isotherms showing the adsorption of any given metal cation to bacterial biomass when (A) there are more ligands than metals, leading to 100% metal adsorption and a steep sorption edge and (B) when there are more metals than biomass and the curve plateaus due to the lack of available ligands. On average, carboxyl groups deprotonate over the pH range 2–6, phosphates pH 5–8, and amino groups pH 8–11.

phosphate groups additionally deprotonate, while at more alkaline pH, amino (pH 8–11) and hydroxyl (pH > 12) groups become increasingly important. Progressive deprotonation reactions result in increasing metal adsorption, up to a point where potentially all the metal is bound. (curve A, Figure 3.19). In addition, the adsorption and desorption of metals reaches the same equilibrium concentration at any given pH value (Fowle and Fein, 2000). Therefore, it does not matter whether steady-state conditions are approached from site undersaturation (no metal associated with bacterial surface) or from site saturation (virtually all metal associated with bacterial surfaces).

Because surface complexation models describe bacterial cell walls as heterogeneous surfaces

with multiple reactive ligands, the adsorption of aqueous ions can effectively be likened to an abiological process, controlled predominantly by the acid–base properties of the exposed ligands and by the affinity of each type of ligand for a specific ion. Indeed, as discussed earlier, most binding sites on the cell are not tailored to capture specific ions from solution, and subsequently they can easily exchange them for others in the bulk fluid phase. This point is driven home by the observation that intact bacteria and their isolated cell wall material exhibit reasonably similar affinities for particular cations, yet that affinity differs between different microorganisms (e.g., Mullen et al., 1989).

Interactions between aqueous metal cations (M^{z+}) and the most common functional groups in bacterial cell walls (i.e., carboxyl, phosphate, hydroxyl) can be represented by reactions (3.21) to (3.23), respectively:

$$M^{z+} + \text{R-COOH} \longleftrightarrow \text{R-COO(M)}^{(z-1)+} + H^+ \qquad (3.21)$$

$$M^{z+} + \text{R-PO}_4\text{H}_2 \longleftrightarrow \text{R-PO}_4\text{H(M)}^{(z-1)+} + H^+ \qquad (3.22)$$

$$M^{z+} + \text{R-OH} \longleftrightarrow \text{R-O(M)}^{(z-1)+} + H^+ \qquad (3.23)$$

The release of protons and the adsorption of metal cations to form a charged complex (e.g., R-COO(M)$^{(z-1)+}$) is quantifiable with the corresponding mass action equation:

$$K = \frac{[\text{R-COO(M)}^{(z-1)+}][H^+]}{[M^{z+}][\text{R-COOH}]} \qquad (3.24)$$

The K value is the experimentally observed metal sorption constant that is related to a true thermodynamic constant ($K_{\text{intrinsic}}$), via activity coefficients, that take into account ionic strength, surface charge and electrical double layers (see below). The equilibrium expression above emphasizes that adsorption of metal cations by microorganisms depends not only on pH and ionic strength, but also on the number and type of functional groups per cell. Since ultrastructural variations exist between different bacterial species, and even within a single strain, observable differences in metal binding capacity are theorized. An additional assumption not immediately obvious from the equilibrium equation, is that all surface functional groups are treated as being structurally and chemically fully equivalent (Buffle, 1990).

In order to consider microbial surfaces as thermodynamic chemical components, the electrostatic interactions between the surface electric field and metal cations can be accounted for using the following equation (Stumm and Morgan, 1996):

$$K_{intrinsic} = K(-\Delta Z F \Psi_0 / RT) \qquad (3.25)$$

$K_{intrinsic}$ represents the equilibrium constant referenced to zero surface charge, F is the Faraday constant, R is a gas constant, T is absolute temperature, ΔZ is the change in the charge of the surface species for the reaction of interest, and Ψ_0 is the surface potential of the cell. Several different SC models (e.g., constant capacitance, diffuse double layer, and triple layer models) were initially proposed to describe the surface electrical field associated with mineral surfaces (see Dzombak and Morel, 1990 for details), and they have now been adapted to microbial surfaces. Such electrostatic sorption models define a mathematical relationship between surface charge and surface potential, but differ in the assumptions they make about where the adsorbed species are positioned in the double layer. The constant capacitance and diffuse-layer models assume that all cations are specifically adsorbed at the shear plane, while the triple-layer model assigns adsorbed species to either the shear plane or a more distant plane (Langmuir, 1997).

3.5.5 Does a generalized sorption model exist?

Despite the apparent simplicity in using SC models to determine microorganism–metal interactions, there are a number of factors that obscure the patterns of metal sorption:

1 *Biomass:metal ratios* – When abundant biomass exists, there are excess anionic ligands present on the bacterial surface compared to the dilute concentration of dissolved metals. This means that equilibration between the metals and the surface complexes they form can easily be achieved at circumneutral pH. The adsorption edge under these conditions is steep, and 100% adsorption is potentially attained using whichever ligand provides the greatest stability for the newly formed surface complex (recall Fig. 3.19). At lower biomass:metal ratios, the available ligands may become fully saturated with metals and the rate with which further adsorption takes place diminishes. Accordingly, the adsorption edge is less steep, and excess metals will remain in solution unless precipitation occurs. What all this implies is that in solute-rich solutions, several ligands may be required to bind available metals, and that the quantity of metals adsorbed at a particular pH increases as the ratio between the total concentration of microbial surface ligands to the total metal concentration increases (e.g., Fein et al., 1997). Clearly, displaying several types of organic ligands on a cell wall is how bacteria ensure metal uptake over a wide range of pH.

2 *Charge modification* – While metal accumulation onto a microorganism is influenced by the surface charge characteristics of the exposed ligands, one of the outcomes of metal binding is that the cell surface progressively becomes less anionic due to charge neutralization within the electric double layer (e.g., Plette et al., 1996). Indeed, with high cation coverage, the surface may even become positively charged. Some studies have revealed that the charge reversal usually occurs in the pH range where the concentration of uncomplexed divalent cations (e.g., Cu^{2+}) decreases and the concentration of their monovalent hydroxylated cations (e.g., $Cu(OH)^+$) correspondingly increases (Fig. 3.20). Over this same pH range, maximum metal adsorption takes place, suggesting that the OH- groups of the hydrolyzed metals play an important role in hydrogen bonding between O-ligands on the bacterial surface and those metals in solution. At more alkaline pH, the cations are neutrally (e.g., $Cu(OH)_2$) or negatively (e.g., $Cu(OH)_2^{2-}$) charged, and the overall charge on the cells once again becomes negative (Collins and Stotzky, 1992). At this stage it is also common

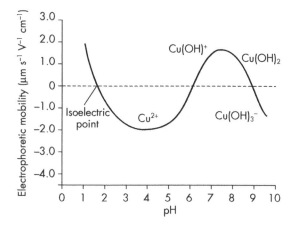

Figure 3.20 The relationship between dissolved copper species and the electrophoretic mobility of bacteria. In the presence of Cu^{2+}, cells remained negatively charged at pH values between their isoelectric points and pH 6. At higher pH values, despite continuing deprotonation of cell wall functional groups, bacteria become positively charged with continued addition of metal. Simultaneously, there is a corresponding change in dissolved copper speciation, from an uncomplexed divalent cation to a monovalent hydroxylated cation ($Cu(OH)^+$), around pH 6. At pH values above 8, the cations become neutrally charged ($Cu(OH)_2$) and then negatively charged ($Cu(OH)_3^-$). These species no longer adsorb to the cell. The cell surface then reverts back to an overall net negative charge, whilst the amino groups deprotonate. (Modified from Collins and Stotzky, 1992.)

to observe a reversal in metal adsorption behavior. Therefore, the pH of the solution not only affects the degree of deprotonation of the cell's functional groups, but it also influences metal speciation, so that different metals present at similar concentrations will be differently adsorbed.

3 *Ionic strength* – It influences metal adsorption by three principal mechanisms. (i) It affects the activities of ions in solution, with higher ionic strengths leading to decreased availability of "free" metal cations in solution: under these conditions, concentration and activity can no longer be considered equivalent. (ii) It governs the thickness of the electrical double

layer between the cell and the ions in the bulk solution. Higher ion availability satisfies the surface charge excesses and results in decreased cell wall surface potentials. By contrast, dilute solutions allow the electric fields to expand outwards from the cell surface. Saturating the cell surface with cations also changes the corresponding isotherms, from those with distinct sorption edges at low ionic strength to a poorly defined sorption edge at high ionic strength (Yee et al., 2004). (iii) Increased ionic strengths lead to competition amongst the various cations for the cell's anionic sites. This, in turn, depends on whether a particular cation bonds to the microbial surface electrostatically as a hydrated, outer-sphere complex or covalently as an inner-sphere complex (Small et al., 2001).

4 *Kinetics* – Metal sorption by microbial biomass often involves two distinct stages. The first, which is passive adsorption to the cell surface, is a rapid process occurring within seconds to minutes after the microorganism comes into contact with the metal (e.g., Hu et al., 1996). When the concentration of cell ligands exceeds dissolved metal concentrations, partitioning can be satisfactorily described by a linear relationship (K_d isotherm) and equilibrium is generally reached within a few hours. During this stage, the contact time of adsorption also exhibits no affect on the kinetics of desorption or on the concentration of the metals bound to the cell (Fowle et al., 2000). The second stage is slower and commonly involves diffusion-controlled, intracellular accumulation. This process occurs over several hours, and if sufficient metals are available to the cell, it can lead to much higher metal accumulations than that of the first stage (e.g., Khummongkol et al., 1982).

5 *Growth phase* – The surface characteristics of any given microorganism, and therefore its capacity to sorb metals, can vary with growth conditions. For example, Chang et al. (1997) reported that Pb^{2+} was most extensively adsorbed by *Pseudomonas aeruginosa* during stationary phase, Cd^{2+} was preferentially adsorbed at exponential phase, while Cu^{2+} was not affected by growth phase. Meanwhile, *B. subtilis* cells growing in exponential phase adsorb 5–10% more Cd^{2+} and Fe^{3+} than cells at stationary phase, which, in turn, is 10–20% more than that adsorbed by sporulated cells (Daughney et al., 2001).

With all these environmental variables to be considered, is it truly possible to apply chemical equilibrium thermodynamics to quantify metal sorption and desorption on bacterial surfaces? The answer remains to be tested in "natural systems", but one thing is likely, certain simplifying assumptions will need to be made. Fortuitously, it is now becoming apparent that metal cations display a similar affinity series for a given group of ligands, regardless of whether the ligands exist on the surface of a microorganism or as an aqueous organic species. This means that common complexes, such as metal–oxalate or metal–acetate, can be used to predict metal–carboxyl surface stabilities of bacteria for those metals whose bacterial adsorption behavior has not yet been measured directly (e.g., Fein et al., 2001). Significantly, this greatly expands the number of aqueous metal cations for which adsorption onto bacteria can be modeled. Recent findings have additionally shown that when biomass:metal ratios are high (i.e., an order of magnitude more ligand sites than dissolved metal on a molar basis), metal adsorption onto the walls of various bacterial species all display very similar sorption edges (Yee and Fein, 2001). Each species can adsorb nearly 100% of the metal cations at similar pH values, suggesting that patterns of metal adsorption may not be too species-specific when abundant biomass is present (Fig. 3.21).

Although the rationale for SCM is that it is possible to describe multiple metals adsorbing onto multiple surface sites by combining equilibrium constants for each specific chemical reaction that occurs, the models are not yet sufficiently developed to predict how lab-based sorption reactions with monocultures compare in an environmental setting with mixed mineral assemblages (metal oxides, clays), multiple organic phases (humic compounds, microbial exudates etc.), multi-elemental pore waters, and a complex mixed microbial community with species in different growth phases held together by EPS and various other extracellular layers of widely different compositions. So, at present it remains to be

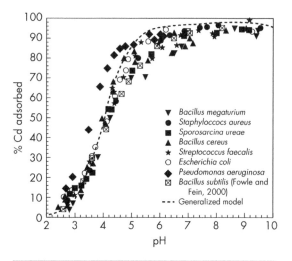

Figure 3.21 Plot showing Cd^{2+} adsorption onto pure cultures of various bacteria. Each point represents individual batch experiments with $10^{-4.1}$ mol L^{-1} Cd and 1.0 g L^{-1} (dry weight) bacteria. The dotted curve represents the modeled adsorption behavior. Notice how all the bacterial species exhibit nearly identical Cd adsorption behavior as a function of pH. (Reprinted from Yee and Fein, 2001 with permission from Elsevier.)

determined whether it is possible to derive a generalized model that can actually be applied to quantify the distribution and concentration of metals in bacteria-bearing water–rock systems.

3.6 The microbial role in contaminant mobility

One of the main motivations for researching microbial–metal interactions is that it has wide-ranging implications for accurately modeling contaminant transport in the environment, and ultimately the design of effective bioremediation strategies (e.g., Bethke and Brady, 2000). The significance of microorganisms in contaminant mobility lies in the fact that they comprise a significant component of the organic fraction in the subsurface, and they possess highly reactive surfaces that allow them to partition metals from solution into their biomass. In sediment and soils, this partitioning capacity is of similar, or even greater, magnitude to some clays and other organic components (e.g., Ledin et al., 1999).

When microorganisms become immobilized onto a solid substratum, or if their movement through an aquifer is inhibited by some form of permeability barrier, they are likely to reduce the transport of contaminants. They do so because coating the original mineral surface with biomass frequently increases the metal binding properties of the substratum, and immobilized cells provide additional surface area to which metals are retained (e.g., Yee and Fein, 2002). At other times, they remain as free-moving particles through the porous media, enhancing the transport and dispersion of sorbed contaminants (e.g., Lindqvist and Enfield, 1992).

3.6.1 Microbial sorption to solid surfaces

Within minutes of a solid being submerged in an aqueous environment, a thin film will collect at the solid–liquid interface due to simple sedimentation and from electrostatic interactions between the solid and the dissolved ions/suspended materials from the bulk aqueous phase (Neihof and Loeb, 1972). This is known as "conditioning" the surface with inorganic and organic compounds for the growth of microorganisms. The actual colonization of this surface involves three steps; (i) transport of the bacteria to the submerged surface, (ii) their initial adhesion via electrostatic interactions, and (iii) their irreversible attachment to the substratum through the excretion of EPS or utilization of surface appendages (van Loosdrecht et al., 1990).

(a) Transport to the surface

In quiescent bodies of water, relatively large cells or aggregates settle to the bottom by sedimentation, whereas smaller cells (radii <1 μm) exhibit a certain degree of diffusive transport due to Brownian motion. In aquifers, the primary driving force for bacterial transport is advection, the process generated by the hydraulic gradients that induce groundwater flow. Under conditions of limited groundwater movement, many bacteria move freely from one location to another by some form of motility, including those that depend on the propulsive action of flagella (swarming and swimming), and those that depend partly on cell to cell interactions (gliding). These modes of transport can be quite fast, with some bacteria showing velocities greater than 10^{-4} cm s^{-1} (Characklis, 1981). Many bacteria can also move chemotactically in response to a chemical gradient (Carlile, 1980).

(b) Initial adhesion – effects of solution chemistry

The initial interaction between a bacterium to a mineral, referred to as reversible adhesion, is an instantaneous attraction by long-range forces holding a bacterium at a small, but finite distance some 5–10 nm from a surface. At this stage there is no direct physical contact between the cell and the solid, and they can readily be removed from the surface by shear forces or the rotational movements of their flagella (Marshall et al., 1971). The extent of interaction can be predicted by colloid chemical theories such as the Derjaguin–Landau–Verwey–Overbeek (DLVO). It describes the magnitude and variation of the potential energy of interaction between a bacterium and a mineral surface as a function of separation distance (see Shaw, 1966 for details). The interaction arises because there is a tendency for surfaces to obtain a minimum Gibbs free energy by satisfying their charges, and one way this can be done is through bacterial adsorption (Absolom et al., 1983).

In its simplest form, if steric effects do not play a role, the total Gibbs free energy is obtained from the difference between the van der Waals attractive energies and the electrostatic repulsive energies (Fig. 3.22). The former are intermolecular forces that result from the formation of temporary dipoles created by fluctuating electron

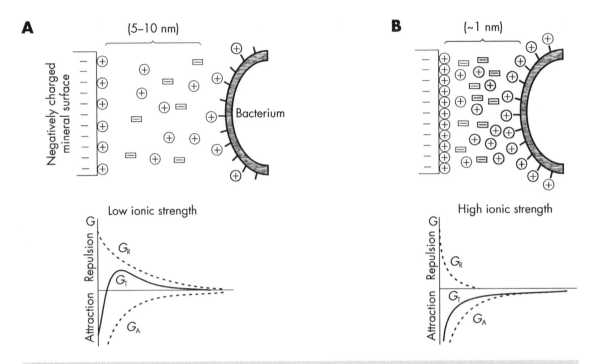

Figure 3.22 Electrochemical interactions between mineral and cell surfaces of like charge according to the DLVO theory. In low ionic strength solutions (A) minimum total free energy is obtained at a long distance (a few nanometers) from the mineral surface where attractive and repulsive forces are equivalent. This means that the cell does not closely approach the mineral surface because G_T constitutes a barrier to adhesion. In high ionic strength solutions (B) attractive forces dominate at all distances due to increasing ion availability and decreased electrostatic repulsion. Since G_T is negative, cation bridging thus brings the cell closer to the mineral surface. G_A, van der Waals attraction; G_R, electrostatic repulsion; G_T, total interaction. (Adapted from van Loosdrecht et al., 1990.)

distributions around atoms in each solid. The latter are due to the overlapping double layers surrounding the mineral surface and cell, which becomes important only when the cell and solid are sufficiently close to one another. According to DLVO theory, the thickness of the double layer is inversely proportional to the square of the ionic strength (van Loosdrecht et al., 1989). Consequently, as the ionic strength is increased, the double layers are compressed, and the surface potential is reduced sufficiently to allow the forces of attraction to exceed repulsion. In terms of total Gibbs free energy (G_T), at low ionic strength, G_T has a positive maximum at short separation distances (~1 nm) that represents an activation energy barrier for adhesion due to the large

electrostatic repulsion between the two solids. At longer distances (several nanometers), the attractive and repulsive forces become balanced. Then, as ionic strength is increased, G_T is lowered to a minimum value due to a reduction in repulsion resulting from increased ion availability and decreased electrostatic interactions (see Stumm and Morgan, 1996). The end result is that cells can approach the mineral surface to shorter separation distances. The type of cation is important here because divalent species (e.g., Mg^{2+}) are more effective at removing bacteria from solutions of identical ionic strength than those fluids containing Na^+ (Simoni et al., 2000).

Once the bacterium overcomes the repulsive forces and gets close to the surface, short-ranged forces, such as hydrogen bonding, then ultimately determine the strength of adhesion. The relationship between ionic strength and bacterial attachment is borne out in a number of experimental studies that have shown an increased number of bacteria on mineral surfaces in high ionic strength solutions, while reduction of ionic strength acts in an opposite manner (e.g., Jewett et al., 1995). The use of atomic force microscopy (AFM), coupled to models that predict the "sticking efficiency" of cells to surfaces, is further providing new insights into the nano-scale interactions that control attractive/repulsive interactions at cell–solid interfaces (e.g., Cail and Hochella, 2005).

(c) Initial adhesion – effects of substratum composition

As might be expected from the discussions on the acid–base properties of cells and metal sorption, the attachment of bacteria to mineral substrata is governed by the surface charge characteristics of both. The effect of pH was recently addressed by Yee et al. (2000), who compared the adsorption of B. subtilis onto the minerals quartz (SiO_2) and corundum (Al_2O_3). They showed that the quartz surface exhibited a negligible affinity for the bacterium because the mineral surface is negatively charged above approximately pH 2, as is the cell surface above pH 2.5. In this case, electrostatic repulsion between the mineral and the bacterium's surface is strong enough to inhibit adsorption. By contrast, the corundum surface is positively charged below pH 9, hence B. subtilis had a positive affinity towards it under normal growth conditions. At very high pH, both the bacterial and mineral surfaces are sufficiently anionic that they repel one another (Fig. 3.23).

However, electrostatic interactions do not fully account for the observed patterns of bacteria-mineral adsorption. In the experiments carried out by Yee et al. (2000) there was significant bacterial adsorption onto corundum even under low pH conditions, where protonated carboxylic acid groups on the cell surface (R-COOH)

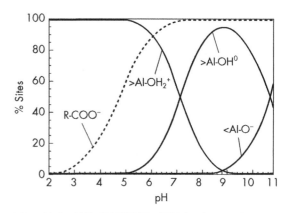

Figure 3.23 Speciation diagram of carboxyl groups associated with a *Bacillus subtilis* cell wall and the Al groups on a corundum surface. (Modified from Yee et al., 2000.)

adsorbed onto positively charged mineral surface sites ($>Al\text{-}OH_2^+$):

$$>Al\text{-}OH_2^+ + R\text{-}COOH \longleftrightarrow >Al\text{-}OH_2\text{-}RCOOH^+$$
$$(3.26)$$

A likely explanation for these particular observations is that the R-COOH sites behave hydrophobically because they are uncharged and not significantly hydrated by water molecules. With higher pH, the bacterium becomes progressively more anionic, increasing the amount of its hydration and hydrophilicity. In turn, this causes the cell to detach until the gradually increasing electrostatic attraction between the mineral and its surface becomes high enough that the bacterium once again becomes adherent. Therefore, attachment is promoted when either hydrophobic cell surfaces interact with uncharged interfaces, or when hydrophilic cell surfaces come into contact with oppositely charged interfaces.

This pattern has an important bearing on how some microorganisms behave in nature. For instance, benthic cyanobacteria tend to exhibit hydrophobic characteristics, while planktonic varieties are more hydrophilic in nature (Fattom and Shilo, 1984). Cyanobacterial hydrophobicity appears to have a genetic basis since some species that are hydrophilic can be made hydrophobic by producing extracellular layers (e.g., sheaths) that are electroneutral, and are thus less likely to interact with water molecules (Phoenix et al., 2002).

The surface charge characteristics of minerals, and hence their ability to attach microorganisms under normal growth conditions, can also be altered by the adsorption of inorganic or organic compounds. With quartz, attachment of cells to the mineral can be dramatically increased by the presence of iron hydroxide coatings. These coatings have an isoelectric point ~8.5 that establishes a positive surface charge at neutral pH so conducive to bacterial adsorption that subsequent exposure of the bacteria/Fe(III)-coated-quartz assemblage to sterile, dilute water does not promote desorption (Mills et al., 1994). By contrast, the addition of anionic phosphate compounds cause the mineral surfaces to become sufficiently negatively charged that they repel the cells (e.g., Sharma et al., 1985). Similarly, when organic matter adsorbs onto Fe-coated quartz, it causes a charge reversal on the mineral surface, making it anionic, and leading to diminished bacterial attachment (Scholl and Harvey, 1992). Meanwhile, the adsorption of organic matter to bacterial surfaces increases their overall negative surface charge and leads to increased attraction to the Fe-coated quartz surface, but a decreased attachment of cells in a quartz-only system (Johnson and Logan, 1996).

Since the initial adhesion of bacteria is usually reversible and relatively weak, surface shear forces and fluid turbulence cause desorption and elevated levels of cellular wash-out (van Loosdrecht et al., 1989). As a measure of protection, microorganisms preferentially colonize easily abraded mineral surfaces with some surface microtopology. Limestones are particularly amenable to surface colonization because of the ease with which the constituent calcite grains degrade. This likely explains why the number of epilithic bacteria on limestone are 10- to 100-fold greater than on harder rock types, such as granite, gabbro, rhyolite, basalt, and quartz sandstone (Ferris et al., 1989). The texture of a mineral also determines the amount of surface area to which cells can attach, with higher cell densities on grains with surface irregularities (DeFlaun and Mayer, 1983).

(d) Irreversible attachment

In the absence of strong shear forces, a bacterium held closely to the mineral surface is ideally positioned to make use of other means to secure a direct and more permanent attachment. This irreversible phase comes about mainly due to the production of EPS that physically bridges the gap between the cell and the solid (e.g., Fig. 3.24). Recent genetic studies have shown that the physical adhesion to surfaces triggers the expression of several genes controlling EPS synthesis (Davies and Geesey, 1995). Correspondingly, biochemical comparisons between benthic and planktonic cells of the same species shows that at least 30% of the membrane proteins are expressed

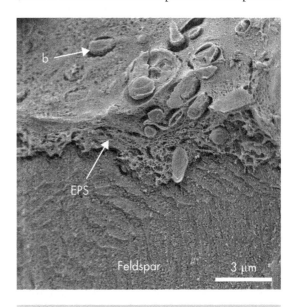

Figure 3.24 Using the technique of cryomicroscopy, the complicated structure of EPS can be visualized in three dimensions without suffering the effects of dehydration during sample preparation. This image shows a plagioclase feldspar grain, with several bacteria (b) residing within the EPS. (From Barker et al., 1997. Reproduced with permission from the Mineralogical Society of America.)

to different extents by cells in these two different modes of growth (Costerton et al., 1995). Other microorganisms employ specific appendages, such as pili, fibrils, or holdfasts, to anchor themselves onto the solid.

A direct correlation exists between the number of attached bacteria and the time allowed for attachment, with increased time leading to a higher number of bacterial collisions with the surface (Fletcher, 1977). Nonetheless, it only takes 10–30 minutes to form a continuous monolayer of cells under laboratory conditions (Characklis, 1973). Cell motility also enhances the likelihood that a bacterium will encounter a surface, with the kinetic energy being important in overcoming the electrical repulsive forces. Once attached, the growth of microorganisms on surfaces is an autocatalytic process, whereby initial colonization increases surface irregularity and promotes biofilm formation (Little et al., 1997).

(e) Effects of cell growth rates

The relationship between cell growth rates and attachment is a complex phenomenon that appears to be species-dependent. For example, a number of bacteria show diminished adhesiveness during exponential phases (e.g., Gilbert et al., 1991), whereas others show the opposite effect (e.g., Fletcher, 1977). This discrepancy is directly related to the specific changes in cell surface charge between different species during optimal growth conditions.

In cyanobacteria, the formation of hydrophilic hormogonia are important for dispersing the species to new environments, yet, as the hormogonia contact new surfaces, they develop back into mature trichomes and concomitantly show increased hydrophobicity (Fattom and Shilo, 1984). This agrees with the common findings in the laboratory, that during continuous culture at high dilution rates, many such microorganisms form flocks or adhere to surfaces in the culture vessel.

During starvation, bacteria show increased levels of adhesion. The "dwarf cells" that are formed produce EPS that enhances their adhesive properties, allowing them to take advantage of the organic and inorganic compounds that accumulate at solid–liquid interfaces (e.g., Dawson et al., 1981). Such tactics may be particularly important in oligotrophic waters where bacteria are exposed to conditions of extreme nutrient limitation. The reduction in size is temporary and can later be reversed on provision of adequate nutrients. Other microorganisms respond to adverse conditions by producing spores. Some spores, such as those of *Bacillus cereus*, have a neutral surface charge that makes them strongly hydrophobic compared with that of the vegetative cell (Rönner et al., 1990).

3.6.2 Microbial transport through porous media

Although bacteria readily attach onto solids, a fraction of them remain mobile in subsurface pore waters. Their ability to move freely through geological material is dependent upon a number of physical parameters, including the system hydrodynamics, permeability, and the magnitude of the clay fraction (Lawrence and Hendry, 1996).

As might be expected, flow rates are an important factor in bacterial dispersion. High flow rates, or the more rapidly the sediment is flushed with groundwater, lead to higher levels of bacterial elution (e.g., Trevors et al., 1990). Under such conditions, bacterial transport rates of over 200 m day^{-1} have been reported (Keswick et al., 1982). Even under no flow conditions, some motile bacteria can move through packed sand cores at rates greater than 0.1 m day^{-1} (Reynolds et al., 1989). Field measurements of bacterial motilities even indicate that bacteria can be transported through porous aquifers faster than chemical tracers because the preferential exclusion of bacteria from smaller, more tortuous pores between sediment particles results in a more direct average path of travel for the unattenuated bacteria (Harvey et al., 1989). In the water-unsaturated (vadose) zone, bacterial movement is instead influenced by gas saturation.

On the one hand, bacteria preferentially accumulate at the gas–water interface, and thus their movement can be impeded (Wan et al., 1994). Conversely, increased rates of cell movement can arise when localized pressure gradients, generated through processes such as fermentation, act to push bacteria through the pore network.

The pore size distribution in soils/sediment or the fracture pattern in rock are also key features governing the spatial distribution of subsurface bacteria. Straining or filtration occurs in unconsolidated material when bacteria are too large to pass through the pore throat aperture, resulting in clogging. Generally if the diameter of the bacterium, or the bacterial aggregates, is greater than the sizes of 5% of the particles in the medium, straining is considered significant (Sharma and McInerney, 1994). In a number of studies, bacteria smaller than 1.0 μm in diameter showed the greatest potential for transport through porous media (e.g., Gannon et al., 1991). However, even if the cells are sufficiently small, aggregation of cells in high ionic strength solutions can cause them to plug available pores (e.g., Jang et al., 1983). Cell shape, as quantified by the ratio of cell width to cell length, also affects the transport of bacterial cells, with spheres moving through porous media more effectively than rod-shaped cells (Weiss et al., 1995). Other experiments have clearly demonstrated how these two properties, size and shape, are related. For example, small spheres (<1 μm diameter) passed easily through coarse-sized sand (1.0 mm), while less than 1% of the larger cells (2 μm diameter rods) passed through the fine-grained sand (<0.3 mm) (Fontes et al., 1991). This effect was largely the result of a significant decrease in the hydraulic conductivity through the fine-grained columns, suggesting that the degree of macropore flow influences the extent of microbial transport (e.g., Fig. 3.25). In rock, conductive fractures have been shown to constitute preferential paths for subsurface bacteria in the water-saturated (phreatic) zone, leading to migrations of several kilometers in distance, whereas small-size fissures dramatically reduced bacterial mobility (e.g., Malard et al., 1994).

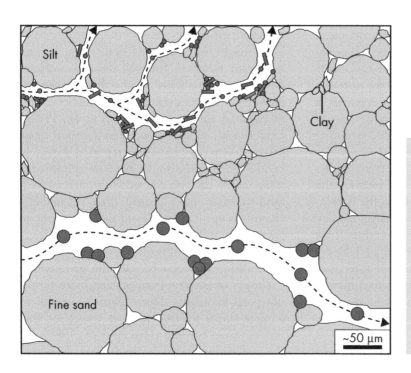

Figure 3.25 Hypothetical drawing of microbial transport through an aquifer. Two flow paths are shown. At the top, small cocci and rods move easily through the silts but are strained by clay particles. The bottom flow path represents larger cocci that are forced to take a route through the larger pore spaces between grains of sand. Note relationship between bacteria and sediment grains are not to scale.

In addition to influencing soil/sediment permeability, clays affect contaminant mobility because they comprise a significant fraction of the solid phase. They also readily form micro-aggregates with organic matter, including microorganisms. Despite the high chemical reactivity of the individual components, aggregation into clay–bacteria composites reduces overall metal binding ability because clays mask or neutralize adsorbing ligands on the cell (Walker et al., 1989). Nevertheless, once metals are bound to clay–bacteria composites, they are difficult to remove. Even strong leaching chemicals, such as nitric acid or EDTA, can remobilize only a fraction of the bound metals. What is particularly noteworthy is that some metals are more difficult to remove from the bacteria–clay composites than from their individual counterparts (Flemming et al., 1990).

3.7 Industrial applications based on microbial surface reactivity

3.7.1 Bioremediation

The increasing societal demands for metals has led to a widescale release of metal pollutants into the environment. Traditional technologies, such as chemical precipitation and sludge separation, oxidation-reduction, evaporation, electrochemical treatment, sorptive resins, and organic solvents have typically been employed in the clean-up. While many of the cheaper processes have become inadequate with progressively stringent regulatory effluent limits, the more effective methods are prohibitively expensive. The need for more affordable technologies has led to the evaluation and design of methods by which the metal binding properties of microbial biomass could be utilized (Eccles, 1995).

Bioremediation is the application of living or dead organisms to degrade or transform hazardous inorganic and organic contaminants. There are several ways in which microorganisms can be utilized in bioremediation strategies (Fig. 3.26). In terms of metal binding, this includes biosorption and bioaccumulation. Biotransformations – the reduction of high valence metals to lower valence insoluble species, or the oxidation that leads to the opposite effect – was discussed in Chapter 2, while biomineralization – the formation of insoluble mineral phases – will be the subject of the next chapter.

Biosorption, defined here as the process whereby microbial biomass acts as a surface upon which metals are passively sorbed, has a major advantage over similar chemical technologies in that large quantities of inexpensive and easily regenerable fungal and bacterial biomass are available from fermentation industry waste, sewage sludge, or the many different types of marine macroalgae that make up seaweed (see Volesky and Holan, 1995; Gadd, 2002 for reviews). Unlike conventional methods, biosorption involves using a nonhazardous material whose application is broad-ranging; it can bind a suite of metals or it can be employed based on the selectivity for binding a specific metal of concern. Moreover, it can be used under a wide range of environmental conditions. Both dead and living biomass can bind metals, but the former is generally preferred because it avoids the problems with toxicity, it can be used under extreme geochemical conditions, and is cheaper to use (see below). Biosorption is most effective as a polishing step where waster-water with low to medium metal concentrations (up to $100 \, \text{mg} \, \text{L}^{-1}$) is purified to drinking-water standard. Treatment of wasterwater with high metal concentrations can lead to rapid exhaustion of the biosorbent material and thus may require larger than desired amounts of biomass. Therefore, pre-treatment of such effluents using other techniques, such as chemical precipitation (which is currently used for 90% of heavy metal removal from industrial wasterwater) or electrolytic recovery, may be more economical. The metal-laden biosorbent is then dealt with in one of two ways: it may be incinerated, with the ash disposed of in landfills, or alternatively, the

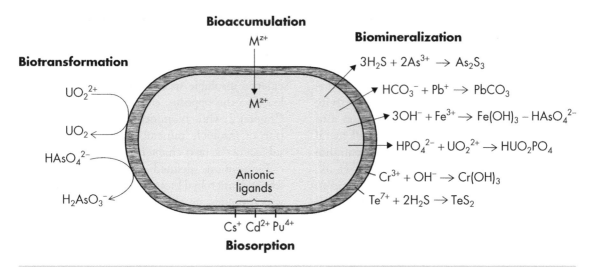

Figure 3.26 Some of the many ways in which natural microbial activity can be used in bioremediation of toxic metals and radionuclides. (Adapted from Lloyd and Macaskie, 2000.)

biomass is regenerated by desorbing the metals from the biomass, yielding a reusable biosorbent and a highly concentrated metal solution (Schiewer and Volesky, 2000).

Bioaccumulation describes absorption of metals by metabolically active cells. Often toxic metals enter the cell as chemical "surrogates," using the transport systems developed by the cell for other elements. Such systems are self-sustaining due to biomass replenishment, and they have the ability to not only absorb high levels of metals, but their excreted metabolic wastes (e.g., H_2S) can also contribute to metal removal. Unfortunately, living biomass present a number of difficulties. First, the final sludge for disposal is of high organic content adding to the cost of transportation to the site of reprocessing or final burial. Second, special care (and associated high costs) needs to be taken to ensure that the growing microbial population is kept uncontaminated by other species and maintained through adequate supply of nutrients, ideal temperatures, and pH buffering. Third, metals accumulated intracellularly are not as easily recovered as those adsorbed to the surface during biosorption, especially if the

metal is bound to metallothioneins or compartmentalized into vacuoles, and as such, require that the cells be physically destroyed (Macaskie et al., 1996).

One area that has received significant attention is the biological removal of radionuclides from low-level nuclear waste processing sites (e.g., Cs, Te, U, Pu, Np). The biogeochemical behavior of these pollutants has become increasingly important due to the issues of their disposal, their long-term containment, and ultimately their movement through the environment. Using uranium as just one example, many studies have shown that several genera of filamentous fungi (e.g., *Rhizopus*, *Aspergillus*, *Penicillium*), yeasts (e.g., *Saccharomyces*), marine algae (*Sargassum*, *Chlorella*), and bacteria (e.g., *Bacillus*, *Pseudomonas*, *Streptomyces*) are very effective scavengers of the radionuclide (see Macaskie and Lloyd, 2002; Kalin et al., 2005 for reviews). Maximum accumulation, with a steep sorption edge, occurs under acidic conditions, between pH 4 and 6, through monodentate adsorption of a cationic uranyl ion (UO_2^{2+}) onto a deprotonated carboxyl group to form the surface complex, $R\text{-}COO\text{-}UO_2^+$ (Figure 3.27). At circumneutral pH, U sorption

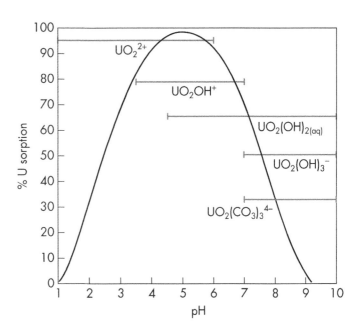

Figure 3.27 Plot showing U(VI) sorption onto *Shewanella putrefaciens* as a function of pH. Under very acidic conditions, uranium is poorly sorbed. It then reaches a maximum at pH 4–5 and remains optimal until pH 6. At higher pH, the extent of sorption diminishes as uranium speciation changes from uranyl cations to anionic hydroxide and carbonate species. (Modified from Haas et al., 2001.)

is best explained by a neutrally charged surface complex on a phosphate site, yielding $PO_4H\text{-}UO_2(OH)$. At higher pH, competition of cell-bound U(VI) with aqueous hydroxyl and carbonate anions reduces the extent of adsorption, resulting in a release of UO_2^{2+} back into solution (Haas et al., 2001). Dead cells often absorb more uranium than their live counterparts, presumably due to an increase in accessible metal binding sites (Volesky and May-Phillips, 1995).

Uranium uptake is commonly greater than $200\ mg\ U\ g^{-1}$ biomass dry weight, but has been reported to reach values in excess of 50% (e.g., Yang and Volesky, 1999). The efficiency of uranium uptake is significant because a total uranium loading capacity of greater than 15% of biomass (dry weight) has been defined as an economic threshold for practical application when compared to traditional technologies (Macaskie, 1991). Given that the amount of uranium removed from solution through biosorption is ultimately governed by the type of available ligands, new techniques in molecular biology are being developed to enhance the sorption capacity of microbial biomass. For instance,

studies have documented how the biochemical composition of *E. coli* cell walls could be modified by inserting sulfur-containing amino acids from other organisms to increase the adsorption of metals that react favorably with the S-ligands (e.g., Sousa et al., 1998). Other studies are instead focusing on coupling the biosorptive abilities of microbial surfaces to the biomineralization of uranyl phosphate, HUO_2PO_4. This has yielded some promising results, with uranium uptake values by *Citrobacter* sp. as much as 900% of the cellular dry weight (Macaskie et al., 1992)! The process involves UO_2^{2+} adsorbing to anionic phosphate ligands in the LPS and then reacting with phosphate excreted by the cell as a result of its enzymatic overproduction of phosphatase on the cell's outer membrane (Macaskie et al., 2000).

One technique of metal sorption by microbial biomass employs freely suspended particles. This permits a high surface area of binding sites, but it suffers from a number of disadvantages. These include small and heterogeneous particle size, low mechanical strength, susceptibility to microbial degradation, and difficult biomass/effluent separation. In the case of the latter, flotation by bubble-generation techniques has been examined as a possible separation process (e.g., Matis et al., 1996). More often, however, the biomass is immobilized in the form of biofilms or pellets. Nonliving microorganisms can also be employed in select wastewater metal removal-treatment systems where extreme aqueous conditions (i.e., low pH, high metal toxicity) are present. Immobilization occurs through supports such as agar, cellulose, silica, alginate, polyacrylamide gels, collagen, and metal precipitates (Brierley et al., 1989).

Based on the various properties highlighted above, several types of reactors have been developed for use in pilot-scale biosorption projects (Volesky, 1990). Briefly, packed-bed columns have the wastewater flowing downwards through a column filled with biomass. This type of reactor offers the advantage of very high effluent quality because the stream exiting the column is in contact with fresh sorbent material. Unfortunately, clogging occurs when significant concentrations of suspended solids are involved. Fluidized-bed reactors have the wastewater passing upwards through the reactor. These reactors avoid the problems of clogging but require more effort to ensure that the flow rate is balanced with the biomass size and density. Stirred tanks contain biomass dispersed throughout the reactor. They provide more contact between the biomass and wastewaters, but more biomass is generally used to achieve the same quality of effluent as the other techniques. A number of novel strategies for improving the biosorption processes are being currently developed, including: the use of pulsed electrical fields to enhance sorption capacity; engineering a spiral bioreactor that minimizes space; chemical or heat pretreatment of the biomass; and growing select, toxic-resistant microorganisms in the form of microbial mats (Lovley and Coates, 1997).

3.7.2 Biorecovery

In contrast to bioremediation, biorecovery is the process whereby microbial biomass is employed to extract either toxic metals/radionuclides from the bioreactors or valuable metals from solutions where their concentrations are below standard recovery levels. In terms of the latter, one such example is gold recovery, because the traditional methods of zinc dust precipitation, carbon adsorption, solvent extraction, or ion-exchange resins are either of low selectivity or extremely expensive. A number of studies have shown that algal cells effectively accumulate gold into their biomass, up to 90% of the gold from solution (e.g., Hosea et al., 1986). During the process, Au(III) is adsorbed onto the cell surfaces, where it is then reduced to Au(1) or Au(0) by some unknown mechanism. Some of the gold even makes its way into the cytoplasm where it forms fine-grained, intracellular colloids (e.g., Southam and Beveridge, 1994). More recently, it has been demonstrated that various mesophilic and hyperthermophilic Fe(III)-reducing bacteria also have the means to precipitate gold epicellularly by reducing Au(III) to Au(0) with H_2 as the electron donor (Fig. 3.28). This process appears to be enzymatically catalyzed, perhaps with specific hydrogenases employed to directly reduce Au(III), although attempts to grow Fe(III)-reducing bacteria with Au(III) as the sole terminal electron acceptor have so far proven unsuccessful (Kashefi et al., 2001).

For biomass to be employed in gold biorecovery, the microbial sorbent would need to feature a high maximum loading curve plateau (in mg Au g^{-1} of biomass), as well as a steep initial portion of the isotherm indicating a high sorption capacity at low equilibrium concentrations. Based on a number of dead marine algae tested, Kuyucak and Volesky (1989a) documented that *Sargassum natans* not only exhibited a desirable steep biosorption isotherm, but it also had a maximum uptake comparable to commercial ion-exchange resins and activated carbon (Fig. 3.29). Batch kinetic experiments further indicate that the time required for full biomass saturation with gold depends on the initial aqueous gold concentration, with dilute solutions requiring only an hour for equilibrium to be achieved, while concentrated solutions can

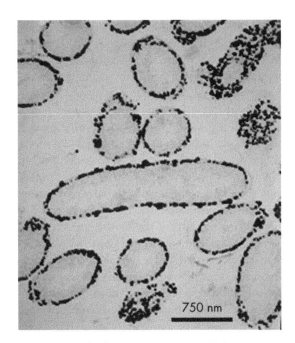

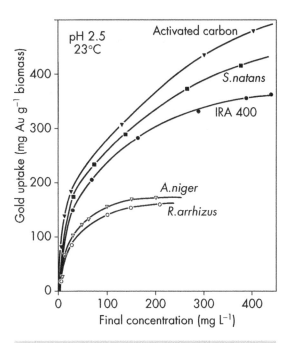

Figure 3.28 TEM image of epicellular elemental gold precipitation associated with *Shewanella algae*. Notice how the gold nicely outlines the cell wall of the intact cells. (Courtesy of Kazem Kashefi and Derek Lovley.)

Figure 3.29 Gold biosorption isotherms for several different types of microbial biomass and industrially-used sorbent materials. Starting gold chloride solutions contained from 10 to 1000 mg L^{-1} Au. The graph shows a steep sorption edge for the brown alga *Sargassum natans*, comparable to the more expensive ion exchange resins (IRA 400) and activated carbon. (Modified from Kuyucak and Volesky, 1989a.)

take several hours. Maximum sorption occurs at pH 2.5, as might be expected from the electrostatic interaction of protonated ligands and the anionic, dissolved gold species used (AuCl$_4^-$). More recently, gold accumulation has been documented for a wide variety of bacteria, fungi, and yeasts (Nakajima, 2003).

In order to use biomass for biorecovery, the elution of the metal sequestered has to be reasonably easy to achieve. The eluting solution should also contain the metal in high concentrations and the regenerated biosorbent must be capable of another uptake cycle. Many types of eluants can be used to desorb metals. Some desorbing agents, such as acids or metal salts, provide cations that outcompete the bound metals for the cell's reactive ligands. Another method is to employ strong organic ligands (e.g., EDTA) that can strip the metals from the biomass, thereby "freeing-up"

the biosorbent for additional metal treatment. In the gold example, Kuyucak and Volesky (1989b) showed that desorption of gold from S. *natans* was achievable through the use of a mixture of ferric ammonium sulfate (to oxidize Au^0 to Au$^+$), and thiourea, which forms soluble complexes with Au$^+$. The elution efficiency was more than 98% effective and the desorption capacity of the eluted biomass remained the same for additional gold biosorption experiments. Darnall et al. (1986) also developed an elution scheme for selective gold recovery from *Chlorella vulgaris*. Most algal-bound metals could be selectively desorbed by lowering the pH to 2. However, to desorb the remaining cell-bound Au(III), the strong ligand mercaptoethanol had to be used.

3.8 Summary

Microorganisms have a variety of surface enveloping layers, including the cell wall and the extracellular structures residing above them, that are directly exposed to diffusible components in the external aqueous environment. At typical growth pH, these layers are studded with organic functional groups that are naturally anionic, wettable and thus highly reactive towards metal cations. Carboxyl and phosphate groups are the most important sites for metal adsorption, and as chemical equilibrium models show, the total number of reactive ligands are a direct function of the architecture and composition of the macromolecules comprising the outermost surfaces. Since considerable ultrastructural variation exists between different bacteria, and can even arise within single species as growth conditions change, the overall metal sorption capacity of microbial biomass can show fundamental variability. In dilute solutions, those metals required for metabolic activities and structural organization are preferentially adsorbed from a range of competing cations, a property that results from the cell possessing specific ligands that favor one metal over another. At other times, metal binding to a cell's surface is largely a nonspecific and reversible electrostatic phenomenon reliant upon thermodynamics, analogous to inorganic systems. In this regard, it is not uncommon for copious amounts of metals to accumulate onto both living and dead microorganisms, at concentrations far exceeding that predicted based on ligand availability, simply as a consequence of them being in a concentrated solution. Living microorganisms can even compensate for this by overproducing protective EPS material that sequesters toxic metals, thereby preventing them from disrupting internal cell functions. Given their ubiquity at the Earth's surface, their rapid rates of metabolism and growth, and their high chemical reactivity, it is clear that microorganisms must play a fundamental role in metal cycling. Indeed, a significant mass of metals in the aqueous environment are intimately associated with cell biomass or tied up as refractory organo-metallic complexes. Moreover, the surface charge properties of microorganisms further facilitate their attachment onto submerged surfaces. This ability has significant implications for contaminant transport because metals sorbed onto attached bacteria show limited dispersion through the environment. Importantly, these same properties have allowed microorganisms to be manipulated for a number of industrial processes, including bioremediation and biorecovery.

4

Biomineralization

Microorganisms are remarkably adept at forming mineral phases. This process, termed biomineralization, can occur in two different ways. The first involves mineral precipitation in the open environment, without any apparent control by the cell over the mineral product. This process was defined by Lowenstam (1981) as "biologically induced biomineralization", with minerals forming simply as a byproduct of the cell's metabolic activity or through its interactions with the surrounding aqueous environment. Simple perturbations, such as the release of metabolic wastes (e.g., O_2, OH^-, HCO_3^-, Fe^{2+}, NH_4^+, H_2S), enzymatic mediated changes in redox state (e.g., oxidation of Fe(II) or Mn(II)), or the development of a charged cell surface can all induce the nucleation of amorphous to poorly crystalline minerals with morphologies and chemical compositions similar to those produced by precipitation from sterile solutions. This is not too surprising considering that biomineralization is governed by the same equilibrium principles that control abiological mineralization processes. By contrast, "biologically controlled biomineralization" is completely regulated, allowing the organism to precipitate minerals that serve some physiological purpose. This process is specifically designed to form minerals through the development of intracellular (within the cytoplasm) or epicellular (on the cell wall) organic matrices, into which specific ions of choice are actively introduced and their concentrations controlled such that mineral saturation states are appropriately achieved. Because the mineralization site is isolated from outside the cell by a

barrier through which ions cannot freely diffuse, minerals form despite external conditions being thermodynamically unfavorable. In this chapter we will review the different types of biominerals formed and examine how the process of biomineralization has affected the geochemical cycling of mineral-forming elements throughout geological time.

4.1 Biologically induced mineralization

4.1.1 Mineral nucleation and growth

The thermodynamic principles underpinning biomineralization, irrespective of whether they are induced or controlled, are the same as those involved in abiological mineral formation. In all cases, before any solid can form a certain amount of energy has to be invested. This energy is required for a number of reasons, including: (i) offsetting the potential repulsive interactions between double layers separating the solid and solutes; (ii) eliminating the hydration shells surrounding dissolved ions, so that a chemical bond can form between them and the surface ligands; (iii) removing organic ligands that have chelated metal cations; and (iv) to subsequently form a new interface between the nascent nucleus and both the aqueous solution and the underlying substratum upon which it is formed. The amount of energy required to do this can be viewed as an activation energy barrier,

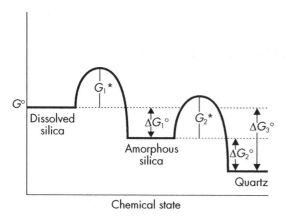

Figure 4.1 Relation between the standard free energy ($G°$) of dissolved solutes (e.g., silica) and the amorphous and crystalline solid phases they form upon supersaturation. The larger the decrease in $\Delta G°$ upon precipitation of the solid, the more stable it will be (e.g., $\Delta G_3°$ versus $\Delta G_1°$). Yet, for the solid to form, a certain amount of activation energy ($G*$) has to be invested. Therefore, the lower the activation energy barrier, the faster the reaction proceeds. This helps explain why amorphous silica nucleates first despite being less stable than quartz. (Modified from de Vrind-de Jong and de Vrind, 1997.)

and for mineral formation to proceed, the barrier must be overcome by the energy released as a consequence of bond formation in the solid phase. The standard free energy ($G°$) of a solid is lower than that of its constituents in solution, so if the activation energy barrier can be overcome, the reaction proceeds spontaneously towards mineral nuclei forming (Fig. 4.1). If instead the activation energy barrier is prohibitively high, metastable solutions will persist until either the barrier is reduced or the concentration of ions are diminished, thereby reducing the thermodynamic driving force towards precipitation.

The first step in mineral formation is nucleation. This process involves the spontaneous growth of a number of critical nuclei of a certain size that are resistant to rapid dissolution. For this to occur, the concentration of ions or atoms in solution

must exceed the solubility product of the solid mineral phase. In other words, a certain degree of supersaturation has to be reached (see Stumm and Morgan, 1996 for details). This can be described in thermodynamic terms, where the free energy of nucleation (ΔG_n) is constrained by the free energy of the bulk solution (ΔG_{bulk}) and the free energy of the developing mineral nucleus (ΔG_{min}):

$$\Delta G_n = \Delta G_{bulk} + \Delta G_{min} \qquad (4.1)$$

The bulk solution free-energy term is, in turn, a secondary function of the degree to which a solution is oversaturated ($\ln\Omega$) (reaction (4.2)), where Ω is a value based on the ion activity product (IAP) of the solution divided by the solubility product of the corresponding mineral phase (K_{sp}): recall activity is the "effective concentration" of the chemical species, which is less than the actual concentration due to ion complexation in solution – for dilute solutions, activity and concentration are essentially equivalent terms. The other terms represent Boltzmann's constant (k), temperature (T, in °K), and the number of ions or molecules in the nucleus (n):

$$\Delta G_{bulk} = -nkT\ln\Omega \qquad (4.2)$$

ΔG_{min} is a product of the surface area of the nucleus (A) and interfacial free energy (also known as surface tension) of the solid phase (γ) (equation 4.3):

$$\Delta G_{min} = \gamma A \qquad (4.3)$$

Accordingly, the overall free energy of nucleation can be written as:

$$\Delta G_n = -nkT\ln\Omega + \gamma A \qquad (4.4)$$

When considering pure solutions in which only the mineral constituents are present, nucleation is said to be homogeneous. In homogeneous reactions, critical nuclei are formed simply by random collisions of ions in a supersaturated solution. Conversely, heterogeneous nucleation involves the development of critical nuclei on the surfaces of foreign solids. The surface can be viewed as a template of ideally spaced ligands that bind and stabilize the nascent nuclei. It essentially acts as a catalyst that reduces the interfacial contributions to the activation energy barrier and thereby increases the nucleation rate.

After critical nuclei are formed, the continued adsorption of ions to them is accompanied by a decrease in free energy. This process is known as mineral growth (if the ions are the same as those of the substratum) or surface precipitation (if the ions are different), and it goes on spontaneously until the decreasing supply of ions becomes prohibitive. The initial mineral phase formed is usually amorphous, characterized by its high degree of hydration and solubility, and its lack of intrinsic structure, compared to more stable, crystalline phases. This pattern arises because even though the surface area of the nucleus increases during hydration, the interfacial free energy between the hydrated surface of an amorphous nucleus and a dissolved ion reduces more rapidly, thereby resulting in lower ΔG_{min}, and hence faster nucleation rates than are possible for crystalline analogs (Nielson and Söhnel, 1971). This means that amorphous phases, such as amorphous silica ($\gamma = 46$ mJ m^{-2}), are kinetically favored if the solution composition exceeds their solubility. By comparison, its crystalline equivalent, quartz ($\gamma = 350$ mJ m^{-2}), has a higher interfacial free energy, is relatively insoluble, and it nucleates slowly at ambient temperatures. Often the transition between amorphous and crystalline phases involves the precipitation of metastable phases.

The nucleation rate also has an important bearing on the size of the critical nuclei formed. If the nucleation rate for an amorphous mineral phase is plotted against the saturation state, a typical curve is obtained. It is characterized by a critical supersaturation value below which the nucleation rate is extremely slow and above which the nucleation rate increases very rapidly (Steefel and Van Cappellan, 1990). What this implies is that at ion activities above the critical

value, new surface area is created mainly by the nucleation of many small grains characterized by high surface area to mass ratios, a regime referred to as nucleation-controlled. At activities below the critical value, surface area increases by the accretion of additional ions to existing grains (i.e., crystal-growth controlled). Now, if the composition of a fluid was to start in the nucleation-controlled regime, the generation of new surfaces by nucleation would rapidly increase, causing the level of supersaturation to collapse to at least the critical value. This means that in nature, a degree of supersaturation above the critical value will not be maintained for lengthy periods of time. Let us turn to silica precipitation as an example. If a concentrated silica solution (10^{-2} mol L^{-1}) was emitted from a hot spring vent, it would be supersaturated with regards to all silica phases, but because amorphous silica has the lower interfacial free energy it nucleates first despite quartz being the more stable phase with lower solubility. As amorphous silica nucleates rapidly it drives the dissolved silica activity down to its critical value, which happens to be below that required to nucleate quartz, i.e., to the left side of quartz's critical value (Fig. 4.2).

Crystalline minerals that would otherwise be difficult or impossible to directly nucleate at low temperatures can circumvent the activation energy barriers by making use of the amorphous precursors as templates for their own growth. Once it begins to grow, the crystal increases its own surface area and, in doing so, controls the proximal free ion activity, driving it down towards its solubility product. When this happens, the saturation state of the solution moves below the solubility of the precursor, causing the latter to dissolve (Steefel and Van Cappellen, 1990).

Although thermodynamics can predict the transformation sequence based on energetics, it cannot determine the kinetics. Sometimes the reactions are relatively quick, such as the formation of magnetite on ferric hydroxide in sediment (see section 4.1.3). At other times, the reaction rates are immeasurably slow over

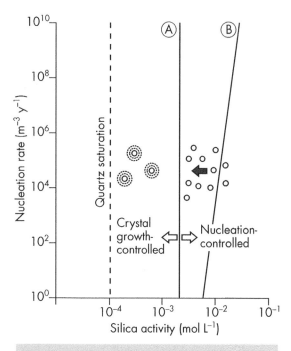

Figure 4.2 Rate of heterogeneous nucleation of quartz and amorphous silica at 25°C, using interfacial free energies of 350 and 46 mJ m^{-2}, respectively. The solid lines reflect the critical silica activities for nucleation of amorphous silica (A) and quartz (B). The graph shows that in a silica supersaturated solution (10^{-2} mol L^{-1}) amorphous silica will nucleate (open circles) in preference to quartz because of its lower interfacial free energy. Accordingly, the SiO$_2$ activity decreases to the saturation state of amorphous silica, which is below the critical value for quartz. This prevents quartz from nucleating in the short term. (Adapted from Steefel and Van Cappellen, 1990.)

hundreds of meters until it too transforms into quartz (see section 6.2.7(c)).

Another important factor influencing the dissolution of the precursor phase is its large surface area and high solubility compared to the newly generated secondary crystals. This feature forms the basis for Ostwald ripening, a process involving the spontaneous redistribution of mass from the more numerous precursor grains in the system to fewer stable crystals that are typically larger, and more evenly distributed, in size (see Baldan, 2002 for details). It can also describe the distribution pattern of dissolved ions into colloids (e.g., Iler, 1979). So, as the smaller precursors act as "seeds" for the larger, secondary phases, the latter grow and the area around them becomes depleted of precursors (e.g., Fig. 4.3). The reason Ostwald ripening takes place is that the larger secondary phases are more energetically

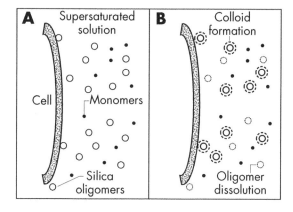

Figure 4.3 An example of Ostwald ripening of silica species at hot springs. (A) As the silica supersaturated solution is discharged, monomeric silica rapidly polymerizes into silica oligomers of various size (dimmers, trimers, etc.). Some of those oligomers then bind to solid surfaces, including microorganisms growing around the vent. (B) Over the course of hours to days, some of those oligomers increase in size to colloidal dimensions by the accretion of monomers and oligomers. As the colloids grow, the monomeric fraction decreases. As a result, some oligomers depolymerize, leading to a bimodal distribution of monomer and colloids.

geological time and the amorphous or metastable phases persist in sediments supersaturated with the thermodynamically most stable minerals. They can show little discernible alteration for tens of millions of years until pressure-temperature changes associated with burial cause the reation sequence to advance to the next stage (Morse and Casey, 1988). For example, amorphous silica shells deposited onto the seafloor slowly dissolve at shallow depths and re-precipitate as cristobalite, which remains stable to depths of favored than the smaller precursors. This might seem a bit contradictory to above, but consider that even though the formation of small grains is kinetically favored in supersaturated solutions, they have large surface area to volume ratios and molecules exposed on the surface are energetically less stable and more reactive than those sheltered at the interior. By contrast, larger crystals (or colloids), with their greater volume to surface area ratios, represent a lower energy state, and are more thermodynamically favored when activation energy barriers of both are overcome.

As will become evident in this chapter, microorganisms contribute significantly to the development of extremely fine grained (often <1 μm in diameter) mineral precipitates. The vast majority are formed passively, and in this regard microorganisms influence mineralization in two significant ways:

1 *Reactive surfaces* – The cell walls and extracellular layers contain an abundance of ionized surface ligands where sorption reactions take place. These sites subsequently lower the interfacial energies for heterogeneous nucleation while simultaneously decreasing the surface area of the nucleus that is in contact with the bulk solution. Furthermore, the spacing of the ligands affects which cations will be bound (recall section 3.3.3), and thus, some microorganisms even have the means to control the structure and orientation of the incipient nucleus. For that reason, microorganisms have been likened to mineralizing templates because the composition and structure of their functional groups are ideal for the passive formation of a number of different types of mineral nuclei. It should be stressed, however, that microorganisms only serve to enhance the precipitation kinetics in supersaturated solutions; they neither increase the extent of precipitation nor do they facilitate precipitation in undersaturated solutions (e.g., Fowle and Fein, 2001).

2 *Metabolism* – Microbial activity can affect mineral saturation states immediately outside the cell through the excretion of metabolites. For example, denitrifying or photosynthetic bacteria promote an increase in solution pH that is supportive of carbonate precipitation; sulfate-reducing bacteria induce the formation of metal sulfides by generating H_2S/HS^-; while the release of Fe^{2+} by Fe(III)-reducing bacteria

may lead to secondary magnetite formation. Other microorganisms enzymatically oxidize reduced metals (e.g., Fe(II)), and facilitate the precipitation of metal oxyhydroxides outside the cell wall. Certainly, the microenvironment surrounding each microbial cell can be quite different from that of the bulk aqueous environment, and as a result the cell surfaces can lead to the development of mineral phases of a type that might not be predicted based solely on our knowledge of the geochemistry of the bulk fluid (Little et al., 1997).

In nature there exists a wide variety of microbial mineral precipitates. This leads to the obvious question – why so much diversity? Fortunately, the answer is simple. Those biominerals formed passively are dependent upon the chemical composition of the fluids in which they are growing, such that a particular microorganism will form a mineral phase from the solutes immediately available to it. Conversely, the same microorganism in a different environment would likely form a different mineral phase altogether. The variations can be as subtle as a change in redox state. For example, it is well known that the anionic ligands comprising a cell's surface can form covalent bonds with dissolved Fe(III) species, which, in turn, can lead to charge reversal at the cell surface. Invariably, this positive charge will attract anionic counter-ions from solution. So, in the sediment, iron staining of a bacterium may lead to the precipitation of an iron sulfate precipitate in the oxic zone, whereas another bacterium may instead form an iron sulfide at depth, where conditions are reducing. In the following section, a number of passively formed biogenic minerals will be discussed, and what will become apparent is that microorganisms interact with the solutes in intimate contact with their surface layers, and in doing so they essentially function as reactive surfaces for mineral precipitation.

4.1.2 Iron hydroxides

The most geologically widespread biomineral is ferric hydroxide (also loosely referred to as ferrihydrite). Its chemical composition is $5Fe_2O_3 \cdot 9H_2O$, but for chemical simplicity it is usually described as $Fe(OH)_3$. Ferric hydroxide has been shown to form in association with microbial biomass in any environment where Fe(II)-bearing waters come into contact with O_2. This includes springs, sediment/soil pore wasters, aquifers, hydrothermal systems, mine wates, and water distribution systems, to name just a few (see Konhauser, 1998 for review). Fossil structures that resemble modern iron-depositing bacteria have also been found in laminated black cherts and Precambrian banded iron formations (BIFs) (Robbins et al., 1987). As will be discussed in section 7.3.2, there is even some circumstantial evidence suggesting that microbial activity was directly involved in the initial deposition of Fe-rich sediment, which later consolidated to make BIF.

When Fe-encrusted cells are viewed in detail under the transmission electron microscope (TEM), it is apparent that mineralization occurs through a series of stages, often beginning with Fe-adsorption to extracellular polymers (EPS) or wall material (recall Fig. 3.12), followed by the nucleation of small (<100 nm in diameter) ferric hydroxide grains, and with sufficient time, the complete encrustation of the cell. These steps have also been demonstrated experimentally and described by Langmuir-type isotherms showing the continuum between metal adsorption and mineral precipitation (e.g., Warren and Ferris, 1998). Not only do bacteria serve as templates for iron deposition, but their organic remains frequently become incorporated into the mineral precipitates during crystal growth such that the sediment ends up with iron–organic composites that may, or may not, retain features of their microbial origins (e.g. Fig. 4.4). Intracellular mineralization typically occurs when the plasma membrane has been breached during cell lysis, yet two recent reports of viable bacteria (*Shewanella putrefaciens*) and photosynthetic protists (*Euglena mutabilis*) that contain intracellular ferric hydroxide granules suggest that the minerals may either serve some unrecognized physiological function (Glasauer et al., 2002) or that they may represent

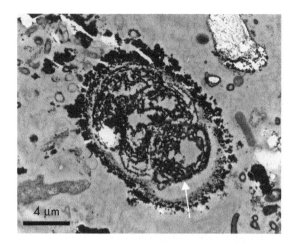

4 µm

Figure 4.4 TEM image of a lysed bacterium in which the cytoplasm (arrow) has been completely replaced by ferric hydroxide. (From Konhauser and Ferris, 1996. Reproduced with permission from the Geological Society of America.)

the cell's way of compartmentalizing and isolating unwanted iron into a localized precipitate (Brake et al., 2002).

(a) Passive iron mineralization

The actual role microorganisms play in ferric hydroxide formation can range from the completely passive to that more facilitated in nature. Yet, by our current definitions, this process is not considered biologically controlled because the microorganisms do not manage all aspects of the mineralization process. In the most passive of examples, dissolved Fe(II) transported into an oxygenated environment at circumneutral pH spontaneously reacts with dissolved O_2 to precipitate inorganically as ferric hydroxide on available nucleation sites. Microorganisms simply act as such sites, and over a short period of time submerged communities can become completely encrusted in amorphous iron (commonly referred to as ochre because of their bright red/brown color) as abiological surface catalysis accelerates the rate of mineral precipitation. While initial microscopic observations of such samples often indicate a paucity of microorganisms, staining the iron-rich sediment with fluorescent dyes for nucleic acids (e.g., acridine orange) often reveals high densities of microorganisms closely associated with the iron precipitates (e.g., Emerson and Revsbech, 1994a). This brings up one important point – it is often quite difficult to delineate microbial versus abiological contributions to mineral precipitation. So, just because a microorganism is associated with a mineral phase does not mean it formed it!

At other time microorganisms are more active in the mineralization process in that ferric hydroxide forms through the oxidation and hydrolysis of cell-bound Fe(II), the binding of ferric ion species (e.g., $Fe(OH)_2^+$; $Fe(OH)^{2+}$; Fe^{3+}) and colloids to negatively charged ligands, or the alteration of local pH and redox conditions around the cell due to their metabolic activity. Indeed, the iron-coatings on cells grown in Fe-rich cultures are sufficiently dense to visualize the bacteria under the TEM without the standard use of metal stains (MacRae and Edwards, 1972). Because of the ubiquity of iron biomineralization in nature, it was suggested that under circumneutral conditions any microorganism that produces anionic ligands will nonspecifically adsorb iron cations or fine-grained iron oxyhydroxides from the surrounding waters (Ghiorse, 1984). This is not unexpected given that the isoelectric point of pure ferric hydroxide is between 8 and 9. Ferric hydroxide also develops on the organic remains of dead cells, implying that iron mineralization can occur independent of cell physiological state. A natural corollary to this is the observation that organic matter commonly adsorbs onto Fe-rich sediment through reactions with surface $>Fe-OH_2^+$ and $>Fe-OH^0$ groups (Tipping, 1981).

(b) Chemoheterotrophic iron mineralization

There are a number of microorganisms, the so-called iron-depositing bacteria, that facilitate iron mineralization by having surface ligands that promote Fe(II) oxidation, although it is not believed that they gain energy from the process (Emerson, 2000). The most common

visible inhabitant of many freshwater, low-oxygenated iron seeps is *Leptothrix ochracea*. This chemoheterotroph frequently forms thick filamentous layers comprising tangled matrices of tubular sheaths encrusted in iron. In an iron seep in Denmark, cell densities range from approximately 10^8–10^9 cells cm^{-3} (Emerson and Revsbech, 1994a). Those high numbers promoted Fe(III) accumulation rates of 3 mm day^{-1}. One interesting observation made was that it was rare to find intact filaments of *L. ochracea* cells inside the sheaths (e.g., Fig. 4.5). This correlates well with experiments that have shown *Leptothrix* continuously abandons its sheath at a rate of 1–2 μm min^{-1}, leaving behind sheaths of 1–10 cells in length that continue to deposit ferric hydroxide. This would seem to indicate that the microorganisms actively prevent themselves from becoming permanently fixed into the mineral matrix (van Veen et al., 1978).

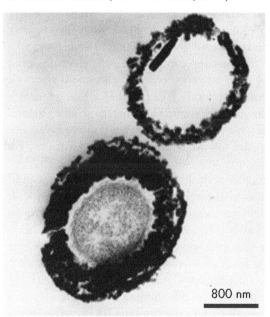

800 nm

Figure 4.5 TEM image of two ferric hydroxide-encrusted *Leptothrix ochracea* cells from an iron seep in Denmark. The cross-section shows one ensheathed cell and one abandoned sheath. (From Emerson, 2000. Reproduced with permission from the American Society of Microbiology.)

Other heterotrophic bacteria, such as filamentous species from the genera *Sphaerotilus*, *Crenothrix*, *Clonothrix*, and *Metallogenium*, as well as unicellular cocci of the *Siderocapsaceae* family, can induce ferric hydroxide precipitation through the oxidation of organic iron chelates. Essentially, they use the organic carbon of such ligands as an energy source, and as a result, the Fe(III) is freed and easily hydrolyzed (Ghiorse and Ehrlich, 1992).

(c) Photoautotrophic iron mineralization

As discussed in section 2.2.3, some anoxygenic photosynthetic bacteria are capable of oxidizing Fe(II) to Fe(III), which then hydrolyzes to $Fe(OH)_3$. This process could be described as "facilitated biomineralization" because ferric iron precipitates as a direct result of the metabolic activity of the microorganisms. These bacteria are phylogenetically diverse and include green sulfur bacteria (e.g., *Chlorobium ferrooxidans*), purple nonsulfur bacteria (e.g., *Rhodobacter ferrooxidans*), and purple sulfur bacteria (e.g., *Thiodictyon* sp.).

Ferrous iron can be used as an electron donor by these bacteria because the standard electrode potential for Fe^{2+}/Fe^{3+} (+0.77 V) is applicable only at very acidic pH, whereas at more neutral pH, the potential shifts to less positive values due to the low solubility of ferric iron cations. For instance, the electrode potential of the Fe^{2+}/Fe^{3+} couple for the bicarbonate–Fe(II) system at pH 7 is approximately +0.20 V, low enough to provide sufficient reducing power to sustain microbial growth (Ehrenreich and Widdel, 1994). Photoferrotrophic growth can also be sustained by the presence of soluble ferrous iron minerals, such as siderite ($FeCO_3$) and iron monosulfide (FeS), but not insoluble minerals, such as vivianite ($Fe_3(PO_4)_2$), magnetite (Fe_3O_4), or pyrite (FeS_2) (Kappler and Newman, 2004).

(d) Chemolithoautotrophic iron mineralization

The formation of ferric hydroxide may also stem from the ability of some chemolithoautotrophic bacteria to oxidize Fe(II) as an energy source (recall section 2.5.5). Although most enzymatic oxidation of Fe(II) occurs at extremely low pH, such as in acid rock drainage environments, the activity of *Acidithiobacillus ferrooxidans* or *Leptospirillum ferrooxidans* generally does not promote *in situ* ferric hydroxide precipitation because the Fe(III) formed remains soluble until more alkaline pH conditions ensue. However, at neutral pH, and under partially reduced conditions, chemolithoautotrophic Fe(II) oxidation by *Gallionella ferruginea* leads to high rates of iron mineralization (e.g., Søgaard et al., 2000). In fact, their extracellular stalk can become so heavily encrusted with amorphous ferric hydroxide that the majority of the dry weight is iron (recall Fig. 2.25). Similar to *L. ochracea*, *G. ferruginea* is a common inhabitant of iron springs, and where it is abundant, the stalk material appears to form the substratum upon which subsequent Fe(II) oxidation occurs. However, actively growing *Gallionella* and *Leptothrix* populations appear to occupy separate microniches, the former preferring areas of sediment with lower oxygen concentrations (Emerson and Revsbech, 1994a).

The rates of iron precipitation by *G. ferruginea* are impressive, with cell densities on the order of 10^9 cells cm^{-3} oxidizing up to 1200 nmol of Fe(II) per hour. This could lead to a hypothetical oxidation rate of 1.1×10^{-11} mol Fe(III) per cell each year (Emerson and Revsbech, 1994b). In the wells, water pipes, and field drains comprising water distribution systems, the large amounts of iron precipitated by *G. ferruginea* has long been recognized as a causative agent of serious clogging problems (e.g., Ivarson and Sojak, 1978).

(e) Hydrothermal ferric hydroxide deposits

Arguably the most persuasive example of ferric hydroxide biomineralization is at marine hydrothermal settings. It commonly precipitates directly on the seafloor from diffuse, low temperature emissions, where subsurface mixing of hydrothermal fluids with infiltrating seawater produces dilute, partially oxidized solutions that range in temperature from near ambient deep sea (~2°C)

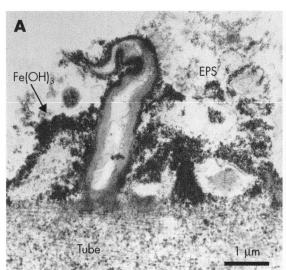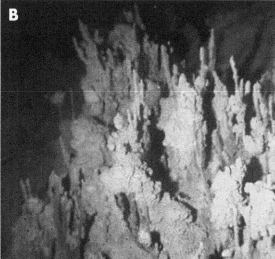

Figure 4.6 (A) SEM close-up of a vestimentiferan tube worm from the vent fields on the southern Juan de Fuca Ridge. The worm is colonized by bacteria that precipitate ferric hydroxide on their cell walls and within the EPS that holds the community together. Eventually the ferric hydroxide grains coalesce between cells and a continuous crust forms in both vertical and horizontal dimensions. This then serves as scaffolding for a new generation of bacteria. (B) Over time, continued iron mineralization leads to the formation of substantial iron deposits that rise as spires from the seafloor. (From Juniper and Tebo, 1995. Reproduced with permission from CRC Press.)

to around 50°C. The deposits themselves range from centimeter-thick oxyhydroxide coatings to more voluminous mud deposits (Juniper and Tebo, 1995). A spectacular example of the former can be seen associated with the unicellular and filamentous bacteria colonizing vestimentiferan tube worms at the southern Juan de Fuca Ridge. (Fig. 4.6).

Extensive Fe(III)-rich mud deposits have been described from a number of sites: (i) the shallow waters of the present caldera of the island of Santorini; (ii) the Red and Larson Seamounts near 21°N on the East Pacific Rise; and (iii) the Loihi Seamount, Hawaii. The Santorini site is perhaps the best cited example of the formation of ferric hydroxide resulting from microbial activity. There, mineralized stalks of *Gallionella ferruginea* occur in such masses that it is more than probable that the bacteria are responsible for iron precipitation (Holm, 1987).

The Red Seamount is characterized by an abundance of Fe-encrusted bacterial filaments, some of which have morphologies reminiscent of neutrophilic Fe(II)-oxidizing bacteria, e.g., twisted ribbons like *Gallionella ferruginea* and straight sheaths similar to *Leptothrix ochracea*. Although direct evidence supporting enzymatic Fe(II) oxidation was never put forth, indications for a microbial role in mineralization comes from the fact that the hydrothermal waters in which these deposits formed were slightly acidic (pH 5–6) and low in dissolved O_2 ($pO_2 = 0.06$ atm). These conditions give a half-time for the oxidation of Fe(II) by O_2 in seawater of approximately 30 years. Coupled with the strong currents on the seamount, it is highly unlikely that spontaneous oxidation of Fe(II) to Fe(III) could occur without biological catalysis (Alt, 1988).

The Loihi Seamount is the newest shield volcano that is part of the Hawaiian archipelago. The impact of high Fe^{2+} (but low H_2S) emissions is readily apparent in extensive deposits of ferric hydroxide that encircle the vent orifices, and in the peripheral regions where visible mats are present. The initial microscopic analyses revealed that both deposits were rich in Fe-encrusted sheaths, similar in appearance to *Leptothrix ochracea* (Karl et al., 1988). Since then, a number of studies have shown that the iron deposits have abundant microbial populations associated with them, up to 10^8 cells ml^{-1} (wet weight) of mat material, and that some of those cells are microaerophilic Fe(II)-oxidizers. It has been estimated that at least 60% of the iron deposited at the Loihi vents is directly or indirectly attributable to bacterial activity. This percentage accounts for the amount of Fe(III) generated through direct catalysis by the bacteria, as well as the proportion of ferric hydroxide that results from Fe(II) auto-oxidation on bacterially bound ferric hydroxide particles (e.g., Emerson and Moyer, 2002).

The recurring observations of bacteria in environments where they are covered in iron suggests that these environments must offer the bacteria propitious growth conditions. Quite possibly the continual supply of trace metals, which adsorb or co-precipitate directly onto iron hydroxides, may serve as an ideal nutrient source in close proximity to the cells (e.g., Ferris et al., 1999). Certainly the high surface reactivity of biogenic ferric hydroxide deposits, often more so than their inorganic equivalents, testifies to the fact that they have very high metal partitioning coefficients. The charge and abundance of the iron hydroxide surfaces will be dependent upon a number of factors, such as salinity, fluid composition, and pH, the latter relating to surface-charge characteristics such as the isoelectric point. The latter, in turn, will be affected by chemical impurities, i.e., silica-containing iron hydroxides have a much lower isoelectric point than pure ferric hydroxides (Schwertmann and Fechter, 1982). Therefore, depending on conditions the amphoteric biogenic iron precipitates can either sorb anions or metal cations. Significantly, given that ferric hydroxides are a common constituent of mid ocean ridge (MORs) venting systems, and that MORs span over 55,000 km on the ocean floor, such metal sorptive properties

must influence the global cycling of trace metals in seawater (Kennedy et al., 2003).

(f) Formation of iron oxides

Once the primary ferric hydroxides are precipitated and incorporated into the sediment, several diagenetic reactions can subsequently alter their surface reactivity, morphological characteristics, and even mineralogy. In most natural systems, ferric hydroxide serves as a precursor to more stable iron oxides, such as goethite (FeOOH) and hematite (Fe$_2$O$_3$). The transformation into more crystalline minerals proceeds through: (i) dehydration and internal rearrangement leading to hematite; and (ii) dissolution-reprecipitation leading to goethite (Schwertmann and Fitzpatrick, 1992). These reactions typically occur without biological participation, yet ferric hydroxide associated with microbial surfaces can similarly undergo these transformations, leading to a cell encrusted in iron oxides (e.g., Fig. 4.7). Experiments have documented that bacterially produced ferric hydroxide can undergo spontaneous dehydration to hematite in an aqueous medium in a

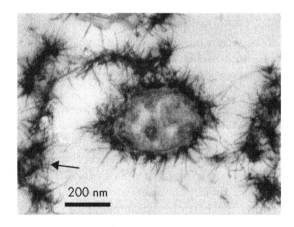

200 nm

Figure 4.7 Precipitation of acicular grains of goethite on an unidentified bacterium collected from a hot spring microbial mat in Iceland. Note how some of the mineral grains have been shed (arrow) from the cell surface. (From Konhauser and Ferris, 1996. Reproduced with permission from the Geological Society of America.)

few hours at 80°C, or in 10–14 days at 40°C (Chukhrov et al., 1973).

The mechanisms by which these transformations occur has been addressed by Banfield et al. (2000), who suggest that crystal growth is accomplished by the elimination of water molecules and the re-assembly of Fe–O–Fe bonds at multiple sites, leading to coarser, polycrystalline material. This, however, requires that some of the particles are not physically adsorbed to the organic ligands because they would constrain the movement and aggregation of surface-bound ferric hydroxide nanoparticles during their natural transformation into an iron oxide.

4.1.3 Magnetite

A great deal of research has focused on the potential for bacteria to contribute to the stable remnant magnetism of modern soils and sediments, and whether biogenic magnetite signals exist in the ancient geomagnetic record (e.g., Kirschvink, 1982). This interest has arisen from the recognition that a number of bacteria appear to form magnetite crystals that are single domain, i.e., grains with a high natural magnetic remanence. These biogenic minerals are known to precipitate under both "biologically controlled" and "biologically induced" conditions.

For the moment, we will concentrate only on those bacteria that "induce" magnetite formation. Dissimilatory Fe(III) reducers such as *Geobacter metallireducens* and *Shewanella putrefaciens* are the most extensively studied species shown to produce magnetite crystals as a byproduct of their metabolism – they oxidize fermentation products and reduce Fe(III) from ferric hydroxide (recall section 2.4.4). The magnetite forms outside the cell and it is not aligned in chains (e.g., Lovley et al., 1987). As a matter of fact, this process is very reminiscent of how G. *metallireducens* forms uraninite from the reduction of uranyl ions (recall section 2.4.5(a)).

Some characteristic features of these magnetite grains are that they are poorly crystalline and they consist of a mixture of round and oval particles that range in size from 10 to 50 nm (Fig. 4.8).

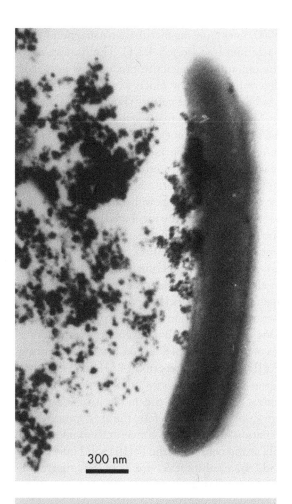

300 nm

Figure 4.8 TEM image of epicellular/extracellular, fine-grained magnetite particles formed as a byproduct of Fe(III) reduction by *Geobacter metallireducens*. (Courtesy of Derek Lovley.)

Most particles (over 95%) are usually found at the lower end of this size range, which means that they fall within the superparamagnetic size range (nonmagnetic behavior), as a diameter greater than 30 nm is required for permanent, single magnetic domain behavior (Moskowitz et al., 1989). Not surprisingly, in wet mounts *Geobacter* does not orient itself in response to an applied magnetic field. Despite this low percentage of single domain magnetite, G. *metallireducens* might still be a major contributor to that size fraction because on a per cell basis, they generate some

5000 times more magnetite than a magnetotactic bacterium (see section 4.2.1). The limitation to the amount of magnetite produced is primarily determined by how much available ferric iron can be added to the culture (Frankel, 1987).

At present, the actual role that Fe(III)-reducing bacteria play in magnetite formation remains unresolved. On the face of it, the abiological reaction of Fe^{2+} with ferric hydroxide should be sufficient to precipitate magnetite. Nevertheless, experimental studies show that magnetite does not form if the cultures are incubated at temperatures too high for growth, if the inoculated medium is sterilized prior to incubation, or if nongrowing cells are added to the experimental solution (Lovley et al., 1987). These observations suggest that the metabolism of the Fe(III)-reducing bacteria must contribute more than just Fe^{2+} to magnetogenesis. One possibility is that magnetite formation is favored by high pH; a condition met during Fe(III) reduction (reaction (4.5)). The Fe^{2+} that forms then adsorbs onto other ferric hydroxide grains, transforming the latter into magnetite (reaction (4.6)). Therefore, the appropriate combination of a high Fe^{2+} concentration and a high pH at the contact of the Fe(III) solid might provide the ideal interface for secondary magnetite formation (Lovley, 1990).

$$CH_3COO^- + 8Fe(OH)_3 \rightarrow$$
$$8Fe^{2+} + 2HCO_3^- + 15OH^- + 5H_2O \quad (4.5)$$

$$2OH^- + Fe^{2+} + 2Fe(OH)_3 \rightarrow Fe_3O_4 + 4H_2O$$
$$(4.6)$$

Magnetite has also been shown to form by microbial reduction of lepidocrocite (γ-FeOOH), a polymorph of goethite (Cooper et al., 2000). In this case, the actual step in magnetite formation proceeds via a ferrous hydroxide intermediate (reaction (4.7)):

$$(\gamma\text{-FeOOH})_2 + Fe^{2+} + H_2O \rightarrow$$
$$(\gamma\text{-FeOOH})_2 \cdot FeOH^+ + H^+ \rightarrow$$
$$Fe_3O_4 + H_2O + 2H^+ \quad (4.7)$$

Recent studies have now additionally shown that magnetite formation does not strictly require the activity of Fe(III)-reducing bacteria. In experiments where Fe(II) was added to cultures of *Dechlorosoma suillum*, with nitrate as the terminal electron acceptor, the bacteria induced the precipitation of a greenish-gray, mixed Fe(II)–Fe(III) hydroxide, known as green rust. This mineral is generally unstable in the environment, and further oxidation led to the formation of magnetite within just 2 weeks (Chaudhuri et al., 2001). Meanwhile, other experimental studies have documented magnetite formation in association with suspended cultures of phototrophic Fe(II)-oxidizing bacteria, through the reaction of Fe^{2+} with biogenic ferric hydroxide precipitates (Jiao et al., 2005).

In modern marine and freshwater sediments, much of the magnetite forms in the suboxic layers where Fe(III) reduction takes place (e.g., Karlin et al., 1987). It has even been found associated with gas seeps and solid bitumen, where its formation appears to be linked to the microbial reduction of iron oxyhydroxides with the hydrocarbons serving as the electron donors (e.g., McCabe et al., 1987). This process is supported by experimental findings of magnetite accumulation during toluene oxidation coupled to Fe(III) reduction by *G. metallireducens* (Lovley and Lonergan, 1990).

Similar processes likely played a role in the geological past. For instance, the isotopically light $\delta^{13}C$ values in Precambrian BIF carbonate minerals and the extensive presence of secondary magnetite in the same sedimentary sequences suggests that Fe(III)-reducing bacteria were important in shaping the mineralogical component of the Fe-rich marine sediments during diagenesis (see section 6.2.4(c)). This respiratory pathway has also been used to explain the general paucity of organic matter in BIFs (Walker, 1984).

4.1.4 Manganese oxides

The development of Mn(III) hydroxides and Mn(IV) oxides occurs in the same types of modern oxic–anoxic interfacial environments where ferric hydroxide forms, but because dissolved Mn(II) is not subject to as rapid a chemical oxidation as Fe(II), it may accumulate to greater concentrations in oxic waters and sediment/soil pore waters. For

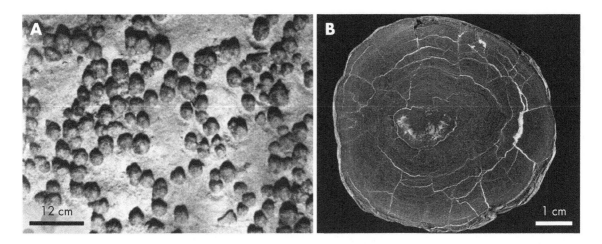

Figure 4.9 (A) Photograph of Fe-rich manganese nodules on the seafloor north of the Puerto Rico Trench. Depth is 5339 meters (courtesy of the Woods Hole Oceanographic Institute, WHOI). (B) Cross-section of a nodule from the Blake plateau, off South Carolina, at a depth of 800 meters, showing concentric laminations. Scale bars are approximate (courtesy of Frank Manheim).

part of Earth's history, chemical stratification might have existed in the water column over much of the continental shelves, where deep anoxic waters with high Fe^{2+} and Mn^{2+} content mixed with shallower, O_2-bearing waters. Not only did this lead to Precambrian BIF deposition, but also some of the world's largest and most valuable manganese deposits (Force and Cannon, 1988).

(a) Hydrothermal manganese deposits

At circumneutral pH, it is generally accepted that most Mn(II) oxidation is due to microbial catalysis (recall section 2.5.6). One environment where this is testable is at some deep-sea hydrothermal vents, where Fe(II) is precipitated at depth as iron sulfides, thus allowing high concentrations of Mn(II) to be released into oxidizing seawater (Mandernack and Tebo, 1993). The most studied hydrothermal manganese oxide deposits are those at the Galápagos rift zone, where actively accreting mounds consisting of todorokite/birnessite (MnO_2) are being formed as a result of the hot, metal-laden fluids percolating up through the sediment (Corliss et al., 1979). Bacterial isolates collected from there suggest that Mn(II) oxidation may be coupled to ATP

synthesis, and thus provides some energy to the cell community (Ehrlich and Salerno, 1990).

Manganese (III/IV) oxyhydroxides are also a major component of metalliferous sediments found on the flanks of ridge crests and where they settle out from hydrothermal plumes. Analyses of the hydrothermal plumes emanating from the southern Juan de Fuca Ridge have shown that the particulate fraction is largely composed of encapsulated bacteria encrusted in iron and manganese, with Fe-rich particles predominating near the vent and Mn-rich particles further off-axis. These observations imply that bacteria scavenge metals from the plumes, but whether they actively oxidize Fe(II) and Mn(II) was left unresolved (e.g., Cowen et al., 1986).

(b) Ferromanganese deposits

It is common for manganese to co-precipitate with iron, leading to what are referred to as ferromanganese oxides. Likely the most recognized examples of ferromanganese precipitation are the laminated concretions and nodules that form in soils, lake sediments, and on the seafloor (Fig. 4.9). The mass of manganese associated with the nodules is truly impressive. In the Pacific

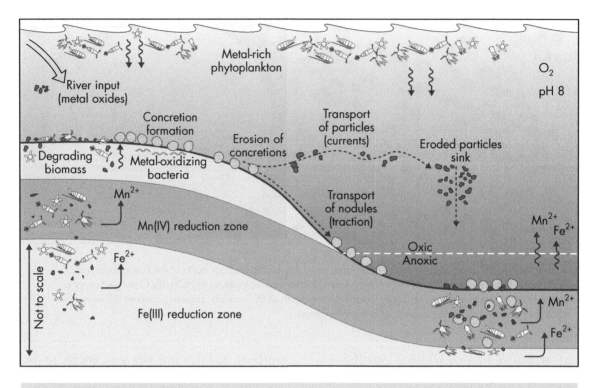

Figure 4.10 Model showing how ferromanganese nodules might form in certain lake sediments. (Adapted from Dean et al., 1981.)

Ocean, it has been estimated that 10^{12} tons of nodules exist, predominantly in pelagic (deep-sea) sediments, with an annual rate of formation of 6×10^6 tons (Mero, 1962). In Oneida Lake, New York, for example, within a 20 km^2 area of the bottom sediment, 10^6 tons of nodules, averaging 15 cm in diameter, have accumulated (Dean and Greeson, 1979).

Nodule formation in marine and freshwater environments involves a series of microbially catalyzed reactions. Based on lake models (Fig. 4.10), the process begins with cyanobacterial and algal plankton concentrating dissolved manganese and iron. Upon their death, the microbial biomass and metals are transported to the bottom sediment and buried, where anaerobic respiratory processes in the suboxic layers release the metals into the pore waters. Concurrently, reduction of riverine particulate Mn(III/IV)/Fe(III) oxyhydroxides occurs. Upward diffusion to the sediment–water interface facilitates microbial re-oxidation and

incorporation of metals onto some form of nucleus, which can be any solid mineral or organic substratum. With continued accretion of iron and manganese to the existing metal surfaces, a nodule forms that may or may not display concentric laminations. In stratified water bodies, some nodules may also be physically transported to the anoxic zone by currents or traction. There the reductive dissolution of the nodules re-liberates Mn^{2+} and Fe^{2+} directly to the water column. Some of the reduced metals make their way back to surface oxygenated waters where the cycle is repeated. Phytoplankton also contribute to metal oxidation by producing high-pH, oxygenated surface waters that are conducive to the re-oxidation and hydrolysis reactions (e.g., Richardson et al., 1988).

In the oceans, nodules can be described in terms of two end-members: (i) those formed from overlying seawater Mn^{2+}; and (ii) those with Mn^{2+} supplied via diagentic processes (similar to lakes). Depending on the source flux of Mn^{2+}, concretion growth rates can range anywhere from 1 mm to tens of centimeters in a million years (Dymond et al., 1984).

Microscopic examination of nodules regularly shows the presence of bacteria on both the surfaces and within the nodules, with population densities of the order of 10^7 cells of bacteria per cubic millimeter of nodule surface (Burnett and Nealson, 1981). Just how important the microorganisms are in terms of the mineralization process itself is unknown, but concretions up to 5 mm in diameter can be produced under idealized laboratory conditions within just 2 months by the Mn(II)-oxidizing bacteria *Metallogenium* sp. (Dubina, 1981).

(c) Desert varnish

Another deposit that incorporates ferromanganese oxides are the so-called desert varnishes. These are the black to orange coatings found on rocks in arid and semiarid environments (see Plate 7). They range in thickness from micrometers to millimeters, and are rich in variable amounts of Mn-Fe oxides and clays, but interestingly, their mineralogy and chemical composition is generally unrelated to the underlying rock substratum. Instead, the main source of Mn and Fe is rainfall or dust.

The predominant microbiological forms associated with desert varnishes are fungi, which are well adapted to the hot and dry conditions. Close examination of varnish shows that fungi hyphae are frequently heavily mineralized and physically embedded in the varnish texture. In media simulating conditions assumed to be similar to those on desert rock, 50% of the fungi studied precipitated Mn(IV) oxides (Grote and Krumbein, 1992). Also present are heterotrophic bacteria, and a large proportion of them are capable of oxidizing Mn(II) to manganese oxyhydroxides (Dorn and Oberlander, 1981). Desert varnish can even be artificially made in the laboratory within months using rock chips, a source of Mn^{2+}, and a mixed inoculum of fungi, heterotrophic bacteria, and cyanobacteria (e.g., Krumbein and Jens, 1981). These results demonstrate that microorganisms

inhabit desert varnish, and that some of those species can contribute to Mn(II) oxidation. However, the question of whether desert varnish is indeed a biological phenomenon, and whether it can only form due to microbial activity, remains unclear.

4.1.5 Clays

Within the past two decades studies in the natural environment have led to the recognition that bacteria mediate the formation of clay-like phases. Some clays form as replacement products from the alteration of primary minerals. For instance, Konhauser et al. (2002a) recently documented that highly altered, glassy tephras within active steam vents at Kilauea Volcano, Hawaii, contained subsurface bacteria with small (<500 nm in diameter), epicellular grains of smectite. They formed from the elements released into the pore waters after the primary glass phase dissolved.

Clays are also significant components of deep-sea hydrothermal deposits, and many of them contain filamentous, organic structures reminiscent of bacteria (e.g., Juniper and Fouquet, 1988). Close microscopic examination of these "biogenic minerals" show intense iron accumulation onto the filaments, upon which silica appears to have subsequently precipitated. Some hydrothermal clay deposits (e.g., nontronite) also comprise intertwining microtube-like structures, thought to be ensheathed, filamentous Fe(II)-oxidizing bacteria. It is believed that the cells not only served as templates for clay precipitation, but also that they may have been instrumental in creating the unique geochemical conditions that favored the formation of nontronite over other minerals, such as ferric hydroxide or amorphous silica (Köhler et al., 1994).

The most frequent observations of biogenic clay phases come from biofilms in lakes and rivers. Ferris et al. (1987) initially described complex (Fe, Al)-silicates on bacterial cells growing in metal-contaminated lake sediment in northern Ontario. These precipitates ranged from poorly ordered and uncharacterized phases to crystalline forms of the Fe-rich chlorite, chamosite

$((Fe)_3(Si_3Al)O_{10}(OH)_2)$. Since then, similar clayey precipitates have been reported from various rivers around the world (e.g., Konhauser et al., 1993). What is particularly remarkable about the riverine clays, irrespective of the chemical composition of the waters from which they were sampled, is that they share a number of similar properties:

1 They are generally amorphous to poorly ordered structures; those crystalline grains attached to cells tend to be detrital in origin.

2 All have grains sizes <1 μm, although the majority are <100 nm.

3 They are commonly attached in a tangential orientation around lightly encrusted cells, while those on heavily encrusted cells have a more random orientation.

4 The grains have a composition dominated by iron, silicon, and aluminum, in varying amounts. With the exception of potassium, no other metals are present in significant amounts. What is particularly interesting is that the most amorphous grains are ferruginous, while the most crystalline phases are highly siliceous, and tend towards illite-like $[(Al)_2(Si_{4-x}Al_x)O_{10}(OH)_2 \bullet K_x]$ compositions (Konhauser et al., 1998).

Based on the observations above, a sequence of events leading to clay biomineralization can be adduced (Konhauser and Urrutia, 1999). In the initial stages, a bacterium adsorbs any number of different Fe cations, e.g., Fe^{2+}, Fe^{3+}, $Fe(OH)^{2+}$, $Fe(OH)_2^+$, depending on solution chemistry and redox potential. If the dissolved iron concentration around the cell surface exceeds the solubility product of ferric hydroxide, then the latter will sorb more iron from solution, leading to the development of small (~100 nm diameter), dense, mineral aggregates on the outer cell surface (Fig. 4.11A).

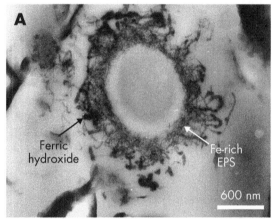

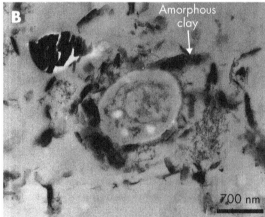

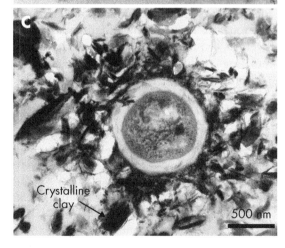

Figure 4.11 (*right*) TEM images of bacteria from a sediment sample in the Rio Solimões, Brazil. (A) Formation of ferric hydroxide aggregates in EPS. (B) Partially encrusted cell with amorphous clays forming on cell wall and within EPS, likely from the precursor ferric hydroxide. (C) Heavily encrusted cell with abundant amorphous and crystalline clay minerals extending several hundred nanometers away from the cell wall. (Reproduced from Konhauser and Urrutia, 1999 with permission from Elsevier.)

Alternatively, bacteria can attract pre-formed nanometer-sized ferric hydroxide particles from suspension (Glasauer et al., 2001), thereby negating the need for the nucleation step.

In most rivers, iron is only found in trace amounts compared to other solutes, particularly silica. Under these conditions, the adsorbed/particulate iron may instead serve as a kinetically favorable site for the development of more complex precipitates of variable clay composition, morphology, and structure. The reason these clays form is as follows. In the pH range of most natural waters, negatively charged counter-ions, or those molecules that are neutrally charged but exhibit residual surface electronegativity (e.g., monomeric, oligomeric, and colloidal silica species), accumulate near the solution–solid interface to neutralize the net positive charge of iron. Two surface species of iron oxide exist in this pH range; $>Fe-OH_2^+$ and $>Fe-OH^0$, but the majority of the surface charge is positive at circumneutral pH (Fig. 4.12). The initial (Fe, Al)-silicate phases then form via hydrogen bonding between the hydroxyl groups associated with the cell-bound iron and the hydroxyl groups in the dissolved silica, aluminum, or aluminosilicate complexes (e.g., Taylor et al., 1997; Davis et al., 2002). Exactly how these reactions occur in nature has

not been ascertained, but we do know that dimeric silica (the species that accounts for more than 99% of the oligomeric silica in natural waters) is highly reactive towards iron hydroxide surfaces (reaction (4.8)), and it exhibits a strong affinity for dissolved aluminum, forming aluminosilicate ions (e.g., reaction (4.9)) that subsequently react with cell bound iron:

$$Si_2O(OH)_6 + >Fe-OH^0 \rightarrow$$
$$>FeSi_2O_2(OH)_5 + H_2O \qquad (4.8)$$

$$Si_2O(OH)_6 + Al(OH)_2^+ \rightarrow$$
$$AlSi_2O_2(OH)_6^+ + H_2O \qquad (4.9)$$

This arrangement of ions forms an electric double layer with iron cations sorbing to the bacterial surface as an inner sphere complex, while the silica-aluminosilicate species attach as more diffuse outer layers. The surface charge of these composites is inevitably dependent upon the solution pH, the ionic strength of the solution, and the time of reaction, such that it becomes progressively more negative as the particles age and more silica sorbs. Indeed, this mechanism of binding Fe to the bacterial cell surface and subsequent reaction with silica (and aluminum) from solution has been confirmed in experimental systems with *Bacillus subtilis* (e.g., Urrutia and Beveridge, 1994).

If the microorganism is subject to sufficiently concentrated solutions, then continued reaction between the solutes and the Fe-bearing cell surface eventually results in the formation of amorphous to poorly ordered clay phases (Fig. 4.11B). Often, these reactions lead to the partial and/or complete encrustation of cells as abiological surface reactions accelerate the rate of mineral precipitation: on some microorganisms, the density of clayey material surrounding them can be so extensive such as to extend hundreds of nanometers away from the cell surface (Fig. 4.11C). Then with time, these hydrous compounds dehydrate, some converting to more stable crystalline forms. Similar steps to this biological model have been observed in the growth of smectite from amorphous Fe-Si-Al precursors

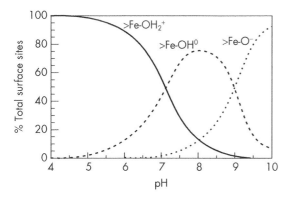

Figure 4.12 A speciation diagram for the surface of ferric hydroxide in water, shown in terms of relative percentages of the three dominant surface site species as a function of pH. (From Fein et al., 2002. Reproduced with permission from Elsevier.)

(Sánchez-Navas et al., 1998) and the glauconitization of precursor smectite phases (Amouric and Parron, 1985).

The reactions above are naturally an oversimplification because a number of inorganic processes are also at play. It is well known that dissolved Al-Si complexes precipitate as poorly ordered Al-silicates when a state of supersaturation is achieved (Wada and Wada, 1980). Moreover, Fe cations can readily be incorporated into those structures, leading to clay-like products (Farmer et al., 1991). All these reactions can conceivably take place in the proximity of the cell wall or within the extensive EPS, particularly since diffusion through the extracellular layers is inherently slow, and a microenvironment can be established that is conducive to mineralization. Additionally, colloidal species of (Fe, Al)-silicates, that either form in the water column or are products of weathering and soil formation, may react directly with the outermost cell surface. It follows that anything that will neutralize or diminish the surface charge of the colloids (e.g., a bacterial wall if colloids are positively charged or adsorbed iron if the colloids are negatively charged) will cause the particles to flocculate out of solution.

The ubiquity of (Fe, Al)-silicate precipitation on freshwater bacteria implies that the latter facilitate this form of biomineralization with relative ease. Perhaps the fact they readily scavenge iron, and that most rivers and lakes typically contain high concentrations of silica, and to a lesser extent aluminum, is all that is required for the formation of authigenic clays. The implications for this mode of biomineralization are, however, quite profound. The sediment–water interface is influenced by: (i) sedimentation and entrainment of metal-rich particulate material; (ii) metal adsorption onto clays, metal oxyhydroxides, or organic material in the bottom sediment; and (iii) precipitation of various authigenic mineral phases (Hart, 1982). The role of microorganisms, in particular biofilms, has seldom been considered an important influencing factor. The thickness of a biofilm may

be only a few millimeters at most, yet when one takes into consideration the large surface area of any given river bed that is colonized by biofilms, the volume of water that falls directly into microbial contact is substantial. In this regard, biofilms dominate the reactivity of the sediment–water interface, and through the adsorption of solutes from the water column, they facilitate the transfer of metals into the bottom sediment. The bound metals may then become immobilized as stable mineral phases that collect as sediment on the river bed, sections of the metal-laden biofilms may be sloughed off by high flows and transported downstream to be deposited in a lake or ocean, or the metals may be recycled back into the overlying water column after microbial organic matter mineralization.

4.1.6 Amorphous silica

Silica precipitation is an important geological process in many modern terrestrial geothermal systems, where venting of supersaturated solutions leads to the formation of finely laminated siliceous sinters around hot spring or geyser vents. When the sinter deposits are examined under the electron microscope, they typically show an association between the indigenous microorganisms and spheroidal, amorphous silica grains that form epicellularly on the sheaths or walls, and intracellularly after the cells have lysed (Fig. 4.13). The silicification process likely begins with the attachment of silica oligomers or preformed silica colloids. These silica species then grow on the cell surface, often reaching tens to hundreds of nanometers in diameter. If silicification is sustained, particles invariably coalesce until the individual precipitates are no longer distinguishable, and frequently, entire colonies are cemented together in a siliceous matrix several micrometers thick.

It was suggested many years ago that the role of microorganisms in silica precipitation is largely a passive process (e.g., Walter et al., 1972). In geothermal systems, waters originating from deep, hot reservoirs, at equilibrium with quartz,

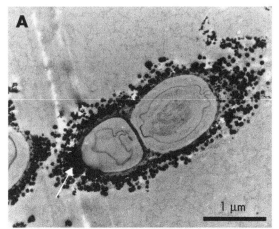

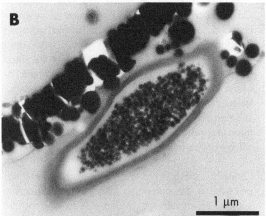

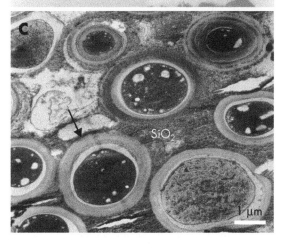

Figure 4.13 (*left*) TEM images of silicified bacteria from a hot spring sinter in Iceland. (A) Two cells completely encrusted in amorphous silica spheres (tens of nanometers in diameter). In some places (arrow) the silica has begun to coalesce into a dense mineralized matrix. (B) A lysed cell with abundant intracellular silica grains. (C) Silica precipitation is so extensive that the entire community of cyanobacteria (*Calothrix* sp.) are encrusted in a dense mineralized matrix. Note how the silicification is restricted to the outer sheath surface (arrow), leaving the cells inside mineral-free. (From Konhauser et al., 2004. Reproduced with permission from the Royal Swedish Academy of Sciences.)

surface, decompressional degassing, rapid cooling to ambient temperatures, evaporation, and changes in pH all conspire together to cause the fluid to suddenly become supersaturated with respect to amorphous silica (Fournier, 1985). Concurrently, the discharged monomeric silica, $Si(OH)_4$, polymerizes, initially to oligomers (e.g., dimers, trimers, and tetramers), and eventually to polymeric species with spherical diameters of 1–5 nm, as the silanol groups (-Si-OH-) of each oligomer condense and dehydrate to produce the siloxane (-Si-O-Si-) cores of larger polymers. The polymers grow in size through Ostwald ripening such that a bimodal composition of monomers and particles of colloidal dimensions (>5 nm) are generated. These either remain in suspension due to the external silanol groups exhibiting a residual negative surface charge, they coagulate via cation bridging and nucleate homogenously, or they precipitate heterogeneously on a solid substratum (Ihler, 1979).

As microorganisms are present in these polymerizing solutions, they inevitably become silicified, much the same as other submerged solids, e.g., pollen, wood, leaves, and sinter. Indeed, Walter (1976a) defined geyserite to mean a laminated, amorphous silica sinter that formed in the proximity of vents and fissures where temperatures in excess of 73°C were deemed sterile except for scattered thermophilic

commonly contain dissolved silica concentrations significantly higher than the solubility of amorphous silica at 100°C (approximately 380 mg L^{-1}). Therefore, when these fluids are discharged at the microorganisms. This temperature exclusion point has since been modified because studies have now shown that microorganisms silicify over a range of temperatures, often in excess of 90°C (e.g., Jones and Renaut, 1996). Moreover, examination of geyserite from Yellowstone and New Zealand indicate that their surfaces are covered with biofilms and that their laminae generally contain silicified microorganisms. Thus, not all geyserite can be regarded as being abiological, and it appears that most siliceous sinters have been constructed, to some degree, around microorganisms (Cady and Farmer, 1996).

Nevertheless, experimental evidence now exists that appears to corroborate the view that the microbial role in silicification is incidental and not limited to any particular taxa. In particular, bacteria have little affinity for monomeric silica, even at high bacterial densities and low pH conditions, where most organic functional groups are fully protonated (Fein et al., 2002). Similarly, under highly supersaturated conditions, the rates of silica polymerization and the magnitude of silica precipitated are independent of the presence of bacterial biomass (e.g., Benning et al., 2003; Yee et al., 2003). Presumably, in concentrated silica solutions there is such a strong chemical driving force for silica polymerization, homogeneous nucleation, and ultimately silica precipitation that there is no obvious need for microbial catalysis. It has also been observed that silicification occurs on dead cells, and continues autocatalytically and abiogenically for some time after their death due to the high reactivity of the newly formed silica. Consequently, silica precipitated in the porous spaces between filaments has the same basic morphology as the silica precipitated on the original filaments (e.g., Jones et al., 1998). These findings certainly support the notion that biogenic silicification at thermal springs occurs simply because microorganisms grow in a polymerizing solution where silicification is inevitable.

With that said, there are species-specific patterns of silicification, because different microorganisms are certainly capable of being silicified with different degrees of fidelity. This is not surprising given that the actual mechanisms of silicification (in solutions where homogeneous nucleation is not possible) rely on the microorganisms providing reactive surface ligands that adsorb silica from solution and, accordingly, reduce the activation energy barriers to heterogeneous nucleation. This means that cell surface charge may have a fundamental control on the initial silicification process.

At present there appear to be three different mechanisms by which microorganisms become silicified (Fig. 4.14):

1 *Hydrogen bonding* – Many bacteria, such as *Calothrix* sp., form sheaths composed of neutrally charged polysaccharides. This can lead to hydrogen bonding between the hydroxyl groups associated with the sugars and the hydroxyl ions of the silica (Phoenix et al., 2002). Although the low reactivity of the sheath gives such cells hydrophobic characteristics that facilitates their attachment to solid submerged substrata, this same property makes them less inhibitive to interaction with the polymeric silica fraction in solution.

2 *Cation bridging* – For microorganisms where the cell wall is the outermost layer, such as *Bacillus subtilis*, silicification is limited due to electrostatic charge repulsion between the anionic ligands and the negatively charged silica species. In order for silicification to proceed, some form of cation bridge is necessitated, whereby metals adsorbed to the cell can act as positively-charged surfaces for silica deposition (e.g., Phoenix et al., 2003).

3 *Direct electrostatic interactions* – Some bacteria, such as *Sulfurihydrogenibium azorense*, produce protein-rich biofilms that contain an abundance of cationic amino groups that adsorb polymeric silica (Lalonde et al., 2005).

One of the more exciting revelations recently has been that silicification may not be detrimental to the microorganism (Phoenix et al., 2000). For instance, when *Calothrix* are grown in silica supersaturated solutions for weeks at a time, and many of the filaments develop extensive mineral crusts up to 5 μm thick, the cells still fluoresce, they continue to generate oxygen, and the

Hydrogen bonding

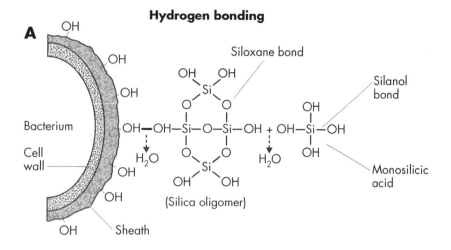

Cation bridging

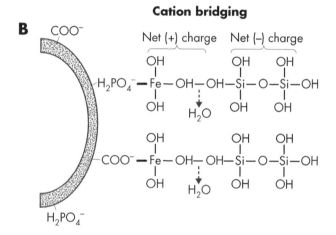

Direct electrostatic interactions

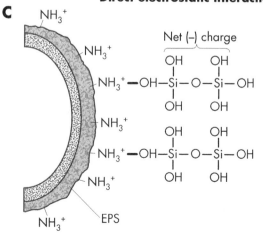

Figure 4.14 Three mechanisms by which microorganisms silicify: (A) hydrogen bonding between dissolved silica and hydroxyl groups associated with some sheaths; (B) cation bridging between silica and negatively charged cell walls; (C) direct electrostatic interactions between silica and positively charged amino groups in some biofilms. Note: stippled arrows show release of H_2O after bond formation.

mineralized colonies exhibit comparable rates of photosynthesis to nonmineralized colonies. Intriguingly, silicification of viable cyanobacterial cells only occurs on the outer surface of the sheath, whereas lysed cells have silica forming within the cytoplasm. This clearly indicates that the sheath is necessary for enabling photosynthetically active cyanobacteria to survive mineralization, by both acting as an alternative mineral nucleation site that prevents cell wall and/or cytoplasmic mineralization, and by providing a physical filter that restricts colloidal silica to its outer surface (recall Fig. 4.13C). Of course at some stage silicification will inhibit diffusional processes. Perhaps their ability to grow upwards within the sheath towards the sediment–water interface, where the magnitude of silica encrustation will be less pronounced than at depth (i.e., where the sinter is older and has been exposed to more silica), is a means by which the cyanobacteria survive in a continuously accreting environment? Biofilm production may be a different version of this defense mechanism.

4.1.7 Carbonates

Microorganisms have played an integral role in carbonate sedimentation since the Archean. The deposits they form are heterogeneous, but the main component is fine-grained, lithified lime mud composed of micrite (1–5 µm crystals of calcium carbonate). It forms as a result of a combination of processes, including mineralization of microbial surfaces, chemical precipitation from supersaturated solutions, and erosion of existing carbonate layers (Riding, 2000). Microorganisms can play both a controlled and passive role in mineral precipitation. The biologically controlled mechanisms will be discussed later in this chapter (section 4.2.4).

(a) Calcium carbonate – mechanism of mineralization

Much emphasis on passive carbonate biomineralization has been placed on the photosynthetic activity and surface reactivity of cyanobacteria

(Merz-Preiß, 2000). The overall reaction that best describes the precipitation process is:

$$M^{2+} + 2HCO_3^- \longleftrightarrow MCO_3 + CH_2O + O_2 \quad (4.10)$$

where M^{2+} represents a divalent metal cation and MCO_3 is a solid carbonate phase. As the cations present in solution can vary from location to location, so too can the different carbonate phases. Consequently, it is not uncommon to see cyanobacteria in close association with a number of carbonate minerals, including calcite/aragonite ($CaCO_3$), dolomite ($(CaMg)(CO_3)_2$), strontionite ($SrCO_3$), and magnesite ($MgCO_3$). Calcite and aragonite are by far the more common carbonate phases, with the concentration of Mg^{2+} determining the more stable form; high Mg^{2+} promotes aragonite precipitation, while lower Mg^{2+} favors calcite precipitation.

The role of cyanobacteria in carbonate precipitation is twofold: metabolic fixation of inorganic carbon tends to increase solution pH and lead to a state of supersaturation, while cation adsorption to the cell surface promotes heterogeneous nucleation (Fig. 4.15). With respect to photosynthesis, in waters with neutral to slightly alkaline pH, cyanobacteria use HCO_3^- instead of, or in addition to, CO_2 as a carbon source for the dark cycle (reaction (4.11)). A byproduct of this reaction, hydroxyl ions, are then excreted into the external environment where they create localized alkalinization around the cell. This, in turn, induces a change in the carbonate speciation towards the carbonate (CO_3^{2-}) anion (reaction (4.12)):

$$HCO_3^- \longleftrightarrow CO_2 + OH^- \quad (4.11)$$

$$HCO_3^- + OH^- \longleftrightarrow CO_3^{2-} + H_2O \quad (4.12)$$

Cyanobacteria also provide reactive ligands towards metal cations and, once bound, they can then react with the CO_3^{2-} anions to form a number of carbonate phases, such as aragonite or calcite:

$$CO_3^{2-} + Ca^{2+} \longleftrightarrow CaCO_3 \quad (4.13)$$

Extracellular layers are particularly favorable sites for nucleation, and cyanobacterial species

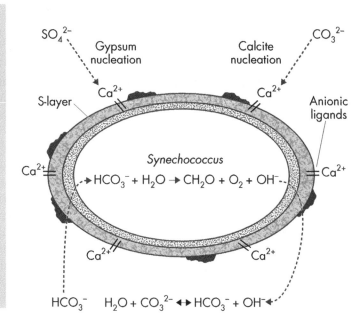

Figure 4.15 Significant insights into the mechanisms underpinning freshwater cyanobacterial calcification have been obtained from experiments replicating the activity of *Synechococcus* sp. communities in Fayetteville Green Lake, New York. When the cyanobacteria were cultured in filter-sterilized lake water (pH 7) they began to precipitate gypsum ($CaSO_4\cdot2H_2O$) on their surfaces within 4 hours of the beginning of the experiment. The biomineralization of gypsum was a two-step process initiated by the binding of Ca^{2+} to the cell's S-layer, followed by reaction with dissolved sulfate. Within 24 hours, an increase in the alkalinization of the microenvironment around the cells pushed the solid mineral stability field towards the formation of calcite. (Modified from Thompson and Ferris, 1990.)

that produce sheaths or EPS generally precipitate more calcium carbonate than those species without such structures (Pentecost, 1978). When calcium carbonate nucleates on the sheath surface it grows radially outwards and, in some cases, this may lead to the complete encrustation of the cell. Conversely, when calcium carbonate nucleates within the intermolecular spaces of the sheath, the latter may become filled with mineral material (Verrecchia et al., 1995). EPS fosters carbonate precipitation by providing diffusion-limited sites that create localized alkalinity gradients in response to metabolic processes, while simultaneously attracting Ca^{2+} to its organic ligands (e.g., Pentecost, 1985). Furthermore, the type of functional groups in EPS affects carbonate morphology and mineralogy, e.g., spherule vs. euhedral calcite or calcite vs. aragonite (Braissant et al., 2003).

Cyanobacteria grown in the presence of various combinations of Sr^{2+}, Mg^{2+}, or Ca^{2+} can precipitate instead strontionite, magnesite, or mixed calcite-strontionite carbonates (Schultze-Lam and Beveridge, 1994). In general, cyanobacteria are equally capable of incorporating Ca^{2+} or Sr^{2+} during carbonate mineral formation, while magnesite is easily inhibited from forming by the prefer-

ential binding of the former two cations over Mg^{2+}. Other studies have documented that cyanobacteria can partition of up to 1.0 wt% strontium in calcite (Ferris et al., 1995). The ability for solid-phase capture of trace metals/radionuclides during biogenic calcification has important implication for bioremediation strategies in calcium carbonate-hosted aquifers because those contaminants can be effectively immobilized from the groundwater flow (Warren et al., 2001).

Much of the foregoing discussion has focused on cyanobacteria. However, a number of studies have described how green and brown algae (e.g., the genera *Chara* and *Halimeda*), that grow as part of marine microbial mats, precipitate aragonite as a result of HCO_3^- uptake during photosynthesis. In most cases, the crystals lack any organizational motif or preferred crystal orientation, they vary in size, and they are not associated with any organic material other than the cell wall (e.g., Borowitzka, 1989). For a limited number of algae (e.g., *Penicillus* sp.), mineral precipitation occurs within a "sheath-like" structure surrounding the cell wall. This sheath appears to serve primarily as a diffusion barrier, aiding in the establishment of a sufficiently large degree of alkalinization.

(b) Calcium carbonate – deposits

Small cyanobacteria (<2 μm in size), known as picoplankton, have been linked to fine-grained calcium carbonate precipitation in both lacustrine and shallow marine environments during times of seasonal blooms. These "whiting events" are believed to be responsible for the bulk of the sedimentary carbonate deposition in some well-described sites, such as Fayetteville Green Lake, New York. There, the unicellular cyanobacterial genus, *Synechococcus*, is the dominant phytoplankton in the surface waters. Under the TEM, *Synechococcus* is frequently shown to be completely mineralized, yet the type of mineralization is seasonally dependent (Thompson et al., 1990). During the cold winter months, when the *Synechococcus* cells are dormant, gypsum ($CaSO_4 \cdot 2H_2O$) crystals develop on the S-layers of nonmetabolizing cells. However, in the spring, as the lake water warms and light intensity increases, the cell population becomes more active in number, the pH increases, and the gypsum becomes unstable and dissolves. Simultaneously, the dominant mineral phase precipitated by individual *Synechococcus* cells changes to calcite (recall Fig. 4.15), which during the warm summer months, falls as a light rain of mineral-encrusted biomass to the lake bottom. Stable carbon isotopic analyses of the unconsolidated carbonate sediment shows that it is enriched in ^{13}C relative to the bulk dissolved inorganic carbon species (Thompson et al., 1997). This isotopic difference is caused by the preferential use of the lighter ^{12}C isotope during photosynthesis, which leaves the organic component depleted in ^{13}C, while the dissolved inorganic carbon, which precipitates as a calcite around the cells, becomes enriched in ^{13}C by as much as 4–5‰.

In the oceans, whiting events can lead to deposits of considerable size and thickness. On the Great Bahamas Bank, for example, satellite imagery has shown some whitings to cover between 35 and 200 km^2 during the summer. Shinn et al. (1989) have estimated that average whitings contain nearly 11 mg L^{-1} of suspended sediment, with $CaCO_3$ settling rates of the order of 35 g m^{-2} h^{-1}. Based on those rates, Robbins et al. (1997) calculated that 1.4×10^6 metric tons of aragonite are suspended each year, and that once sedimented, can account for much of the late Holocene bank-top lime muds on the Great Bahamas Bank. Field studies have suggested that epicellular calcite precipitation, triggered by the fixation of CO_2 by cyanobacterial blooms, may play a role in these whiting events (Robbins and Blackwelder, 1992). Evidence in support of this hypothesis includes the presence of 25% organic matter by weight in the solid whiting material and SEM/TEM images that show individual whiting spheres embedded in an organic matrix, along with the presence of $CaCO_3$ crystals on cyanobacteria surfaces. Just how important picoplankton are to whiting processes will continue to be the subject of examination, but considering that *Synechococcus* blooms are typically around 10^5 cells ml^{-1} and under some conditions can be responsible for 30–70% of the primary productivity of the open ocean (e.g., Waterbury et al., 1979), their biomineralizing abilities might very well be of global importance.

Benthic cyanobacterial communities can form an even wider variety of calcareous deposits. When cyanobacteria growing in biofilms calcify, they can form micritic coatings, crusts, and layers on submerged substrata. For example, "microreefs," consisting of 30% cyanobacteria by weight, have been described forming on submerged limestone gravel in a number of alkaline lakes (Schneider and Le Campion-Alsumard, 1999). Ooids are another such example. These small (<2 mm), concentrically layered, spherical grains are composed of primary calcium carbonate or replacement phases that form where gentle or periodic wave action in shallow marine waters and along lacustrine shores cause equal precipitation on all sides of a cortex of sand, shell fragments, or microbial biomass. Filamentous cyanobacteria, such as *Schizothrix* species, have in particular been heavily implicated in the accretionary process because they produce EPS that binds Ca^{2+}, and their metabolism changes the physicochemical properties at the ooid–water interface (Davaud and Girardclos, 2001). Endolithic varieties, such as *Solentia* sp., also contribute to calcification, but through a multicyclic process of microboring into existing carbonate grains and concurrent infilling of boreholes with aragonite (Macintyre et al., 2000). Processes similar to those described above may account for the high magnesium calcite peloids (elliptical to spheroidal structures 20–60 mm in diameter) that are incorporated in many cemented carbonate deposits in shallow marine and lagoonal settings. They are often characterized by having fine-grained nuclei composed of fossilized clumps of bacteria (Chafetz, 1986).

Lithified carbonate bioherms (also called microbialites) are common in many modern and ancient environments (Fig. 4.16). Freshwater tufa deposits develop at springs and waterfalls in limestone terrains, where loss of dissolved CO_2 due to turbulence and evaporation induces supersaturation and calcite precipitation on available submerged solids. The cyanobacteria, algae, and plants that grow in these moist environments inevitably become incorporated into the precipitating minerals, a process enhanced by the tendency of the carbonates to become trapped in the EPS of the microbial mats (Pentecost and Riding, 1986). A similar process describes the formation of speleothems in caves, where rapid degassing of calcium and bicarbonate-rich groundwaters induces a state of supersaturation (Dreybrodt, 1980). Travertine deposits (a denser form of tufa) are characteristic of a number of thermal spring deposits (see section 6.1.4(a) for details). The primary causes of supersaturation in these systems are the cooling and pressure reduction of the hydrothermal effluent during discharge, and as steam separates from the fluid phase, CO_2 is degassed and the pH correspondingly increases. Similar to sinter formation, the role of the main microbial constituents, the cyanobacteria, may be purely incidental (e.g., Renaut and Jones, 1997). However, the production of EPS by the indigenous community can

serve as an important biological surface upon which authigenic calcite nucleates or particulate grains are trapped (e.g., Emeis et al., 1987).

Thrombolites are macroscopically clotted microbialites that have become increasingly important since the end of the Precambrian. Their formation has been attributed to rapid rates of calcification by coccoid cyanobacteria and, as such, sediment binding and trapping are of minor importance in the overall accretionary process. There are a number of modern thrombolites examples, the largest possibly being the 40 meter high tower-like deposits found in the highly alkaline waters of Lake Van in eastern Anatolia, Turkey (Kempe et al., 1991). Thrombolites can also be formed by green algae in subtidal marine environments. The deepening water, decrease in salinity, and increase in energy and nutrient supply favor algal growth over the cyanobacterially dominated shallow water stromatolites that form with them a laterally gradational biofacies (Feldmann and McKenzie, 1998).

For much of the Precambrian, stromatolites were widespread in shallow marine waters. Although their relative importance has since declined, they are still present in some modern intertidal and subtidal marine environments (e.g., Exuma Sound, Bahamas), seasonally hypersaline embayments (e.g., Shark Bay, Western Australia; see Plate 8), carbonate atolls (e.g., French Polynesia), and shallow coastal lakes (e.g., Lake Clifton, Western Australia). Their mechanisms of formation are discussed in section 6.1.4(b) and their relevance to the Precambrian in section 7.4.2. One of the characteristic features of stromatolites are their laminations. The biological imprint on lamina texture is created by the orientation of the filamentous cyanobacteria, the adhesiveness and abundance of microbial sheath/EPS material, their propensity to facilitate calcium carbonate precipitation, and their growth response to sediment flux and authigenic mineralization (Seong-Joo et al., 2000). Crucially, the microbial mats must be lithified early to strengthen the deposit, and invariably preserve it into the rock record as a stromatolite.

Figure 4.16 Examples of various carbonate microbialites. (A) Speleothems in the Carlsbad caverns, New Mexico (courtesy of Peter Jones/NPS). (B) Travertine deposit from Angel Terrace, Mammoth Hot Spring, Yellowstone National Park (courtesy of Bruce Fouke). (C) Thrombolite mounds from Lake Salda, Turkey (courtesy of Michael Russell). (D) Stromatolites exposed at low tide, Hamelin Pool, Western Australia (courtesy of Ken McNamara).

Fungi are important constituents of lichens, and not only do they excrete large quantities of organic acids that contribute to rock weathering (see section 5.1.2(c)), but they also form authigenic mineral phases, mainly oxalates and carbonates (Verrecchia, 2000). One environment where fungi biomineralize is in calcretes, terrestrial calcareous hardgrounds that are widely distributed throughout the arid and semiarid regions of the world. In such deposits, fungi are often covered

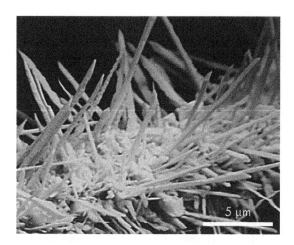

Figure 4.17 SEM image of calcium oxalate crystals on a fungal filament from a calcrete collected at Galilee, Israel. (Courtesy of Eric Verrecchia.)

with calcium oxalate crystals that form from the reaction of cell-released oxalic acid with Ca^{2+} around the cell (Fig. 4.17). The calcium oxalates can then transform into calcite, resulting in the infilling of any available pore spaces and the formation of a hard cement. Fungi (and bacteria) also appear to play a significant role in the transformation of woody tissues in trees to calcite. In the Ivory Coast and Cameroon, some of the trees are actually being calcified *in situ*, and if the quantity of inorganic carbon per tree is extrapolated to account for similar trees throughout tropical Africa, then this biological process could represent a significant long-term carbon sink (Braissant et al., 2004).

(c) Dolomite

The abiological formation of dolomite has proven difficult at room temperature in the laboratory. This is not unexpected given that in nature dolomite commonly forms as a secondary replacement mineral of earlier calcite and/or aragonite. Therefore, it was of great interest to find that the activity of sulfate-reducing bacteria (SRB) could mediate primary dolomite formation under anoxic, hypersaline conditions (e.g., Vasconcelos

et al., 1995; Wright, 1999). Furthermore, SRB have since been shown to experimentally induce the formation of dolomite crystals identical in composition and morphology to those found in the natural systems from which the bacteria were isolated (Warthmann et al., 2000).

The role of SRB in dolomite formation is twofold. First, the process of sulfate reduction overcomes the kinetic barrier to dolomite formation by increasing the pH and alkalinity, and by removing sulfate, which is a known inhibitor to dolomite formation. Since sulfate occurs in seawater as a Mg^{2+}-SO_4^{2-} ion pair, its removal also increases the availability of "free" Mg^{2+} cations in the microenvironment around the cell for dolomite precipitation (van Lith et al., 2003a). Interestingly, only pure cultures of metabolizing SRB form dolomite, and even then some pure strains form high Mg-calcite instead. What this implies is that dolomite formation requires specific environmental conditions, and differences in metabolic activity, salinity or substrate concentration play a role in the establishment of chemical gradients around the cells that sometimes favors dolomite precipitation, whereas at other times it favors the precipitation of different minerals. Second, the cell surfaces of SRB concentrate Ca^{2+} and Mg^{2+} cations around the cell. Because of the relatively large size of the dolomite grains to the SRB themselves, it is likely that the cell material involved in metal binding is the EPS that holds the aggregates of cells (and mineral grains) together (Fig. 4.18). Once bound, these cations subsequently serve as favorable adsorption sites for CO_3^{2-} ions, in a process reminiscent of that for calcite precipitation (recall Fig. 4.15). It would thus appear that the metabolic activity of the SRB and their surface reactivity are complementary in removing all kinetic inhibitors to the formation of a mineral that would otherwise be difficult to precipitate under normal environmental conditions. In fact, the bacteria can be so effective at promoting dolomitization, with rates on the order of 500 mg L^{-1} month^{-1} (van Lith et al., 2003b), that the cells themselves can be completely dwarfed by the product of their labor as the numerous small

Figure 4.18 SEM image showing the relationship between the sulfate-reducing bacteria *Desulfovibrio hydrogenovorans* and the dolomite crystals experimentally precipitated along with them in culture. Not evident from the micrograph is the EPS coating all the dolomite grains. (From van Lith et al., 2003a. Reproduced with permission from Blackwell Publishing Ltd.)

crystals grow in size (via Ostwald ripening) to form the large crystals shown in Fig. 4.18.

Most recently, dolomite has also been shown to form in basalt-hosted aquifers, in association with methanogens (Roberts et al., 2004). Dissolution of basalt yields elevated pore-water concentrations of dissolved Ca^{2+} and Mg^{2+}, which then adsorbs onto the methanogen's surface. When coupled with methanogenic consumption of CO_2, leading to alkalinity generation, a state of localized carbonate supersaturation can easily be attained. What is surprising about this work, however, is that the dolomite grains, only tens of nanometers in size, form directly on the cell surface, at times completely encrusting the cells.

(d) Siderite and rhodochrosite

The formation of siderite is generally limited to sedimentary environments where pore water Fe^{2+} concentrations exceed dissolved H_2S – when sufficient H_2S is produced, the precipitation of FeS and pyrite (see section 4.1.10) never allows ferrous iron concentrations to reach levels suffici-

ent to achieve siderite stability. As a consequence, siderite tends to precipitate in the suboxic layers of freshwater and estuarine sediments, where low dissolved sulfate levels constrain SRB activity (Postma, 1982). Siderite also forms in some anoxic marine sediments, within the zone of methanogenesis, where rapid sedimentation rates lead to subsurface sulfate depletion (Gautier, 1982). One further constraint on siderite formation is that the Fe/Ca ratio of the pore water should be high enough to stabilize siderite over calcite, hence a lowering of Fe(III) reduction rates causes siderite precipitation to cease in favor of calcite or dolomite (Curtis et al., 1986). Rhodochrosite ($MnCO_3$) is formed in a similar environment as siderite, but it can also precipitate in sulfidic sediments because of the high solubility of MnS (Neumann et al., 2002).

Siderite and rhodochrosite can be produced experimentally through the reductive dissolution of ferric hydroxide and MnO_2, respectively (Roden and Lovley, 1993). Both processes involve two steps: the first being the reduction of the metal (recall reactions (2.30) and (2.28), respectively), and the second the reaction of the reduced metals with excess HCO_3^- (reactions (4.14) and (4.15), respectively):

$$Fe^{2+} + HCO_3^- + OH^- \rightarrow FeCO_3 + H_2O \quad (4.14)$$

$$Mn^{2+} + HCO_3^- + OH^- \rightarrow MnCO_3 + H_2O \quad (4.15)$$

The minerals formed experimentally are very similar to those crystals formed naturally, particularly in the case of siderite concretions (e.g. Fig. 4.19). This strengthens the argument that bacterial processes are responsible for early diagenetic siderite precipitation (Mortimer et al., 1997). But, whether the bacteria play a role in their formation beyond supplying the necessary ions remains unresolved.

4.1.8 Phosphates

The formation of phosphate minerals is intimately associated with microbial activity. In

Figure 4.19 SEM image of rhombohedral siderite produced in culture by *Geobacter metallireducens*. (Courtesy of Rob Mortimer.)

modern sediment, phosphogenesis arises from a series of independent biogeochemical reactions beginning with the accumulation of dissolved inorganic phosphate (usually in the form of $H_2PO_4^-$ or HPO_4^{2-}, with a pK_a of 7.2 for the ionization reaction, $H_2PO_4^- \rightarrow HPO_4^{2-} + H^+$) by phytoplankton. Upon death of the cells, the biomass sinks and thus serves as a vehicle by which phosphate is supplied from the water column to the sediments, and eventually released into the interstitial pore waters via heterotrophic degradation (Gulbrandsen, 1969). From there, its fate is multifold: some is readily scavenged by other microorganisms that store it as an energy source; some is adsorbed to minerals phases, such as ferric hydroxide; while the remainder diffuses into the overlying water column. Dissimilatory Fe(III) reduction or reduction of ferric oxyhydroxides by reaction with bacterially generated hydrogen sulfide, also serves as a supplementary source of dissolved pore water phosphate (Gächter et al., 1988).

High localized rates of phosphate release can promote the rapid nucleation of amorphous calcium fluorapatite phases throughout the sediment pore spaces and on the surfaces of pre-existing substrata, including organic matter derived from the microbial cells themselves. As the more stable phase, calcium fluorapatite ($Ca_{10}(PO_4)_{6-x}(CO_3)_xF_{2+x}$), then appears, it causes pore water diffusion of phosphate towards the locus of nucleation, thereby bringing down the bulk degree of supersaturation towards calcium fluorapatite solubility (Van Cappellen and Berner, 1988). To support its growth, calcium fluorapatite then either uses the precursor as an epitaxial (chemically matching) template or it causes the precursor to dissolve and reprecipitate as a more stable phase. At some depth, nucleation and growth of calcium fluorapatite ceases due to rising levels of carbonate alkalinity accompanying the anaerobic decomposition of residual organic matter buried in the sediments (Jahnke, 1984). Given sufficient time, even calcium fluorapatite will eventually transform into either francolite, a highly substituted form of fluorapatite, or the most stable phosphatic phase, that being apatite ($Ca_{10}(PO_4)_6F_2$), with the concomitant loss of CO_2 and fluorine.

Phosphorites are fine-grained, organic-rich sediments containing more than 10% (by volume) phosphate minerals in the form of nodules, crusts, coatings, and pelletal grains. The sites of their deposition tend towards coastal and shelf environments, where upwelling of phosphate-rich, deep ocean waters leads to high phytoplanktonic productivity, while the shallowness of deposition ensures that much of the particulate organic matter reaches the seafloor. Upwelling water also facilitates phosphogenesis because: (i) deep cold water rising towards the surface is heated and tends to lose CO_2 due to a decrease in pressure; and (ii) the phytoplankton fix CO_2 during photosynthesis. Thus, water unusually rich in phosphate moves into a region of increasing pH that should favor deposition of calcium phosphate until the levels of alkalinity become inhibitory (Burnett, 1977). This model is supported by the organic-rich nature of recent phosphorite deposits forming on the continental shelves off the coasts of Southwest Africa and South America (e.g., Bremner, 1980; Glenn and Arthur, 1988). Off the

Peruvian coast, nodules with diameters of several centimeters grow at rates approaching 1 mm yr^{-1}. The resulting phosphatic concretions are resistant to transport by currents, and as a result can be mechanically exhumed and concentrated during periods of sediment reworking.

Microbial structures make up a major part of the modern phosphorite framework, typically comprising filamentous mats of cyanobacteria (Fig. 4.20A) and sulfur-oxidizing bacteria immediately capping the zone of calcium fluorapatite precipitation. Similarly, a close association of benthic microbial activity with the formation of calcium fluorapatite can be widely traced in ancient phosphogenic environments (e.g., Krajewski et al., 1994). The organic matter in fossil phosphorites exhibit features indicative of intense biodegradation of organic matter at, or near, the seafloor, while the microfabrics preserved show that an abundant and diverse benthic microbial assemblage existed at the time of mineralization. Stromatolitic phosphorites are an excellent example of the close spatial association between the activities of ancient microbial mats and the precipitation of phosphatic minerals (e.g. Fig. 4.20B). Furthermore, precipitation of apatitic precursor phases was a common mechanism of bacterial preservation. This has been well documented in the Upper Cretaceous–Lower Eocene Mishash Formation in Israel, where phosphatized mats are preserved as dense apatite overgrowths on remnants of filamentous cyanobacterial sheaths and fungal hyphae, while coccoid cyanobacteria were preserved as apatite infillings (Soudry and Champetier, 1983).

Microorganisms can also play an active role in the mineralization process. One way is through anaerobic respiratory pathways that release metal cations into the pore waters, where they then react with dissolved phosphate. For example, in experimental studies with Fe(III)-reducing bacteria, the ferrous phosphate, vivianite, frequently forms as a secondary product after the metabolic release of ferrous iron into a phosphate-rich medium (e.g., Lovley and Phillips, 1988b). In other studies, bacterial decomposition of

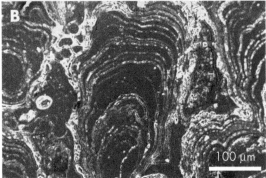

Figure 4.20 (A) SEM image of an experimentally phosphatized microbial mat dominated by filamentous cyanobacteria (*Oscillatoria* sp.). The filaments are coated with thin layers of carbonate fluorapatite, which formed as a result of rapid precipitation of an amorphous calcium phosphate precursor phase. (B) Polished section of a phosphatic columnar microstromatolites from an Upper Cretaceous sequence in the Polish Jura Chain. The microfabric consists of alternating compact (dark gray) and porous (white to pale gray) apatitic laminae. The latter contain remnants of unicellular microorganisms. (Courtesy of Krzysztof Krajewski.)

phosphate-rich organic compounds (e.g., RNA), in the presence of a calcium source (e.g., calcite), produces calcium fluorapatite (Prévôt et al., 1989), while phosphate released through the activity of outer membrane-bound phosphatase enzymes, in solutions containing UO_2^{2+}, has been shown to induce the precipitation of uranium phosphate minerals (Macaskie et al., 2000). Microbial redox processes may further promote chemical gradients and associated pH shifts that help localize calcium fluorapatite precipitation at certain sites in the sedimentary layers (e.g., Van Cappellen and Berner, 1991). In addition, microbial mats may behave as physical barriers by reducing the diffusion of phosphate back into the overlying water column. Empty sheaths and degraded microbial remains would function in a similar manner (Soudry, 2000).

Whether or not there is a direct cellular control over phosphate mineralization is much more ambiguous. In most phosphate-rich sediments, the formation of calcium fluorapatite precursors is a rapid process that takes advantage of any substratum available, and when abundant biomass is present, it can appear as though bacteria are favorable nucleation sites. Yet, experiments specifically designed to test the microbial role have concluded that there is no evidence to suggest that phosphate minerals nucleate preferentially on bacteria; calcium fluorapatite grains were noted to develop on or close to the cell, as well as on solids devoid of bacteria (Hirschler et al., 1990). With that said, detailed microscopic examination of lichens, growing on exposed rock outcrops on Ellesmere Island, in the Canadian Arctic, highlight how ferric iron adsorbed onto cyanobacterial walls and their EPS react with dissolved phosphate (Konhauser et al., 1994). This reaction leads to the secondary precipitation of iron phosphate grains, compositionally similar to strengite ($FePO_4 \cdot 2H_2O$), throughout the biomass (Fig. 4.21). In this particular instance, the microbial community concentrated phosphate within the biofilm by taking advantage of the high adsorptive affinity of Fe(III) for phosphate anions.

4.1.9 Sulfates

(a) Gypsum, celestite, and barite

We have already examined how some *Synechococcus* species directly contribute to the formation of gypsum deposits during the winter in Fayetteville Green Lake. Experimental studies of the cell surface during the initial stages of

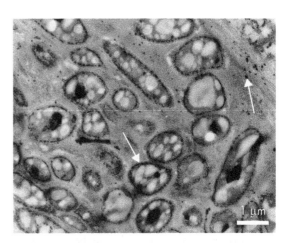

Figure 4.21 TEM image of a lichen scraped off the surface of a granodiorite outcrop on Ellesmere Island, Canada. Arrows indicate the numerous Fe-phosphate grains that are associated with the cyanobacterial cell walls and EPS. The large dark objects within the cells are polyphosphate granules that store temporary excess phosphate, while the large electron-translucent granules inside the cells are polyhydroxy butyrate bodies that function as energy reserves. (From Konhauser et al., 1994. Reproduced with permission from the National Research Council of Canada.)

mineral growth reveal that the S-layer contains small, regularly arranged pores that facilitate the initial nucleation of the gypsum grains (Schultze-Lam et al., 1992). Continued aggregation of the gypsum grains eventually enshrouds the entire cell surface such that the S-layer becomes completely obscured by the growing gypsum crystals. It is interesting that while the cells are still active, the mineralized S-layers are shed from the cell wall into the external environment so that the cells can grow unabated by the biominerals they just formed. Given that the adsorption of dissolved sulfate to the cell-bound calcium is an abiological process, the sloughed off S-layer material then continues to nucleate additional gypsum grains during its descent to, and in, the bottom sediment.

Evaporitic environments, where salinities frequently reach the brine stage, are more typical for gypsum deposition. Modern mats grow abundantly in such concentrated solutions, and it is not uncommon to find laminated gypsum deposits or columnar to conical stromatolitic structures that result from periodically controlled phases of microbial mat development and gypsum precipitation (Fig. 4.22A). There are also a number of ancient stromatolitic gypsum deposits, the most notable being those from the Upper Miocene (Messinian) that circumvent much of the Mediterranean shoreline. Although most of the traces of the ancient microbial communities associated with those deposits are poorly preserved, the relation between the mats and gypsum are still recognizable by the laminations (Fig. 4.22B). Each set of laminae resulted from two superimposed processes controlled by seasonal variations in salinity: (i) growth of cyanobacterial mats during periods of low salinity; and (ii) interstitial crystallization of gypsum when trapped brines reached salinities prohibitively high for the growth of most indigenous microorganisms (Rouchy and Monty, 2000). At times, remains of the microorganisms even became incorporated into the accreting gypsum crystals.

Because a seasonality effect appears to control the ratio of gypsum to calcite precipitated in Fayetteville Green Lake, Schultze-Lam and Beveridge (1994) tested whether cyanobacteria could promote a similar sulfate-to-carbonate transformation when other alkaline earth metals were present in solution, namely Sr^{2+}. In their experimental set-up, the authors inoculated *Synechococcus* cells into artificial lake water with high concentrations of SO_4^{2-} and Sr^{2+}. The precipitates that initially appeared were small grains (tens of nanometers in diameter) composed of the mineral celestite ($SrSO_4$). In time, the celestite grains grew in size until the precipitates completely covered the cells. Then, the mineralogy of the precipitates changed in composition from celestite to strontionite ($SrCO_3$), the carbonate anion being derived by the same alkalinization process described for calcite precipitation. What is intriguing about these results

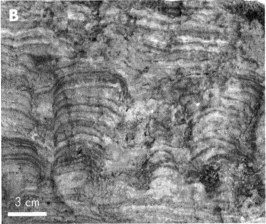

Figure 4.22 (A) Modern gypsified columnar stromatolites from the Ojo de Liebre Lagoon in Baja California (courtesy of Catherine Pierre). (B) Ancient gypsified stromatolites from the Upper Miocene (Messinian) Polemi Basin, Cyprus displaying similar columnar structures as the modern counterparts (courtesy of Jean Marie Rouchy).

is that: (i) the cyanobacterial cell wall avidly bound both Ca^{2+} and Sr^{2+}; and (ii) the minerals that ultimately formed were simply a consequence of the available counter-ions. Along similar lines, the heterotrophic bacterium, *Myxococcus xanthus*, has been reported to form barite ($BaSO_4$), simply by exposing it to a solution rich in Ba^{2+} (González-Munoz et al., 2003).

(b) Iron sulfates

Schwertmannite ($Fe_8O_8(OH)_6SO_4$) and jarosite ($MFe_3(SO_4)_2(OH)_6$), where M may be H^+, Na^+, K^+, or NH_4^+, typically occur as ochreous surface precipitates on stream beds receiving iron and sulfate-rich, acid rock drainage (ARD). When ARD comes in contact with fresh water at an off-site location, the oxidation and hydrolysis of Fe(II) results in a voluminous yellow precipitate, characterized by its high reactivity and efficiency at scavenging other ions from the effluent. At low pH, schwertmannite and jarosite precipitate through anion bridging of ferric iron colloids (reactions (4.16) and (4.17), respectively). At higher alkalinity, and in the absence of appreciable sulfate, the neutralizing effects of relatively unpolluted stream water results instead in the precipitation of either ferric hydroxide or goethite (Bigham et al., 1996).

$$8Fe^{3+} + SO_4^{2-} + 14H_2O \rightarrow$$
$$Fe_8O_8(OH)_6SO_4 + 22H^+ \qquad (4.16)$$

$$M^+ + 3Fe^{3+} + 2SO_4^{2-} + 6H_2O \rightarrow$$
$$MFe_3(SO_4)_2(OH)_6 + 6H^+ \qquad (4.17)$$

Although bacteria are directly involved in the oxidation of sulfidic minerals and the generation of ARD (see section 5.2.2(c)), their involvement in the subsequent precipitation of amorphous iron and sulfur phases is less clear. It is well established that the metabolic oxidation of ferrous sulfate solutions by *Acidithiobacillus ferrooxidans* experimentally leads to the formation of jarosite (e.g., Ivarson, 1973). Other experiments with *Bacillus subtilis* similarly generated ferric sulfate phases of variable stoichiometry depending on the initial Fe(II)/SO_4 ratio used (Fortin and Ferris, 1998). However, only one study has confirmed a direct bacterial role in mineralization, that being of an abandoned coal mine drainage lagoon in West Glamorgan, Wales (Clarke et al., 1997). In the shallow subsurface sediments, a number of unidentified bacteria displayed granular, fine-grained Fe(III)-S precipitates attached to their outer surfaces, while at greater depths, the cells were typically encrusted in a dense mineralized matrix in which individual precipitates appear to have coalesced. What was unexpected was that the Fe:S atomic weight ratio decreased from 3.5:1 at 15 cm to 1.9:1 at 30 cm, highlighting the continued reactivity of the ferric iron for dissolved sulfate as the grains became progressively buried.

4.1.10 Sulfide minerals

The formation of low temperature sulfide minerals (i.e., <100°C) is indirectly linked to the activity of dissimilatory sulfate reduction. As discussed in Chapter 2, SRB couple the oxidation of simple organic molecules to the reduction of sulfate, thereby generating dissolved hydrogen sulfide. It, in turn, reacts abiologically with a number of existing mineral phases within the sediment, including ferric oxyhydroxides. The microbial role is simply to generate the reductant, and as experimental studies showed many years ago, there is no crystallographic differences between iron sulfides formed in the presence of, or absence of, microorganisms (Rickard, 1969). Not surprisingly, in fine-grained anoxic sediments, sulfide minerals are commonly found to be in close association with organic matter (see section 6.2.5(a) for details). In fact, there is generally a good positive linear correlation between the organic carbon and mineral sulfide contents in normal marine shales throughout the Phanerozoic (Raiswell and Berner, 1986). Moreover, patterns of sulfur isotopic fractionations in many sedimentary sulfide deposits are consistent with this form of mineralization, and support a biological origin of reduced sulfur (see Box 7.3 for details).

One of the most prevalent sulfide minerals is pyrite (FeS_2). Although the precise mechanisms by which pyrite forms at temperatures below 100°C remains the subject of debate, it is believed to involve a number of Fe sulfide precursors progressively richer in sulfur (Sweeney and Kaplan, 1973). The process begins with the local precipitation of an amorphous iron

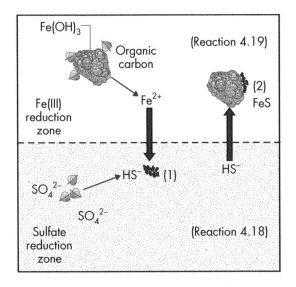

Figure 4.23 Two modes of iron monosulfide formation.

monosulfide phase, e.g., FeS. In sediments, this initial mineralization stage is driven by two separate pathways (Fig. 4.23). One pathway involves Fe^{2+}, produced during biological Fe(III) reduction, diffusing down from the suboxic layers into the sulfate reduction zone, where it reacts with pore water sulfide (in the form of HS^- at marine pH) to form FeS (reaction (4.18)). The second pathway (reaction (4.19)) involves dissolved sulfide, from the underlying anoxic sediments, diffusing upwards where it is removed more slowly, but in greater amounts, by reaction with ferric oxyhydroxide (Canfield, 1989):

$$Fe^{2+} + HS^- \rightarrow FeS + H^+ \qquad (4.18)$$

$$6Fe(OH)_3 + 9HS^- \rightarrow 6FeS + 3S^0 + 9H_2O + 9OH^- \qquad (4.19)$$

Reaction (4.19) is actually much more complicated than written because it involves five steps: (i) inner-sphere surface complex formation between the >Fe(III)-OH ligand with HS^- to form $>Fe(III)SH + OH^-$; (ii) electron transfer from S(-1) to Fe(III); (iii) release of an oxidized

S atom, usually elemental sulfur (S^0), and the concomitant formation of $>Fe(II)OH_2^+$, after reaction with a water molecule; (iv) detachment of Fe^{2+} through a weakening of the bonds between the reduced iron and the O^{2-} ions of the crystalline lattice, thereby exposing a new surface site on the ferric hydroxide; and (v) reaction of Fe^{2+} with HS^- to form FeS (dos Santos Afonso and Stumm, 1992; Poulton et al., 2004a). The first step is usually quite fast unless HS^- reacts instead with a dissolved cation or another mineral phase. In euxinic basins (chemically stratified bodies of water with anoxic waters below the chemocline), the high availability of HS^- and Fe^{2+} leads to rapid monosulfide nucleation within the water column itself (Wilkin and Barnes, 1997).

Once formed, iron monosulfide converts rapidly into mackinawite. In turn, mackinawite can react with any number of intermediate sulfur species with oxidation states between sulfate and sulfide. One such pathway is the reaction with elemental sulfur to form greigite, Fe_3S_4 (reaction (4.20)). The transformation of greigite to pyrite then requires a major crystallographic reorganization of both the iron and sulfur, likely involving a dissolution-reprecipitation pathway (Schoonen and Barnes, 1991b).

$$3FeS + S^0 \rightarrow Fe_3S_4 \qquad (4.20)$$

Based on the rapid formation of pyrite in some sedimentary environments, it has also been proposed that FeS might react with other partially oxidized sulfur phases, including polythionates ($S_xO_6^{2-}$), thiosulfate ($S_2O_3^{2-}$), or polysulfides (S_x^{2-}), and in doing so, avoid the greigite intermediate step (Luther, 1991):

$$FeS + HS_x^- \rightarrow FeS_2 + S_{x-1}^{2-} + H^+ \qquad (4.21)$$

Most controversially, it has been argued that pyrite formation can also proceed at temperatures lower than 100°C under strictly anoxic conditions by reaction of FeS with H_2S (reaction (4.22)) (e.g., Rickard, 1997). Despite experiments demonstrating this reaction pathway (Drobner et al., 1990), they are inconsistent with results obtained from many field and laboratory studies that indicate that the mackinawite to pyrite conversion requires a weak oxidant, not H_2S. Instead, it is widely accepted that the environment for pyritization must be slightly oxidizing (Benning et al., 2000).

$$FeS + H_2S \rightarrow FeS_2 + H_2 \qquad (4.22)$$

The numerous intermediate steps may appear needlessly complicated, but the direct precipitation of pyrite at temperatures below 100°C is unfavorable since its rate of nucleation is slow compared to its formation via the FeS precursor (Schoonen and Barnes, 1991a). This makes sense in light of the fact that the activation energy barrier to pyrite nucleation is so high that in order for this step to occur at a significant rate, the solution must exceed saturation with respect to iron monosulfide. However, once iron monosulfide supersaturation is attained, it will nucleate considerably faster than pyrite and drive the reactant concentrations below the critical value for pyrite. It is only when pyrite begins to grow that it can control the saturation state of the fluid and, therefore, cause the concentrations of Fe^{2+} and S^{2-} to diminish enough that the precursor dissolves. Thus, the precipitation of iron sulfides follows the Ostwald sequence for consecutive reactions, i.e., the thermodynamically least stable phase forms first. Consistent with this is the near universal observation that sediment pore waters are saturated or slightly undersaturated with respect to iron monosulfides, but are always supersaturated with respect to pyrite (e.g., Howarth, 1979).

The most common pyrite textures are clusters of framboids, densely packed mineral aggregates with sizes on the order of tens of micrometers, and possessing an overall raspberry-like appearance (e.g., Fig. 4.24). Framboids have three characteristics: (i) a microcrystalline arrangement that might be an indicator of fast crystal growth and/or magnetic aggregation, the latter

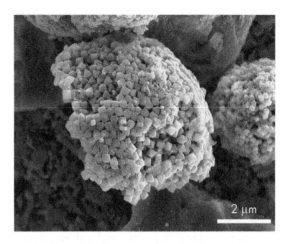

Figure 4.24 SEM image showing the spherical aggregation of individual pyrite crystals to give a typical framboidal morphology. Sample collected from recent sediments in the Black Sea. (Courtesy of Rick Wilkin.)

being important if greigite was the precursor phase; (ii) uniformity in the size and morphology of their microcrystals, suggesting simultaneous nucleation and similar growth rates for the same period of time prior to aggregation; and (iii) overall spheroidicity that might reflect pseudomorphism of a pre-existing spheroidal body, possibly an organism or a microcolony. Certainly, the latter is a possibility when framboids are found associated with extant microbial mats (e.g., Popa et al., 2004). Framboids that form within the water column are a completely different matter because they tend to be smaller and less variable in overall size, features that probably reflect rapid nucleation in the water column with less time for growth (Wilkin et al., 1996). The distinctions in size and morphology between syngenetic framboids (formed in the water column) and diagenetic framboids (formed in sediment) make it possible to determine the redox conditions, bulk C/S values, and the degree of pyritization during deposition of ancient shales (see section 6.2.5(a) for details).

In addition to being primarily responsible for the production of dissolved sulfide, bacteria

can also serve as surfaces for iron sulfide precipitation. For example, in a metal-contaminated
lake sediment in Sudbury, Ontario, mackinawite
was observed directly on the outer surfaces of
bacterial cells and their membranous debris
(Ferris et al., 1987). Some bacteria from the same
samples also precipitated millerite (NiS), indicating that the presence of competing cations
can alter the final product of sulfide biomineralization. Indeed, in the absence of iron, other
metal sulfides, such as galena (PbS) and sphalerite
(ZnS), are associated with microbial biomass in
some black shales, strata-bound and stratiform
base-metal sulfide deposits, and oil reservoirs
(Machel, 2001). These natural deposits corroborate experimental studies suggesting that the
formation of organometallic complexes plays a
critical role in the partitioning of metallic ions
into sulfidic phases. In particular, metals chemically complexed to bacteria are more reactive
towards hydrogen sulfide than when they are in
solution (Mohagheghi et al., 1985).

Bacterial sulfate reduction is likely not an
important process in the formation of hydrothermal massive sulfides because these minerals
are precipitated from solutions containing high
concentrations of geothermally generated H_2S.
However, surface crusts on hydrothermal chimneys at northern Gorda Ridge, for instance,
showed the preservation of bacterial filaments
in fine-grained chalcopyrite ($CuFeS_2$), pearceite
($Ag_{14.7-x}Cu_{1.3+x}As_2S_{11}$), and proustite ($Ag_3AsS_3$)
(Zierenberg and Schiffman, 1990). The bacteria
likely played two roles in biomineralization.
First, they adsorbed Ag, As, and Cu, causing
local concentrations of these metals to exceed
the solubility products of their sulfides, ultimately leading to mineral nucleation. Second, the
bacterial mats may have influenced the physiochemical conditions around the chimneys to
favor metal sulfide precipitation.

It has also been revealed that natural communities of SRB can generate essentially pure ZnS
deposits from dilute groundwater. This extends the
possibility for a biogenic role in low temperature
metal sulfide ore deposits (Labrenz et al., 2000).

4.2 Biologically controlled mineralization

What makes biologically controlled biomineralization different from the processes discussed
above is that the microorganism exerts considerable control over all aspects of the nucleation and
mineral growth stages (Mann, 1988). Initially, a
specific site within the cell is sealed off from the
external environment; this will later become the
locus of mineralization. Two common methods
of space delineation occur. The first involves the
development of intercellular spaces between a
number of cells. The second is the formation of
intracellular deposition vesicles.

Once the cellular compartment is formed, the
next step entails the cells sequestering specific
ions of choice and transferring them to the
mineralization site, where their concentrations
are increased until a state of supersaturation
is achieved. Levels of supersaturation are then
regulated by managing the rate at which mineral
constituents are brought into the cell via specific transport enzymes. Meanwhile, nucleation
is controlled by exposing organic ligands with
distinct stereochemical and electrochemical properties tailored to interact with the mineralizing
ions. These same ligands also act as surrogate
oxyanions that simulate the first layer of the
incipient nuclei (Mann et al., 1993). The
crystals then grow in a highly ordered manner,
with their orientation and size governed by the
overall ultrastructure of the membrane-bound
compartment.

4.2.1 Magnetite

There are a number of microorganisms that
exert significant control over magnetite formation. The best understood are the so-called
magnetotactic bacteria, originally described by
Blakemore (1975). These are a diverse group of
aquatic species (predominantly *Proteobacteria*),
that share three basic features (Bazylinski and
Frankel, 2000):

1 They are microaerophilic, meaning that they exhibit poor growth at atmospheric concentrations of oxygen.

2 Most have bidirectional motility, being able to propel themselves forwards or backwards by rotating one of their polar flagella.

3 They possess a number of intracellular, linearly arranged membrane-bound structures called magnetosomes that house the mineral grains (Fig. 4.25). Most magnetotactic bacteria produce on average 20 or so magnetosomes, although a 9 μm diameter, coccoid bacterium was identified that possessed up to 1000 magnetosomes (Vali et al., 1987).

Unlike the magnetite formed via Fe(III) reduction, the crystals formed by magnetotactic bacteria have unique morphologies (always either cubic, rectangular, or arrow-shaped), they are free from crystallographic imperfections, and chemically they are quite pure Fe_3O_4 (Bazylinski, 1996). Considering that many other metals will be present in their immediate surroundings, this implies that magnetotactic bacteria have the means to exclude nonmagnetite-forming ions from the growing magnetite crystals.

Magnetotactic bacteria also precipitate magnetite within a narrow range of crystal sizes, from approximately 35 to 120 nm. This establishes stable single magnetic domains. A single 40–50 nm magnetosome has a magnetic energy of 3×10^{-14} erg. This energy would be sufficient to align it in the Earth's geomagnetic field were it not for the thermal forces (4×10^{-14} erg) that tend to randomize the cell's orientation in its aqueous environment. However, the magnetosomes are arranged in one or more chains that traverse the cell along its axis of motility, such that the magnetic interactions of a single particle cause its magnetic dipole to orient parallel to the other grains. Thus, the total magnetic energy of the cell is the sum of each of the individual particles; with 20 magnetosomes the cell's magnetic energy is 6×10^{-13} erg. Significantly, this means that the cell is able to align itself passively along geomagnetic field lines while it swims, with the vertical component of the geomagnetic field in

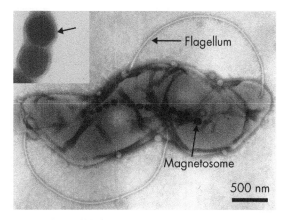

Figure 4.25 TEM image of a magnetotactic bacterium, designated strain MV-4, grown in pure culture. Cells of this strain produce a single chain of magnetite crystals that longitudinally traverse the cell. Inset shows close-up of the magnetosome membrane (arrow) that surrounds each individual particle. (Courtesy of Dennis Bazylinski.)

each hemisphere selecting the predominant polarity type amongst the magnetotactic bacteria (Blakemore and Blakemore, 1990). It is, however, important to stress that the cell is neither attracted nor pulled towards the geomagnetic pole, but merely aligns itself like a compass needle. Thus, dead cells align similar to living cells.

Magnetotactic bacteria have been recovered from a wide variety of environments, where they grow most abundantly at oxic–anoxic interfaces (Fig. 4.26). They are chemoheterotrophic, with oxygen as their usual terminal electron acceptor, and although cells such as *Magnetospirillum magnetotacticum* strain MS-1 produce more magnetite when grown with nitrate, they still require at least 1% O_2 for magnetite synthesis (e.g., Bazylinski and Blakemore, 1983). Other magnetotactic bacteria can use ferric iron, nitrous oxide, and possibly even sulfate as TEAs (Sakaguchi et al., 1993), although the latter has not been confirmed by other studies.

The most unresolved issue regarding magnetotactic bacteria is what is the purpose of possessing magnetic properties? At present we can only

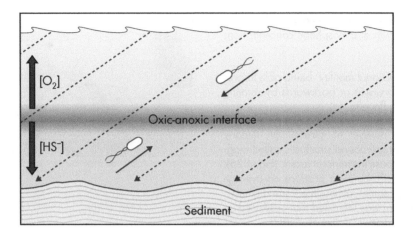

Figure 4.26 Most magnetite-producing magnetotactic bacteria are found at, or above, the oxic–anoxic interface in sediments and stratified bodies of water, where geochemical conditions are appropriate for magnetite formation. They move freely up and down along the inclined geomagnetic field lines (dotted) in response to changing environmental conditions. (Modified from Bazylinski, 1996.)

speculate on some of the advantages, the most probable being that magnetotaxis is a particularly useful navigational tool, increasing the cell's efficiency at locating and maintaining an optimal position in vertical chemical and/or redox gradients typical of sediments and stratified water bodies. Because the magnetotactic bacteria tend to be microaerophilic, their movement above and below the chemocline will have serious repercussions for the health of the cells. So, when a cell inadvertently moves too far upwards, and the concentration of oxygen becomes inhibitory, it reverses direction (Frankel et al., 1997). Similarly, if it moves too far down into the sediment where hydrogen sulfide concentrations are prohibitively high, the cell once again reverses direction and moves back upwards. Like most free-swimming bacteria, magnetotactic bacteria propel themselves forward in their aqueous environment by rotating their helical flagella. However, two questions arise: (i) if knowing which way is up versus down increases a cell's efficiency at finding and maintaining an optimal position relative to the gradient, why then don't all bacteria inhabiting suboxic sediments have magnetic properties; and (ii) at the equator, where the geomagnetic field lines are horizontal, why would bacteria produce

magnetite (Frankel and Blakemore, 1989), especially since the energy expended would surely give them a severe competitive disadvantage compared to nonmagnetic species? To complicate matters more, magnetite has also been found associated with euglenoid algal cells (e.g., Torres de Araujo et al., 1986) and several types of protists, including dinoflagellates and ciliates (Bazylinski et al., 2000). The role of magnetite in these cells is even more of a guess.

One thing is clear, intracellular magnetite must serve a purpose because the processes involved in its formation are complicated and energy intensive (Frankel et al., 1983). Magnetite synthesis involves a series of geochemical steps that begins with the uptake of Fe(III) from the surrounding environment (Fig. 4.27). As discussed in section 3.4.2(a), bacteria commonly rely on iron chelators such as siderophores to facilitate the solubilization and transport of Fe(III) to the cell. Once a specific siderophore has sequestered iron, it then needs to be absorbed by a cell that requires it. This is accomplished by cell synthesis of specific receptor proteins designed to first recognize the Fe(III)–siderophore complex and then, with the aid of other transport proteins, guide the coordinated Fe(III) to the plasma

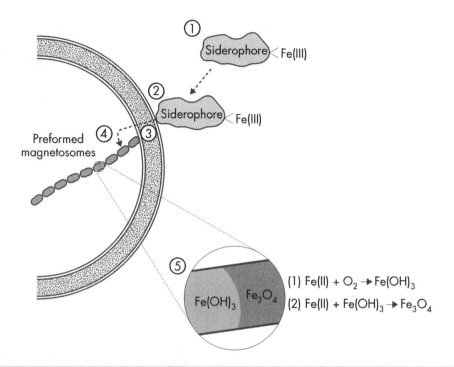

Figure 4.27 The possible mechanisms by which magnetotactic bacteria form intracellular magnetite. There are a number of steps involved, including (1) sequestration of Fe(III) from the aqueous environment via siderophores; (2) siderophore attachment to a receptor site on the outer membrane; (3) transport of the siderophore through the outer membrane to the plasma membrane, where Fe(III) is reduced to Fe(II); (4) transport of Fe(II) to pre-formed magnetosome; and (5) initial precipitation of ferric hydroxide within the magnetosome, followed by conversion to magnetite. Note: size of siderophore not to scale.

membrane (Neilands, 1989). In some species, the siderophore does not penetrate the plasma membrane, but instead donates the iron to a second membrane-bound chelator, while in other species, the entire siderophore is absorbed directly into the cytoplasm (Müller and Raymond, 1984). In either case, the Fe(III) is reduced to Fe(II), and the latter is then shuttled in some form through the cytoplasm into the magnetosome, which appears to be anchored to the plasma membrane. Empty magnetosomes have been observed in iron-starved cells, and recent molecular work has shown that specific magnetosome-associated proteins play a role in vesicle formation prior to biomineralization (Komeili et al., 2004). In the magnetosome, Fe(II) is then re-oxidized to ferric hydroxide, perhaps with O_2

as the electron acceptor. The actual crystallization of magnetite then involves the reaction of the ferric hydroxide with more Fe^{2+}:

$$Fe^{2+} + 2OH^- + 2Fe(OH)_3 \rightarrow Fe_3O_4 + 4H_2O \quad (4.23)$$

The subsequent adsorption of Fe^{2+} on to the ferric hydroxide has been suggested as the possible trigger for magnetite formation, with the solid-state rearrangement manifest as a growing crystal front of magnetite extending into the precursor phase (Mann et al., 1984). This mineralization scenario is, in part, borne out of the observation that some anaerobes, that are capable of dissimilatory Fe(III) reduction, produce a large number of small (30–50 nm in diameter), intracellular grains of ferric hydroxide, as well as lesser amounts of magnetite (Glasauer et al., 2002).

One aspect perhaps not readily obvious is that the mineralization process requires spatial segregation of regions differing in Eh (the redox potential, see Box 6.1 for details) and pH because the necessary conditions to precipitate ferric hydroxide are quite different from those needed to subsequently transform it into magnetite. Add to that the constraints over magnetite morphology and size, it seems clear that the magnetosome must function under precise biogeochemical and genetic control (Gorby et al., 1988).

The characteristic properties of intracellular magnetite are often clearly recognizable in both recent and ancient sedimentary environments. In fact, it has been proposed that biologically controlled magnetite may persist in deep-sea sediments, and thus contribute to the palaeomagnetic record (e.g., Kirschvink and Chang, 1984). Despite the magnetite chains fragmenting upon lysis of the cell, their initial presence can be inferred by observing the morphological/chemical characteristics of magnetically separated fractions of sediment under an electron microscope, and also by using a magnetometer to measure the resistance to demagnetization that distinguishes multidomain from single domain magnetite (e.g., Petersen et al., 1986). Fossil magnetotactic bacteria may even extend as far back as the Precambrian, with magnetofossils extracted from the 2.0 Gyr Gunflint Iron Formation (Fig. 4.28) possibly representing the oldest evidence of controlled biomineralization (Chang et al., 1989).

4.2.2 Greigite

The formation of greigite (Fe_3S_4) proceeds by the same controlled intracellular mineralization process as described for the magnetite-generating magnetotactic bacteria (e.g., Bazylinski et al., 1993). Individual greigite particles are membrane-bound and organized into chains. They are also ferromagnetically ordered, providing the bacterium with properties similar to a magnetite-producing bacterium, although greigite is one-third

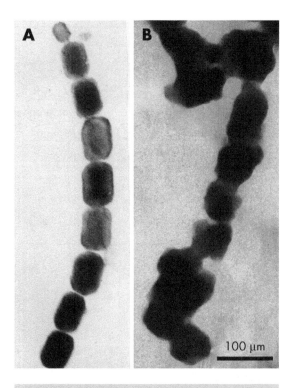

100 μm

Figure 4.28 Comparison of magnetite grains from modern and ancient sedimentary environments. (A) An intact magnetite chain, formed by magnetotactic bacteria, in recent marine sediments of the Santa Barbara Basin, California. (B) Chain of single-domain magnetite grains extracted from limestone within the 2.0 Gyr Gunflint Iron Formation, Canada. (Adapted from Chang and Kirschvink, 1989. Reproduced with permission from the Annual Reviews in Earth and Planetary Sciences.)

as magnetic. The lower magnetism of greigite, however, is compensated for by the fact that the greigite-producing bacteria tend to have many more magnetosome crystals, as many as 100 per cell (Pósfai et al., 1998). Morphologies of greigite include cuboidal and rectangular prismatic crystals in the size range 30–120 nm.

The biomineralization of greigite in magnetotactic bacteria closely resembles the processes of sedimentary sulfide formation, whereby amorphous Fe sulfide transforms into cubic FeS → mackinawite → greigite through a series of solid-state transformations. Similar to above, the magnetotactic bacteria appear to synthesize and align the nonmagnetic sulfides into chains prior to the crystals becoming magnetic. Over time, greigite then converts to pyrite under reducing conditions at neutral pH when excess sulfur is present. In one magnetotactic bacterium, pyrite crystals were observed along with greigite (Mann et al., 1990). Given the lengthy conversion time for greigite to pyrite, it seems unlikely that this process took place during the cell's lifetime. Instead, greigite and pyrite may be biomineralized separately, indicating that the stoichiometry of the metal (Fe) and nonmetal (S) can vary in some magnetotactic bacteria, resulting in different mineral assemblages (Heywood et al., 1990).

While magnetite-producing bacteria prefer microaerophilic conditions, the greigite producers grow below the oxic–anoxic interface, where HS$^-$ concentrations are high. Interestingly, one bacterium, as described by Bazylinski et al. (1995), could produce both minerals, forming magnetite in the oxic zone and greigite in the anoxic zone. This further implies that local oxygen and/or hydrogen sulfide concentrations regulate the type of biomineral formed, but it also hints at the possibility that two different sets of genes control the biomineralization of magnetite and greigite. Although none of the greigite-producing bacteria have as yet been cultured, rRNA analysis has shown that they are associated with sulfate-reducing bacteria (DeLong et al., 1993). Therefore, if these bacteria can reduce sulfate, it then raises the question of whether the sulfide ions present in greigite originate from sulfide present in the aqueous environment, or from sulfate reduction occurring within the cell.

4.2.3 Amorphous silica

Unlike the numerous microorganisms that passively precipitate amorphous silica from supersaturated fluids, some eukaryotes, such as radiolarians and diatoms, exert complete control over the silicification process. The cells are enclosed in siliceous shells (in this context commonly referred

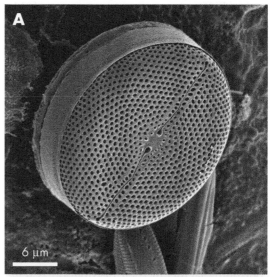

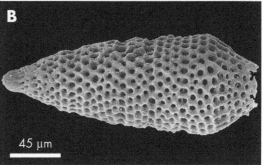

Figure 4.29 SEM micrographs of siliceous eukaryotes. (A) The diatom *Mastogloia cocconeiformis* retrieved from a lagoon in the Grand Cayman, British West Indies (courtesy of Hilary Corlett). (B) Fossil radiolarian *Dictyomitra andersoni* from early Pleistocene sediments of Chatham Island, New Zealand (courtesy of Chris Hollis).

to as opaline silica or opal-A) that tend to have beautifully ornamented structures (e.g., Fig. 4.29). Together, these microorganisms are the major contributors of solid-phase silica fluxes to the seafloor, and it is because of their collective existence that modern oceans (and lakes) are very undersaturated with respect to amorphous silica (Lowenstam and Weiner, 1989). Indeed, prior to the evolution of the radiolarians (and sponges) during the Cambrian, the oceans were probably at equilibrium with respect to amorphous silica (110 mg L^{-1} at less than $25°C$). Thereafter, silica levels began to diminish, and with the evolution of the diatoms around the late Triassic–early Jurassic, and their subsequent proliferation by the middle Cenozoic, seawater silica concentrations progressively declined to modern values of less than 5 mg L^{-1}.

Radiolarians primarily inhabit surface ocean waters, and occupy biogeographical zones comparable with other zooplankton (Racki and Cordey, 2000). Most species are immotile (i.e., they are not capable of movement), and they drift along with currents from one water mass to another. Aside from silica availability, one of the major controlling factors in their distribution is temperature and salinity, with the highest densities found in warm equatorial waters. Radiolarian oozes formed below zones of high productivity can contain as many as 100,000 shells per gram of sediment (Armstrong and Brasier, 2005). One interesting growth strategy radiolarians employ is a symbiotic relationship with algae. When food is scarce, an algal symbiont can provide its host radiolarian with much needed nourishment.

Diatoms are virtually ubiquitous in the hydrosphere, occupying benthic and planktonic niches in both freshwater and seawater. As a group they tolerate an exceptionally large range of temperature, salinity, pH, and nutrient conditions. More than 20,000 modern and fossil species of diatoms are known, 70% of which are marine (Harwood and Nikolaev, 1995). Those marine species, in particular, play an extremely important ecological role, accounting for as much as 40% of the primary productivity in the oceans (Tréguer et al., 1995). In addition, diatoms possess intracellular storage vesicles that acquire and hoard short-term pulses of nutrients while simultaneously depriving competing photosynthetic microorganisms of those essential resources (Tozzi et al., 2004). In this regard, diatoms have periodically been the most significant phytoplanktonic species controlling ocean nutrient cycling, particularly during glacial periods in Earth's history when upper ocean mixing was more vigorous. Their prominence has even been linked to global cooling episodes during Earth's history (Pollock, 1997). Diatoms similarly affect the microbiology of lakes, and mass balance studies have demonstrated that they are responsible for the bulk of silica sedimentation (e.g., Schelske, 1985). As with all autotrophs, their essential environmental requirements include sufficient irradiance to photosynthesize, and, in order to satisfy that demand, continuous residence within the euphotic zone is paramount. In highly productive waters, their shells accumulate in enormous numbers to form a mud known as diatomite.

One of the most interesting paradoxes about radiolarians and diatoms, from a geochemical perspective, is that they expend considerable energy in constructing very elaborate shells composed of a material that is not readily available to them. So two obvious questions arise: (i) how do they form their shells under such seemingly unfavorable conditions; and (ii) why do they not use another mineral, e.g., calcium carbonate, that is easier to form?

The mechanisms underpinning eukaryote silicification are far from resolved (see de Vrind-de Jong and de Vrind, 1997 for details). In the case of diatoms, their cell wall is silicified to form a hard shell, or frustule, comprising two valves, one overlapping the other. New valves are formed within minutes during cell division by the controlled precipitation of silica within a specialized intracellular, membrane-bound silica deposition vesicle, the SDV. To initiate the process, the cells actively pump silicic acid from the external aqueous environment across the plasma membrane and SDV membrane (the silicalemma) with the use of specific transporter proteins (Hildebrand et al., 1997). The energy for this process is driven by photosynthesis (in the light) and glucose respiration (in the dark). Inside the SDV, the silica concentration is increased to a state of supersaturation with respect to amorphous silica. At this stage the monomers polymerize to form nanoscale colloids that adsorb on to the inner face of the silicalemma. It has been estimated that the rate of silicification within diatoms is about 10^6 times higher than abiological formation from supersaturated solutions (Gordon and Drum, 1994).

As the silica is deposited, the SDV increases in size, and there is a concomitant creation of a silica concentration gradient from the borders of the vesicle towards the center as a result of the polymerization process. During this stage, the diatoms exert additional control over silicification because the silicalemma is lined with a mixture of proteins consisting of hydroxyl- and polycationic amino-containing amino acids, such as glycine, serine, and tyrosine (Fig. 4.30). They provide molecular complementarity with polymeric silica such that it sorbs via hydrogen and electrostatic bonding, respectively, to the membrane surface (Volcani, 1983). Actually, the silica binds so strongly to the proteins that only treatment with hydrogen fluoride re-solubilizes the silica. Interestingly, diatoms genetically modify their SDV proteins by inserting more reactive polycationic amino acids when external silica concentrations are low (Kröger et al., 1999). The SDV can take on a number of shapes, and as such, it serves as a template for the manufacture of species-specific shell morphologies. Once a completed valve is formed, a new plasma membrane forms behind it, leaving the old plasma and SDV membranes as an organic casing that protects the siliceous valve against dissolution in the undersaturated waters.

The reason for silica use by diatoms is purely speculative, but it may have its answer in the genetic legacy of when these cells evolved in a more silica-rich hydrosphere and it was energetically "cheaper" to construct a cell wall with silica rather than with organic carbon (Raven, 1983). In any case, there are number of possible benefits to possessing siliceous shells, perhaps the most notable being as armour against predation by zooplankton. For instance, Hamm et al. (2003) have shown that the shells are remarkably strong by virtue of their architecture, and only organisms large enough to ingest them or digest their intracellular contents without opening their shells serve as effective predators. As a result, diatoms typically show lower mortality rates than those of other, smaller algae with similar growth rates. It has also recently been speculated that the silica might play a role in buffering pH, enabling the enzymatic conversion of bicarbonate to CO_2 in waters where the concentration of the latter is less than required for photosynthesis (Milligan and Morel, 2002).

Despite their need for silica, the growth rates of diatoms remain independent of dissolved silica concentrations until they reach 0.1 mg L^{-1} of SiO_2 or less. Silica, therefore, seldom becomes a limiting factor except during intensive diatom blooms, and if this does occur, the diatoms either produce weakly silicified shells or the blooms collapse and they are promptly succeeded by blooms of other nonsiliceous phytoplankton, such as the coccolithophores or cyanobacteria (Schelske and Stoermer, 1971). Furthermore, although ocean surface waters are inherently undersaturated with silica, sufficient quantities are temporarily available at any given time due to a very effective recycling process in which more than 95% of the siliceous shells on their way to, or in, the bottom sediment, are dissolved. This explains how the estimated total present-day silica production by siliceous eukaryotes (2.5×10^{16} g yr^{-1}) is 25 times the input of silica to the oceans from rivers, submarine weathering, and submarine volcanism (Heath, 1974). The efficient recycling process comes about because once the cells are dead, the plasma membrane and silicalemma are degraded by chemoheterotrophic bacteria residing in the water column and seafloor, and the amorphous silica that makes up the shells suddenly finds itself exposed, and in acute disequilibrium with the undersaturated waters (Bidle and Azam, 1999). The siliceous shells rapidly dissolve, with the silica re-circulated to the surface waters by diffusion or upwelling. Zooplankton grazing additionally affects silica re-cycling, as the silica is repackaged into fecal pellets that are transported rapidly to the seafloor.

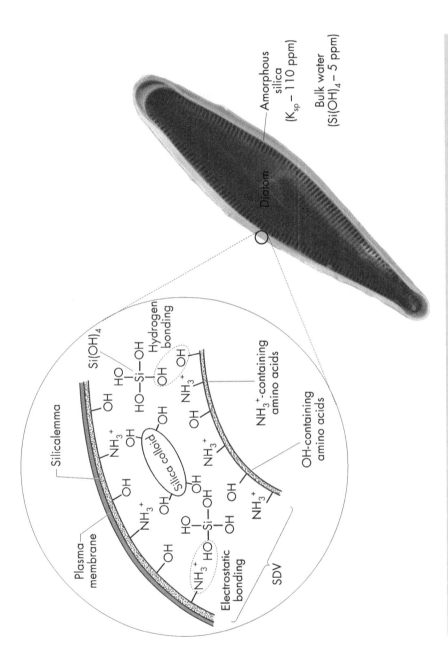

Figure 4.30 Representation of how diatoms form their siliceous shells. Despite living in undersaturated solutions with respect to amorphous silica, diatoms actively extract $Si(OH)_4$ from solution and pump it into an intracellular silica deposition vesicle (SDV) that lines the inside of the plasma membrane. There, the concentration of silica is increased to supersaturation. The SDV also contains hydroxyl and cationic amino-containing amino acids that react with colloidal silica, and thus facilitate the subsequent nucleation and mineralization stages. In this regard, the SDV acts as a template, for silicification.

Notwithstanding the efficient recycling, large areas of the seafloor today (some 15%) are still covered in amorphous silica sediments because silica is less soluble in the colder temperatures associated with the seafloor and there is a progressive loss of reactivity with aging upon sedimentation (Van Cappellen, 1996). These silica-rich deposits are generally associated with regions of active upwelling, where high phytoplankton growth in the surface waters leads to high sedimentation and fast burial rates. Globally, sediments rich in diatom debris are concentrated in the northern and equatorial Pacific Ocean, and around the Antarctic continent, the latter accounting for the majority of the total silica sink (DeMaster, 1981). In turn, some of the diatom and radiolarian shells become rapidly insulated from the undersaturated bottom ocean waters. Not only does this allow the pore waters in this relatively closed system to attain equilibrium with amorphous silica, leading to diminished rates of shell dissolution, but the deposited layers may become sufficiently thick and impenetrable that they prevent irrigation of the surface sediments by benthic infauna, i.e., those animals that live in the sediment (Pike and Kemp, 1999). In the equatorial Pacific today, deposits some 4–6 meters thick may contain over 400 million diatoms per gram (Armstrong and Brasier, 2005).

In the rock record, there are some very significant diatomaceous deposits, such as the several hundred meter thick Belridge Diatomite, within the Monterey Formation, an organic-rich deposit that formed on a Miocene continental margin off the coast of California. Biogenic deposits formed between the Cambrian and Cretaceous contain instead abundant radiolarian shells, i.e., radiolarite deposits (Racki and Cordey, 2000). Their preservation was generally better considering that seawater had higher silica levels in the early Phanerozoic.

4.2.4 Calcite

Calcium carbonate constitutes the largest fraction of known biologically controlled biominerals. The most impressive carbonate precipitation results

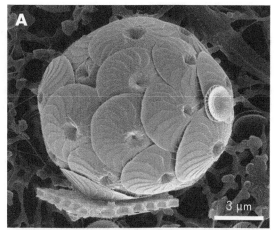

Figure 4.31 SEM micrographs of calcareous eukaryotes. (A) The coccolithophore *Calcidiscus leptoporus* collected off the coast of Namibia (courtesy of Markus Geisen). (B) The foraminifera *Spirillina vivipara* as retrieved from a lagoon in the Grand Cayman, British West Indies (courtesy of Hilary Corlett).

from the coccolithophores and foraminifera. These unicellular algae and protozoa, respectively, remove vast amounts of calcium carbonate (in the form of calcite) from seawater to form their shells, and today they are considered to be the most important carbonate-secreting organisms on Earth (Fig. 4.31).

The coccolithophores have been abundant in the oceans since the Jurassic, yet a population

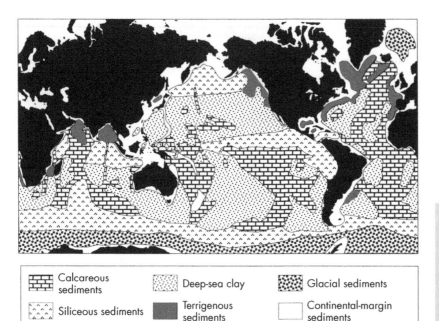

Figure 4.32 The present-day distribution of the principal types of marine sediments. (From Andrews et al., 2004. Reproduced with permission from Blackwell Publishing Ltd.)

Legend:
- Calcareous sediments
- Siliceous sediments
- Deep-sea clay
- Terrigenous sediments
- Glacial sediments
- Continental-margin sediments

explosion in the early Cretaceous saw a massive accumulation of carbonates deposited worldwide. Near the end of the Cretaceous, the coccolithophores suffered a mass extinction; two-thirds of the 50 genera disappeared at that time, though many new groups appeared later in the Palaeocene (Armstrong and Brasier, 2005). The enormous extent to which they have historically precipitated micritic sediment can best be emphasized by the fact that the Cretaceous chalk deposits of north-west Europe are formed almost exclusively from them. Furthermore, over the past 150 Ma (since their first appearance), carbonate sedimentation in the open ocean accounts for 65–80% of the global carbonate inventory (Andrews et al., 2004). These deep-sea deposits, which average 0.5 km in thickness, mantle half the area of the deep ocean (Fig. 4.32).

Modern coccolithophores are abundant at mid to high latitudes coastal areas in waters varying from 2°C to 28°C. *Emiliania huxleyi* is one of the most abundant coccolithophore species, and when nutrients are sufficiently available, blooms with cell densities of 10^8 cells L^{-1} can cover several thousand square kilometers, to a depth where sufficient light exists to support photosynthesis. Blooms of this magnitude can be responsible for the deposition of thousands of tons of calcite (Holligan et al., 1983). Such large masses of suspended calcite crystals also affect the light-scattering properties of the surface waters, and as a result, the blooms can even be detected from space with satellite imagery (see Plate 9). Significantly, the abundance of coccolithophores influences not only the transfer of carbon from the atmosphere to the ocean sediment, as well as the marine calcium budget, but the cells also generate large fluxes of dimethyl sulfide (DMS) that leads to considerable albedo affects over the open ocean, which ultimately may affect global climate change (Westbroek et al., 1993).

The foraminifera play a prominent role in marine ecosystems as micro-omnivores, feeding on bacteria, protozoa, and small invertebrates. They are found in all marine environments, from the intertidal to the deepest ocean trenches, and from the tropics to the poles. Most of the estimated 5,000 extant species live in the world's oceans. Of these, 40 species are planktonic, while the remaining species live on the bottom of the ocean, on shells, rock, and seaweeds, or in the sand and mud (see Murray, 1991 for details). They can be very abundant, comprising over 90% of the deep-sea biomass, with bottom sediments almost exclusively made up of their shells. The oldest foraminifera are from the earliest Cambrian. Much the same as the coccolithophores, they underwent a mass extinction at the end of the Cretaceous, but then experienced a rapid radiation in the Palaeocene (Tappan and Loeblich, 1988). The foraminifera produce ornate calcite shells that range in size from ~30 µm to 1 mm. Because of their intricate morphologies, fossil foraminifera have been widely employed as biostratigraphic markers, i.e., they can be used to establish the relative stratigraphic position of sedimentary rocks between different geographic localities. Moreover, the shells themselves have proven extremely useful as palaeoenvironmental indicators of ancient ocean salinity, water temperatures, surface productivity, and even global climate (e.g., Waelbroeck et al., 2002).

Similar to the silica-secreting eukaryotes, there is still some uncertainty regarding the mechanisms underpinning coccolithophore and foraminifera shell formation. In the case of coccolithophores, they have an internal vesicle that serves as the locus for calcite formation (de Vrind-de Jong and de Vrind, 1997). Import of Ca^{2+} from $CaCO_3$-saturated seawater occurs passively through Ca^{2+} channels into the cytoplasm. However, the Ca^{2+} then proceeds against a concentration gradient to get into the vesicle, requiring energy in the form of ATP. The exact mechanism for transferring Ca^{2+} from the cytoplasm into the vesicle is unknown, but it is hypothesized that specific transport enzymes are involved. To complete calcite formation, HCO_3^- is also introduced into the deposition vesicle:

$$HCO_3^- + Ca^{2+} \longleftrightarrow CaCO_3 + H^+ \quad (4.24)$$

Because the concentration of CO_2 in seawater may be limiting for photosynthesis, coccolithophores have the ability to utilize HCO_3^- instead (recall Fig. 4.15). The net result of this is the generation of OH^-. However, unlike the cyanobacteria that excrete the OH^- ions into the surrounding aqueous environment, it appears that in the coccolithophores OH^- ions are by protons generated by the calcifying vesicle (reaction 4.24). This maintains the pH within the depositional vesicle at an appropriate level for calcification. The subtle differences between cyanobacteria and coccolithophore calcification clearly highlight their differing activities; the former are simply photosynthesizing and inducing calcite precipitation as a byproduct of their metabolism (i.e., they do not apparently use the calcite), while the latter controls the process intracellularly in order to precipitate a crystal with precise size and orientation.

Some algae that comprise reef-building communities also incorporate calcium carbonate into their structures as strengthening agents. These consist of the marine red algae (e.g., *Corallina* sp.) that deposit high-magnesium calcite within their cell walls. They possess sulfated galactans and alginates that preferentially bind Ca^{2+} over Mg^{2+}, thereby creating localized microenvironments that favor calcite precipitation over aragonite. In addition, the cells appear to be capable of regulating the crystallography and orientation of the calcite crystals (Borowitzka, 1989). The importance of the coralline algae in the carbon cycle lies in the fact that, unlike the coccolithophore shells formed in the open oceans, that upon sedimentation into deeper waters (3000–5000 m) re-dissolve back into the water column (due to increased pressures, decreased temperatures, and increased CO_2 concentrations below the calcite compensation depth), the biogenic reef carbonates are semipermanent features in the Earth's sedimentary record, and therefore represent an enormous carbonate sink.

4.3 Fossilization

Most microorganisms lack substantial hard parts and rarely fossilize. Thus, their soft tissues are rapidly degraded and evidence of their existence is wiped away with time. Even among the larger, multicellular groups, the fossil record is nowhere near complete. Despite these inherent shortcomings, the limited microfossil assemblages that exist have proven to be indispensable to our understandings about the evolution of life on Earth because they provide a physical record that represents the geological and environmental conditions of the time when the organisms were living.

As will be discussed in Chapter 7, fossilized prokaryotes are known from very ancient Precambrian rocks, some potentially as old as 3.5 billion years. Considering that there is no evidence to suggest that those microfossils represent species that controlled biomineralization, it suggests that something unique about the species and/or the conditions under which they grew allowed for their preservation (Konhauser et al., 2003). The type of mineralization associated with soft-part preservation of ancient life forms is predominantly in the form of silica. This is unsurprising as the small size of the sorbing silica colloids, relative to the cell, allows for the entire outer surface to be completely enshrouded in a protective mineral coating. Other fossilizing minerals, due to their generally larger crystal size, are of secondary importance in terms of their ability to preserve intact cells.

4.3.1 Silicification

Considering that so many ancient microorganisms are fossilized in silica, there have been surprisingly few studies trying to elucidate the physical changes imposed on microorganisms during silicification. Based on the assumption that Achaean microfossils were cyanobacteria, Oehler (1976) was one of the first to experimentally subject various cyanobacterial genera to colloidal silica solutions over different lengths of time. What he showed was that at temperatures of ~100°C several months were required for complete silicification, and only slight alteration to the cells occurred, while at higher temperatures (165°C) the cells mineralized quickly, but the

filaments fragmented and coalesced, intracellular components were destroyed, and there was a preferential preservation of the sheath and wall material. The importance of the sheath in limiting silicification to the outer surfaces of the cell has already been discussed in section 4.1.6, and it is interesting that when some cyanobacteria, such as *Calothrix* sp., are grown in silica supersaturated conditions, their sheaths double to triple in width, up to 10 µm in diameter (Phoenix et al., 2000). In the case of *Calothrix*, the findings suggest that they genetically respond to high silica concentrations by adapting their surface structure to isolate the cell from the damaging effects of silicification. Crucially, this growth response forms a morphological feature that may be preservable, thereby giving palaeontologists a clue for recognizing their ancient predecessors in the rock record.

Sheaths are not, however, a prerequisite for survival in silica-saturated geothermal waters. *Oscillatoria*, for example, is a cyanobacterium that is either not ensheathed or is thinly sheathed, yet it has been isolated from various hot springs. Studies on unsheathed bacteria have also shown that some produce robust and durable crusts after only a week of silicification, whereas others maintain delicately preserved walls that are only lightly mineralized (Westall, 1997). Only at very high silica concentrations does significant loss of shape and cellular detail occur (Toporski et al., 2002). Recently, exposure of *Sulfurihydrogenibium azorense*, of the order *Aquificales*, demonstrated a new twist on coping with high silica concentrations; it produced protein-rich biofilms that facilitated silicification, but away from the cell surface (Lalonde et al., 2005). By regulating biofilm production appropriately, the *Aquificales* could potentially contribute to or accelerate silicification, though the cells themselves are unlikely to be preserved.

What these studies have collectively shown is that, although the silicification of biomass is an inevitable process in silica-supersaturated solutions, there remains species-specific patterns of silicification, and ultimately different preservation potentials with regards to incorporation into the rock record. Unfortunately, at present only a few microorganisms have been analyzed, and in each study different experimental conditions were used. As a consequence not only do the different studies yield conflicting results regarding the rates and magnitude of silicification, but no comprehensive database is presently available with which to confidently assess what is required for a siliceous microfossil to form.

From what we presently know about silicification, there are at least three main factors that lead to the short-term preservation of intact cell structures.

1 The timing and rate of silicification relative to death of the microorganisms is of paramount importance. When silicification is rapid, recently lysed cells may resist decay, thereby retaining intact morphologies within a relatively impermeable matrix (e.g., Fig. 4.33). Actually, a limited degree of decomposition may help facilitate silicification by exposing cytoplasmic material for hydrogen bonding. Silicification also limits heterotrophic microorganisms from completely degrading the cells prior to their incorporation into the sedimentary record. By contrast, experimental studies have shown that unmineralized cells begin to degrade within days after death (Bartley, 1996). As a result, the remnants of most cells in nature become progressively obscured. This helps explain the general rarity of recognizable microfossils in the Archean rock record, except under conditions of extremely early lithification (see Box 7.5 for details).

2 In Precambrian cherts there is a preservational bias towards cells that had thick cell walls and/or sheaths (Knoll, 1985). This is unsurprising since peptidoglycan and the polymers that comprise the sheath are more resilient to degradation than other cell components, and as long as the constituent autolysins are deactivated, they can persist in the environment long after the cell dies. Furthermore, those structures are more amenable to silicification. Therefore, in terms of preservation potential, fossil assemblages in the Precambrian may be biased towards microorganisms, such as some cyanobacteria, simply because they possessed suitable ultrastructures (Golubic and Seong-Jao,

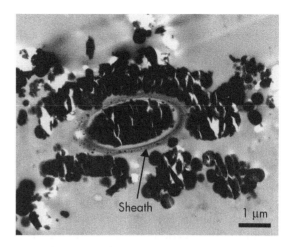

Sheath

1 μm

Figure 4.33 TEM image of a lysed cell, growing as part of a cyanobacterial mat, in silica-rich hot spring waters at Krisuvik, Iceland. This cell exhibits both epicellular and intracellular silicification, with only the sheath and cell wall remaining intact (From Konhauser et al., 2004. Reproduced with permission from the Royal Swedish Academy of Sciences.)

1999). Conversely, other cells with different cellular features degraded and left little evidence of their original organic framework (Horodyski et al., 1992). Aside from the actual preservation of the cell itself, there is some putative evidence to suggest that fossilized EPS is widespread in the rock record (Westall et al., 2000). Whether the structures interpreted in thin section micrographs are indeed biological needs to be verified, but the premise of their existence is reasonable considering that EPS is volumetrically more important than the cells within the biofilm, and as shown above, EPS does provide abundant sites to facilitate silicification.

3 Ferris et al. (1988) showed that the binding of metallic ions, in particular iron to bacterial cell surfaces, was an important contributing factor to the silicification of *Bacillus subtilis*: cells not pre-stained by Fe suffered extensive lysis after several days of aging. This inhibition of cell degradation appears to be related to the ability of some metals to deactivate the cells' own autolytic enzymes. Correspondingly, Walter et al. (1992) suggested that the fossil record is also prejudiced towards those cells that tolerate elevated salinities.

The subtleties of the silicification process are critical because they may control the appearance of the preserved microorganisms and the features that are needed to identify them in terms of extant taxa. Many of the taxonomically critical features of microorganisms are lost during silicification or are concealed by mineral precipitate. Thus, a silicified microorganism analyzed under SEM and/or TEM may display only a few distinct features (e.g., size, general morphology, presence/absence of sheath, septa; and rarely cytoplasmic components) that can be used for identification purposes. Therein lies the problem for microorganism identification by such techniques. For instance, Castenholz and Waterbury (1989) listed 37 characteristics that have been used in the identification of cyanobacteria. Unfortunately, as demonstrated by Jones et al. (2001), silicification may selectively mask and/or destroy some features while preserving others. This can lead to a silicified microorganism that fails to display key features that indicate what it looked like prior to mineralization. A case in point, glass slides left in a silica-supersaturated hot spring pool (with 450 mg L^{-1} SiO_2) at 70°C for only 90 hours showed abundant silicified microorganisms, but silicification concealed essentially all identifiable features that allowed for their recognition (Jones et al., 2004). Simultaneously, silicification can generate artefacts that appear to be microorganisms. In the same hot spring pool, amorphous silica grains (300–400 nm in diameter), that are centred around a mucus strand that is less than 10 nm thick, are morphologically similar in appearance to many of the silicified filamentous microorganisms, yet these "pseudofilaments" are not cellular in origin (Fig. 4.34). Even in the most well-preserved silicified microorganisms only a few of the taxonomically important characteristics can be recognized. It is therefore not surprising that the silicified biota found in hot-spring sinters are usually characterized by low diversities despite the fact that the microorganisms seem to be so well preserved; silicified cells from such settings typically contain less

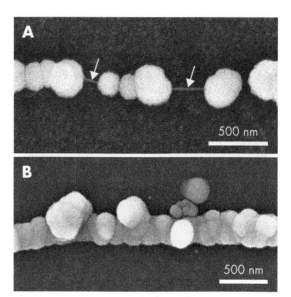

Figure 4.34 SEM images showing the formation of "pseudofossils." These samples were collected from a glass slide left for 90 hours in a New Zealand hot spring where silica concentrations were 450 mg L^{-1}. (A) Nascent silica beads forming on a mucus strand (arrows). (B) The mucus strand is completely covered with silica grains, potentially giving the false appearance of a silica-encrusted filament. (From Jones et al., 2004. Reproduced with permission from the Geological Society, London.)

than 10 morphologically defined taxa; the most taxa yet described from a silicified hot spring sinter is 19 (Jones et al., 2003).

One fundamental point that seems to have escaped close scrutiny is that most of the Precambrian microfossils (as will be described in Chapter 7) are in chert, yet the original microorganisms would have been mineralized by amorphous silica. Therefore, irrespective of how the microorganisms contributed to silicification and the preservation of intact residues, the transformation from amorphous silica to chert would have eliminated most of the morphological evidence of an organic origin. In fact, experiments show that during the transformation from opal-A to opal-CT (cristobalite), most of the detailed wall structures of diatoms are destroyed (Isaacs, 1981). Furthermore, the subsequent phase change from opal-CT to quartz eliminates any remaining wall details, although gross morphology may yet be preserved (Riech and van Rad, 1979).

4.3.2 Other authigenic minerals

In nonsiliceous environments, cell preservation operates in a fine balance between decay and mineralization. On the one hand heterotrophic microorganisms consume the lysed cells, while on the other hand, the metabolic byproducts of their metabolism promote elevated saturation states and mineral encrustation of the cellular remains. As might be expected, given our discussions above, alkalinity generation and phosphate/sulfide release lead to the formation of a wide variety of authigenic minerals.

In experiments with decaying shrimp, where the system was freely open to diffusional-related processes, calcium carbonate forms in a manner typical of biologically induced biomineralization. Although calcite retains the gross morphology of the soft tissue, it completely obliterates the finer detail. Contrastingly, when the system is closed and diffusion limited, the organic material is replicated in a more detailed manner in calcium phosphate; the source of the phosphate being the shrimp itself (Briggs and Kear, 1993). The underlying determinant for which mineral forms appears to be pH; more alkaline values lead to $CaCO_3$, while more neutral values lead to $CaPO_4$. Paradoxically, exceptional preservation of soft tissue in fossils requires elevated, rather than restricted, microbial activity as this leads to anaerobically driven mineral authigenesis (Sageman et al., 1999). It is also important that mineralization occurs quickly, as once the morphology is stabilized by initial mineralization, the potential for preservation of soft tissue record is greatly enhanced.

Bacterial communities themselves are often involved in the preservation of soft tissues

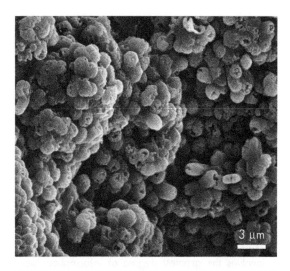

Figure 4.35 SEM image of the fabric in a phosphorite layer from the Triassic Bravaisberget Formation of Spitzbergen, Norway. Aggregates of apatite globules encapsulate remnants of microbial cells. It is presumed that the globules formed as a result of rapid phosphate precipitation on coccoid bacteria that inhabited the surface layers of an organic and phosphate-rich sediment. (Courtesy of Krzysztof Krajewski.)

(e.g., Martill, 1988). They occur as phosphatized replicas replacing the soft tissues, often with detail at the subcellular level. In addition to the presence of characteristic microbial fabrics, in some phosphorites, apatitic molds retain remnants of the original microorganisms (e.g., Fig. 4.35). Examples include the presence of filamentous cyanobacteria in the Mishash Formation, Israel; the reported fungal mat remains in the Tertiary phosphorites of Morocco; and the globule-like clusters of apatite-encrusted coccoid bacteria in the Triassic Bravaisberget Formation of Spitzbergen (Krajewski et al., 1994).

The preservation of soft parts in pyrite is a rarer feature in ancient sediments. At present, the best described examples are associated with the Hunsrück Slate of Budenbach, Germany, the Beecher's Trilobite Bed of New York State, and the Burgess Shale of British Columbia. In general three modes of pyrite fossilization can be recognized (Canfield and Raiswell, 1991):

1 *Mineralized tissue* – Refractory tissues such as cellulose and chitin may be preserved by the precipitation of pyrite in their pore spaces.

2 *Mineral coats* – The preservation of very degradable soft parts most commonly occurs by outline pyritization, but it rarely preserves internal structures because the crystals are too coarse and form too late to replicate the finest details. This typically occurs as a pyritized layer of bacteria that pseudomorph the original structure.

3 *Mineral casts or molds* – This style of preservation involves the greatest degree of information loss, since only the fossil outline is preserved. The casts and molds result from diffuse early diagenetic pyritization in the surrounding sediments.

Pyritization can also preserve shells when it replaces or coats the carbonate minerals (see Plate 10). Replacement under these situations occurs when the solid-phase carbonate is dissolved by H^+ (reaction (4.25)). The loss of protons can then potentially lead to a state of supersaturation with respect to FeS, as reaction (4.26) is driven from left to right:

$$CaCO_3 + H^+ \rightarrow Ca^{2+} + HCO_3^- \qquad (4.25)$$

$$Fe^{2+} + HS^- \rightarrow FeS + H^+ \qquad (4.26)$$

The limited nature of fossil pyritization implies that specific sedimentary conditions must have existed at the site of organic decay during the time of burial (Raiswell, 1997). One constraint is that pyritization took place before compaction of the soft tissue. Another requirement is that the generation of dissolved sulfide, formed at the expense of the decaying organisms, had to proceed at a comparatively slow rate because sulfidic pore waters contain negligible concentrations of iron. In other words, pyritization needs to be limited to the decay site, but if $H_2S \gg Fe^{2+}$, then no ferrous iron would likely be proximal to the organic material. Based on modern sediment studies, such conditions could have been met during the earliest stages of sulfate reduction, at shallow burial depths in the suboxic zone, where Fe(II)-rich pore waters (from Fe(III) reduction) were supersaturated with respect to iron monosulfides (Fig. 4.36). Sulfur isotope data correspondingly suggests that ^{32}S-enriched sulfide, produced at the expense of the organic matter, was unable to diffuse away from the decay site and into the surrounding sediments. Consistent with this, the host sediments must have contained relatively high concentrations of ferric oxyhydroxides at deposition.

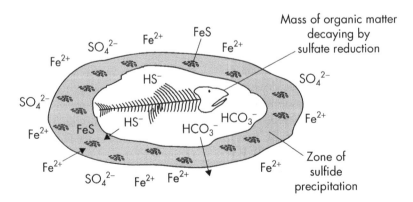

Figure 4.36 Model of organic matter pyritization. Its oxidation by SRB releases HS^- into the pore waters immediately surrounding the degrading biomass. Those pore waters, however, must have sufficiently high Fe^{2+} so that FeS (and ultimately FeS_2) precipitation is confined to the decay site, instead of HS^- diffusing away into the surrounding sediments. (Adapted from Raiswell, 1997.)

4.4 Summary

Microorganisms form an immense variety of authigenic minerals. In the majority of cases, bacterial biomineralization is a two-step process, where cations are initially bound to the anionic ligands of the cell's surface, and subsequently they serve as heterogeneous nucleation sites for mineral precipitation. The biogenic minerals appear identical to those produced abiologically because they are governed by the same thermodynamic principles. As the latter stages of mineralization are inorganically driven, the type of biomineral formed is inevitably dependent on the available counter-ions, and hence the chemical composition of the waters in which the microorganisms are growing. For far fewer microorganisms, biomineralization is a regulated process. The formation of these minerals is a significant drain on their energy reserves, so the mineral formed must be of considerable importance to the functioning of the cell. In either case, the influence biomineralization has on elemental cycling in aqueous and sedimentary environments cannot be overstated. Modern Fe, Mn, Si, Ca, P, C, and S cycles are all affected by biomineralizing processes. Although individual grains are micrometer in scale, if one takes into account the total amount of biomineralizing biomass, it is not difficult to imagine how they can represent a significant geological driver that partitions elements from the hydrosphere into the sediments. In this regard, the precipitation of carbonate minerals by microorganisms is especially relevant because these minerals represent the final products in the weathering of silicate minerals, and a long-term sink for atmospheric carbon dioxide (see section 5.1.6). Importantly, biomineralization has been occurring over geological time, as is evident by the common occurrence of limestone, BIF, chert, sulfide, and phosphorite deposits in the rock record.

5

Microbial weathering

Chemical weathering of the Earth's upper crust includes two major types of processes, mineral dissolution and mineral oxidation. Microorganisms play a fundamental role in both. They attach to exposed mineral surfaces, coat them with extracellular polymers (EPS), and physically disrupt the grains in their attempt to gain access to nutrients and energy in the underlying substrata. At the same time, they create a complex microenvironment at the mineral–water interface, where metabolically catalyzed redox reactions and the generation of acids and complexing agents lead to pH and concentration gradients markedly different from the bulk solution. This often promotes a state of thermodynamic disequilibrium that fosters faster rates of chemical weathering. Microbial EPS further serves as a site for the precipitation of secondary minerals, with compositions and morphologies distinct from those inorganically precipitated in the bulk solution. Aside from its localized effect on degrading individual mineral grains or even entire rock outcrops, microbial weathering has profoundly influenced the Earth's surface environment over geological time. Microorganisms have expedited soil formation since their evolution onto land during the Archean, while the solutes released through chemical weathering have affected the composition of the hydrosphere, from microscale soil pore waters to the enormity of the oceans. The biochemical weathering of some silicate minerals, and the subsequent deposition of calcium carbonate on the seafloor, are also linked through a feedback cycle that impacts atmospheric CO_2 levels, and ultimately the global climate. This chapter will examine how microorganisms influence weathering of the Earth's crust, and then consider how such processes can have both environmental and commercial ramifications.

5.1 Mineral dissolution

5.1.1 Reactivity at mineral surfaces

Chemical weathering rates of minerals are controlled by their composition, morphology, and texture, as well as the geochemistry of the surrounding fluids. The primary determinant underpinning whether a mineral dissolves or not is the competition between: (i) the strength of the chemical bonds holding the crystal structure together, i.e., Coulombic interactions; and (ii) the hydration energy of ions at the mineral's surface (Banfield and Hamers, 1997). If the Coulombic interactions are large but the hydration energy is small, the solid is insoluble, whereas in the opposite case, the solid dissolves easily.

Although these relationships can be used to infer the thermodynamic properties of the mineral in a given solution, they say little about the overall rates of dissolution. Furthermore, during the dissolution of any given mineral there are a number of intermediate chemical steps that take place before an atom or molecule is dissolved from the surface, and each one of those steps has a specific reaction rate (Morse and Arvidson, 2002). Those steps include:

1 transport of reactants through the solution to the mineral surface;

2 adsorption of the reactants to the mineral surface;

3 migration of the reactants on the surface to an "active" site;

4 the actual chemical reaction between the adsorbed reactant and the mineral, which may additionally involve several intermediate steps where bonds are broken and hydrolysis occurs;

5 migration of the hydrated ions away from the reaction site and desorption into solution;

6 transport of the products away from the mineral surface into the bulk solution.

As in any chemical reaction, one of the above steps will be the slowest, the so-called "rate-limiting step". Steps 1 and 6 involve the diffusive or advective transport of reactants and products through the solution, and when either of these steps are rate-limiting, the reaction is said to be transport- (or diffusion)-controlled. Steps 2–5 occur on the mineral surface, and when one of them is slowest, the reaction is surface-controlled. The dissolution of highly soluble minerals tends to be transport-controlled, such that ions are detached so rapidly from the surface that they build up in concentration to form a saturated solution adjacent to the mineral surface. Dissolution is then regulated by the dispersal of those ions into the surrounding undersaturated bulk solution. Crystals dissolved in this manner will exhibit smooth surfaces because ion detachment occurs over the entire surface so quickly that crystallographically controlled surface features, such as etching, do not occur. By contrast, relatively insoluble minerals have surface-controlled reaction rates until very high degrees of disequilibrium are achieved. This means that ion detachment is sufficiently slow that they cannot build up at the minerals surface. Instead, the ions diffuse or advect away from the surface rapidly enough that their concentration at the surface is equal to that in bulk solution

(Berner, 1978). Thus, for most of the minerals discussed in this chapter, steps 2–5 can be viewed as an activation energy barrier that restricts the rate of hydrolysis and the subsequent transfer of the hydrated ions into the surrounding solution. This process costs energy that is recovered once the ion is removed completely.

There are a number of important variables governing the kinetics of dissolution:

1 *Structure of the crystal lattice* – Mineral dissolution rates are related to the strength of the metal–anion bonds. Some minerals (e.g., feldspar) dissolve slowly because they possess an extensively cross-linked structure of silica tetrahedra. In terms of feldspar, the magnitude of dissolution depends on the relative abundance of Al and Si sites at the mineral surface, with Al sites more susceptible to dissolution than the Si sites. As a result, feldspar is subject to selective leaching, though there tends to be a sufficient amount of unreactive bonds left near the mineral surface to maintain integrity once the reactive constituents are removed. Conversely, minerals with poorly cross-linked fabrics, such as olivine, dissolve rapidly and uniformly (Casey and Bunker, 1990).

2 *Orientation of the crystal surface* – Atoms at surfaces always have higher free energy than atoms in a three-dimensional crystal because the former have lower coordination and strongly asymmetric bonding configurations compared with atoms within the bulk crystal. Thus, crystals tend to adopt shapes that minimize their surface free energy (Herring, 1951), that being the surface area and interfacial free energy terms in equation (4.3). Similarly, as the crystal size decreases, its surface reactivity increases because small particles have relatively high surface area:volume ratios.

3 *Defects on the crystal surface* – Not only do particular crystals have different energies, but different locations on an exposed crystal surface have variable energies. For example, most crystal surfaces are not consistently flat, but instead have a stepped topography (Fig. 5.1). Atoms at these step edges have a lower coordination (but higher energy) than those on the flat portions because they have two sides exposed to the solution, and therefore it is easier to form a complete hydration shell around them so as to reduce the activation

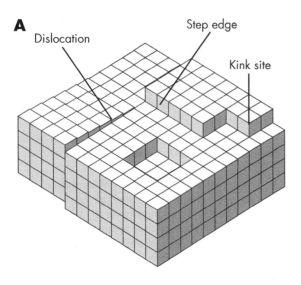
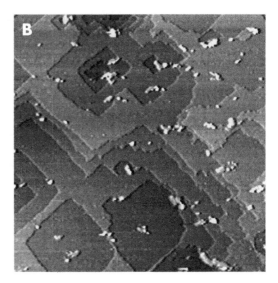

Figure 5.1 (A) Diagram illustrating how mineral surfaces are not perfectly flat, but instead have a number of imperfections, such as step edges, kink sites, and dislocations. (B) Scanning tunneling microscope image of the surface of the mineral galena, highlighting some of those defects. (Courtesy of Steve Higgins.)

barrier to hydrolysis. Kink sites, where step edges turn, are most reactive, with three sides exposed to solution and fewer bonds to adjacent ions. Crystals also have various types of imperfections. Those imperfections include dislocations and planar defects (e.g., stacking faults, grain boundaries), whereby rows of atoms in the crystal are slightly out of place, and hence more energetic than "perfect" surfaces. These then become strongly preferred sites of chemical reactivity, marked by selective etching and growth of pits on the underlying mineral surface (Drever, 1988).

4 *Adsorbed molecules* – Mineral dissolution rates are often diminished if a particular cation or anion is adsorbed that can block access of water to a specific reactive site, e.g., adsorption of phosphate decreases calcite dissolution rates (Berner and Morse, 1974). This is particularly the case at kinks, where because of their excess surface energy, they are preferred sites for the adsorption of ions. By contrast, rates of dissolution are enhanced by the adsorption of protons. They induce a redistribution of the overall electron charge on the minerals' surface, which then fosters the slow rupture of the oxygen–metal bonds in the crystal lattice (Casey and Ludwig, 1995). In Chapter 3,

we covered how the interface between a cell's surface and the surrounding aqueous solution is characterized by an electrical potential that arises from the ionization of surface functional groups. The same concepts hold true for mineral surfaces because most have at least a monolayer of adsorbed water, and their surface functional groups, commonly containing oxygen or hydroxyl ligands, undergo protonation and deprotonation reactions (e.g., reactions (5.1) and (5.2)):

$$>\text{Al-OH} + \text{H}^+ \rightarrow >\text{Al-OH}_2^+ \quad \text{low pH} \qquad (5.1)$$

$$>\text{Al-OH} + \text{OH}^- \rightarrow >\text{Al-O}^- + \text{H}_2\text{O} \quad \text{high pH} \quad (5.2)$$

5 *Reduction or oxidation* – Redox dissolution reactions are important since electron exchange alters the oxygen–metal bond strengths. Iron and manganese oxyhydroxides are subject to reductive dissolution, whereas sulfides and some silicate (e.g., Fe_2SiO_4) and metal oxides (e.g., Cu_2O) are prone to oxidative dissolution. In either case, the concentration of the electron-donating and -accepting species, as well as the activities of H^+ and OH^-, are important parameters in determining dissolution rates (Hering and Stumm, 1990).

5.1.2 Microbial colonization and organic reactions

Immediately upon exposure of a rock at the Earth's surface, a community of bacteria, algae, fungi, and/or lichens attach to the newly available solid surfaces (see Plate 11). For the microorganisms, the minerals in the rock are a rich source of bioessential elements, available only if they can extract them from the crystal lattice. Indeed, in oligotrophic (nutrient-poor) terrestrial environments, mineral solubilization and elemental cycling can be requisite for the microbial communities survival (e.g., Konhauser et al., 1994). To attain those nutrients, colonizing microorganisms do two things: (i) they physically penetrate into the rock causing disaggregation of the mineral; and (ii) they produce organic acids that act as dissolving agents.

(a) Physical processes

As microorganisms colonize rock surfaces, fungal filaments (called hyphae) exploit cracks, cleavages, and grain boundaries to gain access to new mineral resources. In doing so, they cause several alteration features, ranging from simple surface roughing by etching and pitting to extensive physical disintegration of the minerals (Barker et al., 1997). The latter includes detachment, separation, and exfoliation of some constituent grains along cleavage planes (e.g., Fig. 5.2). Minerals without cleavage planes, e.g., quartz, show no such features. Furthermore, grain boundary misfits at the interface between minerals, as well as the new pore spaces created through volume changes associated with the conversion of primary minerals into secondary clay phases, provide sub-nanometer-scale conduits that are exploited as weak points by the fungal hyphae (e.g., Barker and Banfield, 1996).

Bacteria also enshroud all exposed mineral surfaces in EPS (recall Fig. 3.24). From a weathering perspective, the ability of these compounds to retain water helps promote mineral fracturing and it increases the residence time for water to fuel hydrolysis and other chemical reactions (Welch et al., 1999). EPS also serves as a substrate for heterotrophic bacteria, some of which generate acids (see below) that facilitate chemical attack on the underlying minerals (e.g., Ferris and Lowson, 1997).

These biological processes work in tandem with frost wedging, diurnal thermal expansion, and alternate wetting-drying processes to physically break the rock down into the smaller lithic fragments that are more susceptible to dissolution by rain and the effects of organic acids. Then, as the minerals become loosened, macrofauna (e.g., nematodes) accentuate the erosional process through mechanical abrasion as they graze (Schneider and Le Campion-Alsumard, 1999). Eventually, the original rock is transformed into the finer-grained mineral component comprising primitive soils.

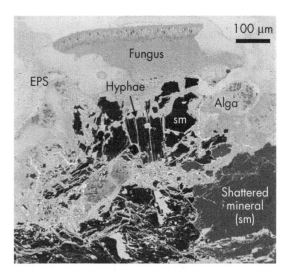

Figure 5.2 SEM image of the interface between the crustose lichen *Porpidia albocaerulescens* and the rock syenite. Notice the extensive nature of mineral shattering and how some of the mineral grains have been exfoliated from the rock by the fungal hyphae (arrow) and subsequently coated in EPS. (Reprinted from Barker and Banfield, 1996 with permission from Elsevier.)

(b) Role of endoliths

Rocks are regularly colonized by endolithic bacteria that grow within the natural cavities and fractures. Those that are photosynthetic, such as the cyanobacteria, are often evident as a distinct blue-green layer 1–10 mm below the rock surface, where sufficient light penetrates (Vestal, 1988). The role of endolithic microorganisms in weathering of limestone and dolomite has been well documented (e.g., Pentecost, 1992). Their main contribution is to actively bore into the host rock by solubilizing cementing mineral grains. This generates more room for their growth, as well as the macrofauna that graze upon their lysed cells. In some carbonate rocks, it has been estimated that endolithic communities average more than half a million cells per cm^2 (Golubic et al., 1970). Such high population densities have the effect of significantly enhancing erosion rates, and along the Adriatic coast, for example, they contribute to an estimated 2 kg m^{-2} of coastline dissolved annually (Schneider and Le Campion-Alsumard, 1999).

Examination of sandstone outcrops (e.g., in South Africa) has shown that endoliths also contribute to the onset of chemical weathering through the process of substratum alkalinization, which involves the cells producing sufficient hydroxyl ions, as a byproduct of photosynthesis (recall section 4.1.7), to increase the pH up to 11 (Büdel et al., 2005). These values are not only high enough to enhance bulk silica dissolution in the endolithic zone, but the associated shift in the carbonate speciation facilitates some minor precipitation of carbonates. As a result of the dissolution process, the upper portion of the rock is loosened and then eroded away by wind and flowing water. These weathering patterns bear a striking similarity to sandstone outcrops elsewhere, such as the Ross Desert of Antarctica (Friedmann and Weed, 1987). Such exfoliative processes not only modify landscape geomorphology, but it has even been proposed that they may be responsible for denuding entire

mountain ranges, albeit over geological time (Büdel et al., 2005).

(c) Production of organic acids

Once bacteria and fungi become established on a newly exposed mineral surface, they immediately begin to accelerate dissolution through the production of organic acids. The majority of organic acids they generate are byproducts of fermentation and/or various intermediate steps of the aerobic respiration of glucose, but some microorganisms further excrete organic acids when growth is limited by the absence of an essential nutrient. Many of the fungal acids contain multiple carboxyl groups that dissociate at circumneutral pH (Berthelin, 1983). For example, oxalic acid has two pK_a values at pH 1.3 and 4.2, while citric acid is a tricarboxylic acid with three pK_a values at 3.1, 4.7, and 6.4. The low pK_a values makes both citric and oxalic acids fairly strong acids. Each lichen also produces a unique suite of compounds, called lichen acids, that are synthesized by the fungi from carbohydrates supplied by the phycobiont (Easton, 1997). Some 300 compounds unique to lichens have been identified.

The organic acids increase mineral dissolution both directly and indirectly. In the first instance, the majority of the organic acids dissociate into organic anions and protons. Some of the protons then react with the ligands of the mineral surface (i.e., protonation reactions), causing a weakening of the metal–oxygen bonds, and ultimately the release of a metal cation from the surface. Concurrently, the organic anions react with metal cations at the mineral surface, similarly destabilizing the metal–oxygen bonds, and promoting dissolution through the formation of a metal–chelate complex. Eventual detachment of the chelate exposes underlying oxygen atoms to further protonation reactions. Therefore, systems with high concentrations of tridentate (three pK_a) or bidentate (two pK_a) organic acids tend to contribute to high levels of ion release, while monofunctional groups, such as acetic acid, have a lesser effect (e.g., Welch and Ullman, 1993). Of all the acids listed above, oxalic acid is the most abundant in natural systems (reaching millimolar concentrations), and it has frequently been observed that oxalic acid production is correlated with high solute availability (e.g., Ca^{2+}) in soils and the reprecipitation of oxalate salts, such as calcium oxalate (e.g., Braissant et al., 2004). In aquifers, concentrations of tens of micromolar have been reported, the lower values indicative of the fact that the organic acids are often not produced *in situ*, but instead migrated in from adjacent organic-rich soils (McMahon and Chapelle, 1991).

Deprotonated organic anions (e.g., oxalate, citrate) indirectly affect dissolution rates by complexing with metals in solution (compared to the mineral's surface as above), thereby lowering the solution's saturation state (e.g., Bennett et al., 1988). EPS acts in a similar manner, particularly those rich in alginate, which contain an abundance of reactive carboxyl groups (Welch et al., 1999). Some organic anions are very strong chelators, and depending on the relative concentration of the anions versus metal cations in solution, pH, and the stability constants of the various complexes, they can effectively partition a metal cation that has dissolved from the mineral into an organo-metallic complex. As an example,

the oxalate anion ($C_2O_4^{2-}$) is a bidentate ligand that can form a four-member chelate ring when it binds to a divalent metal (Fig. 5.3). Trivalent metals that normally form an octahedral six-coordinated complex (e.g., Al^{3+}, Fe^{3+}, Cr^{3+}) can bind three oxalates to form an anionic complex (Gadd, 1999). Such chelation process are important for metal mobility because some, such as Al^{3+} liberated during silicate dissolution, would naturally precipitate as gibbsite ($Al(OH)_3$) at pH values >3, were it not complexed into an organic form (reaction (5.3)). It is only after the oxalate component is microbially oxidized that aluminum is finally precipitated.

$$Al^{3+} + 3C_2O_4^{2-} \rightarrow Al(C_2O_4)_3^{3-} \qquad (5.3)$$

Along similar lines, chelation of Al^{3+} and Fe^{3+} by oxalate anions increases nutrient availability to vegetation because inhibiting gibbsite and ferric hydroxide precipitation permits higher phosphate levels in soil pore waters (Graustein et al., 1977).

Citrates are also strong metal chelators (Fig. 5.3). In soils, this has important implications for Al^{3+} mobility because aluminum–citrate complexes render the metal less toxic to plant roots (Jones and Kochian, 1996). However, citrate's chelating ability may have an adverse environmental impact when it facilitates the leaching

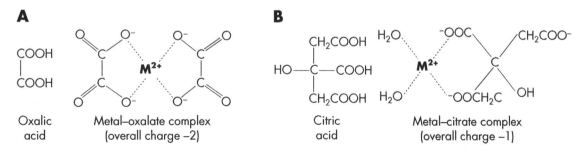

Figure 5.3 (A) Structure of oxalic acid and a metal–oxalate complex. (B) Structure of citric acid and a metal–citrate complex.

and subsequent mobilization of toxic metal contaminants away from soils or waste disposal sites. For instance, uranium forms a complex of two uranyl ions (UO_2^{2+}) and two citrate molecules involving four carboxyl and two hydroxyl groups. This complex is not easily degraded by bacteria, and consequently the radionuclides could enter into public water supplies if left unattenuated (Francis et al., 1992).

(d) Production of siderophores

Siderophores are another example of multidentate organic ligands that form strong complexes with metal cations. They are by definition Fe(III) specific, and show higher formation constants ($\log K_f = 10^{25}-10^{50}$) than low molecular weight organic acids, such as oxalic acid ($\log K_f = 10^8$). They are also found in reasonably high abundance, averaging around 10^{-6} mol L^{-1} in soil pore water (Hersman, 2000). Taking into account a 1:1 binding of siderophore to Fe, and assuming that each siderophore is only used once (although in actuality many are re-used), siderophores alone could remove up to 10^{-6} mol L^{-1} of Fe from solution. This chelating ability, in turn, will invariably affect the dissolution of Fe(III)-bearing oxyhydroxide and silicate minerals.

During dissolution of ferric oxyhydroxides, for example, the coordination of the Fe(III) in the crystal lattice is altered, such that it exchanges its O^{2-} or OH^- ligands for water or an organic ligand. In proton-promoted dissolution, H^+ is adsorbed to the metal surface causing polarization of the neighboring >Fe-OH or >Fe-O groups. In either case, this leads to a weakening of the Fe(III)-anion bond and the subsequent detachment of Fe^{3+} into solution (Stumm and Sulzberger, 1992).

Although siderophores are generally better able to chelate dissolved Fe(III) species because they can form a complete five-member ring, the hydroxamate groups of a siderophore can also bind to Fe atoms on mineral surfaces. In the case of goethite, the dissolution process begins with the adsorption of a single hydroxamate group

to the mineral's surface, followed by structural re-arrangement and dewatering, and eventually detachment of a molecule of Fe(III)-hydroxamate (Holmén and Casey, 1996). Even when a trihydroxamate siderophore, such as deferriferrioxamine, is added to goethite, only one Fe(III) center is coordinated at a time (Cocozza et al., 2002). Siderophores can similarly dissolve hematite at rates comparable to oxalic and ascorbic acids, or to dissolution induced by proton adsorption (Hersman et al., 1995). In fact, siderophores likely work in concert with protons and other organic ligands to promote ferric oxyhydroxide mineral dissolution.

There have also been a number of studies on Fe-silicate dissolution promoted by siderophores or their commercial analogs. Common granitic soil minerals, such as hornblende (a predominantly Fe(II)-containing amphibole with some substituted Fe(III)), are rapidly dissolved in the presence of catecholate-producing bacteria of the genus *Streptomyces*. Experiments have shown that within just a matter of days after inoculation, there was a fivefold increase in Fe release compared to abiological controls (Fig. 5.4), and a doubling of cell mass in hornblende-containing cultures relative to control cultures with *Streptomyces* only (Liermann et al., 2000). Moreover, the bacteria penetrated so deeply into pits and cracks that neither chemical treatment nor extreme heating could fully remove them from the mineral's surface (Fig. 5.5). Other catecholate siderophore-producing microorganisms, including bacteria of the genus *Arthrobacter*, similarly enhanced dissolution rates (Kalinowski et al., 2000b). In both sets of experiments, adding more siderophores only temporarily increased dissolution rates. At some stage the hornblende surfaces became Fe-depleted, and dissolution rates declined nonlinearly, i.e., in a manner similar to the sorption isotherms from Chapter 3. It is also interesting to note that when the commercially available hydroxamate siderophore, desferrioxamine mesylate (DFAM), was used, comparable rates of dissolution occurred, suggesting that it was siderophore

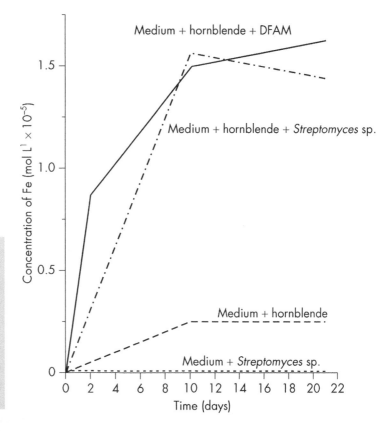

Figure 5.4 Comparison of Fe released during hornblende dissolution experiments as a function of time between buffered solutions, with and without the soil isolate *Streptomyces* sp., and with a commercially available hydroxamate siderophore, desferrioxamine mesylate (DFAM). (Modified from Liermann et al., 2000.)

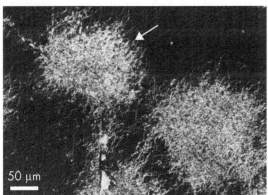

Figure 5.5 Differential interference contrast microscopy image of dendritic colonies of *Streptomyces* sp. on the surface of hornblende. The arrow points to one of the residual colonies still adherent on the mineral after vigorous cleaning with acetone and the enzyme lysozyme. (Reprinted from Liermann et al., 2000 with permission from Elsevier.)

production, irrespective of type, that was driving mineral dissolution.

There is a direct relationship between microbial growth, iron availability, and siderophore/chelate production, as first recognized by Page and Huyer (1984). Using a common N_2-fixing bacterium, *Azotobacter vinelandii*, it was demonstrated that the cells generated different types and amounts of siderophores and organic chelates depending on the iron mineral present: (i) the bacterium solubilizes marcasite (FeS_2) by producing dihydroxybenzoic acid (DHBA); (ii) solubilization of vivianite and olivine (($Mg,Fe)_2SiO_4$) occurred due to the production of the siderophore azotochelin, plus DHBA; (iii) hematite, goethite, siderite, and pyrite induced production of azotochelin and a second siderophore, azotobactin, plus DHBA; and (iv) ilmenite ($FeTiO_3$) and Fe-rich illite caused excess production of azotochelin and azotobactin, plus

DHBA. The sequential production of DHBA and two siderophores would seem to indicate that as the availability of Fe decreased (due to increasing insolubility of the Fe minerals), the microorganisms respond by producing more siderophores. What is clear from that study, and others like it, is that different siderophores are required to sequester Fe from different minerals, and that changing the iron mineralogy can elicit a specific response from the same microorganism.

Although produced in response to Fe stress, siderophores can also inadvertently complex a number of other trivalent metals (e.g., Cr^{3+}) that have a similar ionic potential to Fe^{3+} (Birch and Bachofen, 1990). Despite forming lower stability complexes, divalent metals can also be sequestered by siderophores. In this case, ionic potential is not an issue because only two-thirds of the ligands (e.g., hydroxamate groups) are utilized. Yet, when a trivalent metal is available, all three hydroxamate groups are used, and the siderophore complex must not only be able to wrap around the metal cation, but its ligands must also be properly arranged (Hernlem et al., 1996). Given that many different metals can be chelated, siderophores likely play an important role in accelerating the dissolution of a number of minerals in the environment.

5.1.3 Silicate weathering

Approximately 30% of all minerals are silicates and it is estimated that 90% of the Earth's crust is made up of silicate-based material. During weathering, silicates typically undergo incongruent dissolution, in which most of the easily exchangable (base) cations, such as Ca^{2+}, Mg^{2+}, K^+, and Na^+, and variable amounts of aluminum and silica, are leached out of the crystal lattice, leaving behind a residual clay phase (e.g., reaction (5.4) or metal oxide (e.g., reaction (5.5)):

$$2KAlSi_3O_8 \text{ (K-feldspar)} + 2H^+ + 9H_2O \rightarrow$$
$$Al_2Si_2O_5(OH)_4 \text{ (kaolinite)} + 4Si(OH)_4 + 2K^+$$
$$(5.4)$$

$$2KAlSi_3O_8 + 2H^+ + 14H_2O \rightarrow$$
$$2Al(OH)_3 + 6Si(OH)_4 + 2K^+ \qquad (5.5)$$

What governs the type of secondary mineral formed is: (i) the composition of the primary mineral phase, i.e., felsic versus mafic; (ii) the concentration of dissolved ions at the interface between the leached layer and the intact mineral surface, and the extent to which they are removed from the weathering zone; and (iii) the kinetics of the weathering reaction, which can be affected by the temperature and the amount of water through-flow. In the latter case, notice how the products in reaction (5.5) are more degraded as a consequence of more water on the reactant side of the reaction.

(a) Felsic mineral dissolution

The mechanisms and rates by which feldspar minerals dissolve have received more attention than any other minerals. This is because they constitute some 70–80% of the labile minerals in the upper continental crust, so their dissolution has important bearing on freshwater composition and secondary mineral formation. Experimental studies consistently show that the rate of feldspar dissolution is low, and essentially pH-independent in the range of 5–8, due to their extensively cross-linked structure (Brady and Walther, 1989). However, dissolution rates increase as the acidity increases, and below pH 5, feldspar dissolves by a factor of $a_{H^+}{}^n$, where "n" is the fractional dependence of mineral dissolution on proton activity.

As discussed above, protons have a tendency to adsorb onto mineral surfaces, where they induce rearrangement of charge in the silicate lattice, and concomitantly the hydrolysis of surface Al–O–Si bonds. This has the combined effect of releasing charge-balancing cations and creating a leached zone immediately overlying the intact crystal (e.g., Blum and Lasaga, 1988). The protons then temporarily substitute for the displaced cations, maintaining a local charge balance. Dissolution rates subsequently reach a steady state when cation release is equal to the rate by which they are exchanged for by H^+ (e.g., Helgeson et al., 1984).

Weak inorganic acids, such as carbonic acid, exert minimal effect on silicate dissolution rates because they do not deprotonate completely under circumneutral pH values, and hence generate insufficient protons for chemical attack; they do, however, afford enough acidity to dissolve carbonates (see Box 5.1 and section 5.1.4). Even in soils with high aerobic respiration rates, the carbonic acid generated seldom decreases the pH to values below 4.5. On the other hand, production of sulfuric and nitric acids (from oxidation of reduced sulfur and nitrogen compounds, respectively) causes severe but localized pH changes (e.g., Sand and Bock, 1991). When the concentrations of those acids become sufficient, they may even cause congruent dissolution of the primary mineral phase (reaction (5.6)). In the reaction below, the dissolution of K-feldspar converts a strong acid (sulfuric) into a weak acid (silicic), that can move through a soil or sediment in the undissociated form.

$$KAlSi_3O_8 + 2H_2SO_4 + 4H_2O \rightarrow$$
$$Al^{3+} + 3Si(OH)_4 + K^+ + 2SO_4^{2-} \quad (5.6)$$

Box 5.1 Inorganic carbon speciation

The combination of CO_2 with water forms carbonic acid (H_2CO_3). Despite being a relatively weak acid (it does not readily give up all its hydrogen ions when dissolved in water), it can still accelerate the dissolution of some soluble mineral phases, such as calcite or aragonite. This type of hydrolysis reaction subsequently transforms the carbonic acid to bicarbonate, which is the dominant form of soluble inorganic carbon at pH values between 6.5 and 10.3:

$$CaCO_3 + H_2CO_3 \rightarrow Ca^{2+} + 2HCO_3^-$$

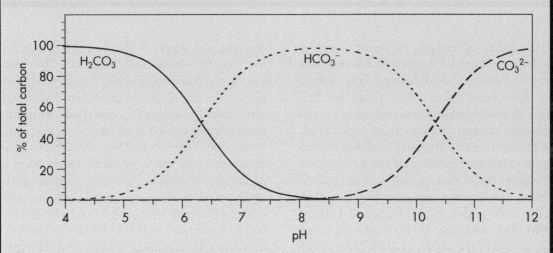

pH-dependent speciation of inorganic carbon in solution.

The ability of carbonic acid to cause carbonate mineral weathering can be exemplified with two examples. At atmospheric p_{CO_2} levels (approximately 3×10^{-4} atm), and at saturation with respect to calcium carbonate, the equilibrium pH is 8.3, and the concentrations of Ca^{2+} and HCO_3^- are $10^{-3.30}$ mol L^{-1} (20 mg L^{-1}) and $10^{-3.00}$ mol L^{-1} (60 mg L^{-1}), respectively. At p_{CO_2} levels of $3 \times$ 10^{-2} atm, characteristic of soil pore waters where biological activity greatly enhances the production of CO_2, the pH at calcium carbonate saturation decreases to 7.0, and the concentrations of Ca^{2+} and HCO_3^- increase to $10^{-2.65}$ mol L^{-1} (90 mg L^{-1}) and $10^{-2.36}$ mol L^{-1} (266 mg L^{-1}), respectively.

continued

Box 5.1 *continued*

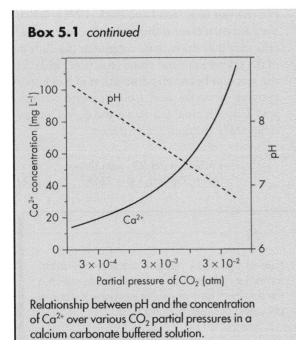

Relationship between pH and the concentration of Ca^{2+} over various CO_2 partial pressures in a calcium carbonate buffered solution.

What this simple example highlights is that an increase in the partial pressure of CO_2 increases the concentration of carbonic acid in solution, causing more calcium carbonate to dissolve, and thus the concentrations of dissolved Ca^{2+} and HCO_3^- to increase. This pattern continues as long as CO_2 in water can be replenished by exchange with the atmosphere. Conversely, a decrease in CO_2 partial pressure (due to photosynthetic activity or degassing) or loss of water (due to evaporation) causes the solution to become supersaturated, resulting in the precipitation of calcium carbonate until equilibrium is restored. The latter process is commonly manifest in the precipitation of carbonate cements in arid soils (e.g., caliche) or the formation of calcite speleothems that line limestone caverns, such as stalagmites and stalactites.

Dissolution of feldspar by organic acids is much more effective. Of importance here are the citric and oxalic acid-producing strains of fungi that have been described in nature as effectively degrading feldspar minerals into various secondary mineral products (e.g., Jones et al., 1981). Experimental studies have further demonstrated that these organic acids can increase rates of feldspar dissolution by orders of magnitude, relative to solutions containing inorganic acids of the same acidity (e.g., Welch and Ullman, 1996). The maximum rate of dissolution occurs near the pH of the organic acid pK_a, when both protons and organic anions are made available to react with the mineral surface. Organic acids also influence the composition of the residual silicate phase because they preferentially solubilize Al-O bonds (Al sites are more prone to organic ligand attack and protonation than Si sites), while in their absence, secondary minerals remain enriched in Al (Stillings et al., 1996).

Physical evidence for microbial weathering comes from surface etch marks that approximate the size and shape of the bacteria colonizing the mineral surface (e.g., Fig. 5.6). To some extent the preferential orientation of etch pits along cleavage planes suggests that dissolution was crystallographically controlled, with the microorganisms taking advantage of structural weak points (Hiebert and Bennett, 1992). Once attached, those microorganisms then create a nanoscale reaction zone where organic acids and metabolites are concentrated on the mineral surface at discrete sites of high reactivity. Microbial colonization, and the extent of etching, also depends on mineral composition. Using *in situ* microcosms, where mineral surfaces were directly exposed to indigenous bacteria within an aquifer, Bennett et al. (1996) showed that after several months, the surfaces of microcline (a K-rich feldspar) were much more deeply weathered than those of albite (a Na-rich feldspar). This pattern of dissolution was attributed to the nutritional requirements of the colonizing bacteria because they were likely potassium limited (K was very low in the groundwaters),

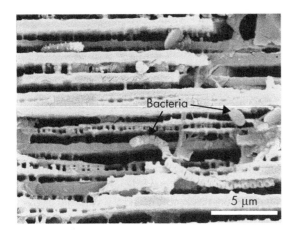

Figure 5.6 SEM image of the surface of a potassium feldspar crystal collected from a peat soil. Note the grooves where etch pits have coalesced. Several different types of bacteria were observed colonizing the mineral surface (arrows). (Courtesy of Martin Lee and Ian Parsons.)

and thus were attacking the K-rich feldspars preferentially.

No discussion on felsic silicate dissolution would be complete without a brief mention of quartz. It is the most stable solid phase of silica, with the highest proportion of unreactive silicate groups. Unlike most other silicate minerals, quartz dissolution is unaffected by acidity except at extremely low pH (<2), where high proton concentrations disrupt silica bonding, or at pH values higher than 8, when deprotonation of surface Si–O–H bonds occurs (Brady and Walther, 1990). As such, it weathers extremely slowly under normal surface conditions, preserving the Si atom in tetrahedral coordination in solution (reaction (5.7)). This insolubility yields concentrations in most surface waters of ~6 mg L^{-1} at 25°C, although most freshwaters have considerably higher dissolved silica concentrations that reflect feldspar hydrolysis.

$$SiO_2 + 2H_2O \rightarrow Si(OH)_4 \quad (5.7)$$

Despite quartz's resistance to dissolution, there is ample evidence from both modern and ancient environments that indicates preferential desilication relative to aluminum or iron, i.e., in tropical soils. The increased mobility of silica is likely brought about by dissolved organic compounds that form soluble silica chelates. This causes a lowering of the silicic acid concentration and a concomitant acceleration of quartz dissolution, as evidenced by quartz grains covered in crystallographically oriented etch pits and solution channels (Bennett and Siegel, 1987). Also, at alkaline pH, citrate forms a bidentate complex with quartz, involving two anionic oxygen ligands interacting with two hydroxyl protons adsorbed onto the quartz surface (Bennett et al., 1988). The interaction might initially take the form of a weak electron donor–acceptor complex, but then the partial electron charge is transferred from the organic anion to the silica molecule, increasing the electron density of the terminal Si–O bond and invariably making them more susceptible to hydrolysis. The combined effect of reducing soluble silica levels and stripping silica from the quartz surface leads to increased quartz dissolution.

(b) Mafic mineral dissolution

The mafic minerals make up a smaller proportion of the continental crust, and their weathering involves both dissolution and oxidation-reduction reactions. Olivine, pyroxene, amphibole, and biotite are enriched in magnesium and ferrous iron, and they weather rapidly in oxic environments as the Fe^{2+} is initially released through congruent dissolution (e.g., reaction (5.8)), and then oxidized and hydrolyzed to ferric hydroxide (reaction (5.9)):

$$Fe_2SiO_4 \text{ (olivine)} + 4H^+ \rightarrow 2Fe^{2+} + Si(OH)_4 \quad (5.8)$$

$$2Fe^{2+} + 0.5O_2 + 5H_2O \rightarrow 2Fe(OH)_3 + 4H^+ \quad (5.9)$$

Unlike feldspar, these minerals display little resistance to weathering because of the relative lack of Si–O–Si cross-linking. Instead, the minerals consist of isolated silicate tetrahedra attached by cation bridges. Therefore, in an acid solution, those silicate groups convert intact to silicic acid as the >Fe–O bonds are protonated; no hydrolysis reaction is needed. Any leached layer of olivine, for instance, will be thin as there are no bridging oxide bonds to maintain integrity once the metal cations are removed (Wogelius and Walther, 1991).

In Chapter 1 it was mentioned that bacteria reside deep within flood basalts, where they eke out a chemolithoautotrophic living from available sources of H_2, some organically sourced and some possibly via water–rock interactions. Direct evidence of microbial involvement in basalt weathering, however, comes from studies of terrestrial lava flows and the pillow basalts associated with seafloor volcanism. On land, weathering is facilitated largely by lichen growth, in which the primary rock-forming minerals undergo oxidation and dissolution as a consequence of respiratory CO_2 production and the excretion of organic and lichen acids. This can lead to variable stages of etching and mineral fragmentation (e.g., Jones et al., 1980). Coincident with oxidative dissolution is cation release, which, in turn, can lead to authigenic mineralization, generally consisting of fine-grained calcium oxalate crystals, along with the formation of ferric hydroxide and various Fe(II)/Fe(III)–silicate assemblages.

Microorganisms are also instrumental in altering the kinetics of dissolution, but with variable results. For instance, chemical weathering rates of recent Hawaiian lava flows colonized by lichens have been reported to be at least 100 times that of bare rock (Jackson and Keller, 1970). At low pH, calculations even suggest that olivine dissolution could support significant populations of Fe(II)-oxidizing bacteria (Santelli et al., 2001). Yet, experimental studies have also shown a decrease in long-term dissolution rates of Fe-rich olivine (fayalite) that is attributed to the precipitation of an unreactive alteration rind on the mineral surface, which eventually limits fluid exchange with the bulk weathering solution (Welch and Banfield, 2002). This is to be

expected at pH values where the ferric iron released reprecipitates on the weathering surface. Clearly, the extent of dissolution is governed by a balance between the net release of products to solution versus the evolution of the mineral surface morphology and reactivity, both of which can be site specific.

Basaltic glass is one of the most abundant and reactive phases in the oceanic crust. Its alteration begins on a very localized scale with an initial loss of cations yielding a leached zone several micrometers thick, followed by a variable degree of dissolution of the silica-rich residues and re-polymerization to form a porous silica network that eliminates nonbridging oxygen atoms (Thorseth et al., 1992). Bacteria are believed to play an important role in the initiation of dissolution reactions because in experiments they rapidly formed micrometer-thick biofilms on fresh glass surfaces (Staudigel et al., 1995). After only weeks to months of colonization, the glass displayed preferential dissolution at points along fractures, that subsequently developed into pronounced etch marks in a years time (Fig. 5.7). It is interesting to note that different bacteria were dominant on

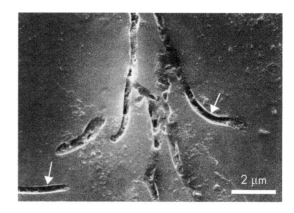

Figure 5.7 SEM image of a polished glass surface left in continuously flowing seawater for 410 days. Removal of the overlying biofilm reveals significant corrosion, as evident from the etch grooves (arrows), some of which exceed 10 μm in length and 0.5 μm in width. (Reprinted from Staudigel et al., 1995 with permission from Elsevier.)

the glass surface at different stages of the experiments, likely a result of the changing pH and redox conditions during dissolution. Consequently, some of the late-stage microorganisms lived within those pits, where they accumulated a range of elements derived from the glass (e.g., Fe, Al, and Si). Often the concentration of elements was sufficiently high to lead to the secondary formation of fine-grained, geochemically heterogeneous material, including palagonite, silicate clays, and (Fe, Al)-hydroxides. Other experiments showed glass dissolution rates on the order of 1 μm annually (Thorseth et al., 1995). If similar rates can be applied throughout the seafloor, then glass dissolution could significantly contribute to the chemical budget of the ocean, particularly for those elements that are selectively leached (e.g., Mg, Na, Ca). It has also been hypothesized that since mafic minerals contain the bulk of transition metals in the upper crust, some of which are bioessential nutrients that have low concentrations in seawater (e.g., Fe), their increased availability through biological weathering may actually have a significant impact on the marine food chain (Staudigel et al., 1998).

5.1.4 Carbonate weathering

Carbonate minerals comprise about 20% of Phanerozoic sedimentary rocks. They are amongst the most reactive minerals found in abundance on the Earth's surface, dissolving congruently at rates that are orders of magnitude faster than that of silicate minerals and at much higher pH values (Morse, 1983). Two important carbonate weathering reactions are the dissolution of calcite and dolomite, respectively:

$$CaCO_3 + H_2CO_3 \rightarrow Ca^{2+} + 2HCO_3^- \quad (5.10)$$

$$\begin{aligned} CaMg(CO_3)_2 + 2H_2CO_3 \rightarrow \\ Ca^{2+} + Mg^{2+} + 4HCO_3^- \end{aligned} \quad (5.11)$$

These two reactions have a number of environmental implications, one of the most important

being that carbonate dissolution directly affects fluxes of Ca^{2+} and HCO_3^- to the hydrosphere. It has been estimated that over 90% of all large world rivers have chemical compositions dominated by limestone and dolomite dissolution (Meybeck, 1979). As such, oceans are chemically buffered against extreme pH fluctuations, in that the addition of excess acid or base has little impact on seawater pH:

$$H^+ + CaCO_3 \rightarrow Ca^{2+} + HCO_3^- \quad (5.12)$$

$$Ca^{2+} + HCO_3^- + OH^- \rightarrow CaCO_3 + H_2O \quad (5.13)$$

Despite reactions that show calcite being dissolved by H_2CO_3, it is usually H^+ that serves as the weathering agent (reaction (5.12)) because the rate of hydration of dissolved CO_2 to form H_2CO_3 is too slow to be as effective as H^+ (Berner and Morse, 1974). In fact, rates of calcite dissolution are transport controlled below pH 4. Furthermore, unlike feldspar dissolution, where the weathering products might accumulate on the mineral surface causing the rate of dissolution to decrease, the carbonate ions that detach from the crystal surface are chemically altered to bicarbonate. Thus, in most cases a state of saturation cannot be achieved at the mineral–solution interface, so calcium carbonate continues to dissolve at low pH. A similar argument can be made at circumneutral pH, but the availability of free H^+ is diminished, leading instead to surface-controlled dissolution rates. Kinks are favored sites for carbonate dissolution, and it has been observed that as a kink ion is removed, a new kink is formed adjacent to the old one. Hence, dissolution can be envisaged as the formation and migration of kinks and the consequent retreat of steps until there is widespread dissolution of the entire surface (e.g., Lasaga and Luttge, 2001).

In nature, prolonged dissolution of limestone and dolomite is evidenced by the pockmarked surfaces characteristic of karst topography. The surficial features include extensive pavement networks with fissures and solution-widened joints,

30 m

Figure 5.8 Spectacular karst topography showing sinkholes in Permian limestone along the Little Colorado River. Scale is approximate. (Courtesy of Louis Maher.)

pinnacles, ridges, canyons, lakes, and sinkholes (Fig. 5.8). Subsurface features include caves and their speleothem deposits. Despite being commonly attributed to abiological dissolution, numerous studies have described the presence of biofilms covering limestone and dolomite, as well as epilithic bacteria and fungi that bore into, and dissolve the underlying carbonates (e.g., Ferris and Lowson, 1997). The extent of karst development, in turn, is a function of mineral hardness, with calcite being more prone to dissolution and endolithic boring than dolomite (e.g., Jones, 1989).

5.1.5 Soil formation

(a) Decomposition of organic matter

Mineral dissolution and organic matter accumulation eventually conspire to form the first layers of soil. The organic fraction is a rather transitory constituent, lasting from only a few hours to several thousand years. This variation occurs, in part, due to the differences in decomposition rates amongst different compounds; starches, proteins, and polynucleotides degrade very quickly; cellulose and chitin have intermediate rates; while lignins degrade very slowly (Table 5.1). Organic decomposition rates also vary according to the

type of environment, with higher rates in the hot and humid tropics versus those in the cold and relatively arid tundra. When fresh organic matter is added to soil, three general sequences of degradation take place (Brady, 2002):

1 Initially, the bulk of the material becomes degraded via the release of hydrolytic enzymes from aerobic bacteria and fungi. As long as there is plenty of fresh organic material available, the number of soil microorganisms remain high (up to 10^9 cells g^{-1}); often the microbial biomass accounts for one-third of the organic fraction in soil (Fenchel et al., 2000). Through fermentation and respiration, the easily degradable polysaccharides are converted into CO_2, some of which volatilizes and ultimately escapes into the atmosphere, while the remainder reacts with water to produce carbonic acid. As discussed above, the carbonic acid then contributes to chemical dissolution of some soil minerals. If the quantity of easily degradable organic matter present is high, its decomposition via aerobic respiration can also lead to a temporary increase in the "biological oxygen demand" (BOD), a term that refers to the quantity of oxygen required to oxidize organic matter.

   Proteins and polynucleotides are simultaneously degraded into their constituent amino acids and nucleic acids, respectively. In turn, these are further broken down into simple inorganic ions such as NH_4^+, NO_3^-, SO_4^{2-}, and $H_2PO_4^-$. Organic compounds also release various cations, such as Ca^{2+}, Mg^{2+}, and K^+. The process that produces these inorganic forms is called mineralization, not to be confused with the formation of minerals. Of these components, nitrate and sulfate are commonly lost due to leaching; phosphate is retained as a calcium fluorapatite phase or other insoluble secondary minerals, while the cations enter into the soil solution, where they are either taken up by roots, become adsorbed onto negatively charged colloids, clays and microbial surfaces, or they are leached from the system, particularly if the soil minerals are protonated (i.e., an acid soil).

2 As soon as the easily degraded carbon is exhausted, cell numbers decline. Cellulose is the most common type of polysaccharide in land plants, yet it is moderately difficult to degrade. Thus, some cellulose tends to remain in the residual organic fraction even after prolonged microbial attack. Chitin behaves in a similar manner.

Table 5.1 Relative degradability of organic compounds. (Data compiled by De Leeuw and Largeau, 1993.)

Organic compounds	Occurrence	Preservation potential
Starch	Vascular plants; some algae; bacteria	−
Fructans	Vascular plants; algae; bacteria	−
DNA/RNA	All organisms	−
Proteins	All organisms	−
Xylans	Vascular plants; some algae	−/+
Pectins	Vascular plants	−/+
Mannans	Vascular plants; fungi; algae	−/+
Galactans	Vascular plants; algae	−/+
Alginic acids	Brown algae	−/+
Cellulose	Vascular plants; some fungi	+
Chitin	Arthropods; crustaceans; fungi; algae	+
Peptidoglycan	Bacteria	+
Teichoic acids	Gram-positive bacteria	+
Sheaths	Some bacteria	+
Cutins, suberins	Vascular plants	+/++
LPS	Gram-negative bacteria	++
Tannins	Vascular plants; algae	+++/++++
Lignins	Vascular plants	++++
Cutans	Vascular plants	++++

The preservation potential ranges from easily degradable (−), intermediate (−/+ , +, ++), to refractory (+++, ++++).

3 After the moderately degradable organic components are reduced, only the complex and refractory materials (e.g., lignin, resin, and waxes) remain relatively intact. Unlike other polymers, lignins have no regular structure to serve as a target for hydrolytic enzymes, and its degradation requires the collective efforts of a variety of nonspecific enzymes (Kirk and Farrell, 1987). Most of those enzymes require O_2, and, in its absence, anaerobic degradation rates of cellulose and lignins are only about 1–30% of aerobic respiration rates (Benner et al., 1984). As a result, these compounds can persist in soils and sediment for many thousands of years, particularly if they are associated with clay minerals that protect them from microbial decay. Rapid burial into anoxic layers (water-logged soils, swamps, etc.) can also lead to very inefficient mineralization rates: in oxic sediment the decay rate is 2–4% per year versus 0.1–0.000001% in the anoxic zone (Swift et al., 1979).

The poorly degradable residues, collectively known as humic substances, are characterized by their black to brown color and their very fine-grained size. On the basis of resistance to degradation and solubility, humic substances have been classified into three groups (Schnitzer and Khan, 1972). Fulvic acids are lowest in molecular weight, lightest in color, and soluble in both acid and alkali. Humic acids are medium in molecular weight and color, soluble in alkali, but insoluble in acid. Humin is highest in molecular weight, darkest in color, insoluble in both alkali and acid, and most resistant to microbial attack. In environments with high rates of cellulose and lignin burial, and where the environment quickly becomes anoxic, refractory organic materials can accumulate to great thicknesses, resulting initially in peat formation, and, if subjected to increased temperatures and pressures that tend to concentrate carbon, they may ultimately be converted into coal through the process of coalification.

Similar to the organic acids discussed previously, humics play a vital role in metal cycling. They contain at their core abundant polycyclic aromatic rings connected by aliphatic chains of different length to form three-dimensional, flexible biopolymers that possess voids capable of trapping other organic and inorganic com-

ponents. Humics also contain an abundance of reactive functional groups, such as carboxyls, that dissociate at normal pH ranges in soils, sediment, and natural waters (Perdue, 1978). As a result of deprotonation, these anionic ligands can efficiently sorb and chelate a variety of metal cations from solution.

(b) Soil profile development

Soil is the ultimate product of mineral weathering, but even as it accumulates, microorganisms continue to shape its mineralogical and geochemical characteristics into distinct soil horizons (Fig. 5.9). The top of the soil, known as the O-horizon, consists of an accumulation of organic litter that is in various states of decay, from just recently deposited and intact to highly degraded with refractory humins. As rainwater

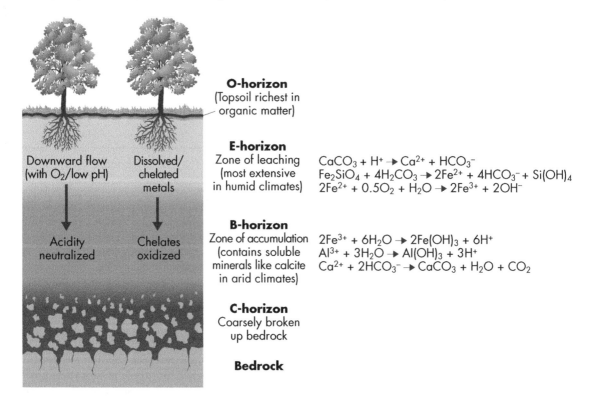

O-horizon
(Topsoil richest in organic matter)

E-horizon
Zone of leaching (most extensive in humid climates)

$$CaCO_3 + H^+ \rightarrow Ca^{2+} + HCO_3^-$$
$$Fe_2SiO_4 + 4H_2CO_3 \rightarrow 2Fe^{2+} + 4HCO_3^- + Si(OH)_4$$
$$2Fe^{2+} + 0.5O_2 + H_2O \rightarrow 2Fe^{3+} + 2OH^-$$

B-horizon
Zone of accumulation (contains soluble minerals like calcite in arid climates)

$$2Fe^{3+} + 6H_2O \rightarrow 2Fe(OH)_3 + 6H^+$$
$$Al^{3+} + 3H_2O \rightarrow Al(OH)_3 + 3H^+$$
$$Ca^{2+} + 2HCO_3^- \rightarrow CaCO_3 + H_2O + CO_2$$

C-horizon
Coarsely broken up bedrock

Bedrock

Downward flow (with O_2/low pH)

Dissolved/ chelated metals

Acidity neutralized

Chelates oxidized

Figure 5.9 An idealized soil profile showing the various horizons. Some of the important dissolution and precipitation reactions are given to highlight the translocation of Fe, Al, and Ca in the profile.

percolates through this organic-rich layer, inorganic and biological processes generate acids of varying strengths that promote intense leaching and removal of the metal cations from the underlying primary minerals. Because the infiltrating waters also contain dissolved O_2, some of those metals are oxidized (e.g., Fe(II) to Fe(III)), and depending on pore water pH, may or may not remain in solution. Organic chelates produced in the O-horizon further aid in the solubilization and transport of metals from the uppermost soil horizon downwards with the infiltrating pore waters. Collectively, these reactions lead to an upper mineral horizon (E-horizon) that becomes progressively enriched in resistant minerals, such as quartz and some metal oxides.

The top two horizons tend to form the bulk of the rhizosphere, the depth to which plant roots extend. This zone has intense microbiological activity adjacent to the plant roots (heterotrophic respiration, N_2 fixation, etc.), and compared to the bulk soil, microbial populations here can be as high as 5×10^9 cells g^{-1} of root tissue (Russell, 1977). In addition to the active microbial communities, the plant roots themselves are important in promoting chemical weathering because they continuously extrude organic acids, and root hairs and their sheaths are rich in organic ligands that sorb metal cations.

Many of the more soluble cations from the E-horizon end up in groundwater, but some (e.g., Al^{3+} and Fe^{3+}) are re-precipitated deeper in the soil, in what is known as the B-horizon, where either the pH is sufficiently buffered to facilitate mineral hydrolysis or the organic chelates (e.g., oxalate, citrate) are oxidized by aerobic microorganisms. In arid soils, calcium carbonate may form as both the loss of CO_2 and H_2O causes the saturation state to increase. This movement of metal cations down the soil profile, and their re-precipitation at lower depths, is one of the main causes for the subsequent differentiation of soils into specific horizons.

The extent of soil profile development depends on a number of variables, including the type of parent material at the time that they were subject

to soil forming processes, climate, topography of the site, and the indigenous vegetation and microbiota (see Brady, 2002). The type of parent material can range from bedrock to detritus transported to the site via rivers (alluvium), ice (till), and wind (loess). The nature of the parent material affects such soil characteristics as composition, mineralogy, texture, and weathering rates. For young soils (just hundreds of years old), horizons will clearly be more distinct in soils formed on granite or basalts than they would if the underlying lithology was sandstone. However, over longer periods of time, the different soils converge to a soil type determined by climate, and once the soil is fully developed, it should be stable indefinitely.

Climate is important in terms of precipitation and temperature: both higher rainfall and temperatures promote greater rates of organic productivity, more organic decay, and ultimately increased organic and inorganic acid generation. Increased temperature and flushing rates also serve to enhance the rates of chemical weathering. Not surprisingly, thicker soil profiles tend to develop in tropical environments, characterized by accumulations of kaolinite, hematite/goethite, and gibbsite. Organic matter accumulation, profile development, nutrient cycling, and structural stability are also intimately tied to the type of vegetation and soil microbiota. Take, as an example, soils formed under grassland versus forests. The organic matter content of grassland soils is generally higher than that of forested areas such that the former are darker in color and have higher moisture-holding capacity, while the acidity associated with coniferous trees will influence soil pore water composition and limits many types of secondary minerals from forming.

5.1.6 Weathering and global climate

A major feedback mechanism controlling atmospheric p_{CO_2} levels is the weathering of some silicates and the subsequent precipitation of Ca-Mg carbonates (Kump et al., 2000). During this process carbonic acid reacts with minerals, such as plagioclase and amphiboles, generating soluble Ca^{2+} and HCO_3^-, and residual clay phases from the incongruent dissolution reactions. The ions are eventually transported to the oceans where they are precipitated either biologically, as calcite or aragonite shells, or abiologically, as a micritic mud when a state of supersaturation is achieved:

$$CO_2 + 2H_2O + CaAl_2Si_2O_8 \text{ (anorthite)} \rightarrow$$
$$Al_2Si_2O_5(OH)_4 + CaCO_3 \quad (5.14)$$

Global changes in atmospheric CO_2 levels are thus determined by the magnitude of the imbalance between the rate of addition of CO_2 to the atmosphere through tectonically induced metamorphic-magmatic decarbonation of limestone/dolomite and sedimentary organic carbon versus the rate of removal by weathering and the incorporation of inorganic carbon into marine sediments and biota (Fig. 5.10). Increased exposure of landmass to surface conditions (through uplift or lowering of sealevel), or higher levels of acid generation, should amplify chemical weathering rates and swing the balance in favor of atmospheric CO_2 drawdown, and potentially glaciation.

Organisms certainly play an integral role in the carbon cycle. Although carbonic acid arises from the oxidation of soil organic matter, the ultimate source of this carbon is atmospheric CO_2 fixed via photosynthesis. Importantly, as Berner et al. (1983) suggested in their seminal paper more than 20 years ago, the real impact of high CO_2 is that it increases Earth's surface temperatures and net precipitation. This, in turn, leads to higher terrestrial biomass production, increased soil biological activity, more organic decay/acid generation, faster chemical weathering rates, and greater solute loads carried by rivers. The increased supply of nutrients to the oceans promotes greater primary plankton productivity that will further reduce atmospheric CO_2 through photosynthesis and the precipitation of calcium carbonate shells. Of course this ends up having a negative feedback because atmospheric CO_2 drawdown inevitably cools global temperatures, such that biological activity and weathering rates diminish, thereby returning the carbonate–silicate cycle to steady state (Walker et al., 1981). Similarly, as nutrient fluxes to the oceans decline, plankton productivity decreases, and less CO_2 is fixed from the

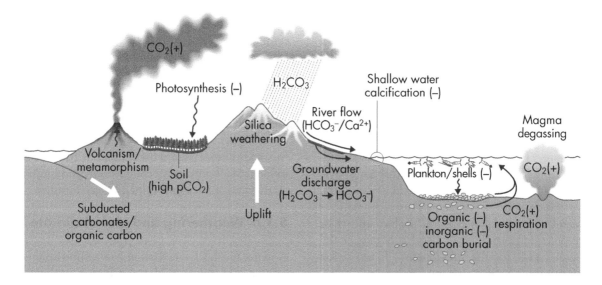

Figure 5.10 Simplified carbon cycle, showing the addition (+) and removal (−) processes in terms of atmospheric CO_2.

atmosphere. Such biological effects on weathering are a prime example of the so-called Gaia hypothesis, in that life helps regulate the Earth's climate to meet its own needs (Lovelock and Whitfield, 1982; Lovelock, 2000).

Implicit in the discussion above is the direct link between the hydrosphere–atmosphere–biosphere subcycle (HAB) and the sedimentary subcycle, that includes both the reservoirs of carbonate minerals and organic carbon. Marine carbonate precipitation is presently dominated by calcareous plankton, and to a lesser extent carbonate-secreting benthic organisms. As indicated above, their net mineralizing capacity, and ultimately their net burial rate, is controlled by the riverine flux of Ca^{2+}/Mg^{2+} and HCO_3^- released by surface weathering, and also by seawater–basalt interactions. For the organic carbon component, the fixation of CO_2 into biomass is achieved principally by photosynthetic organisms, and is shared almost equally in terms of primary productivity between the marine and terrestrial environments (Des Marais, 1997). Of that amount of organic carbon produced, the actual quantity buried into sediments, after respiratory processes, is usually <0.1%, some five times less than the burial rate of carbonate carbon. Despite that small percentage, the sedimentary carbon reserve is much larger than the carbon reserves of the HAB subcycle, and periodic imbalances in the former are potentially large and can have significant effects on the oxidation state of Earth's surface environment (Holland, 1984).

It has been proposed that microbial weathering also played a fundamental role in defining the initial habitability of Earth's terrestrial environment (Schwartzman and Volk, 1989). These authors argue that various thermophiles, possibly including anoxygenic photoautotrophs and chemolithoautotrophs (e.g., methanogens), may have occupied much of the land surface as early as the Archean. These microorganisms would have been effective weathering agents relative to sterile conditions, primarily from their ability (as biofilms) to retain water at the mineral surface

and their transformation of atmospheric CO_2 to the production of inorganic and organic acids. Their presence thus led to CO_2 drawdown, to the extent that surface temperatures could have dropped enough to have facilitated the evolution of mesophilic microorganisms, and ultimately diversification of the terrestrial microbiota. What is not obvious, however, is how such microorganisms could have been prolific given the high UV influxes in the absence of an ozone layer. Perhaps, they relied instead on a shallow sub-surface mode of existence (i.e., as endoliths), where sufficient visible light could penetrate, or they were protected by some other form of UV shield, such as an elemental sulfur smog or a mineral crust (see section 7.3.1)? Irrespective of the mechanism, if sheltered, and if continually moistened in a warm and wet climate, those communities would rapidly have become the primary source of soil CO_2. The later evolution of fungi, and their key role in soil formation, was likely a critical preliminary step for the eventual colonization of land by vascular plants in the Silurian (Schwartzman and Volk, 1991).

5.2 Sulfide oxidation

A number of metal sulfides have the propensity to undergo chemical oxidation when subjected to surface oxidizing conditions. Of those minerals, pyrite (FeS_2) is arguably the most important environmentally because it is an extremely common constituent in coal seams, ore bodies, and shales.

5.2.1 Pyrite oxidation mechanisms

During pyrite's exposure to oxygenated waters, as in the cessation of a mining operation, both its reduced sulfur and iron atoms become oxidized. Three electron acceptors are possible: molecular oxygen, hydrogen peroxide and, under acidic conditions, ferric iron (McKibben and Barnes, 1986). The overall process describing the initiation

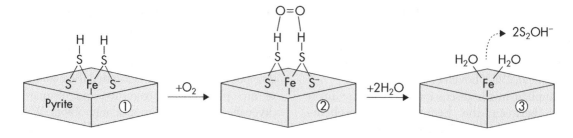

Figure 5.11 Model, based on Goldhaber (1983), of the initial oxidation of pyrite at circumneutral pH by reaction with O_2.

of pyrite oxidation is commonly given by the following incongruent reaction:

$$FeS_2 + 3.75O_2 + 3.5H_2O \rightarrow Fe(OH)_3 + 2H_2SO_4 \tag{5.15}$$

The oxidation and hydrolysis steps shown in the above equation involve the loss of 1 electron by ferrous iron and 14 electrons by disulfide, with the gain of 7.5 electrons by each oxygen per mole of pyrite. All of these redox changes cannot take place in one step; but instead there are a series of electron transfer reactions that need consideration.

(a) Sulfur reactions

The first step in the dissolution of pyrite at circumneutral pH involves attachment of O_2 to the partially protonated sulfur ligands exposed at the mineral's surface (Goldhaber, 1983). The next step requires breaking the O_2 double bond and displacement of S_2OH^- molecules by H_2O (Fig. 5.11). As long as the fluids at the mineral surface are circumneutral, the sulfoxy anions diffuse into the bulk fluid, where they are oxidized to sulfate (reaction (5.16)), via several sulfur intermediates, including thiosulfate ($S_2O_3^{2-}$), polythionates ($S_nO_6^{2-}$), such as tetrathionate ($S_4O_6^{2-}$) and trithionate ($S_3O_6^{2-}$), and sulfite (SO_3^{2-}):

$$S_2OH^- + 3O_2 + H_2O \rightarrow 2SO_4^{2-} + 3H^+ \tag{5.16}$$

Although the progressive oxidation of intermediate sulfur compounds to sulfate is predicted to follow a linear pathway, instead there are a number of variables that make S-cycling much more complex and still incompletely understood (e.g., Xu and Schoonen, 1995). Thiosulfate is the first sulfoxy anion that forms (reaction (5.17)). It is more stable at circumneutral pH than low pH, and slowly disproportionates to elemental sulfur and sulfite in weakly acid solutions (reaction (5.18)). In the presence of pyrite, thiosulfate also oxidizes quickly (with O_2) to form tetrathionate (reaction (5.19)). Tetrathionate is most stable at low pH, and it does do not appear to be significantly oxidized by O_2. By contrast, sulfite is not stable except under alkaline conditions, and rapidly oxidizes to sulfate in the presence of oxygen or any other strong oxidizing agent (reaction (5.20)). Similarly, elemental sulfur is oxidized to sulfate at circumneutral pH (reaction (5.21)). Therefore, in moderately acidic solutions tetrathionate and sulfate predominate as the sulfoxy anions at the expense of thiosulfate and sulfite.

$$S_2OH^- + O_2 \rightarrow S_2O_3^{2-} + H^+ \tag{5.17}$$

$$S_2O_3^{2-} \rightarrow S^0 + SO_3^{2-} \tag{5.18}$$

$$2S_2O_3^{2-} + 0.5O_2 + 2H^+ \rightarrow S_4O_6^{2-} + H_2O \tag{5.19}$$

$$SO_3^{2-} + 0.5O_2 \rightarrow SO_4^{2-} \tag{5.20}$$

$$S^0 + 1.5O_2 + H_2O \rightarrow SO_4^{2-} + 2H^+ \tag{5.21}$$

In reaction (5.16) the complete oxidation of S_2OH^- to sulfate causes the pH to drop. Accordingly, it is sometimes referred to as the initiator reaction, because its leads to the onset of acidic conditions. Simultaneously, the stability of elemental sulfur increases because of its greater insolubility at low pH. This results in the precipitation and accumulation of micrometer-thick agglomerates of elemental sulfur on the pyrite surface (reaction (5.22)), replacing the S_2OH^- molecules (e.g., Sasaki et al., 1995):

$$>S_2^{2-} + 0.5O_2 + 2H^+ \rightarrow 2S^0 + H_2O \quad (5.22)$$

Precipitation of elemental sulfur has the potential to form an inert layer that might inhibit the diffusion of oxidants to the surface, thereby slowing further dissolution. This means that its rate of oxidation to sulfate (reaction (5.21)), relative to its formation from S(−1), can determine the overall dissolution rates of pyrite. In turn, these rates are governed by the transport rates (diffusion, advection) of oxidizing agents, such as O_2 or Fe^{3+}, to the pyrite surface, the pH of the proximal solution, and the presence of S-oxidizing bacteria, as discussed below (Nordstrom, 1982).

(b) Iron reactions

At first, when the pH is still above 4.5, the Fe(II) exposed during the initial reactions spontaneously oxidizes in air to form Fe(III) (reaction (5.23)). Some of that ferric iron dissolves, where it is hydrolyzed and reprecipitated as ferric hydroxide (reaction (5.24)). The remainder is oxidized at the grain surface without going into solution.

$$Fe^{2+} + 0.25O_2 + H^+ \rightarrow Fe^{3+} + 0.5H_2O \quad (5.23)$$

$$Fe^{3+} + 3H_2O \rightarrow Fe(OH)_3 + 3H^+ \quad (5.24)$$

Note that although the initial oxidation reaction consumed protons, and thus led to a temporary rise in pH, the hydrolysis of ferric iron to form $Fe(OH)_3$ inevitably led to more acidity. Because this reaction normally occurs in the presence of sulfate, the ferric hydroxide may convert to the more insoluble minerals, jarosite ($MFe_3(SO_4)_2(OH)_6$), where M may be H^+, H_3O^+, Na^+, K^+, NH_4^+ (reaction (5.25)), or schwertmannite ($Fe_8O_8SO_4(OH)_6$) (e.g., Lazaroff et al., 1982):

$$M^+ + 3Fe(OH)_3 + 2SO_4^{2-} \rightarrow$$
$$MFe_3(SO_4)_2(OH)_6 + 3OH^- \quad (5.25)$$

Reaction (5.23) necessitates that dissolved O_2 serves as the oxidizing agent at circumneutral pH owing to the diminished availability of dissolved Fe(III) at pH values greater than 4.5. However, at this pH range, solid phase or adsorbed Fe(III) can still serve as an effective oxidant of pyrite if it is in direct contact with the mineral surface. This can come about in two ways. First, in the initial oxidation step, Fe(II) diffuses to the surface and becomes oxidized/hydrolyzed to ferric hydroxide. Second, dissolved Fe(II) adsorbs onto pyrite (which has an isoelectric point of 2.5, and hence is anionic at the pH values of most natural waters), where it reacts with, and gives up its electrons to dissolved O_2. In either case, the Fe(III) then rapidly accepts electrons from the pyrite. Adsorbed Fe thus acts as an electron shuttle from Fe(II) in pyrite to dissolved O_2 (Moses and Herman, 1991):

$$Fe(II) + pyrite_{red} \rightarrow Fe(II)\text{-}pyrite_{red} + 0.25O_2 \rightarrow$$
$$Fe(III)\text{-}pyrite_{red} + 0.5H_2O \rightarrow Fe(II)\text{-}pyrite_{ox}$$
$$(5.26)$$

The geochemical reactions described above are borne out in the surface textures of weathered sulfide minerals, including pyrite and pyrrhotite (Fe_7S_8), that show several stages in paragenetic alteration sequences (e.g., Nesbitt and Muir, 1994; Pratt et al., 1994). Initially, after the loss of S_2OH^-, the presence of a thin, featureless ferric hydroxide layer (8–10 nm thick) forms on the mineral surface. Over time it thickens (up to

30 nm) through diffusion of Fe(II) into the surface precipitate. Concomitantly, S(−1) accumulates in the subsurface layers. At some critical stage, the adhesion between the ferric hydroxide and the S(−1) underlayer is weakened, leading to spalling of the ferric hydroxide into solution. Removal of Fe then exposes the reduced sulfur, which in turn becomes sequentially oxidized and eventually released from the pyrite surface as one of the sulfoxy anions. This process becomes to some extent self-sustaining because as the acidity near the pyrite surface increases, the Fe(II) is more easily leached from the surface layer prior to oxidation, thus increasing the exposure of the S-rich sites in the crystal lattice. Thereafter, oxidation of >S(−1) to S^0 becomes increasingly important as the latter becomes stable as a solid phase.

The acid generated by the oxidation of S_2OH^- and S^0 to SO_4^{2-}, as well as Fe(III) hydrolysis, begins low, but given the right conditions, the pH of the waters can drop to values below 4.5.

When that happens, Fe^{2+} becomes stable in the presence of O_2 and its oxidation becomes very slow (Fig. 5.12). Furthermore, unlike equation (2.53), Fe(II) oxidation under acidic conditions becomes independent of pH (Singer and Stumm, 1970), with its kinetic reaction being expressed as:

$$\frac{-d[Fe(II)]}{dt} = k[Fe(II)][O_2] \tag{5.27}$$

where $k = 1.0 \times 10^{-7}\ min^{-1}\ atm^{-1}$ at 25°C. Ferric hydroxide also becomes considerably more soluble at low pH (reaction (5.24) now goes from right to left), and as the Fe^{3+} concentration increases with greater acidity, its role becomes much more important as the pyrite oxidizing agent (Moses et al., 1987):

$$FeS_2 + 14Fe^{3+} + 8H_2O \rightarrow 15Fe^{2+} + 2SO_4^{2-} + 16H^+ \tag{5.28}$$

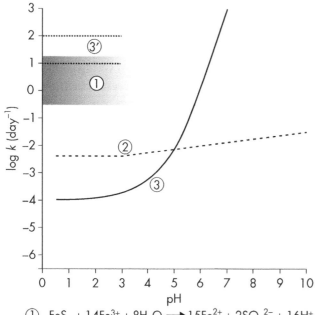

① $FeS_2 + 14Fe^{3+} + 8H_2O \longrightarrow 15Fe^{2+} + 2SO_4^{2-} + 16H^+$
② $FeS_2 + 3.5O_2 + H_2O \longrightarrow Fe^{2+} + 2SO_4^{2-} + 2H^+$
③ $Fe^{2+} + 0.25O_2 + H^+ \longrightarrow Fe^{3+} + 0.5H_2O$
③′ biological oxidation of reaction 3

Figure 5.12 Comparisons of rate constants as a function of pH for (1) the oxidation of pyrite by Fe^{3+}, (2) the oxidation of pyrite by O_2, and (3) the oxidation of Fe^{2+} by O_2 (modified from Nordstrom, 1982). The variation in rate constants for reaction 1 results from different proportions of total Fe^{3+} and FeS_2, as calculated by Singer and Stumm (1969). Reaction 3′ represents the reaction rate enhancement by *Acidithiobacillus ferrooxidans.* (After Lacey and Lawson, 1970.)

Indeed, at pH values lower than 3, Fe^{3+} is the only important oxidizer of pyrite (Fig. 5.12). The increased Fe^{3+} availability also enhances the oxidation of intermediate sulfoxy anions, converting them to sulfate, with the further effect of increasing acidity (e.g., Druschel et al., 2003):

$$S_4O_6^{2-} + 3Fe^{3+} + 2.75O_2 + 4.5H_2O \rightarrow$$
$$4SO_4^{2-} + 3Fe^{2+} + 9H^+ \qquad (5.29)$$

According to molecular orbital theory, reaction (5.28) is initiated by the bridging of Fe^{3+} cations to S_2^{2-} (Luther, 1987). The sulfide moiety is then transformed into more oxidized species, such as $S_2O_3^{2-}$, which may oxidize further in solution to polythionates and sulfate depending on the availability of further oxidizing agents. Concomitantly, the bound Fe^{3+} is reduced to Fe^{2+}, and the bridging complex is eliminated. This binding of Fe^{3+} to the S_2^{2-} ligands (compared to O_2 that cannot bind as easily because of the arrangement of its outer electron shell) further explains why the rates of pyrite oxidation are an order of magnitude faster when Fe^{3+} is available relative to dissolved O_2. The abundance and reactivity of the $S(-1)$ groups for Fe^{3+} likely also explains the different dissolution rates displayed by various iron sulfides, such as pyrite versus arsenopyrite (Edwards et al., 2001).

5.2.2 Biological role in pyrite oxidation

In acid waters, pyrite can reduce Fe^{3+} to Fe^{2+} faster than the latter can be regenerated into Fe^{3+} by O_2. Accordingly, the pyrite will simply reduce all the ferric cations and the reaction will stop. Thus, the oxidation of ferrous iron is considered the rate-determining step in the abiological oxidation of pyrite (Singer and Stumm, 1970). However, as introduced in Chapter 1, acidophilic bacteria use reaction (5.23) as an energy-generating process, and in doing so foster the acidification of their local environment (Fig. 5.12).

(a) Oxidation rate enhancement

Acidophilic Fe(II)-oxidizing bacteria can generate Fe^{3+} some five or six orders of magnitude faster relative to sterile conditions (e.g., Lacey and Lawson, 1970). This increase makes the Fe(II) oxidation rate slightly higher than the rate of the pyrite oxidation by Fe^{3+}. Of course, microorganisms in the environment are always growth-limited by bioessential elements, predators, or some hydrologic condition, hence, their true environmental oxidation rates probably approximate the rate of pyrite oxidation by Fe^{3+} (Nordstrom and Southam, 1997). The Fe^{3+} formed under these conditions, being soluble, is chemically reactive and can effectively scavenge electrons from $S(-1)$ in pyrite, generating Fe^{2+} once again. It is then reoxidized to Fe^{3+} by the bacteria. Because of this re-cycling process, the formation of Fe^{3+} can be viewed as an efficient electron acceptor for sustained lithotrophy, with a progressive, rapidly increasing rate of pyrite oxidation (called the propogation cycle) owing to biocatalysis (Singer and Stumm, 1970). Recall from Chapter 2 that under acidic conditions, very little energy is generated through Fe(II) oxidation. Subsequently, these bacteria must oxidize large amounts of reduced iron in order to sustain themselves, and even a small number of cells can be responsible for extensive pyrite oxidation. Not surprisingly, estimates made in some acid mine environments suggest that the acidophilic bacteria can account for the majority of pyrite dissolution (e.g., Edwards et al., 2000b).

The most widely studied and environmentally important Fe(II)-oxidizing bacteria include the Gram-negative mesophiles, *Acidithiobacillus ferrooxidans* (formerly known as *Thiobacillus ferrooxidans*) and *Leptospirillum ferrooxidans*. The former is rod-shaped, $0.5\,\mu m$ in diameter by $1–2\,\mu m$ long, and possesses a flagellum that enables it to be motile (Fig. 5.13A). It grows best within the pH range 1.8–2.5. Generally more abundant in the latter stages of sulfide oxidation, when pH declines to 1.8 or less, is *L. ferrooxidans*.

It is easily distinguished from A. *ferrooxidans* by its morphology, ranging from helix to curved rods to vibrios, with dimensions of 0.2–0.4 μm in diameter by 1–2 μm in length (Fig. 5.13B). Growing at even lower pH, below 1, is the chemolithoautotroph, *Ferroplasma acidarmanus*. There are many other acidophilic Fe(II)-oxidizing prokaryotes spanning the phylogenetic tree (Fig. 5.14), ranging from mesophiles to thermophiles (Baker and Banfield, 2003). Heterotrophic bacteria coexist with the autotrophs within tailings, where several members of the genus *Acidiphilium* survive by coupling the reduction of elemental sulfur to the degradation of their autotrophic neighbors. Filamentous fungi and protozoa are the most common eukaryotes, where they also function as heterotrophs or grazers, respectively (Johnson and Roberto, 1997).

Not all acidophiles oxidize Fe(II). In fact, considerably more energy is available during the oxidation of reduced sulfur compounds, and A. *ferrooxidans* will preferentially consume sulfide rather than ferrous iron. Studies have even shown that during growth on pyrite, the EPS surrounding A. *ferrooxidans* becomes studded with fine-grained elemental sulfur colloids which are believed to serve as a temporary energy reserve (Rojas et al., 1995). Many of the other Fe(II)-oxidizing bacteria are only capable of using ferrous iron as a substrate (e.g., *L. ferrooxidans*), and in low pH experiments where they are the sole chemolithoautotrophs, a build up of elemental sulfur develops on the pyrite surface (e.g., McGuire et al., 2001). This, in turn, makes life possible for those chemolithoautotrophs that can only oxidize reduced sulfur species (e.g., *Acidithiobacillus thiooxidans*, formerly known as *Thiobacillus thiooxidans*). In the experiments above, A. *thiooxidans* is able to reduce the quantity of elemental sulfur on the pyrite surface to less than 1% observed on samples exposed to Fe(II)-oxidizing cultures only. Notably, the removal of elemental sulfur from pyrite surfaces exposes the underlying minerals to increased oxidative attack. Furthermore, bacterial catalysis of sulfoxy anion oxidation facilitates recycling of sulfur

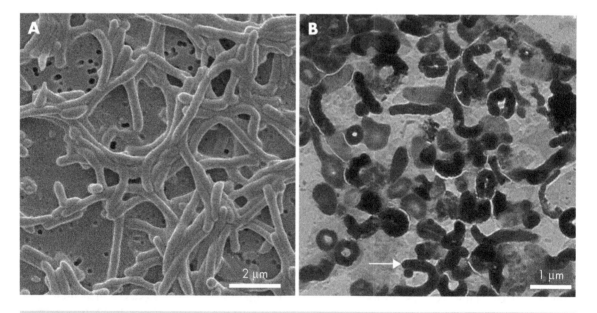

Figure 5.13 (A) SEM image showing a pure culture of *Acidithiobacillus ferrooxidans* on filter paper (courtesy of Aman Haque and Bugscope Project). (B) TEM image of a biofilm dominated by *Leptospirillum ferrooxidans* (arrow) (Reprinted from Rojas-Chapana and Tributsch, 2004 with permission from Elsevier.)

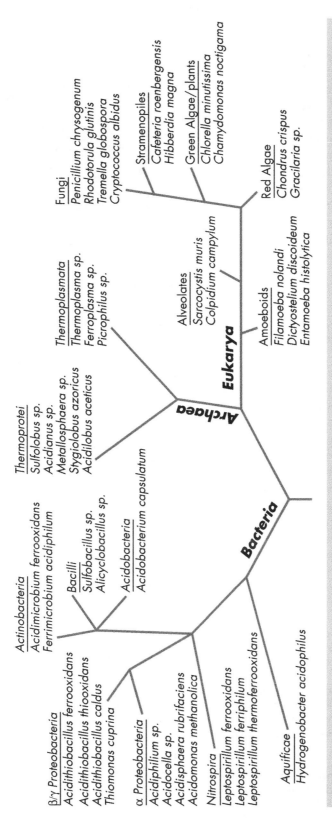

Figure 5.14 Universal phylogenetic tree based on 16S and 18S rRNA sequences, illustrating the wide distribution of microbial species that have been identified in highly acidic waters. The *Bacteria* and *Archaea* are subdivided by taxonomic class, and the *Eukarya* are outlined by their familiar groupings. Representative species are also provided. Note, branch lengths are not to scale.

intermediates, with the net production of increased acidity (e.g., Schippers and Sand, 1999).

Recent studies characterizing microbial diversity in acidic environments have discovered the presence of numerous other S-oxidizing bacteria that appear to be just as numerically important as *A. thiooxidans* (e.g., Bruneel et al., 2003). In view of the specific role each plays in the overall oxidation process, it is not surprising that we find Fe(II)- and S-oxidizing bacteria growing juxtaposed to one another. This also explains the findings that mixed cultures of chemolithoautotrophs increase the rates of sulfide mineral dissolution relative to the actions of a single species growing in isolation (e.g., Lizama and Suzuki, 1989).

(b) Importance of attachment

Bacterial oxidation of pyrite occurs by two mechanisms: indirect and direct. As discussed above, a number of Fe(II)-oxidizing bacteria generate Fe^{3+} from Fe^{2+}, which then reacts abiologically with solid pyrite. This process is considered indirect because the bacteria do not directly oxidize pyrite, and thus can grow attached to nonsulfide minerals as well. By contrast, most of the bacteria discussed above actually grow on pyrite and other types of sulfide minerals (e.g., Fig. 5.15), where they directly oxidize and solubilize the reduced iron and sulfur moieties via enzymatic reactions. The attachment of bacteria to pyrite can be visualized as occurring in four distinct steps:

1 In the first step, bacteria are transported to the pyrite surface after it has already been conditioned with inorganic and organic compounds.

2 Once in the vicinity of the surface, electrochemical interactions between the cell and pyrite surface are initiated; the type and strength being governed by surface properties of the mineral and the bacterium (recall section 3.6.1).

3 The actual physical attachment of the bacterium to the surface occurs by the development of specific structures, such as fibrils or EPS.

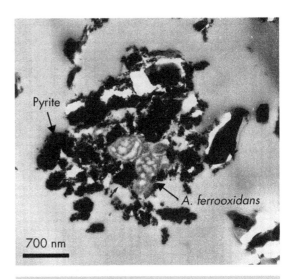

Figure 5.15 TEM image of a colony of *Acidithiobacillus ferrooxidans* cells growing attached to pyrite. (From Southam and Beveridge, 1992. Reproduced with permission from the American Society for Microbiology.)

4 Eventually these reconnaissance bacteria multiply, and form a microcolony directly on the pyrite surface. Given time and sufficient nutrients, the colony expands into a pyrite enshrouding biofilm onto which other species may attach.

Bacterial adsorption onto the pyrite surface is rapid. Experiments have documented that nearly 100% of the total population of planktonic *A. ferrooxidans* cells can adhere to the pyrite within minutes if sufficient surface area is made available (e.g., Bagdigian and Myerson, 1986). The mechanism of attachment is not random, and appears to involve the bacteria colonizing fractures or high surface energy sites, such as dislocations. Aside from free energy gains associated with attachment at dislocations, those sites may afford the acidophiles with a greater flux of reductants – diffusivities along dislocations can be orders of magnitude greater than through pure crystalline solids (Andrews, 1988). Once initiated, the contact sites eventually develop into corrosion pits the size and shape of the bacteria, widening and enlarging until there is a pronounced surface roughening (Fig. 5.16).

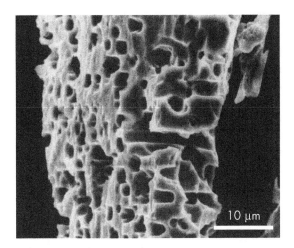

Figure 5.16 SEM image showing extensive pyrite dissolution after only 43 days of oxidation by A. ferrooxidans. (From Mustin et al., 1992. Reproduced with permission from the American Society for Microbiology.)

In turn, the corrosion pits serve as convenient physical recesses and make available newly altered surfaces for colonization by a second wave of bacterial species (e.g., the S-oxidizing A. thiooxidans). Under ideal growth conditions, this process would continue until the pyrite is completely degraded (Mustin et al., 1992).

In the Gram-negative acidithiobacilli, the macromolecule that is responsible for initial mineral binding is the lipopolysaccharide (LPS). Although pyrite is negatively charged above pH 2.5, the functional groups comprising the LPS are still protonated under acidic conditions, thus providing the cells with a neutral to slightly positive charge that allows them to approach and attach onto the pyrite surface. It has even been suggested that variations in LPS chemistry afford acidithiobacilli the means by which to distinguish different atoms in the pyrite lattice (Southam and Beveridge, 1993). This is a useful trait for A. ferrooxidans because it can activate the appropriate oxidative enzymes depending on whether Fe or S sites are exposed. The ability to recognize specific electron donors is even more crucial for strict Fe(II) oxidizers (e.g., L. ferrooxidans) or reduced sulfur oxidizers (e.g.,

A. thiooxidans), because their survival hinges on being able to attach to specific sites on the mineral substrate. Interestingly, the species are able to selectively colonize minerals that have low electrochemical stability and are more reactive and easier to dissolve. This can be relative to matrix materials (sulfide vs. silicates) or even between different sulfide phases, such as the preferential oxidation of arsenopyrite (FeAsS) over pyrite (Norman and Snyman, 1988).

Once attached to pyrite, the production of EPS fixes the cells firmly onto the solid. With time, not only does EPS completely enshroud the exposed mineral surfaces, but under the prevailing acidic conditions, it becomes heavily impregnated with cationic Fe species. This further facilitates the bacterium's electrostatic adsorption onto the negatively charged pyrite by lowering the electronegativity of the bacterium's surface (at pH values where some of the cell's functional groups have already deprotonated) and by reducing any double-layer repulsive barriers (Blake et al., 1994). Moreover, the EPS may actually accelerate pyrite oxidation because the Fe adsorbed to it can potentially serve as an electron shuttle for conveying electrons from the metal sulfide to the cell surface, in a manner reminiscent of reaction (5.26) (Sand et al., 1995). Such a mechanism might be important because the same EPS that aids in surface adhesion could also present a potential inhibitor of Fe^{3+} diffusion away from the cell. Thus, the iron recycling that takes place during the propagation cycle can be envisioned as taking place entirely within the EPS.

The relative importance of direct versus indirect mechanisms in terms of pyrite dissolution rates remains a subject of dispute. It can be argued that rates of oxidation are faster when Fe^{3+} reacts with pyrite versus bacterial oxidation of Fe(II) in the crystal lattice. Furthermore, many of the surface weathering features observed on pyrite can be attributed to abiological reactions with bacterially induced Fe^{3+} (Edwards et al., 2001). Yet, bacteria rapidly attach to pyrite surfaces when they become exposed, and species, such as A. ferrooxidans, do not develop into multiplayer biofilms, suggesting that each cell needs to be in direct physical contact with the pyrite surface in order to grow (e.g., Larrson et al., 1993). Certainly, by adhering to the solid sulfide they ensure that a source of reduced iron and/or sulfur is in close proximity, thereby minimizing the time needed for diffusion of reducing equivalents between the mineral and the bacterium. What is most likely is that both mechanisms work concurrently, and that the more reactive sulfides (e.g., pyrrhotite) are oxidized predominantly by indirect mechanisms, while relatively less reactive sulfides (e.g., pyrite) may require more direct attachment to cause their oxidative dissolution. This could clarify why, on a per cell basis, sulfide mineral dissolution rates appear comparable between attached and planktonic species (Edwards et al., 1999).

(c) Formation of acid mine drainage (AMD)

During coal and metal mining operations, overburden, waste rock, and mill tailings are disposed of in the form of spoil heaps or in tailings ponds. Those waste materials contain residues of pyrite and other sulfide minerals that, upon decommissioning of the mining operation, eventually become exposed to rain and oxygenated surface waters. This places them into chemical disequilibrium, and subject to the oxidation transformations discussed above. Within the vadose zone of the spoil heap, where capillary action on the mineral substrata supplies water for chemolithoautotrophic growth, the sulfide phases provide a source of energy, and the pore spaces allow for the influx of CO_2 and O_2 that serve as the carbon source and terminal electron acceptor (TEA), respectively (see Kleinmann et al., 1981). With all their needs met, sulfide mineral-oxidizing microorganisms can expeditiously establish themselves and begin the biotransformation of fine-grained, pH-neutral, gray sulfide-bearing tailings into the bright yellow and orange stained, leachate-producing residues that are a significant environmental concern (see Plate 12).

The effects of pyrite oxidation are often far removed from the actual mine site. The reasons for this are simple. The highly acidic and sulfate/metal-rich effluent, aptly named acid mine drainage (a specific type of acid rock drainage), is sometimes initially transported by groundwater flow before discharging into surface waters. As long as anoxic effluent is acidic, the Fe^{2+} is stable in the absence of bacteria, but once it comes into contact with O_2 in more alkaline, aerated drainage, oxidation and hydrolysis spontaneously occur. Similarly, Fe^{3+} may be transported away from the site of active pyrite oxidation, without ever having come into chemical contact with any remaining sulfide mineral phases. In the acidic outflow, however, some of it will also precipitate as jarosite, or other ferric sulfate minerals, due to sulfate bridging of ferric iron colloids (recall section 4.1.9(b)).

Mixing of AMD with natural waters in rivers and lakes causes serious degradation in water quality:

1 both the acid and high dissolved metal content (e.g., Fe and trace metals solubilized from the solid-phase sulfides under acidic conditions) are toxic to aquatic life;

2 the acidity changes the dissolved inorganic carbon speciation from HCO_3^- to H_2CO_3, thereby diminishing the autotrophic metabolism of a number of organisms;

3 the ferric hydroxide/ferric sulfates smother benthic species, inhibiting photosynthesis;

4 the acids have a corroding effect on parts of infrastructures along the river course, such as bridges.

If the rate-determining steps are controlled primarily by the activity of A. ferrooxidans, then the oxidation process as a whole depends upon the growth conditions of the bacteria. Aside from the obvious supply of sulfide minerals, one of the most important requirements are that periodic rainwater infiltration provides the needed aeration and removal of oxidation products so that fresh pyrite surfaces are exposed. It has been shown that A. *ferrooxidans* increases acid production for 3–4 days after each rainfall, after which acid generation drops back down to ambient conditions (Kleinmann and Crerar, 1979). A qualifying statement is, however, needed here because excessive rainfall will dilute, or remove, acid build-up, such that the pH may not drop below 4.5, and the onset of the propagation cycle of AMD may never occur. This has been shown to happen during heavy rainfalls at Iron Mountain, California (the most metal-rich and acidic effluent of any abandoned mine reported anywhere in the world), where washout has previously been shown to have reduced the microbial populations from 10^9 cells ml^{-1} to less than 10^4 cells ml^{-1}. It also altered the microbial speciation, such that it took a lag time of 6 months before re-colonization and return to the usual mine microbiota was established (Edwards et al., 2000b). Thus, for AMD to become a significant problem, acid must be allowed to accumulate in the spoil heap pore waters. Such accumulation can often be a seasonal phenomenon. For example, acid flushing into streams is sometimes observed during spring. The underlying cause lies with drainage out of the spoil during winter being prevented by the frozen ground, yet any unfrozen water within the spoil (kept liquid by the heat generated by the oxidative reactions) continues to generate acidity. Then when the ground thaws, the acid is discharged. In a different scenario, if spoils are allowed to dry out completely during the summer months, bacterial numbers and AMD also decline as water becomes limiting for microbial activity (Olson et al., 1981).

Due to the relationship between oxygenation and acid production, it has for many years been perceived that as long as oxygen was excluded from the tailings that pyrite oxidation would cease. However, it has now been recognized that A. *ferrooxidans* is a facultative anaerobe, capable of surviving in the absence of O_2 by using Fe(III) as an electron acceptor, provided that H_2 or a reduced sulfur species serves as the electron donor (e.g., reaction (5.30)):

$$S^0 + 6Fe^{3+} + 4H_2O \rightarrow SO_4^{2-} + 6Fe^{2+} + 8H^+$$
$$(5.30)$$

Cell yields observed during such anaerobic growth are comparable to the cell yields in aerobic, Fe(II)-grown cultures, but two times lower than growth on inorganic sulfur with O_2 as the terminal electron acceptor (Pronk et al., 1992). The practical implications of this are that metal solubilizing activity may take place at the center of poorly aerated ore heaps by using ferric iron that was produced by other bacteria growing at the surface. Indeed, recent findings by Coupland and Johnson (2004) have indicated that A. *ferrooxidans* is the dominant bacterium in both anaerobic and acidic waters from two submerged mines in Wales.

Another major source of controversy regarding acid mine drainage is the origin of the acidity. Although A. *ferrooxidans* plays an active role in pyrite oxidation once the pH has decreased below 4.5, they were not believed to survive at higher pH conditions. Yet, the rather slow kinetics of abiological oxidation by molecular oxygen in air would seem to preclude it as the dominant acidification process. Therefore, it was suggested that the initial steps must be microbially catalyzed, driven perhaps by the activity of neutrophilic chemolithoautotrophs and/or heterotrophs that condition the tailings for subsequent acidophilic populations (e.g., Harrison, 1978; Blowes et al., 1995). More recently, it has become accepted that a succession of neutrophilic and moderate acidophiles microorganisms are not required to generate the needed acidity. In a study of simulated sulfide-rich tailings, it was observed that A. *ferrooxidans* not only survives at pH values of 7, but it was also able to initiate pyrite oxidation and localized acidification within just 2 weeks of colonization (Mielke et al., 2003).

How the bacteria do this is described as follows. First, the bacterium uses cation bridging to fix itself onto high energy sites of the pyrite surface, upon which it then excretes EPS to attain a tenacious bond with the mineral surface. Once attached, it begins directly oxidizing the Fe(II) and S(−1) moieties in the crystal lattice, producing corrosion pits on the surface into which the cells reside. Acidity arises from both the biological oxidation of sulfoxy anions to sulfate and ferric hydroxide precipitation, of which the latter covers the pyrite surface and becomes embedded within EPS. The ferric hydroxide and EPS then act as partial diffusion barriers that maintain H⁺ in a nanoenvironment at the mineral surface (a few nm³). Then, as conditions become more favorable, the bacteria multiply to form micro-colonies that are enshrouded in Fe-rich biofilm. As the proximal pH drops, elemental sulfur

precipitates instead of the intermediate sulfur anions diffusing away, thereby affording S-oxidizing bacteria with an oxidizable substrate. Simultane-ously, ferric hydroxide begins partial solubiliza-tion. The Fe^{3+} formed then reacts with pyrite to initiate the propagation cycle. This extends the acidity to areas away from the immediate surround-ings, eventually affecting the more neutral bulk pore waters (Fig. 5.17). In essence, the micro-bial nanoenvironment exhibits physicochemical conditions conducive to the survival of a particu-lar community of species, even though the bulk pore waters in the tailings sediment is fundament-ally different. Nonetheless, this environmental

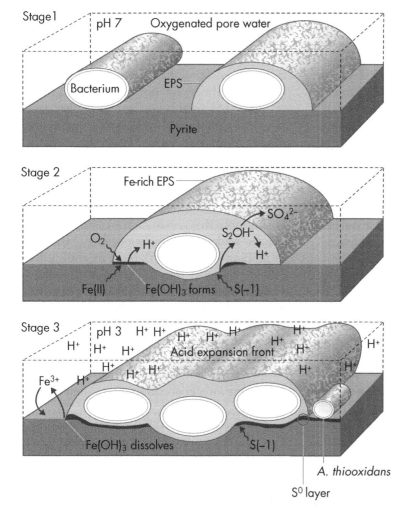

Figure 5.17 Possible model for the colonization of pyrite under circumneutral pH conditions. Stage 1: *A. ferrooxidans* attaches to pyrite surface and generates EPS to firmly attach itself to the mineral surface. Stage 2: Dissolution of pyrite causes release of S_2OH^-, while Fe(II) oxidizes to Fe(III). Oxidation of sulfoxy anions and hydrolysis of ferric iron generate acidity that stays confined to the EPS around the cell. Stage 3: Cells multiply and more acid is created, expanding the acid front. The ferric hydroxide re-dissolves to yield Fe^{3+} that reacts with the pyrite, while elemental sulfur becomes the stable S-phase upon which species, such as *A. thiooxidans*, later attach. (Adapted from model originally proposed by Southam and Beveridge, 1992, and later Mielke et al., 2003.)

modification is transient, and the colonizing microorganisms eventually succumb to the changes beyond their adaptive capabilities, at which point different species begin to predominate and a complete restructuring of the mine waste community ensues. Such changes could include the colonization of even more extreme acidophiles (e.g., Schrenk et al., 1998) or the advent of Fe(III)- and SO_4-reducers, if the tailings become O_2-depleted (e.g., Fortin and Beveridge, 1997).

The exact time associated with the establishment of highly acidic effluent is, at present, still ill defined, but appears to be of the order of years to decades. Despite the rapid colonization and onset of localized acid production in tailings, modeling predictions and studies at a limited number of field sites indicate the peak acid load occurs 5–10 years after mining, followed by a gradual decline over 20–40 years (e.g., Hart et al., 1991). The same study projected very long decay curves for coal refuse (beyond 50 years) before acid leachate is depleted.

Today a number of remediation methods are in place to curb the spread of AMD away from the mine site. They usually develop along two lines: (i) prevention of the actual generation of AMD at the source; or (ii) treatment of the AMD downstream (see Ledin and Pedersen, 1996 for details). In the former case, this may involve: chemical treatment by adding alkalinity (via crushed limestone or $Ca(OH)_2$) to the system before the pH drops below values of 4.5; adding phosphate to inhibit pyrite oxidation; growing vegetation on spoil heaps to consume O_2 and diminish water infiltration; capping the tailings to prevent O_2 diffusion to the sulfides; flooding the tailings so that anoxic conditions inhibit the profusion of *A. ferrooxidans*; or applying biocides to kill off the Fe(II) and S-oxidizing bacteria. The second treatment strategy may involve: diverting AMD to a water treatment plant where chemicals are applied to neutralize the acid and precipitate iron hydroxide (along with the co-precipitation of the trace metals); adding a reactive organic substrate to promote bacterial sulfate reduction and the subsequent immobiliza-

tion of metals as sulfide minerals; or adding living/dead biomass to adsorb iron and other metals (recall section 3.7.1).

Wetlands offer perhaps the best approach because they represent a potentially long-term, self-sustaining system in which both the acidity is consumed and the metals immobilized prior to the effluent being discharged into the regional waterways (Pulford, 1991). Essentially, wetlands remove metals by one of two processes: adsorption/absorption by metal-tolerant plants and FeS precipitation via bacterially mediated sulfate reduction – the SO_4^{2-} coming from the AMD (Fig. 5.18). The sorption processes tend to dominate at the start of wetland construction, but, over time, mineralization becomes the more important process as it converts dissolved Fe into an unreactive form, such as pyrite. As a result, artificial wetlands are now constructed with the view of adding decomposable, organic-rich substrates that facilitate the growth of sulfate-reducing bacteria (SRB) (e.g., Machemer and Wildeman, 1992). Wetlands are also usually low maintenance, involving only periodic dredging of sediment build-up and addition of limestone to treat the acid inflow, although wetland effectiveness has come into question under conditions when high acid loading overwhelms abiological and microbial alkalinity-generating mechanisms (see Wieder, 1993 for details).

5.2.3 Bioleaching

Low grade sulfide ores generally contain a variety of valued metals at concentrations below 0.5% (wt/wt), and their extraction by smelting after milling and ore enrichment is unfavorable because of the high gangue/metal ratio. In order to recover those metals at profit, a number of mining companies have utilized technologies that harness the metal-oxidizing or acid-generating activity of microorganisms (see Hackl, 1997 for details). This process, called bioleaching, places the metals of value in the solution phase, while the solid residue, if any, is discarded as waste material. Bioleaching is now also being used in the

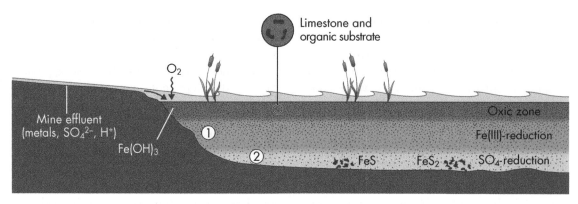

① $8Fe(OH)_3 + CH_3COO^- \rightarrow 8Fe^{2+} + 2HCO_3^- + 15OH^- + 5H_2O$

② $SO_4^{2-} + CH_3COO^- + H_2O \rightarrow H_2S + 2HCO_3^- + OH^-$

Figure 5.18 Model showing how wetlands alleviate acidity and high dissolved metal concentrations from mine drainage. Acid input is initially neutralized by the addition of limestone, causing the hydrolysis of Fe(III) to ferric hydroxide. Secondary alkalinity is then generated through bacterial Fe(III) and SO_4 reduction. Metals are immobilized during burial through reaction of pore water Fe^{2+} and H_2S, with organic compounds supplied to support the dissimilatory bacterial communities.

bioremediation of municipal wastewaters, where the organic sludge produced after treatment often contains high concentrations of heavy metals. Because of the high content of nitrogen, phosphorous, and potassium, sludges have been used as fertilizers in many areas around the world. However, to avoid any potential environmental contamination, the metal contaminants must first be removed (e.g., Tyagi et al., 1990).

(a) Chemolithoautotrophic oxidation

Bioleaching has been used effectively in the recovery of copper, zinc, lead, arsenic, antimony, nickel, and molybdenum from sulfide ores (Table 5.2). There are two common methods used, dump bioleaching and heap bioleaching. Historically dump leaching has been the most widely used method because open-pit mining frequently led to the formation of large piles of waste rock, some on the order of several million tons. During dump leaching, acid is added to the waste rock, and the indigenous bacteria proliferate naturally by oxidizing sulfide mineral residues. In most cases little effort is made to optimize bacterial activity. This process is a slow, inefficient process, with leach cycles measured in years and efficiency at most 50% for metals such as copper. Heap bioleaching is a more sophisticated method. In this process, finely crushed ores are placed on prepared pads, and a dilute sulfuric acid solution (pH ~2) is initially sprayed onto the "heap" to pre-condition the ore for the bacteria. Without the acidification step, bioleaching becomes rapidly ineffectual because waste rocks, made up of silicates, buffer the natural acidity generated through pyrite oxidation before the acidophiles can take hold. Simultaneous aeration of the pile is essential since the microbial leaching process is an aerobic process. The liquid coming out at the bottom of the pile is collected and transported to a collection plant where the metal is re-precipitated and purified. Meanwhile, the Fe(II)-rich liquid, called the lixiviant, is released into an oxidation pond to form Fe(III), and then pumped back to the top of the pile where the cycle is repeated (Fig. 5.19).

Table 5.2 A summary of the reactions involved in the breakdown of various common sulfide minerals. (Data compiled by McIntosh et al., 1997.)

Mineral	Reactions
Chalcopyrite	$4CuFeS_2 + 17O_2 + 2H_2SO_4 \rightarrow 4CuSO_4 + 2Fe_2(SO_4)_3 + 2H_2O$
Covellite	$CuS + 2O_2 \rightarrow CuSO_4$
Chalcocite	$5Cu_2S + 0.5O_2 + H_2SO_4 \rightarrow CuSO_4 + Cu_9S_5 + H_2O$
Bornite	$4Cu_5FeS_4 + 37O_2 + 10H_2SO_4 \rightarrow 20CuSO_4 + 2Fe_2(SO_4)_3 + 10H_2O$
Sphalerite	$ZnS + 2O_2 \rightarrow ZnSO_4$
Galena	$PbS + 2O_2 \rightarrow PbSO_4$
Arsenopyrite	$4FeAsS + 13O_2 + 6H_2O \rightarrow 4FeSO_4 + 4H_3AsO_4$
Stibnite	$2Sb_2S_3 + 13O_2 + 4H_2O \rightarrow (SbO)_2SO_4 + (SbO_2)_2SO_4 + 4H_2SO_4$
Millerite	$NiS + 2O_2 \rightarrow NiSO_4$
Molybdenite	$2MoS_2 + 9O_2 + 6H_2O \rightarrow 2H_2MoO_4 + 4H_2SO_4$

Microorganisms currently used in commercial bioleaching operations are exactly the same as those found naturally associated with exposed sulfidic ore, the only difference is that they may have been selected for rapid growth on the ore of interest. The bacteria should also show versatility in attacking different metal sulfides, and they must be resilient to toxic concentrations of transition metals within the lixiviant (Ehrlich, 2002). *Acidithiobacillus ferrooxidans* fulfills these criteria and, crucially, is ubiquitous in mine tailings. Notwithstanding the environmental concerns posed by growth of A. *ferrooxidans* and the other Fe(II)- and S-oxidizing bacteria present in waste tailings, when those same species are used in the controlled and confined conditions of a processing plant, the undesirable oxidative metabolic processes can serve as catalysts in the metal extraction process.

The rates of oxidative leaching and the efficiency of the process can be predicted based largely on the electrochemical properties of the metals in a mixed ore and on the metabolic abilities of the chemolithoautotrophic species. For example, A. *ferrooxidans* will solubilize Zn from sphalerite (ZnS) much faster than Cu from chalcopyrite (CuFeS$_2$), followed by Fe. By contrast, A. *thiooxidans* can only oxidize the sulfide portion of the ore without preference for particular metals, and its inability to oxidize Fe(II) reduces the electrochemical effect of having high Fe^{3+} concentrations (Lizama and Suzuki, 1988).

The most suitable copper minerals for heap bioleaching are chalcocite (Cu$_2$S) and covellite (CuS) – the major copper mineral of most mine wastes, chalcopyrite, has not been considered economically bioleachable due to the long leach times required. The overall process of copper bioleaching is predicated on the basis of A. *ferrooxidans* being capable of directly oxidizing Cu(I) in chalcocite, removing some copper in the dissolved form (Cu^{2+}), and forming the mineral covellite (reaction (5.31)):

$$Cu_2S + 0.5O_2 + H_2O \rightarrow CuS + Cu^{2+} + 2OH^-$$

$$(5.31)$$

In reaction (5.31), the bacteria utilize Cu(I) as an electron donor before the sulfide (Nielsen and Beck, 1972), although other studies have shown the precipitation of the mineral antlerite

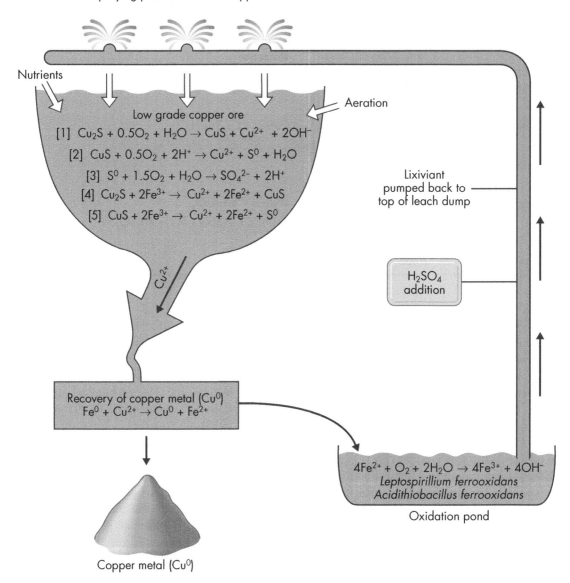

Spraying pH 2 lixiviant on copper ore

Nutrients

Aeration

Low grade copper ore

[1] $Cu_2S + 0.5O_2 + H_2O \rightarrow CuS + Cu^{2+} + 2OH^-$

[2] $CuS + 0.5O_2 + 2H^+ \rightarrow Cu^{2+} + S^0 + H_2O$

[3] $S^0 + 1.5O_2 + H_2O \rightarrow SO_4^{2-} + 2H^+$

[4] $Cu_2S + 2Fe^{3+} \rightarrow Cu^{2+} + 2Fe^{2+} + CuS$

[5] $CuS + 2Fe^{3+} \rightarrow Cu^{2+} + 2Fe^{2+} + S^0$

Cu^{2+}

Lixiviant pumped back to top of leach dump

H_2SO_4 addition

Recovery of copper metal (Cu^0)
$Fe^0 + Cu^{2+} \rightarrow Cu^0 + Fe^{2+}$

$4Fe^{2+} + O_2 + 2H_2O \rightarrow 4Fe^{3+} + 4OH^-$
Leptospirillium ferrooxidans
Acidithiobacillus ferrooxidans

Oxidation pond

Copper metal (Cu^0)

Figure 5.19 The arrangement of a copper heap leaching plant. In the first step, finely crushed ores are placed on prepared pads, and a dilute sulfuric acid solution (pH 2) is initially sprayed onto the "heap" to pre-condition the ore for *Acidithiobacillus ferrooxidans*. Once conditions are ideal for the bacteria, they begin oxidizing Cu(I) and S(−1) through a succession of metabolic reactions. Simultaneously, Fe^{3+}, generated from bacterial oxidation processes, reacts abiologically with the same copper minerals. The Cu-rich liquor is then processed, and the Fe^{2+} from the oxidative and recovery reactions is re-oxidized to Fe^{3+} by *A. ferrooxidans*, in an oxidation pond, and pumped back into the heap to be recycled. (Modified from Madigan et al., 2003.)

$((Cu_3SO_4)(OH)_4)$ in their chalcocite bioleaching experiments (Silver and Torma, 1974). The sulfide in covellite is subsequently oxidized by the same bacteria, initially forming elemental sulfur and Cu^{2+} (reaction (5.32)), but then the sulfur is fully oxidized to sulfate, either by *A. ferrooxidans* or another bacterial species, such as *A. thiooxidans* (recall reaction (5.21)):

$$CuS + 0.5O_2 + 2H^+ \rightarrow Cu^{2+} + S^0 + H_2O \quad (5.32)$$

Because pyrite is a common constituent of metal sulfide ores, its oxidation to Fe^{3+} becomes an additional oxidant of chalcocite and covellite, generating more dissolved Cu^{2+} and Fe^{2+}:

$$Cu_2S + 2Fe^{3+} \rightarrow Cu^{2+} + 2Fe^{2+} + CuS \quad (5.33)$$

$$CuS + 2Fe^{3+} \rightarrow Cu^{2+} + 2Fe^{2+} + S^0 \quad (5.34)$$

This reaction sequence is thermodynamically predictable considering that the standard electrode potential of the Fe^{3+}/Fe^{2+} couple is +0.77, while that of Cu^{2+}/Cu^+ is +0.15, and S^0/S^{2-} is −0.27. This means that ferric iron should act as an oxidant for both Cu(I) and S(−II) (McIntosh et al., 1997). In the presence of O_2, and at the acid pH involved, *A. ferrooxidans* re-oxidizes Fe^{2+} back to Fe^{3+}, thereby regenerating the lixiviant so that it can oxidize more copper sulfide. Thus, copper oxidation is maintained indirectly through bacterial Fe(II) oxidation.

There are a number of variables that affect the efficiency of bioleaching operations, the most important being the maintenance of bacterial growth rates commensurate with the desired cell densities in the overall system (Schnell, 1997). This can be influenced by:

1 *Oxidants* − Ferric iron is considered to be the primary oxidant in the dissolution of copper, and its production is vital for efficient bioleaching. The physical addition of Fe^{3+} is uneconomic, thus bacterially-mediated Fe(II) oxidation is of paramount importance. Because O_2 is required for *A. ferrooxidans*, methods to improve its natural diffusion of into the heap pile must be applied, i.e., through the addition of air injection systems.

2 *Acidity* − To ensure both sufficient Fe^{3+} as a lixiviant and a steady-state population of acidophiles, acidic conditions must be maintained.

3 *Permeability* − The permeability of a heap helps determine the solution distribution and the diffusion of O_2 required for bacterial activity. Good agglomeration, through mixing ore with acid and water to prevent the segregation of fine and coarse material, greatly improves permeability and prevents solution channeling.

4 *Nutrients* − Leaching bacteria require ammonium, phosphate, and potassium, which are supplied as $(NH_4)_2SO_4$ and KH_2PO_4, respectively. Typically, these nutrients are added to a heap with a pH < 2, which maximizes bacterial growth and also prevents the precipitation of ammonium jarosite.

5 *Heat* − *A. ferrooxidans* grows best at 20–35°C, although activity is evident outside these ranges. Temperature is important, and as a general rule of thumb, bacterial activity halves for every 7°C temperature drop, so operations in seasonal environments need to take into account some pore water freezing. Furthermore, many of the reactions that take place within the heap pile are exothermic, and it is not uncommon to record summer temperatures in excess of the optimal conditions of the acidophiles.

(b) Galvanic leaching

An additional reaction mechanism that can have a significant affect on bacterial leaching rates is the process of galvanic leaching. Sulfide minerals tend to be electrically conductive, thus, when two different sulfide minerals are in physical contact, as would be the case in an ore deposit, a galvanic couple is created in which the less reactive mineral acts as a cathode, while the more reactive mineral acts as an anode. The latter preferentially becomes oxidized and dissolved. Minerals can be listed according to their reduction potentials, making it possible to predict how different mineral pairings will interact (Table 5.3).

Table 5.3 Reduction potentials of some Fe- and Cu-bearing sulfide minerals*. (After Rossi, 1990.)

Mineral	Potential (mV)	
Chalcocite (Cu$_2$S)	350	reactive
Chalcopyrite (CuFeS$_2$)	400	
Stannite (Cu$_2$FeSnS$_4$)	450	
Pyrrhotite (FeS)	450	
Tetrahedrite (Cu$_3$SbS$_3$)	450	
Pyrite (FeS$_2$)	550–600	noble

*H$_2$SO$_4$ solution, pH = 2.5. Open-circuit potential is measured against a saturated hydrogen electrode.

A typical example is where chalcopyrite and pyrite are in contact. Chalcopyrite, being the more reactive, acts as the anode (Fig. 5.20). Thus, oxidation begins at the surface of the chalcopyrite crystal, releasing electrons that migrate to the surface of the adjacent pyrite crystal (reaction (5.35)). Oxygen that accumulates at the pyrite surface is subsequently reduced and hydroxyl ions are produced in accordance with the following half reaction (5.36). This process is commonly referred to as an oxygen concentration cell.

$$CuFeS_2 \rightarrow Cu^{2+} + Fe^{2+} + 2S^0 + 4e^- \quad (5.35)$$

$$O_2 + 2H^+ + 4e^- \rightarrow 2OH^- \quad (5.36)$$

These reactions also cause the formation of metal oxyhydroxide and sulfur deposits in the regions around the zone of chalcopyrite pitting. The latter subsequently provides an ideal substrate for S^0-oxidizing bacteria, such as *A. thiooxidans*. This increases the sulfate flux to the aqueous phase. The Cu^{2+} released diffuses away from the surface into the overlying aqueous phase, while the Fe^{2+}

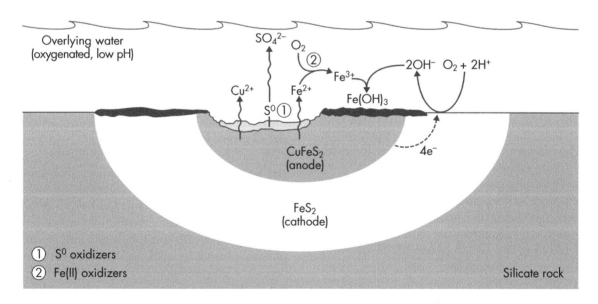

Figure 5.20 Model of a galvanic cell between pyrite and chalcopyrite. Of the products of chalcopyrite oxidation, Cu^{2+} diffuses into the water around the sulfide grains; elemental sulfur accumulates on the chalcopyrite surface, where some of it is biologically oxidized by *A. thiooxidans*; while Fe^{2+} is oxidized by *A. ferrooxidans* and then hydrolyzed to ferric hydroxide on the sulfide surfaces. (Modified from McIntosh et al., 1997.)

released is oxidized by A. *ferrooxidans* (or other Fe(II) oxidizers) to Fe^{3+}, and then hydrolyzed in the more alkaline nanoenvironment above the mineral surface created via the O_2 reduction reaction. The net result of this process is the preferential and accelerated oxidation of chalcopyrite, while pyrite remains relatively unaltered until the chalcopyrite has been completely depleted (Lawrence et al., 1997).

(c) Fungal acid production

Most biological leaching operations rely on acidophile-assisted oxidative processes because the metals of interest are in a reduced oxidation state, housed within a sulfide framework. However, for minerals that contain no redox-active sources of energy, simple dissolution may be all that is required to liberate those metals. In such situations, fungi tend to be highly effective weathering agents because: (i) they tolerate higher pH levels than the acidophiles; (ii) they can survive high metal exposures; (iii) they can be more easily manipulated in bioreactors; and (iv) they produce high concentrations of organic acids that facilitate weathering reactions (Gadd, 1999). Importantly, by altering their growth conditions, fungi can be induced to produce acids on an industrial scale. Citric acid production by the soil fungus *Aspergillus niger* is a case in point. World annual production is estimated at around 400,000 tons (Mattey, 1992). Fungal production can be modified by the concentration and type of carbon source, while withholding certain metals from the growth cultures cause the fungi to increase the amounts of citric acid produced (e.g., Meixner et al., 1985).

There are many examples of laboratory-scale leaching operations. One is the recovery of Ni and Co from low grade laterite ores by species of *Aspergillus* and *Penicillium*. Nearly 60% of the available Ni was leached when fungi were grown in the presence of the ore and the leaching potential was increased to 70% when the metabolic products obtained from cultivation of

fungi were applied at 95°C. The cobalt recovery was almost 50% (Tzeferis, 1994). Not only is *Aspergillus niger* able to solubilize a wide range of insoluble minerals, including phosphates, sulfides, and oxide ores, but it is also able to immobilize those leached metals (e.g., Cu, Cd, Co, Zn, and Mn) by the formation of metal oxalate salts. Many of those metal oxalates are resistant to further solubilization, suggesting that oxalate formation may be a survival mechanism used by the fungi to immobilize potentially toxic metal compounds within their immediate surroundings (Sayer and Gadd, 1997).

Bioleaching can also be employed to remove toxic metals from waste materials. This diminishes problems of disposal and opens up new avenues for recycling metal-rich refuse. As discussed above, chemolithoautotrophs have been employed in sewage sludge treatment, and the use of different acidophiles is often advantageous because they can induce the mobilization of metals at low pH. In organic-rich wastes, heterotrophic bacteria and fungi are more useful because they not only degrade the bulk of the organic carbon, thus reducing the volume of waste, but they also generate acids that effectively leach metals from insoluble mineral constituents. For instance, a strain of *Penicillium simplicissimum* that was isolated from a metal-contaminated site, produced sufficient citric acid to successfully leach 90% of the zinc from insoluble ZnO-containing industrial filter dust (Schinner and Burgstaller, 1989).

5.2.4 Biooxidation of refractory gold

Many ore bodies contain metals, such as gold, that are difficult to extract because the metal is disseminated throughout the host sulfide mineral, such as pyrite or arsenopyrite. For these "refractory" ores, conventional cyanide or bioleaching methods do not work unless the sulfide minerals can first be destroyed by an oxidative pre-treatment to liberate the gold. Two traditional pre-treatment methods for refractory gold ores are roasting and pressure oxidation. Recently, biooxidation has emerged as a viable third alternative. Biooxidation uses similar bacteria as in bioleaching to catalyze the degradation of sulfide minerals, but unlike bioleaching, it leaves the metals of value in the solid phase. To facilitate the biooxidation process, finely ground ores are separated from the gangue and other materials by flotation techniques to produce a concentrate that is then added to a stirred-tank bioreactor, where the chemolithoautotrophic bacteria reside (Lindström et al., 1992).

There are a number of commercial biooxidation processes now in existence. The BIOX® process uses a mixed population of *A. ferrooxidans*, *A. thiooxidans*, and *L. ferrooxidans* to collectively oxidize the reduced Fe and S moieties in arsenopyrite and pyrite (see Dew et al., 1997 for details). In biooxidation, the first step in leaching is oxidation of the sulfide component of the mineral, and the solubilization of Fe^{2+}, As^{3+} (as H_3AsO_3), and the sulfoxy species. The reductants enhance the growth of free-living chemolithoautotrophic bacteria, with the concomitant formation of Fe^{3+}, As^{5+} (as H_3AsO_4), and SO_4^{2-} (see overall reaction (5.37)). The Fe^{3+} subsequently triggers the abiological oxidation of more arsenopyrite and pyrite (reaction (5.38)):

$$2FeAsS + 7O_2 + 2H^+ + 2H_2O \rightarrow$$
$$2H_3AsO_4 + 2Fe^{3+} + 2SO_4^{2-} \qquad (5.37)$$

$$FeAsS + Fe^{3+} + 2.5O_2 + 2H_2O \rightarrow$$
$$2Fe^{2+} + H_3AsO_3 + SO_4^{2-} + H^+ \qquad (5.38)$$

Many bacteria growing naturally on arsenopyrite may be inhibited by the levels of arsenic released. The BIOX® bacteria are, however, tolerant to As(V) concentrations of $15-20\,g\,L^{-1}$. They are less tolerant to As(III), and become inhibited above concentrations of $6\,g\,L^{-1}$, although its rapid oxidation by Fe^{3+} generally maintains very low H_3AsO_3 concentrations. Similarly, other

constituents of the ores, such as antimony and mercury, can become toxic to the bacteria. Other environmental factors that impact on the BIOX® process are pH, temperature, CO_2, and nutrients. The optimal temperature for the BIOX® bacteria is between 35°C and 45°C; the bacteria are not killed at 50°C, but their oxidation rates slow down considerably, and the time required for complete conversion of Fe(II) to Fe(III) increases from 1 day at 40°C to 3 weeks at 50°C.

The BacTech process employs moderate thermophiles in the biooxidation of refractory gold-bearing ores (Miller, 1997). Bacteria of the genera *Sulfobacillus* and *Sulfolobus* grow optimally at temperatures of around 50°C, and when used in biooxidation experiments, they have been shown to provide higher rates of sulfide mineral dissolution than their mesophilic counterparts. Furthermore, the efficient extraction of copper from chalcopyrite concentrates, which cannot be achieved at low temperatures, is a notable potential application of thermophiles in bioleaching.

5.3 Microbial corrosion

The deterioration of a metal by electrochemical reactions with substances in its environment is referred to as corrosion. In most cases the basic process underpinning corrosion involves a flow of electricity between certain areas of a metal surface through a solution that has the ability to conduct an electric current. During corrosion, metal cations develop at an anodic site and the electrons associated with this dissolution are accepted at a cathodic site. The metal that has received the most attention is elemental iron (Fe^0), although copper, aluminum, lead, and silver also succumb to corrosive reactions. The corrosion of elemental iron is best known as rust formation on steel, when in contact with oxygen and water. Such reactions are often manifest as structural damage to buildings and deterioration of machinery, ship structures, cars, and even airplane fuel tanks. In the absence of oxygen, any number of other electron acceptors can be involved in steel corrosion, so that structures underground are similarly susceptible to damage (e.g., pipelines, storage tanks, etc.). Even protons can be reduced because of the very negative electrode potential of Fe^0 (reaction (5.39)). Hence, in principle, protons represent constant potential electron acceptors for anaerobic iron corrosion.

$$Fe^0 + 2H^+ \rightarrow Fe^{2+} + H_2 \qquad (5.39)$$

Under normal circumstances Fe^0 does not corrode completely. Instead, as Fe^{2+} forms and dissolves away from the surface, negative charges are left on the surface. They are strongly reducing and, in the presence of O_2, rust forms, while in anoxic waters, they reduce protons from the dissociation of water to form a protective film of adsorbed H_2. These layers limit continued corrosion by serving to some extent as passivity layers, or barriers, to further corrosion (Cord-Ruwisch, 2000).

Microorganisms facilitate corrosion in a number of ways, one of them being through the mechanism of cathodic depolarization, as initially postulated by von Wolzogen Kuehr and van der Vlugt (1934). In their theory, reaction (5.39) will continue from left to right as long as the H_2 is continually removed from the Fe^0 surface via consumption by various electron accepting molecules, ranging from O_2 to CO_2. Some microorganisms further facilitate the transfer of electrons from Fe^0, by specific catalytic enzymes (e.g., hydrogenases), and in the process capture energy for growth, while others facilitate corrosion simply because they form biofilms. The EPS primarily contributes to corrosion by adsorbing and chelating metals, thereby generating localized concentration gradients that affect saturation states. They also promote the establishment of cathodic and anodic sites on the steel surface, that enhances electron flow

(Beech and Gaylarde, 1991). Several types of microorganisms contribute to metal corrosion, but for simplicity, we will consider them in terms of chemolithoautotrophs, chemoheterotrophs, and fungi.

5.3.1 Chemolithoautotrophs

Most chemolithoautotrophs play some role in metal corrosion. The importance of S- and N-oxidizing bacteria lies in their formation of sulfate and nitrate, and hence sulfuric and nitric acids, respectively. The degradation of concrete sewers or the deterioration of stone monuments (see below) are the most serious problems associated with their growth.

The role of Fe(II)- and Mn(II)-oxidizing bacteria in steel corrosion is based on their ability to form cathodic $Fe(OH)_3$ and MnO_2, respectively. The process is as follows. During the initial stages of steel corrosion, cathodic reduction of O_2 causes an increase in solution pH above the steel surface, which facilitates oxidation and hydrolysis of the Fe^{2+} liberated during corrosion. If sufficient amounts of ferric hydroxide form, then the anodic site may eventually become isolated from the surrounding oxygenated cathode. When that occurs, ferric hydroxide instead serves the cathode, accepting electrons directly from the steel (Little et al., 1997). The extent of the current, in turn, is governed by the Fe mineralization rate because the current becomes self-limiting when the cathode ($Fe(OH)_3$) is depleted via its reduction (to Fe^{2+}), and O_2 is re-established as the cathode. This impasse is overcome by the metabolism of the metal-oxidizing bacterial community that reside on the steel surface and within the dissolution pits. In addition, any soluble Cl^- anions will migrate to the anode to neutralize any buildup of charge, forming heavy metal chlorides that are extremely corrosive (Fig. 5.21). Given the heterogeneity of the corrosive environment, pitting, rather than the even corrosion of the surface, tends to occur.

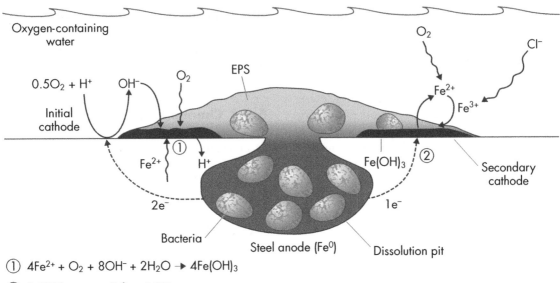

① $4Fe^{2+} + O_2 + 8OH^- + 2H_2O \rightarrow 4Fe(OH)_3$

② $Fe(OH)_3 + e^- \rightarrow Fe^{2+} + 3OH^-$

Figure 5.21 Model illustrating the role of Fe(II)-oxidizing bacteria in metal corrosion. Through their oxidative activity, they generate ferric hydroxide that functions as a cathodic surface, accepting electrons directly from steel. (Adapted from Little et al., 1997.)

5.3.2 Chemoheterotrophs

Many studies have investigated the effects of aerobic bacterial activity on the corrosion of iron. However, because the process also occurs at relatively high rates in the absence of microorganisms, the microbial effect is not easy to monitor or predict. In general, due to the oxygen uptake activity, bacterial biofilms on the metal surface create localized environments of differential aeration that generate cathodic areas (where electrons from Fe^0 reduce O_2) spatially separated from the anodic areas (where ferrous iron dissolves), resulting in a corrosion current and the dissolution of the metal (e.g., Morales et al., 1993).

What is likely the most important group of anaerobic heterotrophs, in terms of their corrosive capabilities, are the SRB (Fig. 5.22). Through their production of H_2S, they indirectly promote corrosion by forming a thin layer of iron sulfide (mackinawite) on the metal surface (reaction (5.40)): strictly, it is not the sulfur atom that accepts the electrons from the corrosion process but the protons that are part of the hydrogen sulfide molecule (Lee et al., 1995).

$$Fe^0 + H_2S \rightarrow FeS + H_2 \qquad (5.40)$$

Mackinawite formation has the effect of accelerating the corrosion process because once electrical contact is established, the steel may behave as an anode, facilitating electron transfer through the cathodic iron sulfide phase, i.e., acting as a galvanic cell (Wikjord et al., 1980). If the Fe^{2+} concentration in solution is low, the mackinawite alters to greigite, as previously covered in section 4.1.10.

SRB can directly cause corrosion via their hydrogenase activity, as follows. Hydrogenase accepts H_2 and releases protons. Those protons then attack the steel surface, causing its oxidation and the release of Fe^{2+}, while simultaneously

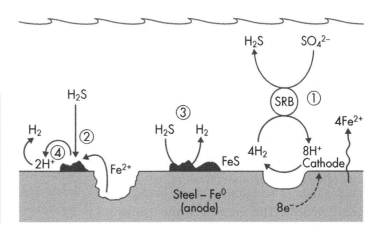

Figure 5.22 Model showing how sulfate-reducing bacteria facilitate the corrosion of steel through (1) cathodic depolarization, (2) anodic depolarization, (3) H_2S production, and (4) indirect supply of H^+. (Adapted from Cord-Ruwisch, 2000.)

accepting electrons from the anode to form H_2 (reaction (5.41)):

$$Fe^0 + 2H^+ \rightarrow Fe^{2+} + H_2 \qquad (5.41)$$

If an appropriate electron acceptor is then available to accept H_2 (e.g., reaction (5.42)), then the hydrogenase once again becomes available to oxidize more H_2 generated at the metal surface; the dissolving process would cease if the H_2 was not removed, for reasons discussed above.

$$4H_2 + SO_4^{2-} \rightarrow H_2S + 2H_2O + 2OH^- \qquad (5.42)$$

By virtue of their resistance to degradation, hydrogenases may remain viable for months after the cell is dead, implying that even lysed cells can continue the corrosion process (Bryant and Laishley, 1989). Thus, in terms of the cathodic depolarization theory, SRB trigger the oxidation of metal by removing the protective H_2 layer and linking the electron flow from the metal to the metabolic reduction of sulfate, with H_2 as the electron donor (Cord-Ruwisch and Widdel, 1986). Additionally, the hydrogen sulfide generated via sulfate reduction reacts with Fe^{2+} to form FeS and H^+ (reaction (5.43)), thereby removing Fe^{2+} from solution and driving reaction (5.41) to form more products. This is anodic depolarization by Fe^{2+} removal. The protons, in turn, repeat the oxidative attack on the steel surface.

$$Fe^{2+} + H_2S \rightarrow FeS + 2H^+ \qquad (5.43)$$

Although not related to metal corrosion, bacterial sulfate reduction can lead to the destruction of gypsum deposits. Their activity is known to limit the preservation of primary gypsum in sediments (Rouchy and Monty, 2000), and considering that a number of antiquities are composed of alabaster (a hydrated form of gypsum), this can have important archaeological implications. On a different note, industry employs gypsum-degrading SRB to convert gypsum sludge, produced during flue gas desulfurization, into marketable calcite and elemental sulfur, while simultaneously oxidizing sewage sludge as their organic substrate (Kaufman et al., 1996).

Under conditions where nitrate serves as the terminal electron acceptor, some bacteria, such as *E. coli*, also promote the oxidation of elemental iron through a similar hydrogenase model as described above (Umbreit, 1976). In recent years, methanogens have similarly been added to the list of microorganisms believed responsible for corrosion. Like many SRB, methanogens consume hydrogen and thus are capable of performing cathodic depolarization-mediated oxidation of elemental iron to produce methane (Boopathy and Daniels, 1991). Because methanogenesis involves proton consumption, the overall reaction (5.44) will be affected by the acidity of the aqueous solution, i.e., at lower pH the reaction becomes more energetically favorable:

$$4Fe^0 + CO_2 + 8H^+ \rightarrow 4Fe^{2+} + CH_4 + 2H_2O$$
$$(\Delta G° = -136 \text{ kJ/CH}_4 \text{ at pH 7 and 37°C}) \qquad (5.44)$$

The above reaction has the same $\Delta G°$ as previously shown for methanogens using reaction (2.46), that is -136 kJ per mol CH_4 at pH 7.

As both the SRB and methanogen examples have shown, anodic dissolution of Fe^0, when coupled only to H_2 production, is not energetically favorable. This means that the oxidative reaction must be coupled to cell growth of a microorganism that consumes the H_2. In this regard, even Fe(III)-reducing bacteria can accelerate the rate of corrosion (Iverson, 1987).

5.3.3 Fungi

Most fungi are capable of producing organic acids that corrode steel and aluminum, as in the highly publicized corrosion failures of aircraft fuel tanks. Another significant corrosion problem is the degradation of cement, and the resulting deterioration of building materials and nuclear waste repositories. Minerals in cement, such as hydrated calcium silicate and portlandite ($Ca(OH)_2$), are readily solubilized and decalcified by fungal-generated organic acids. In particular, the fungus *Aspergillus niger* seems to promote corrosion through the production of acids within the pore spaces physically created by its hyphae (Perfettini et al., 1991). Oxalic acid is also involved in the corrosion of ancient stonework by lichens. Physical and chemical changes in stonework, such as fracturing and encrustation, can lead to biodegradation, with calcium oxalate being a significant chemical component in the surface alteration zone (e.g., Edwards et al., 1994). Concomitantly, carbonic acid is generated through aerobic respiration of lysed lichens, further contributing to chemical weathering.

5.4 Summary

Microorganisms play an important role in accelerating mineral dissolution and oxidation reactions. Through their production of organic acids, they supply H^+ ions to attack metal–oxygen bonds and systematically dissolve the atoms comprising the crystal lattice. Meanwhile, the deprotonated organic anions complex with metal cations, thereby affecting mineral saturation states, promoting even greater mineral dissolution. Other microorganisms produce chelates that act in a similar manner to the organic acids. Even upon death, microorganisms are important agents in weathering because their decay, via the action of respiring heterotrophs, leads to elevated soil CO_2 partial pressures, which, in turn, creates carbonic acid. Collectively, these processes have contributed to the erosion of exposed outcrops, led to soil formation and influenced global climate since the spread of microbial life onto land. Many microorganisms have also evolved the capacity to utilize the energy released by the oxidation of reduced transition metals and sulfur phases. In the deep sea, the microbial oxidation of Fe(II) in basalts likely contributes to the flux of solutes to the bottom waters, while biological sulfide mineral oxidation on land functions as the catalyst for the release of high concentrations of metals, sulfate and acidity into regional waterways. These same reactions have industrial implications. On the one hand, the oxidative and acid-generating properties can be utilized in the biorecovery of economically valued metals from mine wastes or in the remediation of toxic metal concentrations in the environment. On the other hand, these same reactions contribute to the corrosion of various metal structures, necessitating significant financial expenditures on repairs and preventative measures.

6

Microbial zonation

Microorganisms are integral to elemental cycling on Earth's surface. Arguably one of the most important of these cycles involves the transfer of carbon between the atmosphere-hydrosphere-biosphere and the sedimentary environment. In this regard, microorganisms fulfill both the role of primary producers, through the fixation of inorganic carbon into biomass during photosynthesis and chemolithoautotrophy, and as the major degraders of that biomass (as heterotrophs), gaining energy and carbon for growth from its transformation to CO_2 or CH_4. A fraction of that organic carbon also becomes buried into the sedimentary record as kerogen or fossil fuel, where it is ultimately returned to the Earth's surface through long-term geological processes or human activity. Intimately tied to the carbon cycle is the fate of an entire suite of redox-active elements that make up part of the cell and/or function as electron donors or acceptors in metabolic pathways. Elemental cycling occurs within every conceivable environment, from micrometer-thick biofilms to soils and sediments, and even on scales as large as the oceans. It would be far too exhaustive to cover all the environments within this chapter. So, the focus will instead be on two examples particularly relevant to how we interpret the activity of ancient microbial life (the focus of chapter 7), namely microbial mats and marine sediments. Microbial mats are widely distributed in modern environments despite the special circumstances required for their formation. Their propensity to stabilize and modify surficial sediments, and at times induce their lithification into laminated biosedimentary deposits, is one of the few means by which

microbial community activity is imprinted into the rock record. Crucially, morphological and compositional comparisons between modern mineralized mats with Precambrian microbialites has led to significant advances in our understanding of Earth's early biosphere before the onset of multicellular life. Microorganisms are also the driving force behind the geochemical and mineralogical transformations that take place during the burial and lithification of marine sediments. The types of diagenetic minerals formed, and their isotopic signatures, are directly related to the metabolic processes occurring during diagenesis. These, in turn, are influenced by the reactivity of the particulate organic carbon (POC) fraction deposited onto the seafloor and the availability of dissolved and solid-phase oxidants. Considering that the abundance of any given reactant in the sediment column is inevitably related to the oxidation state of the atmosphere and water column, understanding the reactions underpinning marine sediment diagenesis offers perhaps the most valuable information on global paleoenvironmental changes during Earth's history.

6.1 Microbial mats

It is well documented that solids added to aquatic systems are rapidly coated with an organic accretion, and that this adsorbed organic layer constitutes the surface that is actually colonized by adherent bacteria, commonly referred to as "surface conditioning" (Costerton et al., 1981). The adherent bacteria produce microcolonies that initially develop into biofilms composed of only one bacterial type. Given time, and a sufficient supply of energy, undisturbed biofilms thicken into mats, with new species joining the fray to form a complex community of interrelated species. Functionally, a modern living mat is a vertically compressed ecosystem that supports most of the major biogeochemical cycles within a dimension of only a few millimeters. They are largely self-sufficient, in that sunlight or chemical reductants provide the primary source of energy used to convert CO_2 into biomass, while heterotrophy recycles the carbon and reducing equivalents back to other mat microorganisms. Due to differing growth requirements, it is common for mat microorganisms to orient themselves into vertically stratified subsystems that are manifest as distinct, and often, visible layers. The most conspicuous components are the filamentous microorganisms, either photosynthetic or chemolithoautotrophic, that, through entwinement of their filaments and production of extracellular polymers (EPS), provide mechanical cohesion to the mat (e.g., Fig. 6.1).

Mats are generally confined to harsh environmental settings that prohibit widespread disruption by grazing macrofauna. Examples include: subtidal marine environments subject to frequent sediment movement; intermittently exposed intertidal marine settings prone to periodic dessication; hypersaline lakes characterized by elevated salinities; hydrothermal springs where the effluent temperatures exceed the physical tolerance of most organisms; and lakes or marine basins with anoxic bottom waters. In most places the mats are a transient or seasonal phenomena of limited extension, and they grow at most to a few millimeters in thickness. However, under more favorable conditions, and where they are protected for extended periods of time, mats may become more stable structures, and over many years accumulate resilient organic material and mineral deposits. The growth rates of some such mats can be of the order of 1 mm per year, balanced by biomass decomposition in the bottom layers (Fenchel et al., 2000).

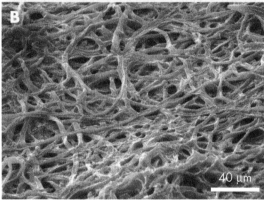

Figure 6.1 (A) Cross-sectional view of a cyanobacterial mat from a barrier island off the coast of North Carolina. Filaments are clearly visible, as are the oxygen bubbles that they produce. The lower layers of the mat are colored black due to iron sulfide precipitation. Quartz sand can be seen below the mat (courtesy of Henry Page and Brad Bebout). (B) Scanning electron microscope (SEM) image showing the top surface of a cyanobacterial mat from Baffin Bay, Texas. Notice how the filaments and EPS completely enshroud the underlying surface (courtesy of Henry Chafetz).

6.1.1 Mat development

In Chapter 3 we outlined how microorganisms attach to submerged solids. The initial colonizers are those species that have both the appropriate cell surface charge and topology to interact with exposed ligands on the solid, and the capability of producing EPS or appropriate appendages to become physically adherent. The subsequent production of a continuous biofilm then arises as a function of cell division within the pioneering communities and new recruitment of similar species from the planktonic phase. As a result, the biofilm consists of single cells and microcolonies of sister cells all embedded together in EPS. As the biofilm gradually occludes the exposed surface, different species begin to colonize the biofilm itself, and over a matter of days, a succession of microorganisms develop into a thicker mat community. This process continues until the supply of nutrients or energy becomes insufficient to support further cell replication or EPS formation (Costerton et al., 1987).

The succession of species colonizing a solid is coincident with the alteration of its surface properties by those initially attached. For instance, the pioneering species may release metabolites, surface active agents, or extracellular polymers, each of which could affect the wettability and adhesiveness of the underlying solid. Under such conditions, the next group of microorganisms adhere to a surface that may be fundamentally different to that encountered by the pioneering group (Fig. 6.2).

With continued accretion, the proportion of EPS to biomass increases significantly; the cell content of biofilms can vary from 90% of the organic matter during the initial stages of formation, to lower than 10% in a well developed biofilm (Christensen and Characklis, 1990). In addition to the benefits to individual cells, as highlighted in section 3.1.2(a), EPS also helps the entire mat community in a variety of ways (see Decho, 2000 for details):

1 The chemical diversity of EPS, caused by possessing different functional groups, allows an assortment of dissolved inorganic and organic species to accumulate within close proximity to the cells, while the hydrolytic enzymes that are located within the matrix break down large organic polymers into their monomeric forms. The diffusional properties of EPS also stabilize geochemical gradients, thereby

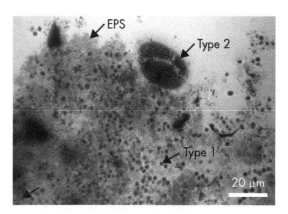

Figure 6.2 Laser scanning microscope image of a complex degradative biofilm community growing on a wood surface (left of arrow) in a bioreactor treating pulp mill effluents. A variety of bacteria are held in the EPS matrix, but it is clear that the small spheres colonized the wood, and they, in turn, provided a surface for the larger cell morphologies. (Courtesy of John Lawrence.)

facilitating unique niches that can be colonized by specific microorganisms.

2 It physically permits mixed communities to live within close proximity to one another, a feature of utmost importance for those microorganisms whose physiological activities require the cooperation of another species. For instance, spatial analyses of cells in mats have shown that they are typically aggregated, and not randomly dispersed, with areas of high cell densities and high microbial activity often adjacent to sparsely populated areas that form little EPS and relatively few cells. The biofilm mode of growth also allows for the pooling and preservation of genetic information and the feasibility of gene transfer.

3 EPS acts as a buffer against environmental extremes, providing protection from short-term pH fluctuations, dessication, and high salt or toxic metal concentrations. It may also provide protection against UV-A/B irradiation. Indeed, the widespread distribution of mats in "extreme" environments may suggest that their formation was a characteristic of life forms that arose amidst the harsh conditions prevalent during Earth's early history (see Chapter 7).

Predictably, attached microorganisms are often more metabolically active than their planktonic counterparts; they assimilate more carbon than planktonic cells, and greatly exceed them in terms of cell densities (e.g., Fletcher, 1985).

Cyanobacteria, diatoms, and other microalgae are dominant in the early stages of biofilm formation. Those initial colonizers tend to include species capable of N_2 fixation, but they may be replaced later on when sufficient organic nitrogen becomes available. The mats also become slowly enriched in a variety of organic compounds due to the overproduction of carbon during photosynthesis and lysis of the primary producers. These compounds, in turn, are degraded by a variety of fermentative and heterotrophic bacteria. The latter include aerobic respirers, denitrifiers, sulfate-reducing bacteria (SRB), and methanogens — their positioning in the mat is governed by the redox potential of the system (Box 6.1). Aerobic heterotrophs are functionally important as their activity leads to O_2 depletion, while SRB provide H_2S for later colonizing representatives of the colorless and purple sulfur bacteria. These two groups of sulfide oxidizers are sandwiched in between the layer of cyanobacteria and SRB, where they are sheltered from high O_2 levels. Additionally, the purple bacteria take advantage

---

Box 6.1 Redox potential

It is generally agreed that pH is the "master variable" controlling chemical reactions in low temperature systems. However, the lack of O_2 in subsurface layers of microbial mats, soils, and sediments can result in a sequence of redox reactions that change the system pH, as well as influence metal cation solubility, the valence of ions (e.g., Fe^{3+} vs. Fe^{2+}), and the nature of the molecules dissolved in pore water solution (e.g., H_2S vs. HS^-).

For any electron-transfer half-reaction, the Nernst equation can be applied to relate the standard electrode potential ($E°$) and solution pH to the electrode potential in any state. When this is done, the symbol "Eh" is used for the measured potential (in volts), also commonly referred to as the oxidation potential or redox potential:

$$Eh = E° + \frac{0.059}{n} \log \frac{[products]}{[reactants]}$$

where n is number of electrons transferred. In natural systems the Eh, like pH, is an environmental parameter whose value reflects the ability of a natural system to donate or accept electrons relative to the standard H_2 electrode that acts only as a reference. Its importance can be high-lighted when we examine the stability limits of water. Recall that liquid water must maintain an equilibrium with O_2 and H_2 gas, and that the oxidation of O in water itself can be viewed as a half-reaction:

$$2H_2O \longleftrightarrow O_2 + 4H^+ + 4e^-$$

In this reaction the O in the water molecule gives up two electrons and acts as an electron-donor that can generate an Eh when it is connected to a H_2 reference electrode. Therefore, we can express the Eh of the "water electrode" by a variation of the Nernst equation:

$$Eh = E° + \frac{0.059}{n} \log[(O_2)(H^+)^4]$$

The concentration of O_2 in equilibrium with liquid water can vary between 1.0 and $10^{-83.1}$ atm. When we apply these limits to the equation above, we obtain for $O_2 = 1$ atm: $Eh = 1.22 - 0.059$ pH, and for $O_2 = 10^{-83.1}$ atm: $Eh = -0.059$ pH.

These equations are straight lines that define the stability field of liquid water in terms of Eh and pH. A positive Eh in the stability field implies that the waters are more oxidizing than the

continued

Box 6.1 *continued*

standard H_2 electrode. Such oxidizing environments contain an abundance of potential electron acceptors, the most common at Earth's surface being O_2. A large reserve of Mn(IV)- and Fe(III)-oxides instead maintain the redox potential at a lower level as reduction of acceptors continues. Environments that have negative Eh values are more strongly reducing than the H_2 standard

electrode, and as such are reducing. They have an abundance of electron donors (e.g., fermentation products, H_2S, etc.). It is worth pointing out that although an environment is "reducing" it may still contain trace amounts of O_2, but the concentrations are exceedingly small and much less than is required to sustain aerobically respiring organisms.

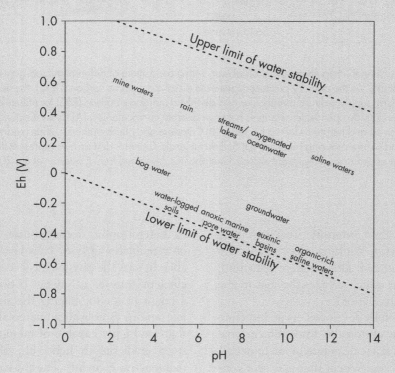

The approximate position of some natural aqueous environments as characterized by Eh and pH. (Modified from Brownlow, 1996.)

How then is Eh related to pH? If we think of Eh as reflecting the abundance of electrons in the environment, then a large number of electrons give a reducing environment, while an absence yields an oxidizing environment. Similarly, we can think of pH as representing the abundance of protons, with a large number of protons generating acidity, and a paucity yielding a basic environment. Since protons and electrons have opposite charges, when we have an abundance of one we would have a shortage of the other. In other words, an oxidizing environment would tend to be acidic, while a reducing environment would tend to be basic (Brownlow, 1996). In fact, were it not for the precipitation of metal carbonate phases (e.g., $MnCO_3$ and $FeCO_3$) that consume the bicarbonate anions generated during anaerobic respiration, the pH of soil/sediment pore waters would go well above 7.

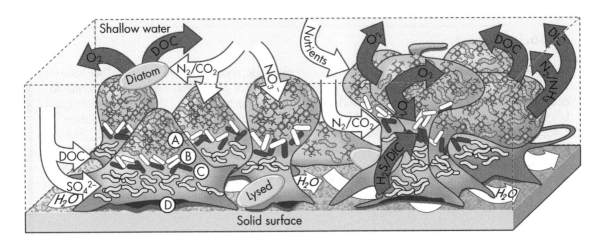

Figure 6.3 Structure of a hypothetical microbial mat, with a consortium of different microorganisms living within a matrix of EPS. As the primary producers generate dissolved organic carbon (DOC), the heterotrophic communities, in turn, supply reduced metabolites and dissolved inorganic carbon (DIC) for utilization by the former. Note the relatively open water channels between discrete microcolonies. (A) Cyanobacteria intermixed with aerobic respirers and nitrate-reducing bacteria. (B) Colorless and photosynthetic sulfide oxidizers. (C) Lysed cells and sulfate-reducing bacteria. (D) Sulfide minerals. Open-ended arrows represent fluxes into the mat, while closed-ended arrows represent excreted metabolites. (Adapted from Costerton et al., 2004.)

of the reduced light intensity. Together, they are responsible for oxidizing more than 99% of the H_2S generated during sulfate reduction, the remainder of which escapes the mat to the atmosphere (Trüper, 1984).

Once a mat is fully developed, its architecture includes an array of irregular, three-dimensional cell aggregates, on a scale of tens of micrometers, separated by a canal system that acts as a conduit for fluid flow between all parts of the mat (Fig. 6.3). In essence, individual mat bacteria enjoy some of the advantages of multicellular life, in that the primitive "circulatory system" delivers nutrients, gases and metabolic reactants from the bulk fluid phase to the microbial community, while simultaneously removing metabolic wastes (Costerton et al., 1994).

Energy production and nutrient acquisition in stable mats are closely interrelated, and usually in steady-state. This involves complex pathways in which primary production of organic carbon and nitrogen fixation by phototrophy and chemoautotrophy in the upper mat are balanced by heterotrophic decomposition below, where only remnants of lysed cells layers are identifiable. In fact, the physiological cooperativity and efficient transfer of energy between species is a major factor in shaping the structure of the mat and in establishing microbial communities that are highly capable of taking advantage of steep gradients in light, Eh, pH, and many redox-active elements. The end result is a microenvironmentally differentiated ecosystem, often only fractions of a millimeter thick. Importantly, many cells within mats are motile and capable of adjusting their position relative to one another, indicating that the ability to move provides cells with the means to both optimize their growth conditions and escape toxic compounds that might have accumulated within certain parts of the mat (Korber et al., 1993).

6.1.2 Photosynthetic mats

Multilayered microbial mats are primarily composed of photosynthetic microorganisms. For them, light has two very important properties, intensity (i.e., light quantity) and the available electromagnetic wavelengths (i.e., light quality).

In the presence of an ozone screen, UV-C (190–280 nm) and UV-B (280–320 nm) radiation are effectively absorbed, although the current decrease in stratospheric ozone by anthropogenic activity has increased UV-B fluxes reaching Earth's surface to levels approaching 1%. UV-A radiation (320–400 nm) and photosynthetically active radiation (PAR, 400–470 nm) are not attenuated by ozone, and thus are available to photosynthetic organisms (Caldwell et al., 1989). Although the typical surface irradiance is about 1000–2000 uEinstein $m^{-2} s^{-1}$, the optimum light intensity for cyanobacteria is between 15 and 150 uEinstein $m^{-2} s^{-1}$, while that for the purple bacteria is between 5 and 10 uEinstein $m^{-2} s^{-1}$ (Stal et al., 1985). When subjected to higher than optimal intensities, photosynthetic activity can become inhibited unless the cells can take appropriate action (Vincent and Roy, 1993).

This can include any number of the following photoresponses:

1 *Motility* – some cells employ vertical migration, using gliding motility to move downwards into the mat where light intensity is reduced (e.g., Castenholz et al., 1991). Others simply change their orientation during peak daylight hours (i.e., from flat-lying to vertical) to reduce their cross-sectional surface area exposed to incident solar radiation, thus minimizing the photon load on their photosystems (Ramsing et al., 2000). Such diel re-orientation is an energy-efficient alternative to vertical migration through an EPS and, possibly, mineral-rich matrix.

2 *Pigmentation* – many mat-forming cells, particularly those incapable of motility, produce compounds that directly absorb UV-A/B. These include extracellular shielding pigments, such as scytonemin (which absorbs UV-A) or mycosporine-like amino acids (which absorbs UV-B), as well as intracellular carotenoids that quench toxic photochemical reaction products (e.g., Quesada and Vincent, 1997). In nature, production of these pigments can change the mat's colour, for example, from orange-yellow (carotenoids) in summer to green (chlorophyll *a*) in winter (Wiegert and Fraleigh, 1972). Models,

however, have shown that cells with radii less than 1 μm cannot rely entirely on sunscreen compounds because the intracellular concentrations needed would be physically unattainable (Garcia-Pichel, 1994).

3 *Synthesis of extracellular layers* – EPS and sheaths provide increased surface area to absorb UV radiation, thus reducing direct exposure of the cells (Ehling-Schulz et al., 1997).

4 *DNA repair* – Many species have the means to restore damaged cellular components by a variety of repair mechanisms (Friedberg, 1985).

In terms of light quality, the majority of photosynthetically active radiation is absorbed by the uppermost meter of water, and in sediments covered by a thin veneer of surface water, it is actually the near-infrared radiation that penetrates furthest. Microorganisms in the top layers of the mat are thus illuminated directly by daylight with a broad spectral composition, while microorganisms in the deeper layers within the photic zone receive only the fraction of light that has been filtered through the overlying pigments. As a consequence of light quantity and quality, photosynthetically dominated mats are frequently stratified, with different species naturally finding their ideal niche at different depths in a mat. This positioning, of course, varies both diurnally and seasonally, with the chemocline shifting upwards and downwards in the same rhythm (Jørgensen, 1989).

(a) Cyanobacterial mats

The most studied microbial mats are those dominated by filamentous cyanobacteria, and in many regards the discussions above on mat development largely pertain to their formation. When viewed in cross-section, the mats typically display a macroscopically visible color banding that can range from micrometers to centimeters in depth, depending, in part, on the extent to which light penetrates (e.g., Fig. 6.4; Plate 13). The upper-most biological layer is often yellow, followed directly underneath by a sequence of layers, beginning with green, red/purple (and

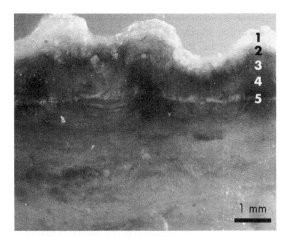

Figure 6.4 Photograph showing the vertical stratification in a very thin slice of microbial mat from Guerrero Negro, Baja California Sur, Mexico. Visible are the layers formed by various types of microorganisms and any accompanying mineral precipitates, including, from top to bottom: (1) gypsum crust; (2) diatoms; (3) filamentous and unicellular cyanobacteria; (4) purple sulfur bacteria – the dark clumps; (5) colorless sulfur bacteria. (Courtesy of Mary Hogan and Tracy Cote.)

Chloroflexus-related bacteria that typically use bacteriochlorophyll *c*. The lower green layer is made up of green sulfur bacteria. These immotile bacteria are obligately phototrophic (using bacteriochlorophyll *b*) and sulfide-dependent, which means that their uppermost limit of growth must coincide with the level of permanent sulfide production in a habitat where it is not exhausted even at a time of maximum daylight oxygenic photosynthesis. The lowermost limit reflects their absolute light limitation. The deepest layers contain anaerobic chemoheterotrophs, usually sulfate reducers and methanogens, that do not require oxygen or light, but merely the microbial remnants of previous surface communities. The presence of SRB is often distinguished by black sediment rich in FeS.

The above zonation is a rather crude representation because the indigenous microorganisms display variable tolerances to light intensity and variations in pore water and gas compositions. As a result, neither oxygenic nor anoxygenic phototrophs are static within a mat. In general, the thickness of any phototrophic-based mat depends primarily on light penetration. Although 1% of surface irradiance is considered to be the lower limit for which cyanobacteria can grow photoautotrophically, viable cyanobacteria have actually been recovered at depths where light levels are below the 1% threshold; the ability to grow photoheterotrophically at low light intensities is common amongst both *Fischerella* and *Calothrix* species, while members of both genera are also capable of slow, chemoheterotrophic growth in the dark (Rippka et al., 1979). Conversely, cyanobacteria undergo biochemical and structural damage when overexposed to high intensity solar radiation, and in response will employ one of the compensating strategies discussed above.

Anoxygenic photosynthesizers similarly suffer light intensity issues, but instead reside deeper in the mat where their bacteriochlorophyll pigments harness the infrared radiation that has penetrated to their depth. In general this is not problematical because they are already restricted to deeper layers where the chemical environment is reducing and H$_2$S is available

occasionally orange), green again, and black (e.g., Pierson et al., 1987; Garcia-Pichel et al., 1994). The top two layers consist of an assemblage of diatoms, algae, and various filamentous cyanobacteria, mainly from the *Oscillatoriales* order (e.g., the genera *Oscillatoria*, *Lyngbya*, *Phormidium*, *Spirulina*, *Microcoleus*, *Schizothrix*) and sometimes other genera, including *Anabaena*, *Fischerella*, as well as the unicellular *Synechococcus*. Being exposed to direct sunlight, these organisms are highly pigmented, with chlorophyll *a*, intracellular carotenoids, and sheath-containing phycobilins. Depending on temperature, the green layer may also include members of the green nonsulfur bacteria (e.g., *Chloroflexus* sp.). The uppermost layers are firmly held together by the intertwined filaments and the production of EPS that bind cells and any sediment grains together. The red/purple layer is dominated by purple sulfur bacteria (e.g., the *Thiocapsa* and *Chromatium* genera), and their bacteriochlorophyll *a* and carotenoid pigments, while the orange layer comprises

(Fenchel and Straarup, 1971). Many also possess carotenoids, which is a useful feature given that radiation between 450 and 550 nm penetrates deepest into water, thus allowing these species to develop mats at the bottom of lakes. Photosynthetic bacteria are, however, more likely to respond chemotactically to O_2 and H_2S concentrations, and as a group differ with respect to their tolerance to O_2 and their need for H_2S (van Gemerden, 1993). For instance, some species can withstand low to moderate O_2 concentrations, whereas for others O_2 is lethal. Accordingly, they move or reorient themselves to a preferred O_2 gradient, and as a result may form specific bands within the mat (Fenchel, 1994). The bacteria further need to respond to fluctuating H_2S concentrations, and interestingly, studies have shown that some species (e.g., *Chromatium minus*) can either increase or decrease their velocity depending on sulfide concentrations (Mitchell et al., 1991).

It is clear from above that photosynthetic mats are complex ecosystems, where at any given time there is an interplay between complementary types of metabolism. This is particularly evident in the linking of the biogeochemical cycles of carbon and sulfur. For instance, during daylight hours, cyanobacterial mats are net producers of O_2, sometimes creating pore water levels exceeding atmospheric saturation by two to three orders of magnitude. In full sunlight, cyanobacteria also photorespire simultaneously with photosynthesis, excreting glycolate that fuels the heterotrophic populations, including SRB and methanogens in the underlying layers (Canfield and Des Marais, 1993). Oxygenation of the upper mat limits the presence of H_2S to deep within the mat, although studies have shown that SRB can populate the mat's surface, presumably if high rates of aerobic respiration limit the O_2 molecules around the cells (Canfield and Des Marais, 1991). By contrast, the methanogens are obligate anaerobes, and hence are excluded from the zone of active oxygenic photosynthesis and carbon cycling, relying instead on a pool of organic carbon not utilized by the SRB (Bebout et al., 2004). Methanogens typically play a minor role in mats, but Hoehler

et al. (2001) have reported significant methane generation in a 4–5 cm thick subtidal, hypersaline mat where H_2 concentrations were an order of magnitude higher than in most other mats.

The accumulation of sulfide below the oxic layers constrains populations of colorless sulfide-oxidizing bacteria (e.g., *Beggiatoa* sp.) and the anoxygenic photosynthetic bacteria (e.g., *Chromatium* sp.) to those depths as well (Fig. 6.5). As long as sufficient long-wavelength light filters through the overlying cyanobacterial layers, and oxygen does not penetrate too deeply, the phototrophs are at a competitive advantage over *Beggiatoa*. Not only do the purple bacteria require less sulfide for their metabolism (photosynthesis only), but the H_2S that diffuses up from the underlying layers comes into contact with them before reaching the lowermost oxic zone where *Beggiatoa* resides (Jørgensen and Des Marais, 1986). *Beggiatoa* is thus forced to oxidize the elemental sulfur they previously accumulated within their cells. Under conditions where the produced oxygen diffuses far below the photic zone, the opposite is true, and *Beggiatoa* will be the dominant sulfur species (van Gemerden, 1993). In darkness, O_2 concentrations drop markedly as

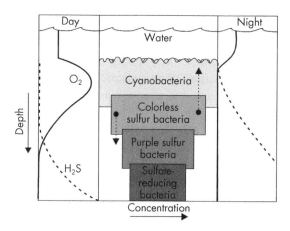

Figure 6.5 Diel fluctuations in O_2 and H_2S in a laminated cyanobacterial mat. The dashed arrows indicate the migration of motile colorless sulfide-oxidizing bacteria to exploit the shifting chemical gradients. (Modified from van Gemerden, 1993.)

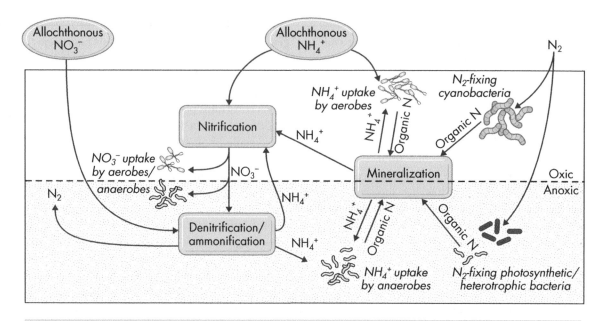

Figure 6.6 Nitrogen cycling within a microbial mat. Primary input of nitrogen comes from N_2-fixation, while losses are attributed to denitrification. In the oxic layers, reduced nitrogenous compounds are ultimately converted to nitrate, while in the anoxic layers, oxidized nitrogenous compounds are reduced to ammonium. Arrows indicate direction of flow.

production ceases and the cyanobacteria/aerobic respirers consume O_2 and release CO_2 – in situations of oxygen deficiency, the cyanobacteria will instead rely on fermentation or anaerobic respiration. As O_2 levels in the mat decline to negligible levels, H_2S diffuses upwards towards the mat surface. The phototrophic bacteria cannot function photoautotrophically in the absence of light, and some species in turn switch to growing chemolithoautotrophically. The lack of light, however, does not stop *Beggiatoa* and other colorless sulfur bacteria from migrating upwards to oxidize H_2S to S^0, changing the mat's surface from green to white (Fenchel and Bernard, 1995).

Within the mat there are also a number of redox reactions that recycle the nitrogenous compounds between reduced and oxidized reservoirs, with the overall system dynamics linked to the diurnal vertical migration of the oxic–anoxic interface (Fig. 6.6). The primary input of nitrogen is through N_2-fixation. During daylight, high O_2 concentrations in the upper mat result in

a general decrease in the rate of N_2-fixation in those species that lack defenses against oxygen (e.g., cells without heterocysts). As a result, N_2-fixation is often limited to anoxic microniches dispersed within the oxic layer and deeper in the anoxic layer of the mat, where some SRB and purple bacteria, e.g., *Chromatium* sp., fix N_2 (Bebout et al., 1987). By contrast, at night, when O_2 concentrations near the mat surface are diminished, the now prevailing anoxic conditions allow for high rates of cyanobacterial N_2-fixation and increased production of organic N for cell biomass. It is quite interesting that the energy required for N_2-fixation is primarily supplied by oxygenic photosynthesis, but the actual process of N_2-fixation requires strict O_2 management. The ability of mat communities to conduct and optimize the intricate C–N cycles is reconciled by two mechanisms: consortial interactions and molecular diffusion (Paerl and Pinkney, 1996). In the first instance, aerobic heterotrophs consume O_2 and create anoxic microniches within which N_2 fixation can occur.

Second, the laminar structure of the mats, and the presence of interstitial pore waters, greatly reduces the molecular diffusion of O_2, ensuring the maintenance of anoxic microniches.

Upon cell lysis, the proteins generated above degrade into NH_4^+ cations:

$$R\text{-}NH_2 + H_2O \rightarrow R\text{-}OH + NH_3 \rightarrow$$
$$NH_3 + H_2O \rightarrow NH_4^+ + OH^- \qquad (6.1)$$

The ammonium may be directly utilized by some microorganisms as a nitrogen source, while the remainder is oxidized by nitrifying bacteria to nitrate. The nitrate supply, in turn, can either: (i) be assimilated by mat microorganisms; (ii) reduced to ammonium via ammonification; (iii) escape from the mat as N_2 via denitrification (e.g., by *Pseudomonas* sp.); or (iv) lost from the local environment due to leaching. Unlike soils or sediment, the limited spatial confines within the mat force the various microorganisms involved in N-cycling (e.g., N_2-fixers, nitrifyers, denitrifiers), as well as the others discussed above, to live within close proximity to one another. Fortuitously, this benefits the overall community because it maximizes diffusional exchange among biogeochemically distinct, yet complementary, groups.

(b) Anoxygenic bacterial mats

The availability of hydrogen sulfide can support two kinds of sulfide-oxidizing bacteria, the purple/green photosynthetic bacteria and the colorless bacteria. The colorless varieties have a competitive advantage when O_2 produced by cyanobacteria supersaturates the pore waters with oxygen, forcing the O_2–H_2S interface far below the compensation depth of the photosynthetic bacterium's photic zone (Jørgensen et al., 1983). On the other hand, when O_2 levels are not inhibitive to the photosynthetic bacteria, sulfide will be exclusively oxidized by them, provided that sufficient long-wavelength light is available. Each of the four broad categories of anoxygenic photosynthetic bacteria discussed in section 2.2.3 form mats where they are the major species.

Many neutral to slightly alkaline sulfide springs in Yellowstone National Park contain mats dominated by the genus *Chromatium*. The mats are usually found above 50°C (inhibitory to cyanobacteria in these sulfide springs) and below 58°C (the upper temperature limit for *Chromatium* sp.). Below *Chromatium* is a thin green layer composed of various species of *Chloroflexus*. These two taxa form an ideal partnership because they have complementary absorption spectra in the infrared region, meaning that they are not in competition for light, and while *Chromatium* relies on hydrogen sulfide utilization, *Chloroflexus* can survive photoheterotrophically on the remains of lysed *Chromatium* cells (Ward et al., 1989). Some purple sulfur bacteria also have the capacity to change their physiology depending on the time of day. For instance, *Thiocapsa roseopersicina* is often associated with cyanobacterial mats, where it is forced to cope with steep gradients of O_2 and H_2S. Unlike many other gradient organisms, this immotile bacterium is capable of switching its metabolism during diurnal cycles, such that it grows photoautotrophically under anoxic conditions and chemolithotrophically in the presence of oxygen by repressing its bacteriochlorophyll pigments (de Wit and van Gemerden, 1987). This ability serves the cells well because during mid-day they are exposed to elevated O_2 concentrations, and since they cannot move away, their utilization of oxygen chemotrophically means that they have a competitive advantage over strictly anaerobic and immotile photosynthesizers, such as the green sulfur bacteria.

Green sulfur bacteria frequently grow beneath the purple bacteria, yet in some New Zealand thermal springs that are moderately acidic (pH 4.5–6), with temperatures in the range 45–55°C, and hydrogen sulfide concentrations above 50 µmol L^{-1}, the unicellular *Chlorobium* sp. forms 0.5–3 mm thick, unlaminated mats directly on top of the sediment (Castenholz, 1988). The intolerance of most cyanobacteria to pH levels below 6 and high sulfide concentrations tends to preclude their growth from the same environments.

In natural habitats dominated by hydrogen sulfide, either from thermal springs or the activity

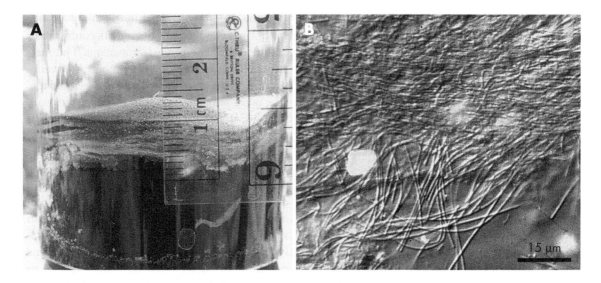

Figure 6.7 (A) Cross-sectional view of a laminated *Chloroflexus* mat, growing at 55°C and pH 6.4, from Mammoth Terraces, Yellowstone National Park. (B) Photomicrograph of *Chloroflexus* filaments that predominate in the upper ~1 mm green layer. (Courtesy of David Ward.)

of sulfate-reducing bacteria, the purple non-sulfur bacteria have a considerable disadvantage relative to the photosynthetic sulfide-oxidizers. However, when the breakdown of organic matter yields an excess of fermentation products beyond those utilized by the heterotrophic communities, purple nonsulfur bacteria find their ecological niche and form viable layers within mat eco-systems. Their continued growth, though, is tied to the overlying production of biomass and the regular supply of readily decomposable organic materials (Pfennig, 1978).

Green nonsulfur bacteria are commonly found in association with cyanobacteria, but they may also be the principal mat constituents at high temperatures (up to 70°C) and when ele-vated concentrations of hydrogen sulfide inhibit cyanobacterial growth (Fig. 6.7). As discussed in section 2.2.3, *Chloroflexus* species are metabol-ically diverse. They often form characteristically orange to green gelatinous, millimeter-thick mats when growing as photoheterotrophs. Yet, despite their name, field studies and experiments have conclusively shown that the addition of hydrogen sulfide stimulates photoautotrophic metabolism

(Castenholz, 1973). Through the use of molecular probes, it is now becoming increasingly obvious that a considerable diversity of *Chloroflexus*-related bacteria exist in hot spring mats, with numerous closely related populations living within a small spatial scale; each species adapted to particular conditions with respect to carbon and light sources (Nübel et al., 2002). For example, at Mammoth Springs, Yellowstone, the uppermost *Chloroflexus* populations grow photoautotro-phically, supplying organic compounds to their underlying photoheterotrophic cousins. Levels of dissolved hydrogen sulfide within the mat often exceed that contributed from the hot spring effluent, suggesting that these mats are also shared by sulfate-reducing bacteria that survive on the photosynthetically generated sulfate.

6.1.3 Chemolithoautotrophic mats

Not all mat communities are dominated by phototrophs. In environments where sunlight is effectively absent, reduced solutes or gases may instead serve as the primary energy sources for chemolithoautotrophy-based mat development.

Many such mats tend to be restricted to a localized source of reductant (e.g., Fe(II) at hydrothermal vents), but several recent studies have now shown that mats based on hydrogen sulfide or methane are much more ubiquitous.

(a) Colorless sulfur bacteria

In Chapter 5 we examined how species such as *Acidithiobacillus thiooxidans* survive by catalyzing the oxidation of sulfide minerals under acidic conditions. The other type of colorless sulfur oxidizers (e.g., *Beggiatoa* sp.) exist under microaerophilic conditions, most frequently forming conspicuous white-colored patches on the seafloor. These bacteria have very high oxygen uptake rates, yet require microaerophilic conditions to exploit the kinetic advantage relative to abiological sulfide oxidation. Indeed, the bacteria's rapid growth rates translate into extremely high hydrogen sulfide oxidation rates, some 10,000- to 100,000-fold faster than in water. Consequently, the cells grow in steep chemical gradients, where their niche is confined to a narrow zone only hundreds of micrometers thick (Fig. 6.8). Within this zone, O_2 and HS^- (the predominant sulfide form in pH 8 waters) overlap by as little as 50 um, with gas concentrations averaging less than 10 μmol L^{-1} and residence times of less than 1 second (Jørgensen and Revsbech, 1983). Exceptions to this exist, and at the interface of sulfide-rich fluids and oxic seawater in the hydrothermal vent fields of Guaymas Basin, Gulf of Mexico, thick (>1 cm) *Beggiatoa* mats proliferate (Gundersen et al., 1992).

Other sulfide-oxidizing bacteria, such as *Thiovulum* sp., are large, spherical bacteria that form fragile "veils" of cells (a few hundred micrometers thick) suspended within the water column itself (Jørgensen, 1982a). Though only loosely attached to the underlying substratum, they are stable enough to withstand slow water movements. Similar to *Beggiatoa* mats, the veils sharply separate the overflowing oxic water from the anoxic, sulfidic water surrounding decaying animals and plants at the sediment–water inter-

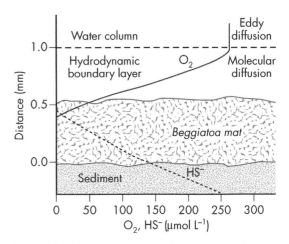

Figure 6.8 Expanded view of the vertical distribution of O_2 and HS^- through a marine *Beggiatoa* mat. The presence of a hydrodynamic boundary layer above the mat, some 0.5 mm thick, creates a steep gradient of O_2 between the overlying circulating waters dominated by eddy diffusion with those in the water immediately above the mat, characterized by molecular (or ionic) diffusion. (Modified from Jørgensen and Revsbech, 1983.)

face, in essence expanding the anoxic zone into the water column itself. In this way, *Thiovulum* can outcompete the sediment-bound *Beggiatoa*, which have sometimes been observed to glide up on the top side of the *Thiovulum* veils.

The formation of mats on solid substrata, and veils in the water column, create a stagnant layer of water just a few hundred micrometers thick. This "hydrodynamic boundary layer" is subject to viscous forces that tends to make it stick to the surface and not participate in the general circulation of water. Therefore, eddy diffusion becomes insignification relative to molecular diffusion, which is the principal mechanism for mass transport across the mat/sediment–water interface (Santschi et al., 1983). From the cells' perspective, this represents a stable microenvironment that positively impacts them in two ways: (i) they can grow under microaerophilic conditions in spite of the oxygenated waters flowing over them; and (ii) it slows the ambient diffusion rates, thereby limiting the amount of abiological sulfide oxidation.

Off the west coast of South America, where high surface-water productivity leads to anoxic bottom waters, centimeter-thick mats of another sulfide-oxidizing genus, *Thioploca*, cover the seafloor. These filamentous bacteria couple the oxidation of HS^- to the reduction of NO_3^-, but what makes this type of metabolism so unique is the fact that the zones where nitrate and sulfide are present do not overlap (Fossing et al., 1995). The cells, however, have developed a strategy to overcome these inherent problems by growing within sheaths that permit their filaments to glide upwards from the seafloor, by as much as 10 cm above the surface, where they accumulate NO_3^- intracellularly within vesicles. They then migrate back downwards and reduce the nitrate with hydrogen sulfide that has diffused upwards from the underlying sediment (Fig. 6.9). Elemental sulfur is the first oxidation product which is stored within the cytoplasm (with N_2 released), until later, when it can be oxidized further to sulfate.

(b) Methanotrophs and sulfate-reducing bacteria

Water column stratification occurs in a number of freshwater and marine systems. In most temperate lakes, stratification is a transient phenomenon associated with the summer season, when the surface waters are heated and the resulting "thermocline" resists mixing by wind until winter cooling causes seasonal overturn. By contrast, meromictic lakes have permanent chemical stratification due to disparate water densities forming a barrier, or pycnocline, to the vertical transport of solutes or gases. A similar situation arises in some marine settings, where a sill obstructs the circulation of deep water, or in coastal basins where relatively fresh water traps an underlying body of normal marine salinity. If, in any of these stratified systems, the surface waters are highly productive in terms of phytoplanktonic biomass, due to excessive nutrient

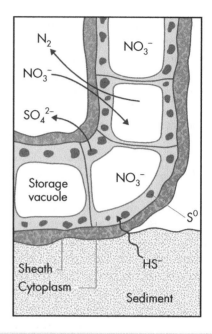

Figure 6.9 Cross-sectional view of a *Thioploca araucae* filament, with each cell about 40 μm wide and 60 μm long. The cells have a large central liquid vacuole in which nitrate is stored at concentrations up to 500 mmol L^{-1}. Hydrogen sulfide, from the underlying sediment, is oxidized intracellularly upon reaction with nitrate, with the initial formation of elemental sulfur grains that are stored within the cytoplasm. The sulfur can subsequently be oxidized to sulfate. (Modified from Fossing et al., 1995.)

input (known as eutrophication), then it is likely that the deep layers will become anoxic and rich in reduced gases, such as H_2S and CH_4, while the surface waters remain oxygenated. This chemical zonation results because aerobic degradation of particulate organic carbon (POC) settling down from the euphotic zone requires more oxygen than can be supplied by turbulent diffusion through the pycnocline (Fenchel et al., 2000). Consequently, large amounts of relatively fresh organic material sinks to the bottom sediments, where it is degraded via anaerobic respiratory pathways (Fig. 6.10). The largest such examples are the euxinic basins represented by the Black Sea, the Cariaco Trench (Venezuela), and the Gotland Basin in the Baltic Sea.

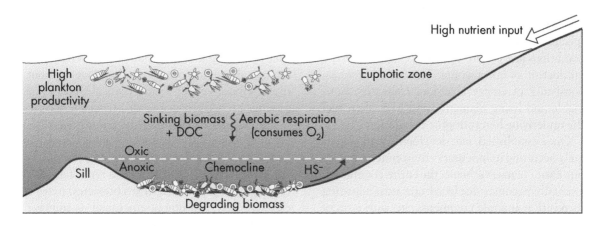

Figure 6.10 Stratification of the water column within a silled marine basin. The upper oxic zone is characterized by high phytoplankton productivity, and as the dead cells sink through the water column, aerobic respirers consume available O_2. Beneath the chemocline, further organic matter oxidation occurs anaerobically, generating dissolved HS^-.

Growing at the sediment–water interface of some of these basins are vast tracts of microbial mats. In the Black Sea, for instance, SRB and methanogens form part of a syntrophic mat community based on the anaerobic oxidation of upwardly diffusing methane (see section 6.2.5(c)). The 10-centimeter-thick mats consist of interconnected, irregularly distributed cavities and channels that enable CH_4 and SO_4^{2-} to be distributed throughout the mat (Michaelis et al., 2002). What is particularly interesting about these mats is that they dwarf their counterparts found in oxic environments due to the lack of macrofaunal disruption. Significantly, they may offer a glimpse into what microbial communities may have looked like on early Earth, when O_2 was essentially absent in deep marine waters and before eukaryotes evolved.

6.1.4 Biosedimentary structures

One of the characteristic features displayed by many mature mats is an overall laminated fabric (e.g., Fig. 6.11). Distinct laminations arise when some form of recurring disruption, due to variations in sedimentation rate, salinity, seasonality, etc., causes a hiatus in the normal accumulation of biomass. In particular, any significant influx of clastic sediments or chemical precipitates that bury the indigenous microbial community will elicit a compensatory response. This usually involves the cells growing or moving upwards over the sediment to colonize the new surfaces and to maintain their optimal growth positions in

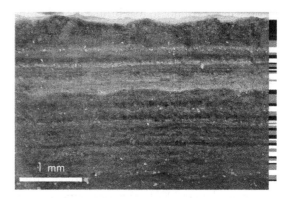

Figure 6.11 Overview of Fig. 6.4, showing the repetitive nature of the laminations, beginning at the top surface of the mat. Color code on right side is a visual aid since the black/white image does not show color variations: black = green cyanobacteria; gray = purple bacteria; white = gypsum precipitates. Notice how the cyanobacterial layers diminish with depth. (Courtesy of Mary Hogan and Tracy Cote.)

the mat (Golubic, 1976). By contrast, immotile species require re-colonization of the surface when conditions prove more favorable, e.g., through the production of hormogonia. The buried biomass, consisting of empty abandoned sheaths, non-viable and lysing filaments, and EPS, then fuel the underlying heterotrophic populations.

Once established, mat development in an actively accreting sedimentary environment follows one basic common scheme: the entire community gradually becomes displaced upwards, adjusting its position at a rate commensurate with rates of sedimentation. These dynamic changes are often manifest in alterations of mineral- and organic-rich laminae, giving rise to what are commonly referred to as microbialites – lithified microbial mats (Burne and Moore, 1987). Each microbialite has an overall morphology, fabric and mineral component that not only is indicative of the environmental conditions under which it formed, but also the growth patterns of the indigenous microbial community, an aspect not lost on geologists trying to ascertain the environment under which early Precambrian stromatolites formed. Recall from Chapter 4 that several mineralizing agents can contribute to mat lithification, but in terms of overall significance for the rock record, two are most important, namely silica and calcite.

(a) Thermal spring deposits

Thermal spring deposits are frequently composed of amorphous silica (sinters) or calcium carbonate (travertines), with the type of mineralization wholly dependent upon the composition of the discharged vent fluids. Using terminology from Renaut and Jones (2000), travertines tend to form in "mesothermal" springs where the bicarbonate-rich effluents have temperatures ranging from 40°C to 75°C. Aragonite is the dominant $CaCO_3$ polymorph at temperatures higher than 40°C. Amorphous silica typically precipitates from neutral to alkaline waters with temperatures in excess of 75°C, but depending on the state of supersaturation, silicification can continue to proceed at lower temperature downstream of the vent. High effluent temperatures make studying thermal springs quite fortuitous because they are often prohibitive to macrofauna, which means that it is possible to ascertain the geochemical-biological interactions that take place without the deleterious effects of mechanical abrasion by grazing.

Individual thermal spring sinters and travertines are architecturally complex and show extensive lateral and vertical variations among the different deposits (e.g., Fig. 6.12). Each type of deposit (also sometimes referred to as a biofacies) may be characterized by a unique microbial assemblage that developed in response to the operative hydrodynamic, geochemical, and temperature conditions (e.g., Renaut et al., 1998; Fouke et al., 2003). Thermal spring mats are usually zoned downstream in response to steep temperature gradients. In waters from 75°C to 100°C, life consists of a few dominant chemolithoautotrophic and heterotrophic Bacteria and Archaea – the more extreme an environment, the fewer the number of taxa, and the microbial community may be shaped by the biological properties of just its dominant member (Golubic, 1991). In shallow channels, where flow rates are high, streamers of various Aquificales become mineral encrusted, forming unique fabrics that preserve the original flow directions. On the discharge aprons, where water temperatures may have cooled below the 73°C threshold, thick photosynthetic mats often form, containing species of Synechococcus and Chloroflexus. In waters cooler than 65°C, filamentous cyanobacteria appear, including species of Oscillatoria and Phormidium, as well as some purple and green sulfur bacteria (see Plate 14). In the most distal parts of the drainage system, where temperatures are lower than 40°C, other cyanobacterial genera, such as Spirulina, Fischerella, and Calothrix, begin to dominate. Green algae, diatoms, and fungi also become important mat constituents at the lowermost temperature regimes. Calcite is the primary $CaCO_3$ phase in this part of the drainage system.

At similar temperatures, the determining factor governing species distribution is the composition of the geothermal fluids and gases. For instance, Calothrix prefers to grow in alkaline water; fungi dominate areas with acidic waters

Figure 6.12 An example of the complexity and textural variations of travertine depositional biofacies, within a distance of only 4 m, in the drainage system at Angel Terrace, Mammoth Hot Springs, Yellowstone National Park. See Fig. 4.16B for an overview of the spring system. (A) Filamentous strands of aragonite-encrusted *Aquificales* that project from the apron into the pond. (B) Shrub-like clusters of aragonite on the floor of the upstream end of the pond, where the predominant species are β-*Proteobacteria*. (C) Dendritic clusters of aragonite on the downstream end of the pond, dominated by α-*Proteobacteria*. (D) Aragonite microterracettes on the slope. Predominant species are β-*Proteobacteria* and cyanobacteria. (Adapted from Fouke et al., 2003. Reproduced with permission from the National Research Council of Canada.)

(pH < 5.0), while *Fischerella* appears to be restricted to waters that contain little or no sulfur. Clearly there are many more such examples of microbial niches in thermal spring settings, and hence, it should not come as a surprise that variations in microbial communities give rise to textural variations in sinters or travertines simply because different microorganisms interact uniquely with the waters in which they are bathed. The actual role of the indigenous microorganisms in mineral precipitation was covered in Chapter 4, but it is fair to suggest that they become more important as their abundance and diversity increase with distance from the vent, and critically, as the chemical forces driving solutions towards equilibrium are decreased. Yet, when the geothermal waters become undersaturated with respect to $CaCO_3$ and amorphous silica, travertine and sinters no longer form, irrespective of whether microorganisms are present.

It is also important to point out that there is significant textural variation between the minerals that encrust microorganisms and other organic surfaces (e.g., wood, leaves) with the isopachous cements that line and/or fill pores and cavities in the primary deposit. Mineralization of biomass, and other organic materials, is driven by surface water interactions, while precipitation of isopachous cements is mediated by subsurface water flow through the older, buried parts of the deposit (e.g., Jones et al., 1997).

Most sinters and travertines are characterized by laminations that reflect variations in features, such as the types of microorganisms they contain (as in above), systematic temporal changes in the depositional environment, in particular the amount of mineral-forming solutes delivered by flow or spray, and whether the deposit is submergent or emergent. These features can be best illustrated by considering stratiform stromatolites, a common sinter found associated with hot spring discharge aprons. This sinter consists of two types of alternating laminae. One lamina is dominated by filamentous microorganisms that lie parallel to subparallel to the depositional surface (P-lamina). There is relatively little porosity in P-lamina because the filaments are densely packed and the intrafilament spaces are often filled with amorphous silica. The other type of lamina consists of composite pillars, many micrometers in diameter, that are formed from the merger of subvertical to vertical silicified filamentous microorganisms, typically species of *Calothrix* or *Phormidium* (U-lamina). These laminae often have high porosity that is formed in the open spaces between the filament pillars, between the filaments themselves, as well as the hollow tubes that were once occupied by the actual filament (Jones et al., 1998). Frequently the P- and U-laminae are formed of the same filamentous microorganisms. Indeed, in many cases it is possible to find individual flat-lying filaments in the upper part of the P-lamina that, with growth, adopted an upright growth habit (Fig. 6.13A).

For the stratiform stromatolites, as well as many other biofacies, the cyclic alternation between

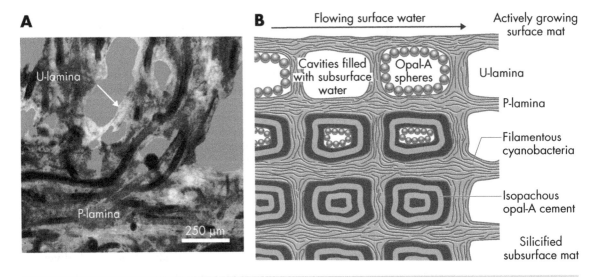

Figure 6.13 (A) SEM image of the change in growth attitude of the filamentous cyanobacterium, *Phormidium* sp., at the transition from P-lamina to U-lamina in a stratiform stromatolite from Ohaaki Pool, New Zealand. Arrow denotes silica precipitation. (B) Representation of the structural and fabric evolution in the same stratiform stromatolite, showing how large cavities between pillars in U-lamina fill with isopachous amorphous silica cement that is precipitated from subsurface water flowing through the porous sinter. (Modified from Konhauser et al., 2004.)

different laminae has been attributed to variations in growth patterns between the summer and winter months (e.g., Chafetz et al., 1991). Caution though must be attached to the whole-hearted acceptance of this idea because some microorganisms are known to change their growth attitude and positioning in response to daily phototactic controls. The classically described example of this is at Octopus Springs, Yellowstone National Park, where mats of *Synechococcus lividus* and *Chloroflexus aurantiacus* alternate their positions in the mat diurnally. The cyanobacteria grow at the surface during the day, generating O_2 that limits H_2S to the deeper layers where *Chloroflexus* populations can grow either photoautotrophically, or photoheterotrophically on organic compounds extruded by the cyanobacteria. At night, *Chloroflexus* moves to the top due to low light levels (which cannot support the cyanobacteria) and positive aerotaxis (Doemel and Brock, 1977). The laminae that result are tens of micrometers thick. Nevertheless, in most thermal springs, filamentous microorganisms tend to adopt an upright attitude during the summer that contrasts sharply with their prone attitude during the winter months (Walter, 1976b).

In some extremes, such as those in Iceland, the laminations consist of alternating layers, each ~250 μm thick, of cyanobacteria and pure silica. The cyclical pattern arises from active cell growth towards the sediment surface during spring/summer when the microorganisms can keep pace with silicification, while during their natural slow growth phase in the dark autumn/winter months silicification exceeds the cyanobacterium's ability to grow upwards and the cell becomes completely buried in amorphous silica (Fig. 6.14). When conditions once again become favorable for growth, recolonization of the solid silica surface by free-living hormogonia occurs (Konhauser et al., 2001). For those cells that have become buried, they can remain viable as long as there is sufficient light for cell maintenance, but in time, they lyse and become part of the degradable organic pool utilized by the underlying chemoheterotrophic populations.

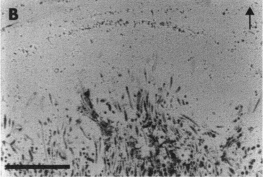

Figure 6.14 SEM images from a section of microstromatolite (overall view in Fig. 6.15A) at 5 mm below the sinter apron surface. (A) Sharp contact between an inorganic silica layer (bottom left) and the predominantly filamentous, vertically aligned microbial layer dominated by *Calothrix* sp. (top right). (B) The gradational upper surface of the microbial layer. Note the preferred vertical orientation of cyanobacteria towards the sediment–water interface (arrows) and isolated filaments projecting into silica layer above. There is no continuum between cyanobacteria of one layer with those of an overlying microbial layer hundreds of micrometers above. Scale bars = 100 μm. (From Konhauser et al., 2001. Reproduced with permission from Blackwell Publishing Ltd.)

The U- and P-laminae commonly form couplets that are stacked one on top of the other. As the couplets accrete vertically through time, in direct response the continual seasonal cycles of growth, they become progressively buried by younger mats. Once below the surface of the discharge apron, the laminae are placed under the influence of subsurface waters that may, or may not, be related to the surface waters. Most of these laminae are characterized by such high porosity and permeability that subsurface waters will easily pass through them, particularly the open compartments in the U-lamina. If the fluids are sufficiently silica supersaturated, isopachous cements will then infill any remaining porosity (Fig. 6.13B), giving the sinter greater rigidity, and possibly enhancing its long-term survival.

Understanding modern sinter (or travertine) formation has important implications for the rock record. As the pioneering work by Walter et al. (1972) suggested, it may be possible to interpret some distinctive ancient stromatolite morphologies from modern thermal spring analogs, such as extant coniform stromatolites, with the Precambrian stromatolitic form *Conophyton* (e.g., Fig. 6.15). Modern coniform stromatolites, dominated by *Phormidium* species, are restricted to continuously submerged, low energy geothermal pools. This environmental setting has also been proposed for the origins of *Conophyton* stromatolites, suggesting that similar cyanobacterial species may have been involved in their formation (Jones et al., 2002). For this to be useful, however, a general framework illustrating the association between the principal types of thermal spring biofacies and the dominant mat-forming communities still needs to be developed. Unfortunately, this is where gaps in our knowledge emerge. It is well known that cyanobacteria exert dominant control on fabric development at temperatures below 73°C, but how anoxygenic phototrophic mats influence sinter development has not yet been ascertained. Hyperthermophilic *Archaea* and viruses are probably also present in some geyserite, but as yet have not been identified with confidence, so their potential roles in microbialite formation are unknown (e.g., Barns et al., 1996; Rice et al., 2001).

(b) Intertidal and subtidal stromatolites

In modern marine intertidal mats, lithification occurs within the top few centimeters of the depositional interface by the precipitation of calcium carbonate cements to form hard, current-

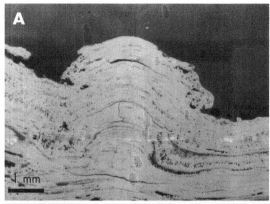

Figure 6.15 Comparison of modern siliceous stromatolite formation in hot springs with some Precambrian stromatolites. (A) SEM image of an approximately 7 mm thick section of siliceous coniform stromatolite, collected from the main hot spring vent at Krisuvik, Iceland. Note the dome-like structure (on the order of 2 mm in diameter) and the underlying alterations between the cyanobacterial layers (dark) and the pure amorphous silica (white) (from Konhauser et al., 2001. Reproduced with permission from Blackwell Publishing Ltd). (B) Small, *Conophyton*-like stromatolites, preserved in chert, from the 2.7–2.8 Gyr rocks at Kanowna, near Kalgoorlie, Western Australia (courtesy of Ken McNamara).

resistant structures. Contrary to what we might predict, the role of cyanobacteria in stromatolite formation is principally by trapping and binding processes, in which the intertwined filaments incorporate detritus within their EPS and sheaths to form a cohesive mat-like structure. Much more calcium carbonate is actually precipitated within the aphotic zone on cyanobacterial remains than in the euphotic zone dominated by living cyanobacteria (Paerl et al., 2001). Moreover, the morphology of that natural calcium carbonate is nearly identical to that produced in the laboratory by heterotrophic bacterial cultures without the presence of cyanobacteria (Chafetz and Buczynski, 1992). This implies that the activity of anaerobic heterotrophic bacteria may be the driving factor in the formation of calcareous marine microbialites (Fig. 6.16).

The underlying reason why cyanobacterial biomineralization is not important during the overall lithification process in marine settings is as follows. In the surface layers daytime photosynthetic activity causes $CaCO_3$ precipitation in the surface layers, along with O_2 generation:

$$2HCO_3^- + Ca^{2+} \rightarrow CH_2O + CaCO_3 + O_2 \quad (6.2)$$

As a result, the key enzyme rubisco switches from predominantly CO_2 (carboxylation) to O_2 (oxygenation) utilization. This results in photorespiration and production of organic compounds (e.g., glycolate), a fraction of which is excreted and available for fermenters and chemoheterotrophs. Aerobic oxidation of that organic carbon yields H^+, decreasing the pH and favoring $CaCO_3$ dissolution (reaction (6.3)). Therefore a balance of these processes results in little or no net lithification at the surface (Visscher et al., 1998).

$$CH_2O + O_2 + CaCO_3 \rightarrow 2HCO_3^- + Ca^{2+} \quad (6.3)$$

In addition to these opposing metabolic processes, some cyanobacteria may actively prevent their calcification by producing amine-rich proteins that inhibit electrostatic interactions with Ca^{2+}. This then limits the mineralization process to the anionic extracellular layers, or to those species that do not form such proteins (Gautret and Trichet, 2005).

The cementation that does take place is induced by the heterotrophic bacterial populations that utilize the cyanobacterial remains buried into the underlying anoxic layers. The key chemoheterotrophic processes include nitrate ammonification (reaction (6.4)) and sulfate reduction (reaction (6.5)). These forms of metabolism contribute to localized increases in alkalinity, which coupled to the release of EPS/sheath-bound Ca^{2+}, leads to a state of supersaturation with respect to calcium carbonate. This, in turn, induces the precipitation of micritic cements that lithify the sand grains together, and trap both living and dead microorganisms in a carbonate matrix, often to the ultimate demise of those entombed microorganisms (Castanier et al., 2000).

$$2(CH_2O) + NO_3^- + Ca^{2+} \rightarrow CaCO_3 + CO_2 + NH_4^+ \quad (6.4)$$

$$2(CH_2O) + SO_4^{2-} + Ca^{2+} + 2OH^- \rightarrow \\ CaCO_3 + HS^- + HCO_3^- + 2H_2O \quad (6.5)$$

Sandwiched in between the oxic and anoxic layers is a zone of active sulfide oxidation, consisting of populations of colorless sulfur chemolithoautotrophs and purple photosynthetic bacteria that utilize the HS^- generated by the SRB below. The former thrive at the O_2/HS^- interface, where their metabolism leads to a decrease in pH and additional $CaCO_3$ dissolution:

$$HS^- + 2O_2 + CaCO_3 \rightarrow SO_4^{2-} + Ca^{2+} + HCO_3^- \quad (6.6)$$

Considering that at night, when oxygenic photosynthesis ceases, and HS^- diffuses upwards to the mat surface, this process (in addition to aerobic respiration) could be responsible for dissolving any residual carbonates formed earlier during the day, leading to a net absence of calcium carbonate in the surface layers. Meanwhile, the SRB also move upwards at night, and as they temporarily populate more of the subsurface mat, their increased activity promotes $CaCO_3$ precipitation in those layers.

The purple bacteria typically grow under the colorless sulfide-oxidizers because they require the complete absence of O_2. Their fixation of CO_2 similarly drives equilibrium towards carbonate

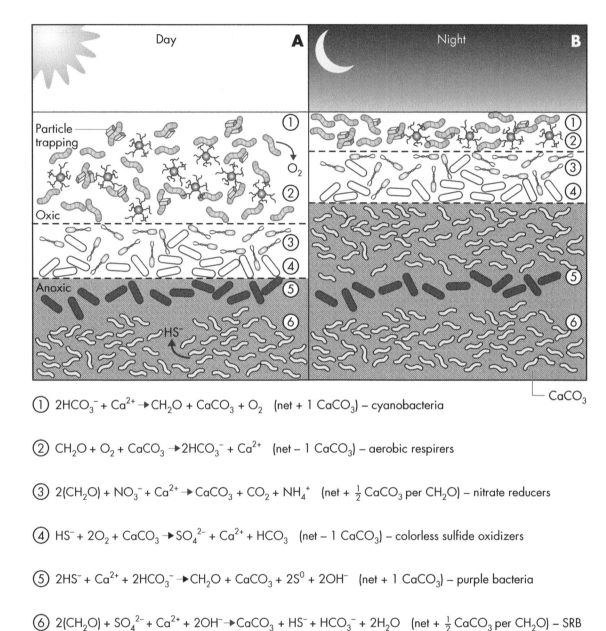

① $2HCO_3^- + Ca^{2+} \rightarrow CH_2O + CaCO_3 + O_2$ (net + 1 $CaCO_3$) – cyanobacteria

② $CH_2O + O_2 + CaCO_3 \rightarrow 2HCO_3^- + Ca^{2+}$ (net – 1 $CaCO_3$) – aerobic respirers

③ $2(CH_2O) + NO_3^- + Ca^{2+} \rightarrow CaCO_3 + CO_2 + NH_4^+$ (net + $\frac{1}{2}$ $CaCO_3$ per CH_2O) – nitrate reducers

④ $HS^- + 2O_2 + CaCO_3 \rightarrow SO_4^{2-} + Ca^{2+} + HCO_3$ (net – 1 $CaCO_3$) – colorless sulfide oxidizers

⑤ $2HS^- + Ca^{2+} + 2HCO_3^- \rightarrow CH_2O + CaCO_3 + 2S^0 + 2OH^-$ (net + 1 $CaCO_3$) – purple bacteria

⑥ $2(CH_2O) + SO_4^{2-} + Ca^{2+} + 2OH^- \rightarrow CaCO_3 + HS^- + HCO_3^- + 2H_2O$ (net + $\frac{1}{2}$ $CaCO_3$ per CH_2O) – SRB

Figure 6.16 Model showing the key microbial reactions influencing calcium carbonate cementation in a modern marine intertidal stromatolite. (A) During the day, there is a fine balance between the metabolic processes that induce calcium carbonate precipitation and those that would lead to its dissolution. Only in the anoxic layers, where SRB and purple photosynthetic bacteria are active, does calcium carbonate form (shaded, stippled region). (B) At night, O_2 production ceases and HS^- diffuses upwards, from the anoxic layers, towards the bottom of the oxic zone. That, in turn, allows the colorless sulfide oxidizers to move upwards as well. This impacts calcium carbonate stability in the surface layers because they generate acidity and, hence, promote dissolution of any residual cement. Yet, increased SRB activity acts in an opposing manner, and at the depths where they temporarily reside, they help lithify loose sand grains together.

precipitation during the day when they are most active (reaction (6.7)):

$$2HS^- + Ca^{2+} + 2HCO_3^- \rightarrow$$
$$CH_2O + CaCO_3 + 2S^0 + 2OH^- \qquad (6.7)$$

On balance, cyanobacterial photosynthesis, anoxygenic bacterial photosynthesis, sulfate reduction, and ammonification account for $CaCO_3$ precipitation, while aerobic respiration and aerobic sulfide oxidation cause dissolution.

Models accounting for the development of modern subtidal and intertidal carbonate microbialites additionally need to take into account the physical processes of sedimentation and the microbial responses to burial beneath clastic and chemical sediments (e.g., Reid et al., 2000; Seong-Joo et al., 2000). For example, in intertidal settings, the incipient stage of stromatolite formation is characterized by the activity of highly motile pioneer communities that can withstand wave agitation (Fig. 6.17A). These species produce

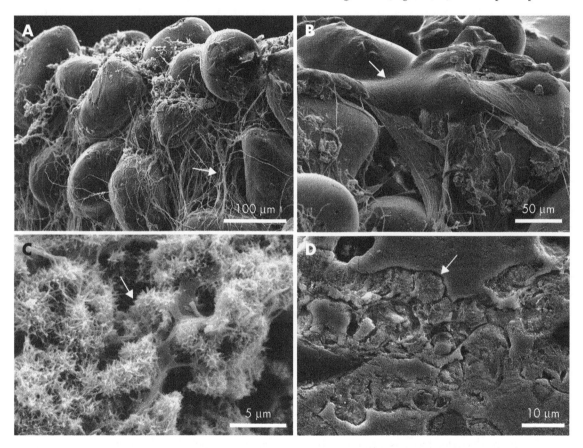

Figure 6.17 SEM images showing some of the stages of intertidal stromatolite formation. (A) A pioneer community of filamentous cyanobacteria, *Schizothrix* sp. (arrow), that bind carbonate sand grains and initiate the process of sediment stabilization. (B) The cyanobacteria generate a continuous sheet of EPS (arrow) to hold the community together. (C) As the cyanobacteria move upwards, they leave behind lysed cells and abandoned sheaths that serve as organic substrates for heterotrophic populations (e.g., SRB). This activity, in turn, increases alkalinity and facilitates carbonate precipitation, in this case aragonite needles (arrow). (D) Climax community where endoliths (*Solentia* sp.) cross existing carbonate grains at point contacts, infilling their bores with aragonite (arrow), fusing grains together. (From Reid et al., 2000. Reproduced with permission from Nature Publishing Group.)

copious amount of EPS that trap sediment grains (Fig. 6.17B), and as the cells move upwards towards the sediment surface, they leave behind abandoned sheaths and EPS-coated sediment grains. Some cyanobacteria (e.g., *Phormidium* sp.) produce daily laminae as a phototactic response, with their filaments predominantly upright (U-lamina) during the day to flat lying (P-lamina) during the night. Different cyanobacteria (e.g., *Schizothrix* sp.) change their growth and movement orientation in response to daily burial by sediment, such that the P-lamina forms on the sediment surface, but the U-lamina forms when the cells become vertically oriented while assuming escape positions. Less mobile species remain buried by the sediment, where they are killed off and excluded from the uppermost mat surface, whereas others respond by sending hormogonia upwards to settle in and between the EPS to mature when conditions once again become favorable.

The above stage alternates with periods of relative quiescence during which time sedimentation ceases and the surface cyanobacterial community rapidly spreads out laterally to develop a continuous mat of flat lying cells. These mats, subsequently, support aerobic heterotrophic activity that degrades the EPS, sheaths, and lysed cells in the topmost layers. The resulting O_2 depletion stimulates the development of anoxygenic phototrophs beneath the cyanobacteria, and below, fermentative and anaerobic heterotrophs (e.g., SRB) that utilize the buried organic remnants of the overlying community. Metabolites released by the SRB then serve as reductants for chemolithoautotrophic bacteria that straddle the oxic–anoxic interface. Concurrently, the alkalinity generated in the anoxic layers, and the release of Ca^{2+} from the degraded biomass, promotes the nucleation and precipitation of thin micritic crusts directly underneath, and in contact with, the surface mat (Fig. 6.17C). Therefore, it is during the quiescent periods, when heterotrophic metabolisms dominate, that the bulk of calcification occurs. Conversely, during the active accretionary stages of stromatolite growth, lithification of the microbial community is minimized.

During prolonged quiescent periods, climax communities with more benthic diversity develop, including both immotile and slower growing cyanobacteria (e.g., species of *Calothrix* and *Scytonema*), diatoms, algae, foraminifera, as well as endolithic coccoid cyanobacteria (e.g., *Solentia* sp.). The latter modify the sediment through boring and infilling (Fig. 6.17D), leading to micritized grains that display complex internal patterns (Fig. 6.18). Although destroying the original granular texture, microboring plays a constructive role in stromatolite formation by fusing loose grains together at grain boundaries into a lithified and rigid horizon.

Stromatolites can also experience changes in species composition simply due to seasonal variations in water chemistry (e.g., Monty, 1976). For instance, at the peak of the summer season, evaporation may cause increased hypersalinity, and potentially carbonate and gypsum precipitation. This forces some of the filamentous cyanobacteria to glide downwards into the microbial mat for protection, leaving a surface population of more resistant species (e.g., the coccoid cyanobacterial genera, *Entophysalis* and *Chroococcus*, or green algae, such as *Dunaliella* sp.) that are better adapted to the high insolation and evaporative conditions at the surface. The lithification of the mat surface is often followed by colonization by an epilithic–endolithic community, with the stromatolite becoming subject to microbial erosion. Following seasonal lowering of salinity and light intensity, the filamentous cyanobacteria glide upwards, through the overlying communities and sediment, to re-establish themselves as a new surface mat. Microorganisms that grow in intermittently to permanently highly saline solutions have had to make a number of physiological adaptations, including the overproduction of EPS to retain water and protect against osmotic stresses and extreme temperatures (Gerdes et al., 2000). Too much daylight can be another potential crisis, but some cyanobacteria, such as *Oscillatoria limnetica*, resolve this difficulty by moving deeper into the anoxic layers of the mat, assimilating HS^- as their electron donor by photosystem I alone (Cohen et al., 1975). Fortunately for the

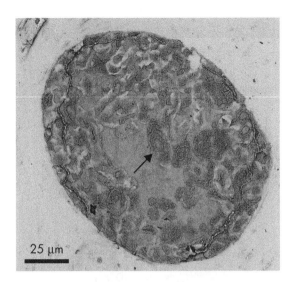

25 µm

Figure 6.18 An etched SEM thin section of the interior of a fine-grained carbonate grain from Exuma Cays, Bahamas. It is almost completely re-worked through microboring and aragonite in-filling of the bore holes (arrow) by the endolithic coccoid cyanobacterium *Solentia* sp. Note how only the interior of the grain has remained intact. (Courtesy of Ian Macintyre.)

cyanobacteria, the translucency provided by the abundant EPS and gypsum allows light to penetrate deeply into the mineralized mats, thus extending the euphotic zone to greater depths (Canfield et al., 2004).

As the discussions above imply, the carbonate stromatolite is a dynamic ecosystem in which the interplay between various microorganisms leads to an upwardly accreting and lithifying structure, with a concomitant downward-progressing decomposition of the microbial remains. In the long term, the overall morphology of any given carbonate microbialite may systematically change with fluctuating environmental conditions. For example, columnar carbonate structures in the tidal channels off Lee Stocking Island, Bahamas, were originally interpreted as high-relief stromatolites that formed from trapping ooid and pelletal carbonate sand and synsedimentary precipitation of carbonate cement (Dill et al., 1986). How-

ever, closer inspection of those structures revealed three discernible internal structures: (i) cyanobacterial (mostly *Schizothrix*) stromatolites consisting of alternating layers of clastic sediment and fine-grained micrite that are linked to microbial activity; (ii) algal stromatolites (chlorophytes, diatoms) similar to above; and (iii) thrombolites displaying their characteristically irregular, clotted fabrics that are formed by bacteria, algae and macrofauna. It has been suggested that the cyanobacterial stromatolites represent forms that began to develop in an intertidal setting with the Holocene flooding of the Great Bahaman Bank. The thrombolites instead formed under the present, normal subtidal conditions, while the eukaryotic stromatolites represent an intermediate facies. It therefore appears as though a gradual change from stromatolite to thrombolite might be explained simply on the basis of rising sealevel, such that with deepening waters there would be a decrease in salinity and an increase in energy, all factors that favor thrombolite growth (Feldmann and McKenzie, 1998).

6.2 Marine sediments

Nearly 70% of the Earth's surface is occupied by marine sediments. Newly deposited sediment essentially comprises three main components; inorganic detritus, organic matter, and water and its solutes. The land-derived (terrigenous) detrital content generally varies from coarse-grained sands and silts along the coast to increasingly finer-grained particles (e.g., clays) in the open ocean, although turbidity currents transfer coarse sediments to deep-sea (pelagic) sediments. In addition to the main mineral constituents, quartz, feldspars, clays, and residual lithic fragments of various source rocks, comes a suite of metal oxides, dominated by iron, manganese, and aluminum. Many marine phytoplankton also form mineralized skeletons, and upon their death comes a significant flux of biogenic silica and calcium carbonate (often referred to as oozes) from the water column to the seafloor. The

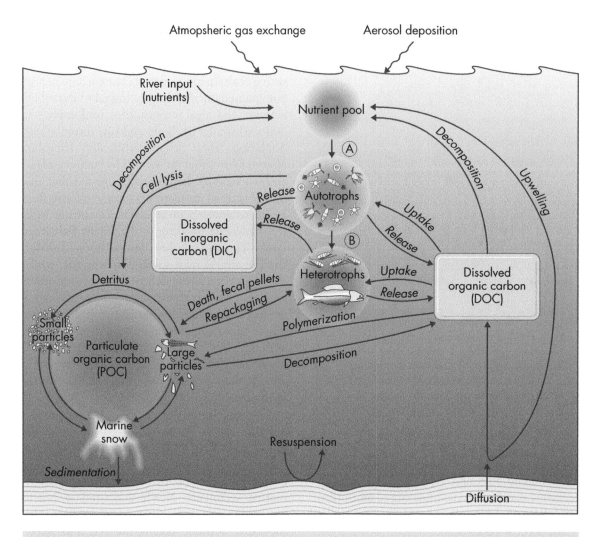

Figure 6.19 Simplified model of the organic matter cycle in the oceans. (A) primary production by autotrophic phytoplankton; (B) degradation and ingestion of autotrophs by heterotrophic bacteria and zooplankton/animals, respectively. (Modified from Wakeman and Lee, 1993.)

composition of the particulate organic carbon similarly varies from a predominantly terrestrial origin nearshore to almost entirely planktonic in the open ocean, the latter accounting for over 90% of the ocean's carbon production. The composition of the uppermost sediment pore waters largely reflect ocean bottom waters, characterized by high concentrations of Cl^-, Na^+, SO_4^{2-}, Mg^{2+}, Ca^{2+}, K^+, HCO_3^-, and Br^-, in descending order of concentrations (Drever, 1988). Generally, the only reducing agent is organic matter.

6.2.1 Organic sedimentation

In the oceans there is a light but constant rain of allochthonous and autochthonous particulate organic material, and their degradation products, sinking from surface waters to the seafloor (see Fig. 6.19 for marine cycling overview). Primary productivity is two to five times greater in nearshore environments (e.g., lagoons, estuaries, coasts, shelf) than the open ocean because of greater nutrient supply from land, but the more important

Table 6.1 Global distribution of carbon burial in marine sediments. (Data from Middleburgh et al., 1993.)

	Depth range (m)	Areal coverage (%)	Global carbon (%)
Coastal sediments	<200	9	87
Continental slope and rise	200–4000	33	9
Deep-sea sediments	>4000	58	4

factor governing the flux of POC to the seafloor is the depth of the water column traversed by the sinking particles, and hence the time over which the particles can be degraded. For example, the fraction of POC reaching 100 m depth is approximately 50 times greater than the fraction sinking to 5000 m. Indeed, the most readily decomposable POC in the open ocean is oxidized aerobically within the upper few hundred meters of the water column, leaving less than a few percent of the material to make its way through deeper waters and onto the seafloor (Suess, 1980). Of that amount, less than 0.1% is buried into pelagic sediments, compared to 5% in nearshore sediments (Chester, 2003). This helps explain why only around 4% of the global distribution of carbon is buried into pelagic sediments, despite these sediments covering more than 50% of the Earth's surface (Table 6.1). Of the particulate fraction settling through the water column, over 95% consists of small (<20 μm) particles that sink very slowly (<1 m day^{-1}), and are thus recycled in the upper water column. Conversely, the large particle pool (>200 μm), comprising algal cells, fecal pellets and debris, zooplankton carcasses, and amorphous aggregates (known as "marine snow") are relatively rare, but due to their rapid sinking rates of hundreds of meters per day, they constitute the bulk of the vertical flux of material to the seafloor (Wakeman and Lee, 1993).

Within most of the ocean water column, the oxidation of organic matter is not sufficient to use up all of the available oxygen. As a result, bulk ocean waters remain oxygenated, with the exception being in some coastal waters where nutrient upwelling supports significant phytoplanktonic productivity (i.e., oxidation of the dead biomass consumes available O_2). The availability of O_2 in the water column promotes extensive alteration of the organic fraction by aerobic chemoheterotrophs, yielding sedimentary POC with chemical compositions markedly different from that of the material originally synthesized in the surface waters. For instance, the most labile compounds (e.g., amino acids and phospholipids) are preferentially degraded, with the residual organic fraction being enriched in carbon relative to nitrogen and phosphorous (e.g., Wefer et al., 1982). This becomes particularly clear when we compare the average elemental composition of marine phytoplankton sinking through the water column, with an initial 106C:16N:1P ratio (referred to as the Redfield ratio) to their debris arriving at the deep seafloor, with a C:P ratio as high as 900:1 and a C:N ratio as high as 30:1 (Knauer et al., 1979). These ratios reflect the greater ease with which N- and P-containing compounds are hydrolyzed (and the nutrients consumed) versus the C-containing structural components (cellulose, lignin, chitin, etc.).

Protozoans and zooplankton are also important catalysts in POC alteration. Phytoplankton population size is kept in check by heterotrophic protozoan ingestion (e.g., the flagellates), that, in turn, are preyed upon by zooplankton, and then larger animals. It has been estimated that some 45% of phytoplankton or detrital particles ingested by zooplankton goes into the growth of the organisms, 27% is lost as dissolved organic carbon (DOC), 24% is respired to form dissolved inorganic carbon (DIC), while the remaining 4% is excreted as fecal material. The latter tend to have very different organic compositions compared to those of the animal's original diet (Copping and Lorenzen, 1980).

When the heterotrophs die, their carcasses are attacked by other aerobically respiring micro-organisms, releasing some DOC that is readily scavenged by phytoplankton (if in the euphotic zone) and other heterotrophic populations. In other words, there is a "microbial loop" that involves the partial recycling of nutrients and organics through the food chain, with only a fraction lost from the water column as it settles out as POC to the seafloor (Azam et al., 1983). Similar re-cycling of organic material occurs at depth. It has been estimated that bacterial chemolithoautotrophy in some bottom waters can generate organic carbon at a rate equivalent to about 1% of the primary production in the surface waters. Energy for this process may, in part, be provided by NH_4^+ derived from the sinking flux of POC, as indicated by high densities of nitrifying bacteria at depth (Karl et al., 1984). In fact, employment of rRNA-targeted fluorescent probes have revealed *Bacteria* and *Archaea* cell densities as high as 10^4 cells ml^{-1} (Karner et al., 2001). This microbial biomass can then support the numerous protozoans, zooplankton, and fish that inhabit the deep waters.

On the seafloor, recently deposited organic matter is subject to grazing, burrowing, and particle reworking by relatively large animals moving over the surface of the sediment, and the more numerous small organisms (the infauna) living within the top few centimeters of the sediment. Collectively these processes go by the name bioturbation. Studies have also shown that a significant amount of settled POC is microbially respired at the sediment surface, with estimates as high as 70% (e.g., Mayer, 1993). Some of the DOC and nutrients generated through these metabolic processes will eventually return to the overlying water column (via diffusion and upwelling), thereby linking biological productivity in the surface waters with biogeochemical processes in sediments (Wollast, 1993). Another fraction of

DOC and nutrients will be resynthesized into sedimentary microbial biomass. The re-cycling efficiency is poorly constrained, but values as high as several tens of percent have been reported. The residual POC, not respired at the seafloor, is then buried into the sedimentary column, where it ultimately serves as the primary reducing agent for subsurface microbial communities (see below).

6.2.2 An overview of sediment diagenesis

After sediment is deposited, a combination of physical, chemical, and biological processes transform the unconsolidated material into sedimentary rock. These changes go by the name diagenesis, and include: compaction; mineral transformations; organic matter degradation; generation of hydrocarbons; changes in pore water composition; and cementation of unconsolidated sediment. At the sediment–water interface, the freshly deposited material is porous and exposed to bottom waters and the dissolved gases (O_2, CO_2) that they contain. Below a depth of a few centimeters, the porosity is reduced by compaction, and contact with bottom waters is essentially lost. At still deeper levels, the older sediment is affected by reactions induced by increasing pressure and temperature, and by the infiltration of reactive pore waters. The kind of reactions taking place will change with depth: oxidation near the sediment–water interface and reduction at lower levels if organic matter is still present. At even greater depths, pore spaces are no longer continually filled with water and physicochemical reactions begin to fuse the sediment grains together, leading to what is known as lithification. Diagenesis, in effect, can be viewed as the slow equilibration of surface sediment to a new set of conditions associated with burial.

Throughout the sediment column, up to the temperature threshold of life (below the boiling point of water), microbial activity plays an integral role in diagenesis. It is widely recognized that through their various chemoheterotrophic pathways, microorganisms are ultimately responsible for the conversion of organic carbon to CO_2 and CH_4. Some bacteria use hydrolytic enzymes to break down complex organic compounds into simple monomers, such as sugars, amino acids, and fatty acids, that they can then make use of, while others are restricted to simple fermentation products, of which acetate is probably the most utilized (Lovley and Chapelle, 1995). Typically, the more labile materials are degraded in near-surface sediments on timescales of days to years, more refractory materials are broken down deeper in the sediment on timescales of hundreds to thousands of years, while the most resistant materials, precursors to fossil fuels, are transformed only on timescales of millions of years. The amount of refractory material left after microbial attack will depend on the relative proportions of terrestrial- versus marine-derived biomass, and on the degree of processing that has taken place in the river catchment area or oceanic water column, respectively. In the end, less than 1% of the original organic material buried into sediment may ultimately contribute to the sedimentary organic geochemical record (e.g., Emerson and Hedges, 1988).

Associated with the decrease in organic matter with burial is a progressive change in its composition, perhaps most evident in the residual C:N:P ratio. Although the C:N:P ratio of the organic matter input will depend on its source and the amount of alteration during descent in the water column, there are still recognizable trends with depth in a given sediment. For example, the preferential stripping of N from amino acids into solution causes the C:N ratio of the sediment to increase with depth, with pore water amino acid levels virtually vanishing below 10 cm (Macko et al., 1993). The loss of POC nitrogen is partially counteracted by the assimilation of some of the NH_4^+ into growing bacterial cells. Correspondingly, the C:N ratio of organic matter mineralized in the upper sediment actually decreases in shallow sediment, but once again increases with depth as the organic matter undergoes more advanced degradation and N is lost through denitrification (Blackburn, 1980). This serves to illustrate how microbial cells themselves

should be viewed as an integral component of the sedimentary package. Although there is clear evidence for changes in the C:N ratio during early diagenesis, there is debate as to whether or not preferential stripping of P also occurs, thereby increasing the C:P ratio (Ingall and Jahnke, 1997), or whether this is partially counteracted by the active assimilation of P into cells when excess amounts are available (Ramirez and Rose, 1992). The actual retention of P in sediments is a fine balance between the rates of precipitation (as diagenetic calcium fluorapatite phases), its adsorption (as phosphate) onto ferric oxyhydroxides, and biological uptake versus surface mixing and diffusive losses to the overlying water column.

Pore water and mineralogical changes during diagenesis are also directly related to the bacterial reduction of dissolved (O_2, NO_3^-, SO_4^{2-}, CO_2) or metal oxyhydroxides in the sediment. The terminal electron accepting process (TEAP) that occurs at any given depth depends on what oxidants are available and, in the situation when multiple electron acceptors are present (as in the uppermost sediment layers), on the free energy yield of the specific reaction. Thus, the decomposition of freshly deposited organic material in sediments proceeds in a continuous sequence of redox reactions, with the most electropositive oxidants being consumed at, or near the surface, and progressively poorer oxidants being consumed at depth, until the labile organic fraction is exhausted and the deeper sediments are left with a composition very different from the sediments originally deposited. Inorganic byproducts of chemoheterotrophy (e.g., HCO_3^-, Mn^{2+}, Fe^{2+}, NH_4^+, NO_2^-, HS^-, HPO_4^{2-}, and CH_4) seldom accumulate to very high concentrations. Instead, diffusion and macrofaunal processes will cause the net transport of these reduced species from deeper sediment towards the seafloor, where residing chemolithoautotrophic bacteria use them as reactants in their metabolism. Alternatively, their presence in pore waters may trigger important abiological reactions between the solid and dissolved phases, leading to secondary mineral precipitation.

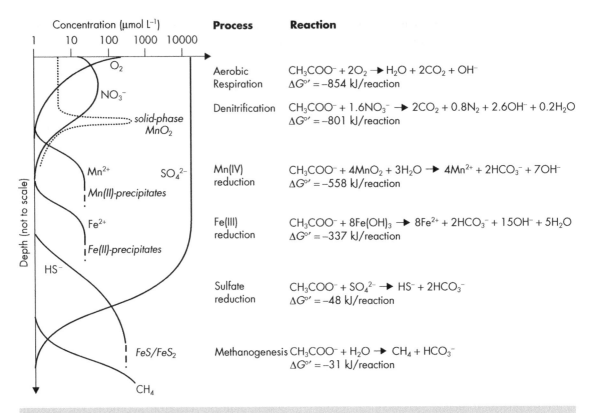

Figure 6.20 Idealized pore water and solid-phase profiles based on the successive utilization of terminal electron acceptors during the decomposition of marine sedimentary organic matter. (Adapted from Froelich et al., 1979.)

The sequence of redox reactions is manifest as a biogeochemically and mineralogically zoned sediment that reflects the dominant microbial community growing at a particular depth, although in many cases they may overlap (Fig. 6.20). In actuality, the changes in organic and inorganic chemistry are complex, transient, and involve numerous recycling reactions that can complicate our full understanding of what microorganisms are metabolically active at any given depth. Nonetheless, sediments are broadly defined as "oxic," "suboxic," and "anoxic" by the levels of dissolved O_2 (and HS^-) in the pore waters, which, in turn, are directly influenced by microbial respiration.

Evidence for the existence of distinct biogeochemical zones has been known for more than half a century, but a formalized depth-related scheme describing the main diagenetic reactions occurring in each zone was not developed until the 1970s (e.g., Froelich et al., 1979). During the last 20 years, this sequential scheme has formed the framework with which to successfully describe the degradation of organic matter in all aquatic environments, from groundwater to eutrophic surface waters to the oligotrophic open ocean. Although the depth over which these zones occurs may vary from a few millimeters or centimeters in coastal sediments, with high organic carbon fluxes, to several meters in the deep ocean, where carbon sedimentation rates are very slow, they occur in any system where the supply of labile organic matter outpaces diffusion of oxygen into the sediment (Jørgensen, 1983).

6.2.3 Oxic sediments

The sediment–water interface experiences the most intense heterotrophic activity. Aerobic respiration (reaction (6.8)) is the first used metabolic pathway for the degradation of organic matter, not only because it generates the largest free energy yield per mole of organic carbon oxidized, but also the aerobes are equipped with a full suite of enzymes capable of degrading complex organic polymers into simpler substrates.

$$C_6H_{12}O_6 + 6O_2 \rightarrow 6CO_2 + 6H_2O \qquad (6.8)$$

It is estimated that on a global scale, the majority of all POC deposited is aerobically respired. Naturally, this percentage varies amongst different marine environments, with aerobic respiration accounting for more than 90% of the organic carbon oxidized in pelagic sediments (e.g., Bender and Heggie, 1984), but <50% in coastal marine settings (e.g., Jørgensen, 1980). Such great variability in importance helps explain why oxygen consumption rates typically range from 0.001 to 1 mmol cm^{-2} yr^{-1} in marine sediments (Canfield, 1993).

There are several factors that primarily control the sediment depth to which localized oxic conditions prevail:

1 *Sedimentation rate* – The rate with which sediment is deposited directly influences the amount of time organic matter decomposes in oxygenated waters, at or near the sediment–water interface. Under the low sedimentation rates associated with the open ocean (<0.01 mm yr^{-1}), small amounts of organic matter (<0.001 g C cm^{-2} yr^{-1}) are deposited onto the seafloor (Henrichs, 1992). This organic fraction will remain exposed to oxygenated waters for thousands of years, leading to almost complete oxidation by O$_2$, with only the most refractory fraction available for other terminal electron acceptors (TEAs). In pelagic sediments, the oxic zone can extend from several centimeters to more than a meter in depth (Murray and Grundmanis, 1980). By contrast, under the high deposition rates characteristic of coastal environments (1 mm yr^{-1}), abundant organic matter is deposited (>0.01 g C cm^{-2} yr^{-1}). With high rates of accumulation, O$_2$ diffusion is very shallow and the oxic zone may only extend down a few millimeters in depth, although physical stirring of surface sediments by waves and currents, circulation of overlying water into sediment through macrofaunal burrows and tubes (known as irrigation), and particle remobilization through grazing may locally increase that depth to several centimeters (e.g., Hammond et al., 1985). With high accumulation rates, freshly deposited organic matter passes through the oxic zone in anywhere from one year to tens of years, leaving a significant amount of labile carbon intact to support anaerobic heterotrophic communities. Based on the formula: C = 4(SR)$^{0.68}$, where C is the aerobic carbon oxidation rates (in mmol C cm^{-2} yr^{-1}) and "SR" is the sediment deposition rate (g cm^{-2} yr^{-1}), it has been calculated that carbon oxidation rates are reduced by a factor of 4.8 for every factor of 10 increase in sedimentation rate (Canfield, 1993).

2 *Reactivity of POC* – Plankton-derived organic matter, rich in lipids and cellulose, exhibits significantly higher reactivity than terrestrial-derived organic material consisting of vascular plants rich in lignins (Hedges et al., 1988). Consequently, the presence of easily degradable organic compounds on the seafloor promotes high rates of aerobic respiration, and the rapid consumption of available O$_2$. Therefore, the "quality," and not just quantity of organic material influences O$_2$ utilization.

3 *Oxygen concentration in the overlying water* – The rate of oxidation of organic carbon by aerobic respiration is dependent on the concentration of dissolved O$_2$, with 3–10 µmol L^{-1} being a critical value below which the rate of aerobic respiration diminishes significantly, and stops when virtually no O$_2$ is left (Van Cappellen and Gaillard, 1996). Oxygenation of bottom waters is not uniform throughout the oceans; bottom seawater can range anywhere from 0 to 300 µmol L^{-1} O$_2$, the former representing redox stratification in euxinic marine basins. Of significance for the bulk of marine sediments is the diffusion of O$_2$ from the bottom waters, through the hydrodynamic boundary layer, and into the underlying sediment. The impact of this boundary layer is that it has the potential to inhibit O$_2$ fluxes across the solid–water interface, particularly under high biological uptake rates. Therefore, sediments and aerobes normally exposed to oxygenated water could find themselves temporarily under anoxic conditions, while in a similar manner, microaerophiles (e.g., *Beggiatoa*) can grow as mats directly on the seafloor, protected from the fully oxygenated waters. Only by reducing

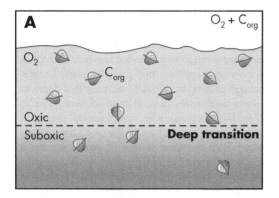
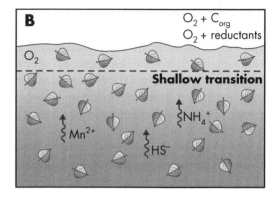

Figure 6.21 Possible positions of the oxic-suboxic interface when (A) sedimentation rates are slow, and most of the O_2 is used in aerobic respiration or (B) when sedimentation rates are rapid, and much of the O_2 is used instead in the abiological or chemolithoautotrophic oxidation of reduced metabolites, such as NH_4^+, Mn^{2+}, or HS^-.

the thickness of the boundary layer, increasing the bulk O_2 concentration, or reducing metabolic O_2 consumption can the aerobic populations regain their vigor (Jørgensen and Revsbech, 1985).

4 *Physical regime of the environment* – Sediments differ markedly from the water column in terms of their transport properties. Unlike the bottom waters, the presence of a solid matrix eliminates large-scale fluid flow and turbulent mixing (Van Cappellen and Gaillard, 1996). Instead, the exchange of solutes between sediments and the overlying water column takes place through the interstitial pore waters, where rates of diffusion are considerably less than in open water because a large proportion of the cross-sectional area through which the solutes have to move is occupied by solid particles (reducing porosity) and the path for diffusion is not a straight line (high tortuosity). For instance, most freshly deposited marine sediments have fairly high porosities, ranging from 60% to 95%, but those porosities decrease by as much as 20% from the sediment–water interface down to depths of a few centimeters (Ullman and Aller, 1982). Similarly, sediment grain size has a bearing on the chemical exchange with bottom waters because fine-grained silts and clays have lower permeability than coarse-grained sands. Thus, diffusive influx of a bottom-water component, say O_2, may quickly diminish with depth, leading to the establishment of steep pore water gradients in the subsurface sediments.

5 *O_2 consumption by pore water reductants* – Recall from Chapter 2 that a vast number of chemolitho-autotrophic bacteria use O_2 as their preferred electron acceptor. Therefore, as reduced metabolites from underlying suboxic and anoxic layers diffuse upwards, a fraction of the oxygen is diverted away from aerobic respiration, towards their re-oxidation (Fig. 6.21). As a consequence, the depth of O_2 penetration, and hence the oxic zone, correspondingly decreases. In fact, some diagenesis models show that the amount of O_2 partitioned to organic carbon oxidation can be less than 1%, with nitrification or hydrogen sulfide making up the major O_2 sinks (e.g., Blackburn and Blackburn, 1993).

6.2.4 Suboxic sediments

(a) Nitrification-denitrification

Nitrogen cycling in aquatic sediments is relatively complex, with overall nitrate pore water concentrations balanced between nitrification and denitrification in the oxic/suboxic zones – nitrate diffusing downwards into the sediment from the bottom waters is generally not considered a significant source of input (Seitzinger, 1988). The amount of nitrification is, in turn, largely governed by the concentration of pore water NH_4^+, which itself depends on: (i) the C:N ratio of the sedimentary organic matter being oxidized, with high POC loading resulting in increased rates of nitrification; and (ii) the relative amount of nitrate ammonification. In most sediments higher rates of nitrification are also associated with deeper O_2 penetration (Blackburn and Blackburn, 1993). As previously discussed in section 2.5.7, in the presence of O_2, ammonium is first oxidized to nitrite by ammonium-oxidizing bacteria, and the nitrite is then further oxidized to nitrate by nitrite oxidizers (see overall reaction (6.9)). Collectively, these processes cause a subsurface peak of nitrate in the pore water (recall Fig. 6.20). In coastal sediments with high carbon input, this nitrification peak is often very close to the sediment–water interface (Mortimer et al., 1999).

$$NH_4^+ + 2O_2 \rightarrow NO_3^- + H_2O + 2H^+ \qquad (6.9)$$

In many sedimentary environments, however, the average distributions of O_2 and NO_3^- are not consistent with nitrification exclusively in oxygenated sediment. Instead, sediment pore water profiles indicate the presence of a suboxic nitrate maxima that may be the result of bacterial nitrification coupled to the reduction of MnO_2 (reaction (6.10)). This reaction is energetically feasible only below pH values of 7.8 (Luther et al., 1997), but if it is prevalent in marine sediments, it implies that nitrification and denitrification (see below) can occur simultaneously without vertical stratification. The most likely environment for such a reaction is where suboxic sediments are more or less constantly mixed with Mn(IV) oxides from the oxidized surficial layers, e.g., coastal settings (Hulth et al., 1999).

$$NH_4^+ + 4MnO_2 + 6H^+ \rightarrow 4Mn^{2+} + NO_3^- + 5H_2O \qquad (6.10)$$

The opposing process in marine N-cycling is denitrification. Denitrifiers most closely resemble aerobic respirers in that they are capable of completely degrading complex organic matter to carbon dioxide:

$$2.5C_6H_{12}O_6 + 12NO_3^- \rightarrow$$
$$6N_2 + 15CO_2 + 12OH^- + 9H_2O \qquad (6.11)$$

In coastal sediments, these bacteria reduce nitrate to negligible levels within 1–10 cm of the pore water profile. No nitrate occurs below this depth except where burrows, lined with nitrifying bacteria, irrigate oxygenated water to depths well below the oxic–suboxic boundary (Hansen et al., 1981). In pelagic sediment, the vertical zone of all TEAs are expanded, and for nitrate, concentrations may not become negligible until depths of greater than 1 m.

Rates of denitrification range over several orders of magnitude in marine sediments, between values as low as 4×10^{-5} mmol cm^{-2} yr^{-1} in pelagic sediments (Bender and Heggie, 1984) to over 0.1 mmol cm^{-2} yr^{-1} in continental shelf sediments (e.g. Devol, 1991). It has been estimated that denitrification in marine sediments may oxidize $12–16 \times 10^9$ mol day^{-1} of organic carbon, an amount equivalent to about 3% of the sedimentary carbon oxidation by aerobic respiration (Christensen et al., 1987). One of the main factors controlling the relative significance of denitrification is the NO_3^-:O_2 concentration ratio. In estuarine sediments, where water column NO_3^- levels exceed dissolved O_2 levels, denitrification is responsible for an equal amount of, or even more, organic carbon oxidation than aerobic respiration (Jørgensen and Sørensen, 1985). High rates of N_2 loss would be quite problematic for sedimentary life were it not for the continuous supply of oxidizable organic-N, and its cycling within the sediments between oxidized (NO_2^-/NO_3^-) and reduced (NH_4^+) reservoirs via nitrification and nitrate ammonification, respectively (Tiedje et al., 1982).

The denitrifyers, however, are in direct competition with other metabolic pathways, and even some abiological reactions, for the available supply of nitrate. For instance, some chemolithoautotrophic bacteria couple nitrate reduction to sulfide oxidation (reaction (6.12)). Unlike the mat-forming *Thioploca*, which cannot grow on sediments subject to high fluidity and instability, other species, such as the 750 μm, balloon-shaped *Thiomargarita namibiensis*, obtain nitrate during occasional sediment re-suspension events, at which time they accumulate enough to last them until the next such event occurs (Schulz et al., 1999).

$$5HS^- + 8NO_3^- \rightarrow 5SO_4^{2-} + 4N_2 + 3OH^- + H_2O \tag{6.12}$$

N_2 may also be formed through reduction of NO_3^- by upwardly diffusing Mn^{2+} (reaction (6.13)). This reaction, coined chemo-denitrification, is often discernible in pelagic sediments by an absence of an overlap between the distributions of Mn^{2+} and O_2, i.e., there is a nitrate buffer zone (e.g., Sørensen et al., 1987). Similar reactions (6.14) have also been shown to occur when significant amounts of Fe^{2+} diffuse upwards towards the chemocline (Sørensen, 1987):

$$5Mn^{2+} + 2NO_3^- + 4H_2O \rightarrow 5MnO_2 + N_2 + 8H^+ \tag{6.13}$$

$$10Fe^{2+} + 2NO_3^- + 24H_2O \rightarrow 10Fe(OH)_3 + N_2 + 18H^+ \tag{6.14}$$

The flux of N_2 escaping from sediments is often higher than the predicted biological or abiological denitrification rates based solely on pore water nitrate concentrations. This means that in the sediments, excess N_2 is being formed independent of a nitrate intermediate. One way this occurs is through the anammox process described in section 2.5.7, where shunting nitrogen directly from NH_4^+ to N_2 (reaction (6.15)) can promote ammonium deficiencies in sediments where this process plays a key role in nitrogen cycling (Thamdrup and Dalsgaard, 2002). This reaction has since been shown to be an important source of N_2 in stratified water columns as well, where NH_4^+ diffusing upwards from deep anoxic waters is consumed by anammox bacteria higher up in the water column (e.g., Kuypers et al., 2003).

$$NH_4^+ + NO_2^- \rightarrow N_2 + 2H_2O \tag{6.15}$$

Ammonium can also react with MnO_2 to form N_2 instead of nitrate (reaction (6.16)). The Mn^{2+}

formed in this reaction is subsequently utilized by Mn(II)-oxidizing bacteria, with O_2 as the TEA (see reaction (2.55)), producing more reactive MnO_2 to continue the oxidation of ammonium (Luther et al., 1997). The same authors have further observed that this reaction can outcompete the direct oxidation of NH_4^+ by O_2 (forming NO_3^-) in Mn-rich coastal sediments, and in doing so, potentially account for 90% of N_2 formation. In essence, this process short-circuits the traditionally considered nitrification–denitrification cycle.

$$2NH_4^+ + 3MnO_2 + 4H^+ \rightarrow 3Mn^{2+} + N_2 + 6H_2O \tag{6.16}$$

Neither aerobic respiration nor denitrification (and nitrate ammonification) have a significant impact on sediment mineralogy. On the other hand, the terminal electron accepting processes described below give rise to most of the important geochemical and mineralogical transformations occurring during early diagenesis. As discussed in section 2.4.7, anaerobic decomposition of organic matter is not generally accomplished by a single microorganism, but instead by a community of interdependent species. Most anaerobes are very limited in the types of organic compounds that they can oxidize. That is why in suboxic to anoxic sediments, much of the organic matter released from the hydrolysis of complex organic matter is first metabolized via fermentative microorganisms. The latter do not completely oxidize organic matter to CO_2 or CH_4, making it possible for other types of heterotrophic microorganisms to coexist with them (Senior et al., 1990). In fact, without the anaerobes, fermentative microorganisms would struggle to survive since the accumulation of their waste byproducts would make their metabolic reactions thermodynamically unfavorable.

(b) Manganese cycling

Detrital mineral phases that survive transport and deposition are for the most part an unreactive component of the bottom sediment. The oxides and hydroxides of manganese and iron are the exceptions. These phases are stable under the oxygenated conditions encountered in the water column, but they quickly become unstable after burial below the oxic layers of sediment. Following denitrification, reduction of MnO_2 to dissolved Mn(II) becomes the most energy-efficient bacterial respiratory process:

$$CH_3COO^- + 4MnO_2 + 3H_2O \rightarrow$$
$$4Mn^{2+} + 2HCO_3^- + 7OH^- \qquad (6.17)$$

In most sediment, the limited amount of metal oxyhydroxides deposited, as well as their slower rates of burial, compared to the downward diffusion of O_2, NO_3^-, and SO_4^{2-}, generally makes their reduction of minor importance (<10%) in terms of the amount of total organic carbon oxidized (Burdige, 1993). There are, nevertheless, exceptions to this generalization. One of the well-described sites is the Panama Basin, where nearly 100% of carbon oxidation is linked to Mn(IV) reduction (Aller, 1990). These hemipelagic sediments are an intermediate class of deep water deposit (a mixture of terrigenous and pelagic materials) where high rates of Mn(IV) reduction are driven by a combination of hydrothermal manganese inputs from the East Pacific Rise and active macrofaunal activity in the upper 30 cm of sediment that mix suboxic sediment with oxygenated bottom waters, thereby re-oxidizing Mn(II) to Mn(III) hydroxides (e.g., MnOOH) and Mn(IV) oxides (e.g., MnO_2). Not only does this facilitate another round of reduction that simultaneously consumes more organic carbon, but the O_2 flux is largely diverted to Mn(II)-oxidation, not aerobic respiration.

In typical sediments, the concentration of pore water Mn^{2+} is negligible in the surface sediments, but at some depth between the NO_3^- peak and zero NO_3^-, it begins to increase in concentration towards its maximum value. A characteristic feature of Mn cycling in marine sediments is the upward convexity of the pore water Mn^{2+}

profile and the discrete layer of MnO_2 forming at the base of the oxic zone (recall Fig. 6.20), suggesting that the TEA can be either nitrate or oxygen (Froelich et al., 1979. In the absence of significant sediment disruption, the depth of this manganese "spike" is governed by the balance between O_2/NO_3^- diffusing downwards and Mn^{2+} diffusing upwards, with all Mn^{2+} usually completely consumed within the spike. If there is little Mn^{2+}, O_2/NO_3^- would diffuse downwards to oxidize Mn^{2+} deeper in the sediment, while if there is abundant Mn^{2+}, it would diffuse upwards to be oxidized at even more shallow depth (Burdige and Gieskes, 1983). It is unlikely that Mn^{2+} would actually diffuse above the MnO_2 layer because the oxides effectively adsorb dissolved manganese, in essence acting as a cap preventing Mn^{2+} escape, although in some instances Mn^{2+} does indeed diffuse across the sediment–water interface to contribute to ferromanganese nodule formation. In a steady-state system, the concentration of Mn in the MnO_2 layer will increase until the sedimentary input of MnO_2 is balanced by the efficiency of its reduction and remobilization as it passes through the nitrate reduction zone. Under conditions of changing sedimentation and associated fluctuations in redox conditions, it is not uncommon to find multiple MnO_2 layers within the same sediment, each one reflecting its relict pore water composition. In some coastal sediments, it has been estimated that each molecule of manganese (and iron) is recycled between 100 and 300 times before being finally buried in the sediments (Canfield et al., 1993).

Downward diffusion of Mn^{2+}, and its reaction with HCO_3^-, most commonly results in the formation of manganous carbonates, such as rhodochrosite ($MnCO_3$), or mixed Mn- and Ca-carbonates, such as kutnahorite and manganoan calcite (Middleburg et al., 1987). The precipitation of these mineral phases (e.g., reaction (6.18)) are largely responsible for constraining the concentrations of Mn^{2+} in sediment pore waters and for limiting pH rises much above 7 (Aller and Rude, 1988):

$$Mn^{2+} + Ca^{2+} + 4HCO_3^- \rightarrow$$
$$MnCa(CO_3)_2 + 2CO_2 + 2H_2O \qquad (6.18)$$

Unlike their carbonate counterparts, manganese sulfides are rather soluble. Consequently, Mn^{2+} is commonly observed diffusing out of sulfide-rich layers in anoxic marine sediments, the exception being the Black Sea where dissolved sulfide levels are sufficiently high that supersaturation with regards to minerals, such as wurtzite (MnS), are attained (Suess, 1979). In most instances, the reaction of MnO_2 with HS^- yields Mn^{2+} and elemental sulfur (reaction (6.19)). Bacterial disproportionation of S^0 to HS^- and SO_4^{2-} then regenerates HS^- to potentially repeat the cycle (Thamdrup et al., 1993). Interestingly, reaction (6.19) has been implicated in supplying more Mn^{2+} to the chemocline in stratified marine waters, such as the Black Sea, than the reduction of MnO_2 via the dissimilatory reductive pathway (Burdige and Nealson, 1986). No comparison was made with the abiological reduction of MnO_2 by Fe^{2+} (reaction (6.20)):

$$MnO_2 + HS^- + H_2O \rightarrow Mn^{2+} + S^0 + 3OH^-$$
$$(6.19)$$

$$MnO_2 + 2Fe^{2+} + 4H_2O \rightarrow 2Fe(OH)_3 + Mn^{2+} + 2H^+$$
$$(6.20)$$

Manganese (IV) reduction can also take place by reaction with solid-phase sulfides (FeS) that are brought into contact during physical or biological reworking (Aller and Rude, 1988). These mixtures are thermodynamically unstable and will decompose through reactions such as (6.21):

$$4.5MnO_2 + FeS + 7CO_2 + 4H_2O \rightarrow$$
$$4.5Mn^{2+} + SO_4^{2-} + 7HCO_3^- + FeOOH$$
$$(6.21)$$

The potential catalyst for the above reaction is chemolithoautotrophic sulfur oxidation, with the bacteria using manganese oxides as their terminal electron acceptor.

(c) Iron cycling

Below the zone of dissimilatory Mn(IV) reduction, and at the depth of complete nitrate removal from pore waters, is where Fe(III) reduction takes place (reaction (6.22)). Ferric hydroxide is the most reactive and easily reducible solid-phase iron mineral, and rates of Fe(III) reduction in sediment decline rapidly with depth as it becomes depleted (Lovley and Phillips, 1986).

$$CH_3COO^- + 8Fe(OH)_3 \rightarrow$$
$$8Fe^{2+} + 2HCO_3^- + 15OH^- + 5H_2O \qquad (6.22)$$

Despite the lower reactivity of crystalline phases (e.g., hematite, goethite, magnetite), in sediment where ferric hydroxide has either been depleted or where crystalline iron oxides are inherently more abundant, the latter can serve as a significant source of reducible iron that couples the oxidation of buried organic matter.

With respect to the percentage of total organic carbon oxidized, the importance of Fe(III) reduction, compared to other terminal electron accepting pathways, varies according to the environment, but in general it is not nearly as significant as sulfate reduction (see next section) because of the limited amount of ferric iron minerals initially deposited. One exception, however, is along the Amazon Shelf, where the Fe(III) reduction zone is up to a meter in thickness (Aller et al., 1986). Conditions that lead to such extensive Fe(III) reduction are: (i) inner shelf muds rich in ferric oxyhydroxides from the highly weathered continental shield regions; and (ii) intense physical reworking of surface deposits by shelf currents and waves brings subsurface FeS and FeS_2 into contact with oxygenated bottom waters, causing their oxidation. In this manner the $Fe(OH)_3$ is regenerated but the reactive organic reductants are limited by fixed inputs from primary production and terrestrial POC (Fig. 6.22). This translates into early diagenetic processes dominated by Fe cycling, leaving a S-poor mineral composition more characteristic of a freshwater than a marine environment.

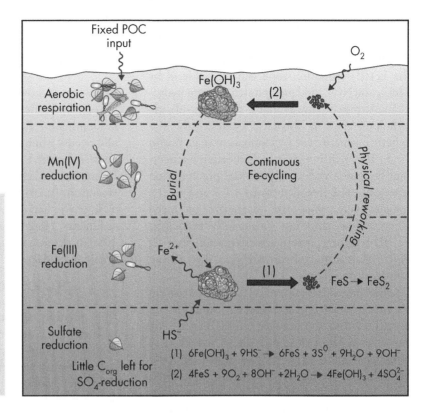

Figure 6.22 How physical reworking of sediment regenerates Fe(OH)$_3$, while a fixed supply of organic carbon is progressively depleted in sediments due to various respiratory pathways. The loss of organic carbon diminishes the importance of sulfate reduction, while the reductive Fe(III) pathway dominates.

The reduction of ferric oxyhydroxide minerals produces an increase in the concentration of Fe^{2+} in suboxic sediment pore waters, with a peak in concentration at the boundary between the Fe(III) and sulfate reduction zones. Some of this ferrous iron may diffuse upwards to be re-oxidized to Fe(OH)$_3$ abiologically by reaction with either MnO$_2$ or NO$_3^-$ (e.g., Myers and Nealson, 1988b). The importance of manganese oxides is particularly evident from the general observations that Fe^{2+} does not accumulate in pore waters above the Fe(III) reduction zone until the reduction of MnO$_2$ is complete, and from the simple fact that Fe^{2+} diffusing up from lower depths contacts MnO$_2$ before nitrate (Lovley and Phillips, 1988a). The reduction of MnO$_2$ by Fe^{2+} (reaction (6.20)) may also help explain the clear separation of the zones of Fe(III) reduction from either nitrate or Mn(IV) reduction, regardless of the actual overlap between these suboxic processes.

Pore water Fe^{2+} is also removed by reaction with HS$^-$ (reaction (6.23)) produced by the underlying populations of SRB (Thamdrup et al., 1994).

This forms metastable iron monosulfide minerals, such as mackinawite, that are precursors to pyrite, provided that adequate supplies of intermediate sulfur species are present:

$$Fe^{2+} + HS^- \rightarrow FeS + H^+ \qquad (6.23)$$

In normal marine sediments, where dissolved sulfate (around 28 mmol L^{-1}) and ferric oxyhydroxide minerals are abundant, organic matter appears to be the major control on pyrite formation. By contrast, pyrite formation in freshwater sediments is severely limited by the low concentration of sulfate (100 μmol L^{-1} or less), while in euxinic settings, HS$^-$ builds up in both the bottom sediments and water column, but iron is insufficiently available for the formation of FeS (Berner, 1984).

When abundant reactive iron minerals are present in marine sediment, pore water HS$^-$ concentrations may be kept low enough that Fe^{2+} concentrations exceed the solubility product of a number of other minerals, including mixed valence oxide, carbonate, phosphate and silicate phases, depending on the chemical conditions of the particular sediment (Zachara et al., 2002). As shown above, Fe^{2+} diffusing upwards from the Fe(III) reduction zone can be oxidized by MnO_2 and NO_3^-, and precipitated as ferric hydroxide coatings on any available solids. In suboxic sediments there is also clear evidence for diagenetic magnetite (Fe_3O_4) formation (e.g., Karlin et al., 1987), although its actual origin is uncertain because a variety of potential precipitation reactions exist (recall sections 4.1.3 and 4.2.1). Irrespectively, since the magnetite produced is crystalline, and not particularly reducible, Fe(III)-reducing bacteria effectively convert part of the labile iron oxyhydroxide pool into a more refractory one (Lovley, 1991).

Other minerals that have been observed to form are siderite and vivianite. Siderite formation has been linked to the activity of Fe(III)-reducing bacteria since they generate both Fe^{2+} and HCO_3^-, although some sulfate reducers have similarly been shown to reduce ferric oxyhydroxides when the latter are more available (e.g., Coleman et al., 1993). In Fe-rich sediments, the general availability of HCO_3^-, in suboxic and anoxic pore waters, tends to cause Fe^{2+} to precipitate quickly as a cementing agent. By contrast, when sulfate reduction rates exceed Fe(III) reduction rates, enough HS^- is produced to react preferentially with any ferrous iron to precipitate iron monosulfide minerals instead (FeS has lower solubility than $FeCO_3$). Therefore, the precise juxtaposition of Fe(III) reduction and sulfate reduction in marine sediments controls whether siderite and/or iron sulfides may form (e.g., Coleman, 1985). The formation of vivianite in many ways resembles that of siderite, namely Fe^{2+} concentrations must exceed those of HS^-. It also requires that dissolved phosphate is made available through oxidation of organic matter, the dissolution of phosphate-bearing minerals, or through reduction of phosphate-adsorbing ferric oxyhydroxides. Precipitation is extremely slow and is often inhibited by the presence of various mineral and organic phases that preferentially adsorb phosphate from solution (Krom and Berner, 1980).

6.2.5 Anoxic sediments

(a) Sulfate reduction

Below the suboxic layer is a highly reducing, oxygen-free zone under which sulfate reduction predominates. Recall, sulfate reduction typically occurs when all other available TEAs have been exhausted, with the exception of CO_2. In marine sediments, the reduction of SO_4^{2-} (reaction (6.24)) accounts for approximately 50% of the carbon oxidation in coastal marine sediments (Jørgensen, 1982b), but less than 10% in deep sea and freshwater environments (Bender and Heggie, 1984; Jones, 1985):

$$CH_3COO^- + SO_4^{2-} \rightarrow HS^- + 2HCO_3^- \quad (6.24)$$

The relative importance of this metabolic pathway in coastal marine environments stems from the higher concentrations of sulfate at the sediment–water interface, some 50 times greater than the combined sum of all other electron acceptors with higher electrode potentials. Moreover, external electron acceptors that yield more energy than it typically disappear within the first few centimeters of sediment depth, leaving it as the dominant TEA for most of the sediment column (D'Hondt et al., 2002).

The rates of dissimilatory sulfate reduction are proportional to the quantity and reactivity of organic matter entering the anoxic zone. This relationship was first described over 40 years ago, in the first-order (one-G) model developed by Berner (1964). In this model it was assumed that reactive organic matter decomposes at an overall rate directly proportional to its own concentration (i.e., first-order kinetics):

$$\frac{dG}{dt} = -kG \quad (6.25)$$

where G is the concentration of the organic carbon that can be degraded by the SRB community, k is the first-order rate constant, and t is time. The downfall of this model is that it does not take into account that the pool of decomposable organic material actually consists of various groups of organic compounds with vastly differing reactivities. This realization has led to the "multi-G model," whereby each compound is assumed to degrade via first-order kinetics:

$$\frac{-dG_T}{dt} = \Sigma k_i G_i \qquad (6.26)$$

where k_i and G_i refer to individual fractions, each with a different reactivity, and G_T represents total decomposable organic matter (Westrich and Berner, 1984).

What these models essentially indicate is that not only is the overall sedimentation rate important, but so too is the extent to which this organic matter was first degraded in the oxic zone; the greater the amount of aerobic decomposition, the more refractory the residual material becomes, leaving less labile organic matter available for sulfate reduction. As a result of natural variations in the above, sulfate reduction rates in marine sediments can vary by six orders of magnitude, with the highest (>1 mmol cm^{-2} yr^{-1}) in rapidly deposited lagoonal and coastal sediments, where it occurs just below the sediment–water interface, and the lowest in pelagic sediments, where rates

may be <10^{-5} μmol cm^{-2} yr^{-1} (Canfield, 1993). Euxinic basins also have high sulfate reduction rates because no aerobic decomposition can occur in the sediment, hence labile organic materials are abundantly available for SRB (Fig. 6.23). Quite clearly, sulfate reduction can occur over vast tracts of the oceans, from within the water column in euxinic basins, to just below the sediment–water interface in coastal environments, to hundreds of meters in depth in pelagic sediment. But, in all of these sedimentary environments, rates of sulfate reduction decrease with depth, as the most readily metabolizable organic fraction is progressively degraded and as pore water SO_4^{2-} concentrations decrease due to lack of exchange with overlying seawater.

The typical product of sulfate reduction is hydrogen sulfide, an extremely toxic compound even to the SRB because it combines with the iron of cytochromes. Therefore, when H_2S/HS^- concentrations exceed 16 mmol L^{-1}, dissimilatory sulfate reduction ceases (Reis et al., 1992). Yet, a fortuitous detoxification mechanism for sulfide exists in most sediments – its reaction with extracellular iron, and the formation of insoluble iron monosulfides, and eventually pyrite (recall section 4.1.10). Pyritization can occur on scales

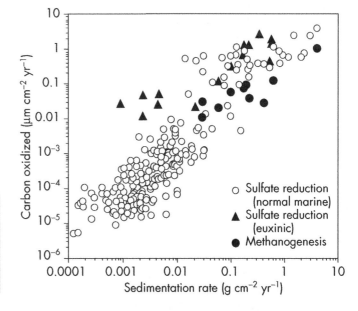

Figure 6.23 Rates of sulfate reduction in normal marine and euxinic sediments compared to methanogenesis, plotted as a function of the rate of sediment accumulation. (Modified from Canfield, 1993.)

that range from microscopic aggregates and framboids, to mesoscale fecal pellets, to macroscale burrows, shell infillings, fossils, and concretionary layers (e.g., Canfield and Raiswell, 1991). Similar processes do not occur in freshwaters as hydrogen sulfide is seldom produced in sufficient abundance to alter the iron chemistry of the sediments. Reactions of HS^- and MnO_2 in marine sediments are of lesser importance in regulating the concentration of the former because manganese oxides are present in much smaller concentrations than ferric iron minerals, and there are no comparably insoluble manganese sulfides.

The amount and reactivity of the detrital iron minerals dictates how much pyrite will form, commonly referred to as the degree of pyritization (Raiswell et al., 1988). In slowly accumulating sediments, little labile organic matter is buried, sulfate reduction rates are slow, less HS^- is available for reaction with ferric oxyhydroxide, so less of the latter are eventually converted to pyrite. Since the solubility of iron monosulfides controls the pore water concentrations of dissolved Fe^{2+} and HS^-, limited hydrogen sulfide means that pore water Fe^{2+} will increase in concentration. By contrast, in rapidly accumulating sediments, more labile organics are buried, near-surface sulfate reduction rates are high, HS^- is abundant, and more of it is available for reaction with dissolved Fe^{2+} and ferric oxyhydroxide minerals (Berner and Raiswell, 1983). Nonetheless, as long as ferric oxyhydroxides are present in relatively high amounts in shallow, suboxic sediments, HS^- is rapidly removed from the pore waters to form iron monosulfide. Its concentration in suboxic sediment is thus kept negligible, even in the presence of active sulfate reduction. Accordingly, Fe^{2+} is able to accumulate in the pore waters due to the dissimilatory reduction of ferric oxyhydroxides and abiological reactions with HS^-. At some stage, when those ferric iron minerals become consumed with deeper burial, HS^- generation overwhelms Fe^{2+}, and the former accumulates in the pore waters.

Despite the generalizations above, not all ferric iron phases are pyritized. Thus far, discussions have assumed reactions with ferric oxyhydroxides, but in reality there are a number of other ferric iron minerals deposited. Some of those phases react very slowly with HS^-, and complete pyritization becomes impossible in the timespan over which sulfate reduction occurs in the sediment. In other words, a fraction of the detrital ferric iron minerals pass through the sulfate reduction zone faster than they can react with HS^-. Studies have shown that the half-lives of ferric iron minerals towards HS^- falls into three groups (Canfield et al., 1992; Poulton et al., 2004a). Ferric hydroxide and lepidocrocite have very short half lives (<1 day), and on a geological timescale, can be considered highly reactive towards dissolved sulfide (recall Fig. 6.22). This reactivity stems largely from higher surface area and lower interfacial free energies associated with these phases. Goethite, magnetite, and hematite are increasingly less reactive, with half-lives on the order of tens of days. The common sheet silicates (clays, biotite) and some primary silicates (pyroxenes, amphiboles, garnet) contain iron that is poorly reactive towards dissolved sulfide, with reaction half-lives approaching 100,000 years (Table 6.2). Such dramatic differences in reactivity help explain why 75–80% of the iron for pyrite formation comes from near-surface reactions with ferric hydroxide through to hematite, and after their consumption, HS^- builds up in deep pore waters. Then, very little additional pyrite formation takes place due to the poor reactivity of the residual iron silicates. At depth, HS^- concentrations decline significantly due to: (i) decreased rates of sulfate reduction; and (ii) subsequent reactions with the unreactive Fe pool. This causes Fe^{2+} concentrations to temporarily increase until addition of HS^-, from the thermal desulfurization of organic matter, occurs (Raiswell, 1997).

Only a fraction of the HS^- that forms via bacterial sulfate reduction is actually precipitated as pyrite sulfur (Berner, 1982). Instead, some 90% of the HS^- is re-oxidized to sulfate. Although O_2, NO_3^-, MnO_2, or Fe(III) are the ultimate oxidants in sedimentary systems, there also exists a dynamic S-subcycle that involves a number of sulfur intermediate products, with

Table 6.2 Half-lives of sedimentary iron minerals with respect to their reaction with dissolved sulfide. (Data from Canfield et al., 1992; Poulton et al., 2004a.)

Iron mineral	Half-life
Ferric hydroxide	5 min to 12 h
Lepidocrocite	11 h
Goethite	63 d
Magnetite	72 d
Hematite	182 d
"Reactive" silicates	230 yr
Sheet silicates	84,000 yr
Ilmenite, garnet, augite, amphibole	>84,000 yr

thiosulfate appearing to be the most significant. By the use of ^{35}S-tracer studies, Jørgensen (1990) demonstrated that thiosulfate constitutes between 68% and 78% of the immediate HS^- oxidation products, but it is subsequently: (i) oxidized to SO_4^{2-} by any of the oxidants listed above, including O_2 (reaction (6.27)) or ferric oxyhydroxide (reaction (6.28)); (ii) reduced back to HS^- by SRB metabolism (reaction (6.29)); or (iii) disproportionated to HS^- and SO_4^{2-} (reaction (6.30)). The relative importance of the three pathways varies with depth, from predominantly oxidative to reductive, but disproportionation is most important overall.

$$O_2 + 0.5S_2O_3^{2-} + 0.5H_2O \rightarrow SO_4^{2-} + H^+ \quad (6.27)$$

$$8Fe(OH)_3 + S_2O_3^{2-} + 14H^+ \rightarrow \\ 2SO_4^{2-} + 8Fe^{2+} + 19H_2O \quad (6.28)$$

$$CH_3COO^- + S_2O_3^{2-} + H^+ \rightarrow \\ 2HS^- + 2CO_2 + H_2O \quad (6.29)$$

$$S_2O_3^{2-} + OH^- \rightarrow SO_4^{2-} + HS^- \quad (6.30)$$

Much of the HS^- that ultimately escapes the thiosulfate subcycle provides a useable source of reducing power for various chemolithoautotrophs, most of which gain energy from oxidizing sulfide, with O_2 as the TEA (recall reaction (2.49)). Oxygen-sulfide reaction zones can be classified according to the degree of overlap between the

O_2 and HS^- profiles (Boudreau, 1991). In the first category, there is discernible overlap in the two species, a situation most frequently encountered in the water column of stratified bodies of water, where O_2 and HS^- mix along the chemocline. Second, as the zone of oxygen-sulfide overlap decreases in thickness, these species begin to disappear along a thin interfacial zone. This situation is typical of bacterial mats (e.g., *Beggiatoa*) growing on the surface of rapidly accreting sediments, i.e., where there are high levels of sulfate reduction in the sediments facilitating HS^- upward diffusion into oxygenated bottom waters. In the third category, O_2 and HS^- are separated by a buffer zone consisting of redox-sensitive intermediate species, such as Fe and Mn. In such systems, the oxidized solid forms of these metals are reduced (recall reaction (6.19) and Fig. 6.22), along with the oxidation of HS^- to S^0 under completely O_2-free conditions (e.g., Aller et al., 1986).

What the reactions above highlight is the close interrelationship between the Fe and S cycles, and how Fe(III) and sulfate reduction may work in concert within the same sedimentary layers. Significantly, the burial of organic matter and pyrite in sediments exerts a fundamental control on C-S-O cycles over geological time. Although they cover only 9% of the seafloor, sediments underlying shallow, but highly productive near-shore regions of the world oceans accumu-late nearly 90% of the total organic matter buried in sediments annually (recall Table 6.1). In these "normal marine" environments, where there is adequate detrital ferric oxyhydroxide minerals, and the amount and reactivity of organic matter limits the degree of pyritization, there tends to be a positive correlation between the % C and % pyrite incorporated into the marine sediments. In essence, a constant fraction of the initially deposited organic matter is degraded during sulfate reduction and the amount of this degradation dictates how much pyrite can be formed and how much organic matter is buried with it. This cor-relation varies in euxinic basins where a greater amount of pyrite is formed per unit of organic carbon buried because HS^- is abundant every-where beneath the chemocline, and pyrite can form from sedimenting detrital iron minerals even within the water column (Raiswell and Berner, 1985). As a result, organic carbon is not needed at any given location for pyrite formation, and this leads to the common observation of high pyrite sulfur values accompanying low organic carbon contents. The limiting factor in pyrite formation in these environments is iron availability.

Based on these variations, it is possible to show that the worldwide ratio of carbon to sulfur burial has varied considerably over geological time. From Fig. 6.24 we notice a lower C:S burial ratio in the early Palaeozoic as compared to today. This likely reflects more widespread bottom water anoxia, i.e., euxinic-type con-ditions, which is reflected in the rock record by the greater abundance of black shales from that time. By contrast, the large peak in C:S ratio for the Permian-Carboniferous might be ascribed to the rise in vascular land plants, which, because of their lower reactivity than marine plankton, coupled with limited sulfate availability in freshwater, would lead to increased organic matter burial in freshwater swamps and lakes, lower rates of sulfate reduction and pyrite formation, and ultimately a higher sediment C:S ratio (Raiswell and Berner, 1986). Collectively, these changes had important implications for global climate since for every mole of organic carbon or half-mole of pyrite sulfur buried, one mole of photosynthetically produced O_2 is left behind to react with rocks or remain in the atmosphere or oceans (e.g., Kump and Garrels, 1986). To complete the redox cycle, the pyrite and organic carbon are eventually uplifted and reacted with atmospheric O_2 to form dissolved SO_4^{2-} and CO_2, which are ultimately returned to the ocean and atmosphere.

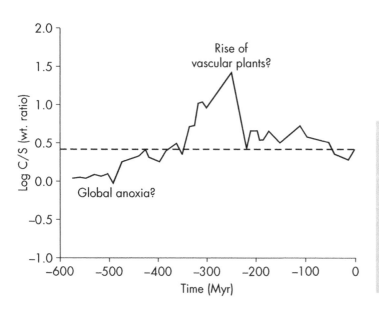

Figure 6.24 Plot of log C/S ratios versus time for the Phanerozoic. The C/S ratio refers to the mean worldwide weight ratio of organic carbon to pyrite sulfur at the time of burial for all sediments, not just those buried in normal marine environments. Dashed line represents modern C/S ratios for comparison. (Modified from Raiswell and Berner, 1986.)

(b) Methanogenesis

The terminal step in the anaerobic degradation of organic material is methanogenesis. On average, methanogenesis is responsible for 5–10 times less organic carbon degradation than sulfate reduction, and it is usually limited to sediments of rapidly depositing coastal environments, where the accumulation of labile organic material (i.e., fermentation products) exceeds sulfate availability (recall Fig. 6.23). This situation can come about when subsurface activity of SRB is higher than the rates of downward SO_4^{2-} diffusion, causing its concentration to diminish enough that methanogens develop a competitive advantage. Indeed, methane does not accumulate in sediments until more than 90% of the dissolved sulfate has been reduced (Martens and Berner, 1974). This is unsurprising given that the growth of methanogens is inhibited by SRB, because the latter have a higher affinity for the oxidizable substrates than do the former (Lovley et al., 1982).

The CH_4 generated by methanogens in the underlying sediment diffuses upwards towards the sulfate reduction zone, where it subsequently serves as an electron donor (see below for details):

$$CH_4 + SO_4^{2-} \rightarrow HCO_3^- + HS^- + H_2O \quad (6.31)$$
$$\Delta G^{\circ\prime} \text{ at pH } 7 = -17 \text{ k}$$

Although the above reaction is thermodynamically favorable, to date no single bacterium capable of the anaerobic oxidation of methane (AOM) has been isolated, and it appears as though AOM is the result of a syntrophic relationship between SRB and certain *Archaea* (recall section 2.5.3).

Recent studies of some marginal marine sedimentary environments have calculated that nearly 100% of the upward CH_4-flux can be oxidized anaerobically with sulfate (e.g., Niewöhner et al., 1998). Crucially, if these estimates can be demonstrated at a number of other sites, it might mean that anaerobic methane oxidation is overall a very important sink for sulfate in marine sediments, perhaps more so than that linked to organic carbon oxidation (D'Hondt et al., 2002).

The diffusive flux of CH_4 upwards is, however, usually inadequate to completely reduce pore water sulfate, and hence, accumulate above the methane–sulfate transition zone. Yet, it has been observed in some rapidly depositing shallow water sediments, where sulfate is depleted near the sediment–water interface (e.g., Cape Lookout Bight, North Carolina, and with sediment accumulation rates of 10 cm yr^{-1}), that methane saturates pore waters and ultimately escapes the sediment by bubble ebullition (Martens and Klump, 1984).

(c) Methanotrophy

Some of the hydrocarbon and nonhydrocarbon gases, generated in marine sediments through the bacterial (and thermogenic) degradation of buried organic material, also discharge from the seafloor at vents, mud volcanoes, and the so-called "cold seeps." Under high pressure (>5 MPa), low temperature (<10°C), and gas concentrations that exceed solubility, hydrocarbon gases can condense to form ice-like crystalline deposits, referred to as "gas hydrates", or methane hydrates when the natural gas is more than 99.9% methane. Individual gas hydrate deposits range from small grains and nodules irregularly disseminated in the pore spaces of sediments, to vein in-fillings in hemipelagic muds (Fig. 6.25) and fractured rock, to massive mounds that occasionally break through the sediment at seeps and vents. The largest fraction of the gas hydrate reservoir is buried beneath 200–300 m of sediment on continental slopes and subduction zones, where they can comprise 20% of the pore volume. Lesser amounts occur in the upper few meters of seafloor sediment. Collectively, gas hydrates represent an enormous sink of methane, estimated crudely at 10^4 Gt C, where Gt = 10^{15} g. To put this amount into perspective, the amount of carbon in hydrates is some two times that of conventional fossil fuel reserves on the planet (Kvenvolden, 1988).

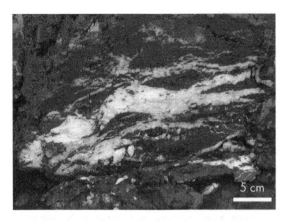

Figure 6.25 Photograph of pure white gas hydrate layers (containing >97% methane) infilling clayey sediment from Hydrate Ridge, on the Cascadia convergent margin off Oregon. (Courtesy of Erwin Suess and Antje Boetius.)

Gas hydrate stability depends largely on the right combination of pressure and temperature on the seafloor. Any decrease in pressure, through uplift or sealevel fall, or an increase in bottom water temperature, through global warming or altered ocean circulation, would cause the decomposition of shallow buried hydrates. This would not only destabilize the sediment by creating abnormally high porosity, but it could also lead to sediment slope failure or collapse and the sudden transfer of methane to the water column. Today, methane production in anoxic marine sediments is largely balanced by oxidation within the sediments themselves, and to some extent in the water column, with the result that on a global scale the ocean is a relatively minor source of methane to the atmosphere (Frea, 1984). However, if methane release from marine gas hydrates were to overwhelm oxidation, it would flood the upper oceanic and atmospheric carbon reservoirs, and major alterations to global climate might ensue. In this regard, decomposition of gas hydrates over broad regions of the oceans (and on land in regions of permafrost) may well have been responsible for dramatic changes in past global temperatures, as marked by the postglacial episode

of the Neoproterozoic (e.g., Kennedy et al., 2001) and during various times in the Phanerozoic (e.g., Kennett et al., 2000).

A potential buffer to methane efflux is microbial methane oxidation (methanotrophy). Within vast regions of the pelagic sediments, with low nutritive and POC input, hydrocarbon-based chemolithoautotrophy is the primary form of metabolism. Sure enough, recent studies have revealed that gas hydrates and cold seeps play host to a robust and diverse chemosynthetic community of mat-forming methanogenic, methylotrophic, sulfate-reducing and sulfur-oxidizing bacteria, with cell densities on the order of 10^9 cells cm^{-3} (Orcutt et al., 2004). As discussed above, the bulk of methane oxidation tends to lie at the base of the SO_4^{2-} reduction zone. On the seafloor this can be anywhere from beneath the microbial mats covering exposed hydrate mounds to several meters deep in the sediment. In euxinic basins, such as the Black Sea, the reaction zone can extend throughout much of the anoxic water column (e.g., Wakeham et al., 2003). Rates of methane-induced sulfate reduction can be extremely high, up to 600 times higher at Gulf of Mexico seeps than nonseep environments. The importance of this reaction pathway is highlighted by the strong sulfur isotope fractionations, with $\delta^{34}S$ values of residual sulfate up to +71‰ (Aharon and Fu, 2003). Interestingly, when rates of sulfate reduction are compared to methane oxidation, it is has been observed that the former are higher by at least an order of magnitude, suggesting that methanotrophy is likely CH_4-limited, and SRB are additionally fueled by the oxidation of crude oil or C_2–C_5 gases (e.g., ethane, propane, butane, and pentane; Joye et al., 2004).

Where ample amounts of hydrocarbons diffuse or advect upwards into shallow oxic sediment and bottom waters (or above the chemocline in stratified waters), the gases undergo aerobic oxidation by a variety of benthic and planktonic methanotrophs, respectively. These microorganisms can effectively drive methane to near undetectable concentrations (e.g., Valentine et al., 2001).

The oxidation of methane (and other hydrocarbon gas) under anoxic conditions leads to an increase in alkalinity (recall reaction (6.31)), and the subsequent precipitation of aragonite and Mg-calcites as: (i) isopachous cements that line cavities formed by bedding expansion or relict hydrates; and (ii) as carbonate nodules that tend to show repeated zonation with framboidal pyrite. The precipitation of carbonates at cold seeps can even produce topographic features on the seafloor with as much as several meters of relief (e.g., Fig. 6.26). Collectively, these precipitates not only serve as a major inorganic sink of carbon, but they also help stabilize the microbial community by cementation of the soft sediments and formation of a hard, solid substratum that can be used for the attachment of symbiotic macrofauna, such as tubeworms, clams and mussel colonies. Methane oxidation coupled to sulfate reduction further generates HS^-, and provides an energy source for sulfur-oxidizing bacteria, such as the genera *Beggiatoa*, *Thiothrix*, and *Thioploca*. Any excess HS^- reacts with dissolved or solid-phase iron to precipitate as iron monosulfide, and it is not uncommon to observe pyritized remains of bacteria (Sassen et al., 2004).

The diagenetic carbonate phases associated with hydrocarbon oxidation are typified by negative $\delta^{13}C$ values, reportedly as low as −60‰ (see Peckmann and Thiel, 2004 for details). The variable isotopic signatures reflect the mixing ratio between ocean DIC (at 0–2‰), highly negative DIC generated by the various metabolic pathways that preferentially oxidize ^{12}C- from ^{13}C-depleted biogenic gases, e.g., methane (as low as −110‰), and DIC from the thermogenic pool (−30‰ to −50‰). Some of the ^{13}C-depleted bicarbonate is also incorporated into various benthic and planktonic chemolithotrophic species, yielding biomass with $\delta^{13}C$ values of −100‰ (Orphan et al., 2001). Importantly, from a palaeoenvironmental perspective, some foraminifera are much more depleted in $\delta^{13}C$ than plankton that do not incorporate methane-derived DIC, and upon their sedimentation they yield horizons indicative of elevated methane levels in the water column (Thomas et al., 2002).

Relict methane-rich point sources have been identified in rocks ranging in age from Quaternary to Devonian. Limited data is available for hydrate or methane seep oxidation in the Precambrian, although distinctive negative

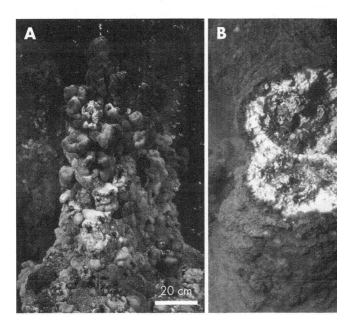

Figure 6.26 Microbial reef structures found at methane seeps in the Black Sea. (A) Tip of a chimney-like structure, with methane gas emanating into the water column. (B) Broken chimney about 1 m high showing gray outer microbial mat, with an inner pinkish layer of porous aragonite and calcite (seen as white), with a green interior core comprising carbonate and more microbial biomass. (From Michaelis et al., 2002. Reproduced with permission from Nature Publishing Group.)

isotopic excursions and sedimentary features in some cap carbonates suggest that methane hydrate decomposition might have induced a short-lived period of global warming that ended the "Snowball" glaciations of the late Neoproterozoic (e.g., Jiang et al., 2003).

6.2.6 Preservation of organic carbon into the sedimentary record

Although most organic carbon is continuously recycled within the water column and unconsolidated sediment, a fraction does become incorporated into sedimentary rocks. There it remains relatively inaccessible to microbial attack until re-exposed to bulk pore waters or atmospheric conditions through natural causes or human intervention. The quantity of organic carbon associated with sedimentary rocks is estimated to be as high as 2.5×10^{22} g C (Fenchel et al., 2000).

(a) Organic carbon preservation

Previous explanations for the enrichment of seemingly labile organic carbon in some marine deposits (black shales, hydrocarbon source rocks) were tied to the belief that increased carbon preservation occurred when POC was deposited under anoxic conditions. In other words, anaerobic decomposition of organic matter was intrinsically slower than aerobic decomposition. Despite being true for certain refractory substances (e.g., lignins, waxes, etc.), several experimental and field-based studies have now shown that differences between decomposition rates in oxic and anoxic sediments are relatively small (see Henrichs and Reeburgh, 1987 for details). In fact, this is what one might expect considering that anaerobic respiration rates typically exceed aerobic respiration rates because the anaerobes must oxidize larger amounts of substrate simply to gain the same amount of usable free energy and produce the same amount of biomass as their aerobic counterparts.

A more accepted view today is that organic preservation is linked to surface plankton pro-

ductivity, in that the rates of organic accretion under highly productive waters is so fast that a significant fraction of the organic matter is passed through the oxic zone before complete degradation (e.g., Pedersen and Calvert, 1990). Recall our previous discussions on rates of aerobic respiration – the slower the sedimentation rate the more time for oxidation, resulting in less preservation, while the opposite is true under faster sedimentation rates. It thus appears that the time organic matter resides in oxic sediments, rather than the total time since deposition, influences preservation. In support of this view is a compilation of organic preservation data suggesting that when sedimentation rates are greater than 0.04 g cm^{-2} yr^{-1} similar preservation is observed for deposition in both the presence and absence of O_2 (Fig. 6.27). Conversely, at sedimentation rates lower than 0.04 g cm^{-2} yr^{-1}, enhanced preservation is only observed in low-O_2 and euxinic sedimentary environments; at sedimentation rates lower than 10^{-3} g cm^{-2} yr^{-1}, less than 1% carbon

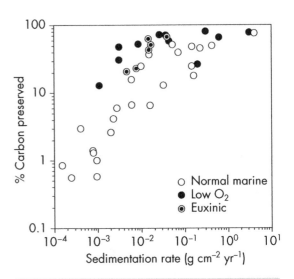

Figure 6.27 Percent organic carbon preservation as a function of sedimentation rate. Sediment types include: normal marine (>20 μmol L^{-1} bottom water O_2); low O_2 (0–20 μmol L^{-1} bottom water O_2); and those deposited in euxinic (0 μmol L^{-1} bottom-water O_2). (Modified from Canfield, 1994.)

is preserved (Canfield, 1994). There is, however, a caveat to these rate calculations. The higher preservational efficiencies (i.e., the ratio of carbon buried versus carbon deposited) in euxinic basins may have a significant estimation artefact because it is generally assumed that sulfate reduction is the predominant anoxic respiratory pathway, even though methanogenesis may be as, or even more, important in the total anaerobic mineralization budget. Consequently, the amount of organic matter mineralized in anoxic systems is typically underestimated (Middleburg et al., 1993). As a matter of fact, calculations suggest that carbon burial efficiency in Black Sea sediments is not significantly different from that in oxygenated environments with similar sedimentation rates (Calvert et al., 1991).

Another possible cause for preservation lies in the observation that the bulk of organic matter in sediments is physically associated with the fine-grained mineral fraction, either as lysed cells within their siliceous and calcareous skeletons or in association with detrital metal oxyhydroxides and clays. The latter are particularly reactive towards organic compounds, and it has been suggested that clay-organic aggregates facilitate the burial of intact, labile organic residues into the anoxic zone, presumably with the minerals functioning as protective coatings (Hedges and Keil, 1995). Moreover, upon compaction, some of the organic fraction becomes further insulated from the bulk pore waters, making them inaccessible to the action of many large hydrolytic enzymes (Mayer, 1993). This theory, however, needs to reconcile the fact that essentially all the fine-grained sediment deposited under oxygenated bottom waters is under macrofaunal influence (see below), which would likely disrupt the organic-mineral assemblage (Berner, 1995).

A third explanation is that there is an absence of macrofauna in anoxic sediments. In oxygenated waters these organisms can have a number of effects on deposited organic matter. First, grazing causes disintegration of intact organic remains, leading to higher surface areas for microbial oxidation. Second, by burrowing

they mix fresh organic matter with older, more decomposed materials, causing enhanced degradation of the latter. Third, ingestion of refractory organic carbon may lead to excretion of more labile fecal pellets (Lee, 1992). The depth in the sediments where animals are found is primarily limited by the magnitude of organic detritus and bottom water O_2 availability, with macrofaunal activity ceasing when O_2 levels reach values lower than 0.2 mg L^{-1}. The benthos are also inhibited in oxygenated environments when: (i) the sediment is too fine-grained and compacted to provide sufficient space for them to live; and (ii) when strong ebb and current flows redistribute sediment too quickly for burrows to become established (e.g., Van Cappellen and Gaillard, 1996).

Irrespective of above, once buried and subjected to increased temperatures and pressures, those organic remains are slowly transformed into nonhydrolyzable (insoluble) degradation products referred to as kerogen. Kerogen can be either labile (sapropelic) or refractory (humic), depending on the nature of the original organic material. Sapropelic kerogens that are lipid and protein-rich, i.e., sourced from marine biomass with high H:C ratios, are particularly important economically because they convert to various liquid and gaseous fossil fuels at a depth of several kilometers and a temperature between 50°C and 100°C (de Leeuw and Largeau, 1993). The most labile kerogens break down to form heavy hydrocarbons (e.g., oils), while more resistant kerogens break down to form light hydrocarbons (e.g., gases). The most refractory kerogens form graphite.

(b) Hydrocarbon-microorganism interactions

Microorganisms play a critical role in hydrocarbon formation during the early stages of diagenesis, as described above. In addition, those microorganisms indigenous to hydrocarbon reservoirs, through deposition with the original sediment, and that have survived ever since (predominantly fermenters, SRB, and methanogens), as well as those imported with groundwaters (aerobes), can continue to induce changes in hydrocarbon composition if circumstances favor their growth (Machel and Foght, 2000). For instance, aerobic respirers, usually restricted to the shallow subsurface, generate a series of compositional and isotopic modifications on crude oil, with gas and gasoline-range compounds preferentially oxidized to CO_2, while the oil fraction becomes more viscous, higher in density, and richer in sulfur and nitrogen. These transformations lead to heavy oil, asphalt, and tar, which often form a special class of unconventional resources known as "tar sands." Meanwhile, the injection of anoxic seawater to enhance secondary oil recovery stimulates growth of thermophilic sulfate reducers in the reservoir due to the high concentrations of sulfate artificially introduced into what was essentially a closed system (Nilsen et al., 1996). A similar situation occurs from the use of methanol as an oilfield solvent, with some acetogenic bacteria converting it into acetate, a substrate for the SRB communities. From the consideration of an oil company, SRB generation of HS^- is a potential problem for a number of reasons: (i) it causes corrosion of steel alloys in oil wells and in oil- and gas-processing systems; (ii) it contaminates hydrocarbons, changing "sweet crude" into "sour crude" over a period of weeks to months; and (iii) it leads to precipitation of secondary iron sulfides or calcium carbonates (via alkalinity generation), thereby reducing the permeability of the "reservoir" rock (e.g., Cord-Ruwisch et al., 1987).

The reduction of reservoir permeability inevitably causes the recovery yields to diminish to unprofitable levels. At this stage, tertiary recovery processes may be employed, involving steam injection to reduce viscosity, possibly coupled to biological methods aimed at improving hydrocarbon mobility. In this regard specific bacteria might be injected into the well to help liberate oil from rock surfaces by dissolving carbonate and sulfate minerals to which oil may adhere or by generating gases whose pressure could help force the hydrocarbons though less permeable strata (ZoBell, 1952). EPS production by a number of bacteria can also be used beneficially in many subsurface engineering problems including microbially enhanced oil recovery. In the oil industry, one area of concern is the undesirable flow of injected water through unconstrained pathways. As a remedy, EPS-producing bacteria are injected into oil-bearing reservoirs, where they are then supplied with nutrients for a finite period of time in order to grow and eventually form a plug. Due to this practice, the injected water is diverted into more desirable channels, pushing the oil out through production wells instead. One of the benefits of using such bacteria is that after the nutrient supply is terminated (when a desired permeability is attained) certain species are still able to survive starvation conditions and, hence, maintain the biological plug (e.g., Kim et al., 2000).

The association of microorganisms and fossil fuels extends to the bioremediation of surface oil spills and polluted aquifers. A variety of bacteria and fungi are able to grow directly on viscous oil films, where they oxidize the aliphatic and aromatic components (e.g., Fig. 6.28), and in doing so contribute to its decomposition and dispersal. The addition of nutrients (N, P) to oil spills has been shown to effectively increase the rates of bioremediation, such as in the case of the 41-million-liter crude oil spill from the supertanker *Exxon Valdez* in Prince William Sound, Alaska (Bragg et al., 1994). Most hydrocarbon-degrading bacteria tend to be aerobic respirers since the aliphatic fraction is not fermentable, and hence cannot be directly attacked by the majority of anaerobes (Atlas, 1981). Volatile hydrocarbons are similarly oxidized under oxic conditions, although as discussed above, anaerobic oxidation of methane and other gaseous hydrocarbons is also a significant process in marine sediments. In other instances, the ability to attack hydrocarbons stems not so much from the microorganisms being able to directly use those compounds as their source of carbon and energy, but instead the hydrocarbons are oxidized in a process known as co-metabolism. It necessitates that another compound, which

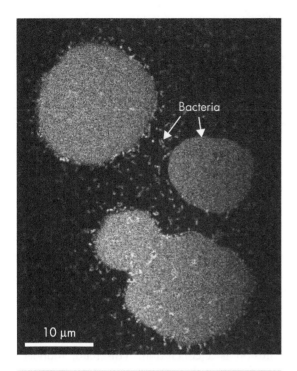

Bacteria

10 µm

Figure 6.28 Confocal micrograph showing fluorescent cells of *Rhodococcus erythropolis* (arrows) completely surrounding oil droplets. (From Dorobantu et al., 2004. Reproduced with permission from the American Society for Microbiology.)

may be quite unrelated, is the carbon and energy source, but its presence in some way permits the simultaneous oxidation of the hydrocarbon (Horvath, 1972).

A well publicized fossil fuel-related issue is groundwater contamination by benzene. It is a major environmental concern due to its toxicity and high solubility, thus significant attention has focused on microorganisms that can attenuate its dispersion into drinking water supplies. Fortunately, a number benzene-degrading aerobes have been identified, notably *Pseudomonas* species which can comprise the majority of gasoline-degrading bacteria in contaminated aquifers (Ridgway et al., 1990). Similarly, benzene degradation has been demonstrated in sediments with nitrate, ferric iron, sulfate, and carbon dioxide as alternate elec-

tron acceptors (e.g., Lovley et al., 1994), thus extending the possibility for remediation over a broad range of Eh conditions. The metabolic versatility displayed by hydrocarbon degrading bacteria lends itself well to genome-based remediation strategies, where attempts are being made to identify the genes that regulate the bioremediation reaction. A global analysis of which genes are being expressed under various conditions in a contaminated aquifer could reveal the metabolic status of the individuals in that community, and possibly by the addition of a trace nutrient, manipulate the population dynamics in favor of the species that might best accelerate the bioremediation process (Lovley, 2003).

6.2.7 Diagenetic mineralization

In the previous sections we outlined the major chemical and mineralogical changes that occur with depth through successive biogeochemical zones. However, both carbonate/phosphate mineral formation and amorphous silica dissolution-reprecipitation may occur in any biogeochemical zone, and these are worth describing separately.

(a) Calcium carbonates

The saturation state of sediment pore waters with respect to various carbonate minerals is directly related to the metabolic process responsible for organic matter oxidation. Aerobic respiration typically results in the complete oxidation of organic carbon to CO_2, thereby promoting the dissolution of biogenic carbonate both above and below the calcite saturation horizon (Archer et al., 1989). In fact, aerobic respiration has been shown to result in the dissolution of a large fraction of the calcium carbonate raining onto the seafloor from the calcifying plankton above (Emerson and Bender, 1981). Anaerobic respiration, on the other hand, releases HCO_3^- into the pore water system. Increased alkalinity can cause the equilibrium ion activity product for calcium carbonate to be exceeded, leading to the precipitation of early diagenetic carbonate

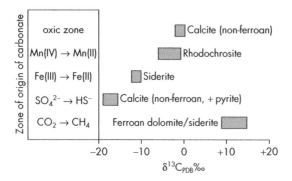

Figure 6.29 Carbon isotopic composition of various diagenetic carbonates formed during progressive burial in marine sediments. (Modified from Coleman, 1985.)

minerals that are relatively stable once formed, and not subject to rapid recycling by redox reactions in the same way as sulfides and oxides (Irwin et al., 1977). Such processes become most predominant at depth, in particular within the sulfate reduction zone.

A characteristic sequence of different terminal electron accepting processes can be distinguished on both mineralogical and stable isotopic grounds in environments where precipitation of carbonate minerals is possible (Fig. 6.29). Although aerobic respiration induces calcite dissolution, other aerobic processes can actually facilitate the precipitation of nonferroan calcite (no dissolved iron is present under oxic conditions), with a stable carbon isotopic composition ($\delta^{13}C$) of 0‰. For instance, chemolithoautotrophic oxidation of NH_4^+ and H_2S, at the base of the oxic zone, actually leads to a rise in pH because these reactions effectively remove protolytic species from solution, resulting in a pH higher than predicted from the reaction stoichiometries (Boudreau and Canfield, 1993). Nitrate reduction does not produce carbonate minerals with characteristic chemistry. Mn(IV) and Fe(III) reduction produce rhodochrosite and siderite, respectively, with more negative $\delta^{13}C$ values (typically −2‰ and −10‰, respectively) due to the incorporation of an increasing component of biogenic carbonate (CO_3^{2-}) that has a $\delta^{13}C$ signature of −20‰ to −30‰. Sulfate reduction produces nonferroan

calcite because any available iron reacts preferentially with sulfide instead. These calcites have an even more negative $\delta^{13}C$ signature (typically −15‰). Finally, below the zone of sulfate reduction, methanogenesis leads to the production of ferroan carbonates with a characteristic positive $\delta^{13}C$ signature due to the preferential removal of $\delta^{12}C$ in methane (Curtis et al., 1986). What is not shown in the figure are the isotopic signatures for carbonates associated with methane oxidation (up to −60‰, as described above).

During burial, the carbonate minerals formed may be subject to further alterations. Pore waters experience changes in temperature and CO_2 concentration with time, so that the precipitates are in contact with alternately unsaturated and supersaturated solutions. Thus ions of the mineral can move from one grain to another, permitting large crystals to grow at the expense of more soluble small ones. One product of such diffusion is a concretion. This hard, rounded mass forms by the precipitation of a cementing agent around some form of nucleus (Fig. 6.30A). Siderite cements are most prominent in freshwater and brackish environments, where alkalinity is generated in part by sulfate reduction, but Fe(III) reduction rates are higher to preclude sulfide mineralization. Conversely, pyrite and calcium carbonate are more often encountered in marine sediments. Not infrequently, a foreign object can be found at the center of a concretion – a fragment of shell, bone, plant material and, most unusually, old gun shell casings from World War II in estuaries along the Norfolk coast in England (e.g., Pye et al., 1990).

Once the nucleus is established, further growth proceeds when the ions comprising the concretion diffuse from bulk sediment pore waters towards the reactive site by either dissolution of existing minerals within the nearby sediment (diagenetic redistribution) or transportation from a more distant source. Depending on the supply of critical ions, rates of carbonate concretion formation can range from tens to thousands of years (see Berner, 1980 for details). It is also important to point out that the chemical composition, and even mineralogy, of the concretion

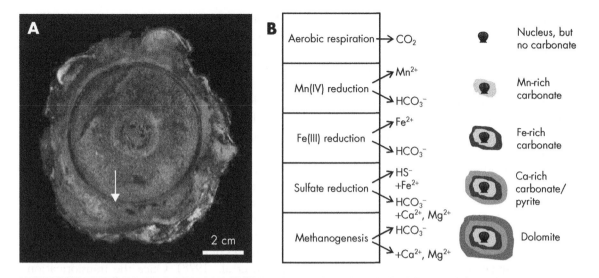

Figure 6.30 (A) A siderite concretion formed around an unexploded munitions shell (arrow) from intertidal marsh and sandflat sediments, north Norfolk, England (courtesy of Stuart Burley). (B) Idealized diagram of concretion formation, beginning with the presence of nuclei in the oxic zone (shell fragment, grains, etc.). As the concretion is buried through the successive biogeochemical zones, changes in pore water chemistry alter the concretions' composition and, often, their mineralogy (modified from Curtis et al., 1986).

can change with burial, or within the same sediment layers, as the concentrations of the constituent ions change (see Fisher et al., 1998 and Al-Agha et al., 1995 for variations based on freshwater and marine systems, respectively). For example, as Fe(III) reduction ceases due to consumption of available ferric oxyhydroxide minerals, Ca^{2+} cations may subsequently replace Fe^{2+}, leading to a Ca-rich siderite or calcium carbonate precipitation instead (Fig. 6.30B).

Water moving through the unconsolidated sediment may additionally carry dissolved ions capable of reacting with $CaCO_3$ grains. One such cation is Mg^{2+}, which under some conditions can react with calcite to form dolomite. Collectively, these processes gradually fill in the intergrain spaces and enlarge or replace existing crystals, leading to lithification (Krauskopf and Bird, 1995).

(b) Phosphates

During many of the respiratory processes, phosphate is released from the organic fraction into the pore waters. This does not, however, necessitate high dissolved phosphate concentrations. Instead, if its concentration in the sediment pore waters (usually in the form of HPO_4^{2-}) becomes sufficiently high, some form of diagenetic phosphate minerals will precipitate. The chemistry of phosphate in many ways resembles that of the carbonates, in that the parental acid, phosphoric acid (H_3PO_4), is a fairly weak acid (1st pK_a 2.1), and that the reaction of the dissociated bases, e.g., $H_2PO_4^-$ and HPO_4^{2-} (2nd pK_a 7.2) with metal cations, under acidic and alkaline pH conditions, respectively, leads to the formation of insoluble mineral phases. One such instance is where Ca^{2+} is added to a phosphate-rich solution, and either calcium fluorapatite or its metastable precursors may precipitate, depending on the pore water chemistry and state of supersaturation. Calcium fluorapatite has also been reported to form by the replacement of calcium carbonate by reaction with dissolved phosphate. The latter process is strikingly illustrated by solid-phase phosphate beds consisting of shelly fragments, with the original calcareous shell material being now entirely converted to fluorapatite (Glenn and Arthur, 1988).

At circumneutral pH, HPO_4^{2-} is a comparatively strong base, such that it reacts with carbonic acid (H_2CO_3) to form HCO_3^- and $H_2PO_4^-$. This increase in alkalinity leads to the paragenetic formation of calcium carbonate. Because sulfate reduction also generates alkalinity, pyrite formation in marine sediments is frequently coincident with the precipitation of carbonate and phosphate minerals.

Another important sink for dissolved phosphate is through reaction with ferric (and aluminum) oxyhydroxide minerals. They effectively adsorb HPO_4^{2-} from solution (e.g., reaction (6.32)), so the existence of such mineral phases in the uppermost sediments can act as a trap for upwardly diffusing phosphate (e.g., Froelich et al., 1988):

$$Fe(OH)_3 + HPO_4^{2-} \longleftrightarrow Fe(OH)\text{-}HPO_4 + 2OH^- \tag{6.32}$$

In part, this leads to increased sediment residence times for recycled phosphate relative to recycled nitrogen or carbon, and in some cases, the presence of surficial ferric oxyhydroxides may cause a net uptake of phosphate by sediment. Of course, this iron trap is not permanent, and as soon as the ferric oxyhydroxides are buried into the Fe(III) reduction zone, phosphate is re-liberated into the pore water system (Krom and Berner, 1981).

(c) Amorphous silica

Unlike the mineral authigenesis processes above, the Si-cycle in marine sediments is more based on a series of dissolution-reprecipitation reactions that take place during the sedimentation of amorphous silica shells from the overlying euphotic zone (Hurd, 1983). As discussed in section 4.2.3, over 95% of the shells settling through the water column and depositing onto the seafloor are recycled. The remaining shells are then incorporated into the sediment where they are diagenetically modified, beginning with their dissolution (Fig. 6.31A). If the rate of opal-A dissolution rapidly exceeds the slow rate of direct quartz nucleation, and if dissolved silica diffusion out of the sediment pore waters is minimized, silica concentrations in the interstitial waters begin to increase until they reach values in excess of the next most soluble phase, that being the metastable form of disordered cristobalite, known as opal-CT (Williams et al., 1985), or in the presence of aluminum and iron, clay minerals (e.g., Michalopoulos and Aller, 1995). In the first instance, as the shells are buried and dissolved, and subjected to slowly increasing temperatures (>30°C), eventual supersaturation with respect to opal-CT causes it to nucleate and crystallize in the form of dense spherical masses of the deposit known as porcellanite (Hein et al., 1978). Once the transformations are initiated, almost all of the biogenous opal-A shells will be converted to opal-CT (Fig. 6.31B), sometimes retaining the outline of the original diatom or radiolarian.

Under conditions close to the Earth's surface opal-CT may persist for millions of years because the silica polymers contain covalent bonds that have to be broken and reformed upon crystallization. However, at the elevated pressure and temperature conditions of deep burial (hundreds of meters in depth), the dissolution-reprecipitation process may proceed to the stage of quartz formation, often manifest in the form of bedded and/or nodular cherts (e.g., Hesse, 1990). Exceptions to the above transformations are, however, numerous, suggesting that temperature, pressure, and time are not the only important contributing factors. For instance, the overall pathway is strongly affected by the composition of the pore waters and of the host sediment, with clayey sediments promoting opal-CT, while in carbonate sediments quartz is more common (see Kastner et al., 1977 for details). As an interesting aside, similar patterns of silica re-crystallization have also been observed in freshwater, where the dissolution and reprecipitation of diatom shells, that grow as biofilms on submerged trees, leads to the in situ petrification of the outer wood surfaces whilst the trees are still living (Konhauser et al., 1992).

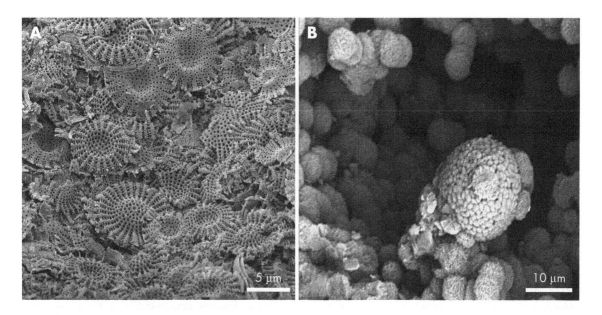

Figure 6.31 (A) SEM image of partially degraded planktonic diatoms (*Cyclostephanos* sp.) from Miocene laminated shales, near Kamloops, British Columbia, Canada (courtesy of Alex Wolfe). (B) SEM image of opal-CT lepispheres from chalk deposits at Dalbyover, Denmark (courtesy of Kåre Røsvik Jensen).

6.2.8 Sediment hydrogen concentrations

As discussed above, the dominant TEAs in sediments are generally segregated into distinct zones based on the potential free energy yields of the various metabolic processes. Yet, on a purely thermodynamic basis, reactions yielding less energy should also take place as long as they are energetically favorable. This implies that the segregation might instead be more accurately explained by competition between different types of microorganisms for electron donors, such as the fermentation products H_2 and acetate (Lovley and Klug, 1986). This premise is, to some extent, borne out by the observation that the rate-limiting step in dissimilatory metabolic pathways is the generation of the fermentative byproducts. Both H_2 and acetate have rather short turnover times (minutes to hours), which translates into low sediment pore water concentrations; H_2 concentrations are measured in nanomoles,

while acetate concentrations exist in the micromolar range (e.g., Novelli et al., 1988).

Given the limited availability of fermentation products at any given time, the key to a microorganism's survival then seems to lie in its ability to utilize the reduced substrates as quickly as possible. In this regard, it is noteworthy that those species with a thermodynamic advantage also tend to have a kinetic advantage (Postma and Jakobsen, 1996). In other words, the maximum potential rate of H_2 utilization, as an example, is equally important as the affinity for H_2 in determining the outcome of competition. Therefore, microorganisms utilizing electropositive TEAs effectively maintain H_2 concentrations below the capabilities of microorganisms using TEAs that yield less energy, and it is only when the electropositive TEA becomes depleted that the next most electropositive TEA controls the H_2 concentrations in the pore water system (Lovley and Goodwin, 1988). In sediments, this translates into distinct H_2 concentration gradients for each particular metabolism and a threshold level below which H_2 cannot be further metabolized. When nitrate and/or Mn(IV) reduction are the dominant terminal electron accepting pathways, pore water H_2 concentrations are extremely low (<0.05 nmol L^{-1}), with increasing levels through Fe(III) reduction (0.2 nmol L^{-1}), sulfate reduction (1–1.5 nmol L^{-1}), and methanogenesis (7–10 nmol L^{-1}). Acetate concentrations follow a similar pattern (Chapelle and Lovley, 1992).

As might be expected, this pattern of H_2 and acetate competition can be ascribed to the physiological capabilities of the microorganisms growing in the sediments. Microorganisms that use ferric hydroxide as their TEA can metabolize H_2 or acetate to concentrations lower than those that can be utilized by sulfate reducers, and the sulfate reducers can metabolize the same substrates to concentrations below those usable by methanogens (Lovley and Phillips, 1987a). These findings are consistent with observations of sediments where sulfate is not reduced, and methane is not produced until the reducible ferric hydroxide fraction has been converted to dissolved Fe^{2+}. Accordingly, when the availability of ferric hydroxides do not limit the rates of microbial Fe(III) reduction, sulfate reduction and CH_4 production from H_2 or acetate are inhibited. Crystalline iron oxides (hematite, magnetite) do not permit Fe(III) reduction to effectively compete for electron donors in sediment. Given the pattern above, it is not surprising that the availability of NO_3^- or MnO_2 diminishes the magnitude of Fe(III) reduction in suboxic sediment, and under such conditions some of the Fe(III)-reducers may even switch over to nitrate or Mn(IV) reduction when H_2 becomes limiting for them under their normal mode of growth (Lovley and Phillips, 1988a; diChristina, 1992).

6.2.9 Problems with the biogeochemical zone scheme

It is now widely accepted that a sequential scheme of biogeochemical zones is an oversimplification. This realization has come from a number observations, some of which are listed below (Fig. 6.32):

1 Natural aquatic environments contain complex microbial communities that exhibit uneven distributions, rather than a simple, vertically stratified, downward succession of "pure cultures." This means that as the populations aggregate into specific microniches, there is potential for overlap of the processes described above, leading to a mosaic of biogeochemical zones within a relatively homogeneous background. Take as examples the simultaneous occurrence of Fe(III) and sulfate reduction in the same sedimentary layer because of the wide range of iron oxide stabilities that results in different kinetic effects (Postma and Jakobsen, 1996). Similarly, active SRB communities exist within anoxic microniches of large organic particles and fecal pellets lying in oxic sediments (Jørgensen, 1977). An even more dramatic demonstration is the observation of active SRB communities in the oxic sediments (Jørgensen and Bak, 1991). The potential for high rates of sulfate reduction under oxic conditions is likely explained by their ability to not only tolerate oxygen for prolonged periods of time, but to carry out their anaerobic reactions even in the presence of oxygen (Dillig and Cypionka, 1990).

2 Bacteria show incredible metabolic plasticity, being able to completely switch their TEAs when their preferable substrates are consumed or as environmental conditions change. Some bacteria can use TEAs that are more energetic when available, i.e., sulfate reducers that utilize Fe(III), while other bacteria can "downgrade" when their preferred TEA is exhausted, i.e., nitrate-reducing bacteria switching to Fe(III) reduction.

3 Many of the microbial and abiological reactions also occur within close proximity, making it very difficult to discern the effects of each reaction from bulk chemical analyses, especially when their net effects cancel out one another. Recall the complex interactions between the Mn and N cycles in coastal sediments, whereby upwardly diffusing Mn^{2+} reduces NO_3^- to N_2, and the MnO_2 that forms is then rapidly re-reduced by NH_4^+ to re-generate more NO_3^- or N_2. Aside from the anomalously high N_2 flux measured, the actual formation of MnO_2 might be completely obscured without extremely detailed study (Luther et al., 1997). Moreover, anoxic NO_3^- production during Mn(IV) reduction indicates the

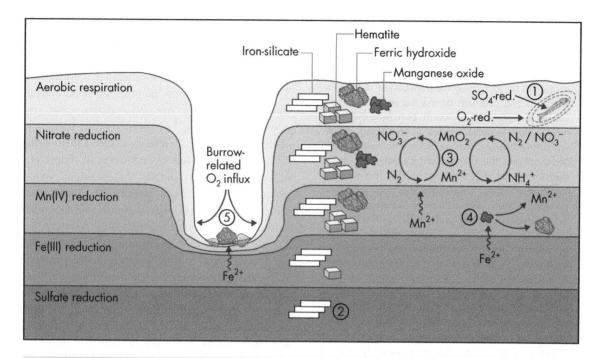

Figure 6.32 Some of the many ways in which the idealized one-dimensional vertical succession of TEAPs in a marine sequence can be complicated by physical and biological disruptions. This includes (1) aggregates of microbial communities growing in juxtaposition to one another, such as SRB in the oxic zone; (2) metabolic pathways occurring outside their "normal" depth, i.e., Fe(III) reduction in anoxic sediments due to the poor reactivity of some of the Fe minerals; (3) subcycles within a given sedimentary layer that obscure the actual processes taking place, e.g., the Mn- and N-cycles; (4) abiological reactions, for instance, the reduction of MnO_2 by upwardly diffusing Fe^{2+}; and (5) bioturbation increasing the flux of oxidants (e.g., O_2) into suboxic-anoxic layers of sediment that facilitate reactions, e.g., Fe(II) oxidation that would not normally take place because the upwardly diffusing Fe^{2+} would be consumed by reaction with MnO_2. It also mixes fresh organic matter into deeper sediment.

existence of a subsurface internal nitrification–denitrification cycle, with the small net production of NO_3^- leading to an underestimate of true denitrification rates from NO_3^- profiles alone (Hulth et al., 1999).

4 Along similar lines to point 3, it is very difficult to untangle the biological versus the abiological reactions taking place (e.g., Canfield et al., 1993). As discussed earlier, the reductants generated from anaerobic respiration (HS^-, Fe^{2+}, Mn^{2+}, NH_4^+) can each independently diffuse upwards and reduce an overlying electron acceptor. So, any sulfate-reducing bacteria will be able to indirectly reduce available ferric oxyhydroxides to Fe^{2+} via upwardly diffusing HS^-, with the sulfide being converted into SO_4^{2-}. Some of the Fe^{2+}, in turn, diffuses upwards to

reduce MnO_2; Mn^{2+} then reduces NO_3^-; and NH_4^+ ultimately is oxidized by O_2. Thus, a cascade of redox reactions can abiologically take place through the entire sediment column by a combined "electron shuttle," beginning simply with the independent action of a community of SRB in deep anoxic sediment. In sediments lacking the full suite of biogeochemical zones, HS^- may directly reduce any of the overlying oxidants. In terms of pore water analyses, it might then be possible to measure low dissolved HS^-, despite active sulfate reduction. In a similar manner, low Fe^{2+} concentrations are not necessarily indicative of low Fe(III) reduction rates since the Fe^{2+} readily becomes incorporated into sulfide, carbonate and phosphate mineral phases, as well as being re-oxidized back to $Fe(OH)_3$ by MnO_2 or nitrate.

5 Physical and biological disruption of the surface
sedimentary layers not only affects the organic
properties (as discussed above), but it also alters the
normal sequence of solid-phase and pore water
properties by increasing the transport of diagenetic
reactants and products across the sediment–water
interface. Irrigation of worm burrow tubes with
oxygenated seawater results in increased downward and lateral diffusion of O_2 into surrounding
sediment and the concomitant oxidation of pyrite
and enhanced diffusive loss of HS^- (Berner and
Westrich, 1985). Subsurface flushing with oxygenated waters also influences nitrification, and
the oxidation of Fe(II) and Mn(II). In the case of
Fe cycling, as long as ferric oxyhydroxides persist,
phosphate remains adsorbed to the mineral surface and is thus unavailable to the pore waters.
Consequently, surficial sediments cannot be considered a homogeneous medium dominated by
one-dimensional vertical diffusion, but rather they
are heterogeneous and intensively mixed, with a
convoluted spatial distribution of biogeochemical
zones that are impossible to accurately resolve
during sampling (Aller, 1980).

Our understanding of sediment biogeochemistry, and the intricate subtleties of microniches, is advancing rapidly with the advent
of new geochemical and molecular techniques.
The traditional means of pore water sampling
through sectioning core and centrifugation are
now, when suitable, being replaced with microelectrode profiling. These glass electrodes, with
tips as small as a few micrometers in diameter,
make it possible to analyze the chemical environment (e.g., O_2, H_2S, H_2, pH, Eh) and some
microbial processes, such as photosynthesis and
aerobic respiration, in dimensions relevant to
the life of the microbial community. Nonspecific
electrodes, that employ scanning voltammetry,
have also proven useful for measuring Fe and
Mn concentrations on small spatial scales. Such
methods, for example, have revolutionized our
understanding of how the hydrodynamic boundary layer and bioturbation impact the diffusion
of O_2 across the sediment–water interface (e.g.,

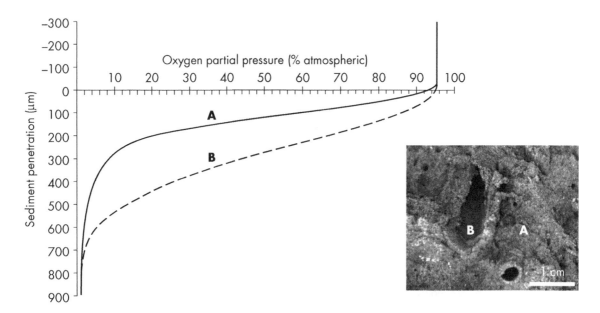

Figure 6.33 Oxygen microsensor profiles illustrating the average oxygen penetration into a split core of anoxic, highly bioturbated sediment obtained from the tidal flats of Willipa Bay, Washington, USA. Inset shows the unlined burrow of a *Maldanid* worm, with labels A and B marking the points where the profiles where obtained from the surrounding sediment and burrow wall, respectively. The corresponding profiles show the increased diffusion of oxygen into the burrow wall relative to the surrounding matrix. (Courtesy of Marilyn Zorn and Stefan Lalonde.)

Fig. 6.33), which then affects the oxidation of redox-active elements. Similarly, development of polyacrylamide gels in the 1990s now permits the measuring of pore water ions *in situ*, with diffusive equilibria established within minutes of the thin film of gel being inserted into the sediment (Davison et al., 1991).

In terms of the microbiology, the routine use of signature lipid biomarkers and stable isotope probing have made it possible to distinguish

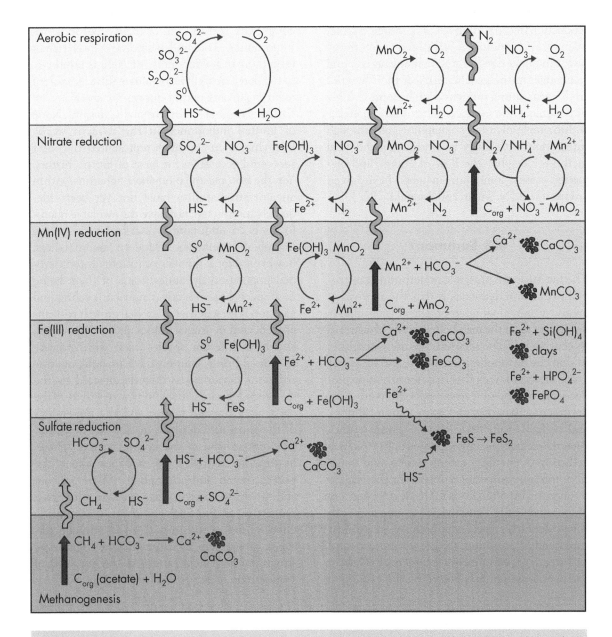

Figure 6.34 Some of the complex elemental cycling that takes place with marine sediments. Notice how within each biogeochemical zone there is a continuous drive towards reducing the overall Eh of the system, but with the reduced metabolites diffusing upwards to regenerate their oxidized counterparts.

in situ microbial biomass and metabolic activity (e.g., White, 1993), while nucleic acid probes are being used to more fully characterize community structure (e.g., Boetius et al., 2000). The most promising developments, however, are coming from the application of functional genomics and proteomics to assess gene expression and protein production (recall section 1.4). For instance, DNA microarray technology is being used to determine how (and how rapidly) microorganisms respond to changes in the geochemistry of their habitat. In the near future, it might even become feasible to document the gene content of all organisms within simple microbial communities, and through integration with high resolution geochemical methods, decipher the flow of energy and elements between community members in any given environment (Newman and Banfield, 2002).

6.3 Summary

The fundamental principle underpinning microbial population dynamics is that the survival of an given individual microorganism is ultimately dependent upon the metabolic activity of others in its ecosystem. Some relationships are synergistic, in that the metabolic waste of one species serves as the substrate for another (e.g., fermentation produces H_2 for anaerobic respirers). Others are more detrimental, whereby the wastes are inhibitory, and perhaps even toxic (e.g., O_2 from cyanobacteria excludes anaerobic phototrophs). As each cell requires energy, a carbon source, and nutrition, intense competition ensues for the limited resources. This often leads to the development of biogeochemically stratified environments, with a few dominant species effectively controlling the chemical conditions (redox, saturation state) within any given microenvironment, but where the overall system is balanced by the cycling of

elements between oxidized and reduced reservoirs (e.g., Fig. 6.34). Stratification can develop on widely different scales, from millimeter-thick microbial mats to marine sediments that can be measured in terms of tens to hundreds of meters. Recent advances in analytical techniques are now making it possible to better observe the internal complexity within each of these environments. Importantly, as our understanding of modern mat formation and sedimentary diagenesis improves, so too does our ability to evaluate the role ancient microorganisms played during the evolution of Earth's biosphere. For instance, despite being of limited importance in the modern world, lithifying microbial mats represented the most successful ecosystem for most of Earth's history. Yet, the isotopic and biomarker signatures within ancient stromatolites have not yet been adequately interpreted, nor have the available microfossils been undisputedly ascribed to a specific microbial assemblage. Modern microbialites, however, offer a number of insights into microbial population dynamics, some of them being how specific species affect nutrient cycling and electron flow, whether microbial growth patterns are reflected in unique fabrics and morphologies, and how different species cope with growing in mineralizing solutions. Meanwhile, marine sediment diagenesis is directly affected by the materials deposited, including the reactivity of the particulate organic matter and the mineralogy of the inorganic detritus, as well as the chemical composition of the bottom waters. These, in turn, directly reflect the oxidation state of the water column and atmosphere. Therefore, our ability to relate modern pore water and solid-phase properties to specific microbial communities offers us the opportunity to assess through isotopes, biomarkers, etc., the ancient community structure and paleodepositional environment of sedimentary rocks prior to lithification.

7

Early microbial life

Under what conditions did prebiotic organic monomers form and how did they polymerize? When did the first life form evolve, what was it, and how did it metabolize? How has biological innovation through time impacted the composition of the Earth's surface? These are just some of today's many unresolved questions. What we do know, however, is that life must have begun under conditions considerably different from today. Imagine an environment with surface temperatures approaching 100°C; a steamy atmosphere with over 1000 times the present level of carbon dioxide, high levels of sulfur dioxide, elemental sulfur smog and carbon monoxide, virtually no oxygen, and hence little ozone that could attenuate incoming UV radiation; an expansive ocean with abundant ferrous iron, silica, and ammonium; less landmass than present and no continents in the modern sense; and a surface subject to heavy bombardment by extraterrestrial material. This was the Hadean Earth between 4.4 and 4.0 billion years ago (Ga), when quite likely the beginnings of life took place. In this chapter we will consider how the physical and chemical conditions on the early Earth gave rise to the emergence of life, and the evolution of the biosphere. Supporting evidence for this journey back in time comes from an extremely limited suite of ancient geological samples, and unfortunately, analyses based on the chemical and isotopic composition of their minerals and organic remains, the few microfossils, and the fabric of some potential organosedimentary structures has led to interpretations fraught with controversy. Coupled to the physical evidence is the extrapolation of biochemical and molecular information from modern microorganisms. But there again, incorporating those findings into phylogenetic trees has proven equally contentious, and has led to highly ambiguous results. As a result, various aspects of the model outlined below are necessarily speculative and should not be viewed dogmatically, but instead as the favored consensus of what the majority of the scientific community believes presently.

7.1 The prebiotic Earth

The Earth formed by accretion some 4.5 billion years ago. In the first 100 million years thereafter a differentiated planet was forged with a liquid iron core, a highly convective mantle, and a primitive crust. During this transient phase the mass of mantle cycled to the near surface was comparable to the total mass thus cycled during all subsequent Earth history (Stevenson, 1983). Along with the eruption of molten material at the surface, extensive outgassing of volatiles ultimately led to the formation of the primordial atmosphere. As the accretionary energy input waned, surface temperatures dropped to below 100°C and water condensed to form the primordial oceans. The early atmospheres and oceans may have been reprocessed several times by episodic bombardment, with lost volatiles continuously replenished by mantle outgassing and the addition of volatile-rich meteoritic and cometary material. Some estimates suggest that the Earth may have acquired as much as 10^{21} L of water between 4.5 and 3.8 Ga from cometary collisions alone, essentially equivalent to all the water in modern oceans (Chyba, 1987). It was only when the energy flux from impacts became sufficiently diminished that a relatively cool surface with a stable atmosphere and ocean could persist. Based on noble gas isotopic ratios from mid-ocean ridge basalts, the current atmosphere may be as old as 4.4 billion years (Allègre et al., 1987), while oxygen isotopic ratios in 4.4 billion year old zircons suggest that liquid water was also present at that time (Valley et al., 2002).

7.1.1 The Hadean environment

As a consequence of rapid tectonic recycling, large amounts of mantle carbon were released to the Hadean atmosphere. The atmosphere immediately after accretion was reducing with $CO > CH_4 > CO_2$ and abundant H_2 (Chang, 1994). This reflected chemical equilibrium with reduced iron phases in the mantle, e.g., iron–wustite assemblage, and the pyrolysis of accreted organic compounds from extraterrestrial impacts. With such a reduced mantle redox state, as much as half of the subducted water could have been converted to H_2 prior to being outgassed. Most of the hydrogen would subsequently have been lost to space, while the O_2 left behind would have oxidized minerals in the newly formed crust and been carried down towards the mantle through subduction (Kasting et al., 1993). As the mantle became progressively oxidized, perhaps within the first 200 million years after accretion (Delano, 2001), the gases released by volcanic activity would have reflected a more neutral redox state, that being $CO_2 > CO > CH_4$, similar to the order of gases released from volcanoes today.

(a) The atmosphere

Carbon dioxide partial pressures may have been as high as tens of bars immediately after accretion, and remained up to several bars until the early Archean, some 4.0 Ga. Taking into account that solar luminosity at that time was an estimated 30% lower compared to today, such a CO_2-rich

atmosphere would have produced a mean global surface temperature of around 85°C (Kasting and Ackerman, 1986). Evidence in support of a hot Archean atmosphere has since been provided in the form of $\delta^{18}O$ compositions of 3.5–3.2 Gyr cherts in the Barberton greenstone belt, South Africa (Knauth and Lowe, 2003). Carbon monoxide was also likely present at higher levels than today. Despite its propensity for being photochemically oxidized, CO could have been produced in copious quantities early in Earth's history by vaporizing CO- and organic-rich comets or Fe^0-containing chondrite meteorites in a CO_2-dominated atmosphere (Kasting, 1990). Methane would initially have been degassed at substantial rates as long as the primitive mantle was more reducing than today, but after core segregation, any accumulation thereafter would have been dissipated by photochemical reactions in the atmosphere in less than 10^4 years. Thus, in the absence of methanogenesis, methane concentrations would have remained negligible (Kasting et al., 1983).

Sulfur would have been emitted to the atmosphere from volcanoes, mostly as H_2S or SO_2. Both gases could then have been photochemically converted to a number of other sulfur species, one of them being elemental sulfur (Kasting et al., 1989). Importantly, if the surface temperature of the early Earth was considerably warmer than today, that sulfur may have been in the form of a vapor which could have acted as a UV shield.

Nitrogen gas would have degassed due to its low solubility in magmas, and on the early Earth, the partial pressure of N_2 in the atmosphere may have been approximately 1 bar (Wen et al., 1989). Some of this N_2 was fixed as nitric oxide in lightning, while some may have reacted with hot magma below subaerial volcanoes and underwent thermally catalyzed reactions producing similar N-species (e.g., Mather et al., 2004). What happened to the ammonia released during early outgassing is of greater uncertainty. Photochemical kinetics suggest that ammonia would have been rapidly photo-oxidized to N_2 within decades (Kuhn and Atreya, 1979).

Conversely, the sulfur vapors may have served to absorb UV radiation and allowed ammonia to persist for much longer times (Kasting, 1982).

Oxygen would also have been produced in the upper atmosphere through various photochemical reactions (Canuto et al., 1982), although its concentrations would have been kept low due to reactions with reduced atmospheric gases and the Earth's crust (Walker et al., 1983).

(b) The oceans

The initial composition of the oceans was dominated by seawater circulation through deep-sea hydrothermal systems. This process generated reduced, solute-rich waters that would have been replenished on a timescale shorter than 10^7 years, the time needed for the entire ocean to pass through the vents (Wolery and Sleep, 1976). The hydrothermal inputs would have been complemented by atmospheric contributions. An atmosphere with several bars of CO_2, and perhaps as much as 1 mol L^{-1} HCl, would have acidified rainwater to a pH < 1. This, in turn, would have led to vigorous crustal weathering, the release of a significant flux of solutes, and eventually the formation of a saline ocean with Na > Ca > Mg > K (Garrels and Mackenzie, 1971). Much of the CO_2 would also have been buffered through ocean crust carbonatization (i.e., replacement by carbonates), resulting in high bicarbonate concentrations, and possibly elevating the pH to values as high as 6 by the Early Archaen (e.g., Grotzinger and Kasting, 1993). Even after significant amounts of continental crust had formed, the presence of hydrothermal carbonates in Archean greenstone belts indicate that ocean crust carbonatization was still an important CO_2 sink (Walker, 1990).

Some of the volcanogenic SO_2 would have dissolved in the oceans, yielding bisulfite (HSO_3^-) and sulfite (SO_3^{2-}), a fraction of which may have undergone secondary photochemical oxidation in the surface waters or disproportionation to sulfate. Elemental sulfur would similarly have been rained out of the atmosphere upon transformation into sulfate by photochemical reactions.

Despite these sources, sulfate concentration would have been low by modern standards because of the absence of river-borne fluxes, coupled with its loss via reaction of seawater with hot basalt in seafloor hydrothermal systems (Walker and Brimblecombe, 1985).

If the ocean pH was circumneutral, then much of the nitrogen oxides rained out into the oceans could have been reduced by dissolved Fe^{2+} to yield high NH_4^+ concentrations (Summers and Chang, 1993). It has also been proposed that NH_4^+ could have been produced by reduction of N_2 in crustal and hydrothermal systems at temperatures between 300 and 800°C (Brandes et al., 1998). Certainly, in the absence of O_2 in the Hadean water column, NH_4^+ must have been the dominant dissolved N-species.

(c) The landmass

The internal tectonics and degassing mechanisms on the early Hadean Earth were unique from the rest of its history. During the first 10^8 years, heavy meteorite bombardment and vigorous mantle convection prevented the emplacement of a large, continuous, solid lithosphere, thick enough to support material differentiated from the mantle. Estimates place the rate of mantle overturn on the order of 10^5 years during that time, so even if a large volume of crust had formed, it was rapidly recycled back into the mantle (Stevenson, 1983). It was only when the final vestiges of very hot material in the mantle erupted that plate tectonics became the dominant mode of heat transfer from the Earth's interior. The crust produced during this stage must have been primarily komatiitic (formed of ultramafic lava), with sialic crust (rich in silica and alumina) formed later via fractional crystallization from extensive magma chambers and by the re-melting of the komatiites. The principal land masses consisted of low-relief mafic volcanic islands developed along spreading centers and subduction zones, and over intraplate thermal plumes or hot spots (Lowe, 1994).

As the Earth continued to cool, recycling slowed, microcontinental blocks with a more rigid and buoyant crust formed, and a stable tetonic system began to take shape. This had occurred by at least 4.1 Ga, as reflected by the oldest known rocks in existence, the Acasta gneiss from the Northwest Territories, Canada (Bowring et al., 1989), although the chemical composition of recently discovered 4.4 Gyr detrital zircons confuses the story because it implies that differentiated crust might already have been in existence during the early Hadean (Wilde et al., 2001). The intermediate to granitic continental rocks subsequently provided platforms for the deposition of sediment, the purported oldest sedimentary rocks being the 3.8 Gyr Isua-Akilia greenstone belts of SW Greenland, that consist of volcaniclastic material, clastic turbidites, and chemical precipitates such as banded iron formations (see below) and chert (Nutman et al., 1997).

7.1.2 Origins of life

Beginning in 1871 with Charles Darwin's infamous allusion to life's beginnings in a "warm little pond," it has subsequently been argued that the uppermost few hundred meters of the Hadean ocean were a rich mixture of organic monomers available for the ascendancy of life, the so-called "Organic Soup" model. Their origins were diverse, including those that formed on the primordial Earth, from chemical reactions between simple molecules in the atmosphere and hydrosphere,

as well as those delivered as components of extraterrestrial material (Fig. 7.1). In the past twenty years, a new hypothesis has gained wider acceptance, that being life began at hydrothermal vents. This view is supported by the apparent antiquity of hyperthermophilic microorganisms and the theoretical ease with which organic compounds form at the interface between warm to hot, reducing hydrothermal fluids and cooler, relatively oxidizing seawater, especially in the presence of catalyzing mineral surfaces that abound at such sites. A third theory, that of Panspermia, was briefly mentioned in Chapter 1, but given its limited support, it is not covered in any more detail here.

(a) The Organic Soup model

UV was potentially the largest source of energy for organic synthesis on the primordial Earth (Table 7.1), but its effectiveness was limited since most photochemical products would be formed in the upper atmosphere only to be decomposed by long wavelength UV before reaching the oceans. By contrast, simple organic compounds synthesized by electrical discharges would have formed close to the Earth's surface, allowing them to potentially reach the oceans before being photodissociated (Miller and Orgel, 1974). The first successful synthesis of amino

Table 7.1 Comparison of various potential sources of organic compounds on Earth around 4.0 Ga.

Source	Organic production: reduced atmosphere (kg yr^{-1})	Organic production: neutral atmosphere (kg yr^{-1})
Lightning	3×10^9	3×10^7
Ultraviolet light	2×10^{11}	3×10^8
Atmospheric shocks from meteors	1×10^9	3×10^1
Atmospheric shocks from postimpact plumes	2×10^{10}	4×10^2
Interplanetary dust particles	6×10^7	6×10^7
Hydrothermal synthesis	2×10^8	2×10^8

Note: Hydrothermal synthesis data from Shock (1992), all other data from Chyba and Sagan (1992).

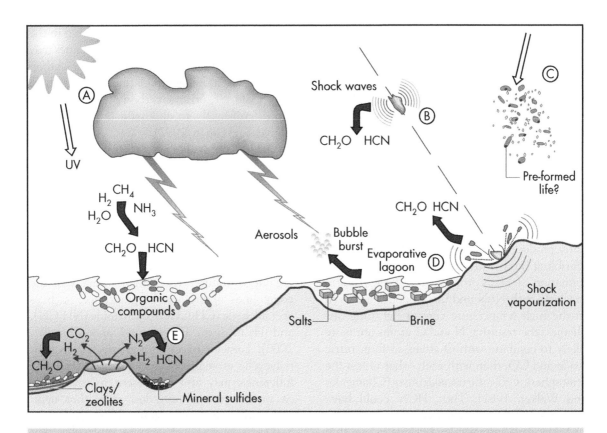

Figure 7.1 The various mechanisms by which organic compounds may have arisen on Earth. (A) Key prebiotic monomers (CH_2O and HCN) were formed when reduced gases in the atmosphere were subjected to UV irradiation and electrical discharges. They then condensed to form more complex compounds, such as amino acids. (B) Meteorites passing through the atmosphere at hypersonic speeds could also form CH_2O or HCN either through shock waves that ionize the reduced atmospheric gases or from shock vaporization after impacting land. (C) Extraterrestrial micrometeorites, carbonaceous chondrites, comets, and interstellar dust falling to Earth could have brought extraterrestrial organic compounds, while some hypothesize that they may even have included pre-formed life, i.e., Panspermia. (D) Evaporation of monomers may have led to polymerization, while bubble burst sent spray into the atmosphere, where the nuclei were concentrated by evaporation and altered by irradiation and lightning. (E) Reduced gases, emitted at hydrothermal vents, reacted with more oxidizing seawater to form HCN, CH_2O, and some more complex organic compounds. Mineral substrata could have served as templates for further polymerization reactions. Note, horizontal not to scale.

acids, under what was once considered plausible prebiotic conditions, was accomplished by subjecting a reduced gaseous mixture of CH_4, H_2, H_2O, and NH_3 to electric discharges (Miller, 1953). The amino acids were synthesized by what is known as the Strecker condensation. In this reaction two key prebiotic compounds, hydrogen cyanide (HCN) and formaldehyde (CH_2O), both of which are formed in electrical discharges as well, condense with each other, in the presence of ammonia, to sequentially form amino nitriles and amino amides, that upon hydrolysis would yield amino acids (Miller, 1957):

$$RCHO + HCN + NH_3 + H_2O \rightarrow$$
$$RCH(NH_2)COOH + NH_3 \qquad (7.1)$$

Since those early experiments, several other studies have shown alternate pathways for amino acid synthesis, as well as the feasibility of forming some of the requisite components of nucleic acids, i.e., the formation of ribose ($C_5H_{10}O_5$) from CH_2O, and adenine ($C_5H_5N_5$) from HCN (see Oró, 1994 for review).

If atmospheric gases were the source of pre-biotic organic compounds, synthesis would have been favored in a highly reducing atmosphere because it theoretically generates a wider variety, and greater yield, of organic compounds than mildly reducing and neutral atmospheres, i.e., 2×10^{11} kg yr^{-1} versus 3×10^8 kg yr^{-1} (Chyba and Sagan, 1992). Some key components in organic synthesis, such as HCN, are also not efficiently produced under neutral redox conditions (Stribling and Miller, 1987). For example, formation of HCN from N_2 and CO_2 requires breaking both N≡N and C=O bonds, and even in the high temperature core of lightning discharges, the resulting N and C atoms are more likely to combine with O atoms to form nitric oxide and CO_2 than with each other unless the atmospheric C:O ratio exceeds unity (Chameides and Walker, 1981). Thus, HCN could have been produced quite efficiently in a CH_4- or CO-dominated atmosphere but not in one that consisted mostly of CO_2.

Another potential energy source on the early Earth was the shock waves of incoming extra-terrestrial bolides. Theoretically, the passage of hypersonic meteorites at velocities greater than 7 km s^{-1} can generate temperatures up to 10,000°K. This is sufficiently high to ionize atmospheric gases and yield a variety of secondary organic compounds (Gilvarry and Hochstim, 1963). Significant amounts of HCN and CH_2O could have been generated from just a few large impacts every 1–5 years (Fegley et al., 1986). Experiments have since formed small amounts of HCN, aldehydes, and amino acids from shock vaporization of meteorites in a reduced gas mixture of CH_4, C_2H_6, NH_3, and H_2O (e.g., Mukhin et al., 1989). Although the energy from ultraviolet radiation is estimated at being around 1000 times more abundant than shock waves, the latter were perhaps 10^6 times more efficient at producing amino acids (Bar-Nun et al., 1970). In the

reducing atmosphere that initially existed in the early Hadean, shocks by postimpact plumes may have produced up to 2×10^{10} kg organic carbon yr^{-1}, but in more neutral atmospheres, where $[H_2]/[CO_2] < 1$, organic carbon production rates drop significantly to approximately 400 kg yr^{-1} (Chyba and Sagan, 1992).

The low yields of organic material synthesized under nonreducing, CO_2-rich atmospheric conditions has led to the suggestion that extra-terrestrial material may instead have provided a more suitable source of organic compounds (Oró, 1961). Well over a hundred different molecules have been spectroscopically identified in interstellar regions, most of which are organic in nature. These include many of the compounds of biological significance, such as acetaldehyde (CH_3CHO), cyanamide (NH_2CN), and HCN (e.g., Ehrenfreund and Charnley, 2000). Larger carbon-bearing species, such as polycyclic aromatic hydrocarbons (PAHs) and fullerenes, may also be present in interstellar gas or incorporated onto dust grains that, upon UV irradiation, could potentially transform into a diverse mixture of organic molecules that include ethers, quinones, and alcohols (Bernstein et al., 1999). Comets and meteorites impacting Earth may have been another source of extraterrestrial carbonaceous material (e.g., Maurette, 1998). Significantly, the discovery of abundant amphipathic compounds within samples of the Murchison carbonaceous chondrite, some of which could self-assemble into vesicular membranes with diffusive properties, offers a viable solution to the difficulties in generating fatty acids on a prebiotic Earth (Deamer, 1985). Conservative estimates propose that (at 4.0 Ga) approximately 10^8 kg carbon yr^{-1} from interstellar dust, 10^5–10^6 kg carbon yr^{-1} from comets, and 10^3–10^4 carbon yr^{-1} from meteorites were delivered to Earth (Chyba et al., 1990). This bombardment declined in intensity by several orders of magnitude, nearing its present comparatively low level by 3.5 Ga. If all the organic products were fully soluble in oceans of contemporary extent and depth, then the steady-state organic abundance in the oceans at 4.0 Ga could have been approximately $1 \, g \, L^{-1}$ for a reducing atmosphere and $10^{-3} \, g \, L^{-1}$ for a neutral atmosphere (Chyba and Sagan, 1992).

Irrespective of source, the organic monomers would have necessitated some means to prevent rapid re-mixing with seawater, and hence their dilution. In the "drying lagoon" scenario of Robertson and Miller (1995), prebiotic components were concentrated and then polymerized in evaporitic environments. In a comparable model, Lerman (1986) proposed that particulate material scavenged from the ocean–atmosphere interface by bubble burst could have undergone multiple stages of evaporative dehydration and rehydration, whilst continuously exposed to radiation and lightning. The aerosols that formed subsequently returned to the ocean waters enriched in organic compounds, phosphate, and various biologically important metals.

One aspect, however, that the Organic Soup-type models do not take into account is the oxidizing effect of ultraviolet light on redox-sensitive elements in the upper water column. Studies have shown that dissolved Fe(II), which was likely abundant in the prebiotic oceans, could have been photo-oxidized to Fe(III) by UV radiation in the 200–400 nm wavelength range (Cairns-Smith, 1978; Braterman et al., 1983):

$$2Fe^{2+} + 2H^+ \xrightarrow{h\nu} 2Fe^{3+} + H_2 \qquad (7.2)$$

where h is Planck's constant and ν is the frequency of the radiation. As long as the pH was greater than 4, this reaction would have led to the formation of ferric hydroxide particles that were extremely reactive towards organic matter, and with their deposition, would systematically remove the monomers from solution. Therefore, photochemical reactions could not only have stripped the ocean surface of iron, but indirectly the vital organic molecules necessary for origin of life reactions (Cairns-Smith et al., 1992).

(b) The hydrothermal model

Giant impacts would undoubtedly have constrained the timing of life's origin at the Earth's surface. Postimpact plumes of vaporized rock could have enveloped the planet in a cloud of dust that increased surface pressures to above 100 bars. The rock vapor atmosphere may have radiated downwards onto the oceans with a temperature of 2,000°K, high enough to effectively evaporate large volumes of ocean water and sterilize much of the planet for several thousand years (Sleep et al., 1989). Certainly Darwin's pond would have been lost, as well as the upper illuminated, euphotic zone of the oceans, where most of the plankton in the oceans live today. Although the heavy bombardment continued until sometime between 4.4 and 3.8 Ga, life may yet have arisen during this unlikely time within a sheltered environment. A number of such sites have been proposed, but based on life's necessity for a steady supply of usable free energy and bioessential nutrients, the diffuse vents and black smokers associated with mid-ocean ridge systems, or the subsurface fractures associated with terrestrial thermal springs, may have offered the most favorable conditions for early life (Nisbet and Sleep, 2001).

The potential for prebiotic organic synthesis at hydrothermal systems stems from the thermal and chemical disequilibrium that arises when hot, buoyant and reducing fluids, containing H_2, N_2 and CO_2 (or HCO_3^-), circulate upwards through fractures in the crust, quenching rapidly upon contact with relatively cooler, more oxidizing seawater (Shock and Schulte, 1998). Mixing can occur either in the subsurface or upon exposure at the seafloor, thereby creating any number of temperature gradients and flow rates. Aqueous fluids with contrasting compositions and oxidation states are thermodynamically unstable when they mix, and because of kinetic barriers that inhibit stable equilibrium in the C–H–O–N system (i.e., the formation of H_2O, CH_4, and NH_3), the metastable solutions that form instead may yield organic compounds. Dissolved hydrothermal H_2, derived from high-temperature reactions of seawater with mafic volcanic rocks, provides the reduction potential and the thermodynamic drive for organic synthesis. In turn, the H_2 content is a function of the oxidation state of the mineral assemblages hosting the hydrothermal system, that being quartz–fayalite–magnetite (QFM) or pyrite–pyrrhotite–magnetite (PPM) in shallow and deep parts of the circulation system, respectively.

Hydrothermal fluids with oxidation states buffered at, or below, the QFM assemblage show the greatest potential for organic synthesis. Formaldehyde and HCN are two such compounds that theoretically can form by Strecker synthesis at vent systems (Schulte and Shock, 1995). More complex organic compounds are also predicted to form. Ketones and alkenes would be favored at high temperatures; alcohols generally require mid-temperature ranges; and organic acids dominate the distribution of carbon compounds at low temperatures (Fig. 7.2). It is even possible that organic acids could account for nearly 100% of the carbon at temperatures lower than 100°C, while inorganic carbon is favored at both the highest temperatures (>250°C as CO_2) and lowest temperatures (<30°C as HCO_3^-) (Shock et al.,

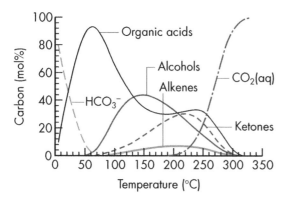

Figure 7.2 Calculated distribution of carbon amongst organic and inorganic forms at metastable equilibrium during the mixing of anoxic seawater with hydrothermal fluid. The H_2 fugacity is set by the QFM buffer assemblage. (Copyright (1998) From Shock et al. Reproduced by permission of Routledge/Taylor & Francis Group, LLC.)

1998). The thermodynamic models above are supported to some extent by experimental data showing amino acid production by reacting formaldehyde with ammonia or HCN at temperatures as low as 150°C (Hennet et al., 1992).

Fischer–Tropsch type reactions (FTT) may have occurred as well. Gaseous mixtures of CO, NH_3, and H_2, that are at equilibrium at elevated temperatures (>800°C) and pressures, have experimentally yielded fatty acids essential for the formation of lipid bilayer membranes, as well as nucleotide bases and other nitrogen-containing molecules (Hayatsu and Anders, 1981). Although the experimental temperatures are much higher than encountered in hydrothermal systems, the FTT reactions proceed at lower temperatures in the presence of mineral catalysts (Ferris, 1992).

Based on the thermodynamic predictive models, fluids circulating through the flanks of submarine hydrothermal systems (at temperatures <250°C) could generate organic carbon with yields in excess of 10^8 kg yr^{-1} (Shock, 1992). Although there is more thermal contrast between the ocean and black smoker vents, it is the off-ridge seepages that are further from electrochemical equilibrium. This is because CO_2 is stable in high temperature fluids buffered by the PPM assemblage, and accordingly, no reduction of CO_2 to organic compounds is expected. However, at lower temperatures CH_4 is more stable, and given the appropriate kinetic constraints, CO_2 may become reduced to an organic form in the immediate vicinity of the vent. The enrichment of komatiites in the Hadean crust, relative to modern oceanic basalt, would have accentuated this contrast by generating even more reducing hydrothermal fluids (Shock and Schulte, 1998). It is worth mentioning that part of the organic production in the flank system would be offset by destruction at the black-smoker vents, as advocated by Miller and Bada (1988), but only by an amount estimated at 5% of the production rate.

Hydrothermal vents additionally represent localized environments where a continuous source of free energy and nutrients are released to sustain a microbial community (recall Fig. 1.12). Some hydrothermal gases (H_2, H_2S) and solids (FeS, S^0) were prerequisites to prebiotic organic synthesis, and later they provided the redox potential for primitive metabolic pathways (e.g., Russell et al., 2005). Off-ridge seepages can also be extremely reduced and alkaline through the serpentinization of the ultramafic crust (e.g., reaction (7.3)), and the subsequent dissolution of a fraction of the metal hydroxide and serpentine (e.g., Kelley et al., 2001):

$$2Mg_{1.8}Fe_{0.2}SiO_4 \text{ (olivine)} + 2.933H_2O \rightarrow$$
$$Mg_{2.7}Fe_{0.3}Si_2O_5(OH)_4 \text{ (serpentine)} +$$
$$0.9Mg(OH)_2 + 0.033Fe_3O_4 + 0.033H_2 \quad (7.3)$$

This leads to vent fluids enriched in a number of major ions and transition metals. Ammonium was probably released at hydrothermal vents through its extraction from basalts and high temperature N_2 fixation (Lilley et al., 1993). Iron and phosphate, though absent from alkaline hydrothermal fluids, would have been present in the Hadean ocean as a result of discharge from the more acid black smoker vents (Russell and Hall, 1997).

7.1.3 Mineral templates

A major issue in both the Organic Soup and hydrothermal models is, given the low concentrations of the organic products and their dispersive loss to the bulk ocean, how did the monomers polymerize to form more complex species in aqueous solution? For instance, the formation of an amino acid, and its further polymerization into a peptide, requires condensation reactions. In the absence of enzymes it is difficult to conceive how these reactions may have occurred. Thus was born the "Surface Metabolism" model, whereby organic compounds were repeatedly synthesized by autocatalytic reactions between simple inorganic components (e.g., CO_2, $H_2PO_4^-$, and NH_4^+) on mineral sorbing surfaces as the circulating fluids passed up through the overlying altered basalts (see Cody, 2004, for details).

In the original hypothesis of Wächtershäuser (1988a), the first metabolic pathway reduced CO_2 into an anionic carboxylate group, using energy from the reaction between FeS (mackinawite) with dissolved H_2S (reaction (7.4)). This reaction is exergonic at 100°C, yielding pyrite as its byproduct (Drobner et al., 1990).

$$FeS + H_2S \rightarrow FeS_2 + H_2 \quad \Delta G^{0\prime} = -39 \text{ kJ mol}^{-1} \quad (7.4)$$

In the presence of Fe^{2+}, and at pH values up to 5, pyrite has a positively charged surface (Bebie et al., 1998). On the prebiotic Earth, that meant pyrite could have adsorbed inorganic anions and negatively charged organic molecules from the hydrothermal fluids. These accumulated and polymerized into more complex compounds directly on the mineral surface, some of which may have been highly unstable, converting into more stable products faster than they could detach and decompose (e.g., Fig. 7.3(a)). Every once in a while a new organic compound may have formed that not only provided a novel catalytic reaction, but was autocatalytic towards its own biosynthesis from the surface-bonded constituents, thus ensuring that it became an inheritable feature of the "surface metabolist" (Wächtershäuser, 1994). In time, some of those compounds might have become enveloped in a lipid membrane and detached from the pyrite surface to take on an independent existence (Orgel, 1998). Other compounds that were not capable of surface bonding would have disappeared into the surrounding waters. This means that pyrite may have had a self-sorting effect, perhaps explaining the polyanionic nature of so many of the constituents of central metabolism, such as polycarboxylates, anionic peptides, polyanionic coenzymes, and even RNA and DNA. The central role of iron in those early biochemical reactions may also explain its later incorporation into a number of enzymes, such as Fe-S proteins and cytochromes (Wächtershäuser, 1988b).

Surface bonding additionally mitigates the problem of phosphate limitation because inorganic

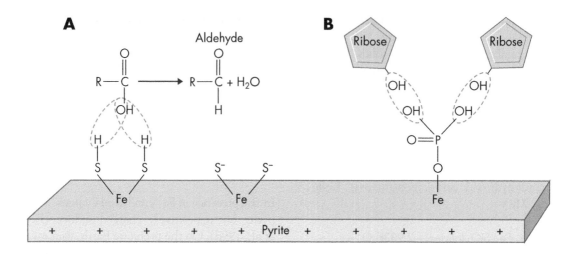

Figure 7.3 Representation of how pyrite surfaces might have facilitated some chemical reactions relevant to the origins of life. (A) The reduction of a carboxylic acid via a thiol, and the subsequent formation of an aldehyde (modified from Wächtershäuser, 1988b). (B) The adsorption of phosphate, and its secondary reaction with ribose (adapted from Gedulin and Arrhenius, 1994).

phosphate and anionic phosphorylated organic molecules (e.g., adenosine monophosphate) readily concentrate and assemble into macromolecules on pyrite surfaces (Bebié and Schoonen, 1999). Adenosine on its own is not significantly bound, but requires phosphorylation to adenosine monophosphate, thereby necessitating a pyrite–phosphate–adenosine ternary complex. This leads to the question of whether the adsorption of phosphate onto pyrite surfaces may have contributed to surface-induced phosphorylation of organic compounds on the prebiotic Earth? Certainly such a molecule could link two separate ribose molecules (each recall a component of nucleotides) via hydrogen bonding, while still remaining ionized and reactive towards the mineral surface (Gedulin and Arrhenius, 1994) (Fig. 7.3b).

Experimental studies attempting to confirm reactions between FeS and H_2S under prebiotic conditions, however, have shown that solutions containing CO_2 and H_2S yield organic sulfur compounds in preference to carboxylic acids (Heinen and Lauwers, 1996). Furthermore, the electron transfer from FeS to CO_2 is hindered by a high activation energy, even though the reac-

tion is thermodynamically favorable (Schoonen et al., 1999). This implies that other inorganic carbon molecules may have been a more important source for primary organic material. One such molecule might have been carbon monoxide which, when reacted with methane thiol (CH_3SH), can be transformed via a series of steps (using NiS and FeS surfaces as catalysts), first into thioacetic acid (CH_3COSH) and subsequently into either a thioester acetyl methylsulfide (CH_3COSCH_3), or its hydrolyzed product, acetate (CH_3COO^-):

$$CH_3SH + CO + OH^- \rightarrow CH_3COO^- + H_2S \quad (7.5)$$

CH_3SH is thought to form as an intermediate product by reduction of CO or CO_2 with FeS and H_2S. Its ensuing carbonylation (adding CO) to form thioacetic acid, and then its conversion to a thioester, is akin to the reductive acetyl-CoA pathway: the key enzyme in this pathway is the Ni-Fe-S-containing acetyl-CoA synthase (Huber and Wächtershäuser, 1997). Thioacetic acid could, hypothetically, also be fed into a reductive citric acid cycle, driven by the high CO_2 partial pressures at that time, to form pyruvate,

$C_3H_3O_3^-$ (Wächtershäuser, 1988b). Whether these reactions were significant as a means of pre-biotic organic carbon formation relies, in part, on the abundance of the precursors on the early Earth: CH_3SH has been detected in volcanic gases and in fluid inclusions of Archean origin, both FeS and NiS are common sulfide minerals around hydrothermal vents, and CO may have been abundant in Hadean–Archean hydro-thermal vent waters. Other studies have since shown that pyruvate can be synthesized from methane thiol at 250°C and elevated pressures, through a series of reactions that mimic the flow of hydrothermal fluids through an iron sulfide-containing crust (Cody et al., 2000). If the subsequent carboxylation of pyruvate to oxaloacetate, followed by reduction to malate, can be demonstrated, then a potential pathway for the origin of the reductive citric acid cycle around hydrothermal vents may be at hand.

The potential importance of FeS extends beyond simple catalysis. Based on laboratory experiments, it has been suggested that thin membranes of mackinawite could have pre-cipitated spontaneously at the interface where exhaling, hot (100°C), alkaline (pH 10), sulfide-rich submarine seepage waters mixed with acidic (pH 5–6), Fe-bearing, cooler (20°C), and mildly oxidizing Hadean ocean water (Russell et al., 1993). Disequilibrium between the two fluids would have been acute, and as the sulfide mem-branes grew into bubble-like vesicles, about 10–100 μm across, the membranes could have served to encapsulate, and hence separate, the hydrothermal waters inside the vesicles from external seawater. Crucially, such semipermeable catalytic boundaries might have established an electrochemical potential of similar magnitude to the protonmotive force (pmf) that drives ATP generation in extant cells (recall section 2.1.4(b)). Yet, unlike most modern cells, the pmf was not generated from the segregation of electrons and protons in the plasma membrane, but instead it originated from the naturally occurring concentra-tion of protons in the Hadean ocean. The energy associated with the pmf may then have been used

to drive specific molecules onto, or across, the mackinawite membrane (Fig. 7.4). These include H_2, H_2CO_3, NH_4^+, a variety of simple organic compounds (e.g., HCN), as well as dissolved metals and phosphate. It has been proposed that the membrane might also have possessed catalytic properties, perhaps a forerunner to the Fe-S enzymes used in chemiosmosis (Russell et al., 2005). Quite possibly, as the mackinawite membrane conducted electrons from inside to outside, towards an external terminal electron acceptor (TEA) (such as the ferric iron from photo-oxidation), the mackinawite progressively became oxidized via an intermediate ferredoxin stage, to greigite. During this transition, the ferredoxin might have catalyzed the reduction of the CO_2 (as carbonic acid at low pH) to produce organic compounds, such as CH_3SH.

Once inside the membrane, several possible reactions, as described above, could have taken place. It has further been postulated that some of those organic products, the carboxylic acids, might subsequently have become aminated on reaction with hydrothermal ammonia, and pos-sibly HCN, to form amino acids (Russell and Hall, 1997). In turn, those amino acids could have been converted into polypeptides on the surfaces of (Ni,Fe) sulfides as long as pH values inside the membranes were between 7 and 10 (Huber and Wächtershäuser, 1998). Thermodynamically, the formation of peptides at high temperature is endergonic, but it can be made possible by coupling it to an exergonic reaction, such as the formation of acetate from CO and CH_3SH.

As organic anions were generated within the vesicles, increasing osmotic pressure might have caused distension, budding, and reproduction of additional iron monosulfide bubbles at the grow-ing surface of the sulfide mound. Contiguous compartments could have maintained fluid mix-tures of slightly different Eh and pH conditions, and harbored different reactants and products. As these new organic compounds began to poly-merize, they could have assembled themselves on the interior of the inorganic membrane, where continued build up eventually may have weakened

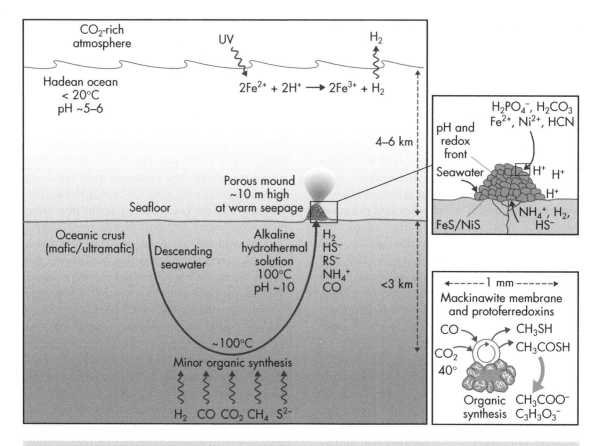

Figure 7.4 Model environment for the origins of life on the seafloor at an alkaline hydrothermal vent, sometime before 4.0 Ga. (Modified from Russell, 2003.)

the iron sulfide membrane, allowing the organic polymers to take over the role of keeping the two fluids apart (Russell and Hall, 1997).

Other minerals may have similarly catalyzed prebiotic organic formation. By their very nature, clay surfaces are a composite of anionically and cationically charged sites. The negative charges arise from incomplete ionic substitution in the crystal lattice, while the positive charges may result from either the binding of metal cations or the hydrolyzation of aluminum to form highly electropositive ligands. This means that they could adsorb metal catalysts to their surfaces, while simultaneously binding phosphate elsewhere on the surface for use in biosynthesis. It is thus conceivable that, under certain conditions,

clays served as primitive biochemical templates, whereby "genetic information" was stored as a distribution of charges along the mineral edges (Cairns-Smith, 1982). Once organic monomers were bound, they could have been converted into more complex oligomers with a prescribed structural orientation and size. Indeed, experiments have shown the formation of oligopeptides at the surfaces of illite and oligonucleotides at the surface of montmorillonite (Ferris et al., 1996). Once oligomers are formed, they, in turn, serve as the catalysts for further polymerization reactions (Lahav and Nir, 1997).

Zeolites are highly porous minerals that, like clays, have variable surface charges, which is why they are often used as ion exchange resins (e.g., used in water softeners). Experiments have shown that they can bind amino acids, and in doing so, dramatically accelerate the formation of high molecular weight oligopeptides (Zamaraev et al., 1997). In this regard, metal hydroxides, such as $Mg(OH)_2$ (brucite), $Mn(OH)_2$ (pyrochroite), and $Ca(OH)_2$ (portlandite), might play a similar role. Replacement of a fraction of the divalent ion with a trivalent cation (i.e., Fe^{3+}, Al^{3+}) introduces excess positive charge on the hydroxide sheets, which is enough to adsorb anions such as glycoaldehyde phosphate. Once bound, they condense into a mixture of more complex sugar phosphates, which are important precursors in prebiotic synthesis models (Pitsch et al., 1995). Highly weathered feldspars offer similar features, with their regular distribution of etch pits and grooves providing a natural catalytic surface for monomer formation, a source of nutrients within the crystal lattice, and shelter from dispersion into dilute solutions (Parsons et al., 1998).

7.2 The first cellular life forms

As the various biochemical reactions above worked synchronously in the prebiotic oceans, we can envision a rich mixture of simple organic compounds that, in the absence of living microorganisms, would have accumulated. The next step in prebiotic evolution would have necessitated the assembly of monomers and oligomers into high molecular weight macromolecules. They principally included: (i) lipids as self-sealing cell envelopes that protected the cytoplasmic contents from becoming randomly dispersed; (ii) nucleic acids for replication and transfer of genetic properties to progeny; (iii) proteins as catalysts for energy transduction; and (iv) carbohydrates as the structural backbone of the cell wall (Trevors, 2003). This, however, is where real gaps in our knowledge emerge, because at present we still have not been able to synthesize the basic building blocks of life, let alone assemble together something even remotely resembling a living entity (Shapiro, 1999; Szostak

et al., 2001). As the arguments made for the various mechanisms likely to have underpinned this critical stage in prebiotic synthesis are beyond the scope of this book, we will instead skip ahead in time, to the introduction of the first primitive cell-like form, the progenote.

7.2.1 The chemolithoautotrophs

The first progenote was undoubtedly a simple entity that possessed minimal metabolic capability, perhaps requiring only a few proteins to drive its bioenergetic machinery. According to the Organic Soup model, those life forms were anaerobic chemoheterotrophs that depended on abiologically synthesized organic molecules as both carbon and energy sources (Ehrlich, 2002). As cell populations increased and intense competition for available substrates developed, the community became carbon-limited. Thus, an evolutionary step was needed to make them self-sufficient. This was the emergence of autotrophy, and the energy for this process could have come from either the oxidation of an inorganic compound (chemotrophy) or from the transformation of radiant energy in sunlight (phototrophy). In the Surface Metabolism model, the progenote began chemolithoautotrophically, with its energy coming from redox-driven reactions occurring naturally at hydrothermal vents. This viewpoint is supported by: (i) the identification of enzymes, including those involved in autotrophic carbon fixation, that are shared amongst all living organisms (Olsen and Woese, 1996); and (ii) the fact that all of the deepest branches for *Bacteria* and *Archaea* within the universal phylogenetic tree are occupied by hyperthermophiles and thermophiles (Fig. 7.5) that gain their energy chemolithoautotrophically from anaerobic reactions. However, it should be pointed out that the placement of hyperthermophiles on the universal phylogenetic tree is controversial. Some suggest that their deep-branching may be due to methodological artefacts when using single gene sequences (e.g., Philippe and Laurent, 1998). Others propose that the hyperthermophile

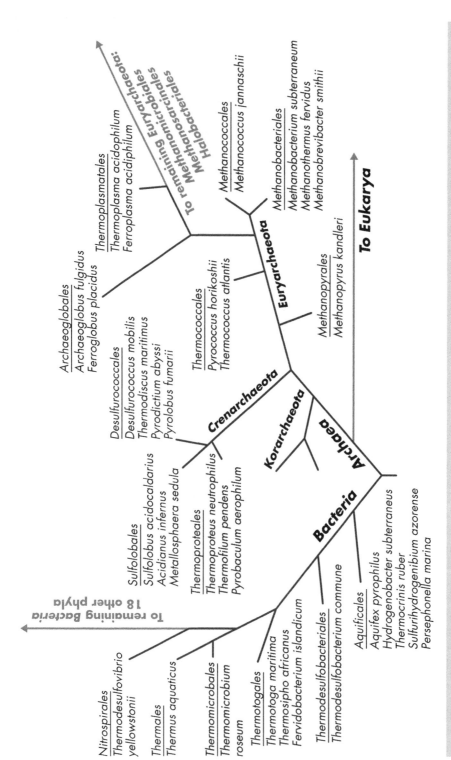

Figure 7.5 Universal phylogenetic tree of the hyperthermophilic-thermophilic prokaryotes, based on 16S rRNA sequence comparisons, illustrating their deeply rooted position. They are grouped by taxonomic order, and example organisms are provided to indicate the variety of genera therein. Archaeal phylogeny is highly contested, therefore the evolutionary distance and branching pattern is, in part, approximated from the work of Matte-Tailliez et al. (2002).

lineages are simply a consequence of increased selective pressure acting on their rRNA molecules, and instead they may have descended from an even earlier mesophilic ancestor (Galtier et al., 1999). Indeed, this leads to the concept referred to as the "impact bottleneck," where some microorganisms found refuge from early bolide bombardment and were selectively preserved (e.g., the ancestors of extant thermophilic *Bacteria* and *Archaea*), whilst those species (possibly mesophiles) inhabiting the surface environment were extinguished (Gogarten-Boekels et al., 1995). Notwithstanding, based on the general consensus favoring a thermophilic origin, the following section sets the stage for a possible sequence of steps that take us from the progenote's metabolism at a hydrothermal vent to the eventual diversification of microbial life throughout the geosphere.

We begin our discussion with early progenotes continuing to take advantage of the reaction between H_2S with FeS (reaction (7.4)). It is sufficiently exergonic to facilitate the formation of energy-rich phosphate compounds, such as pyrophosphate, $P_2O_5(OH)_2^{2-}$ (PP_i): the potential role of pyrophosphate, not ATP, in early biological energy conversion stems from its capabilities as both an energy and phosphate donor, its comparatively uncomplicated structure, and its ability to form under abiological conditions from volcanic magma (Baltscheffsky and Baltscheffsky, 1994). The reaction would also have provided the reducing power for primitive CO_2 reduction and organic synthesis, possibly using either the reductive citric acid cycle or the reductive acetyl-CoA pathway. The former has been suggested as being the oldest since it operates in phylogenetically diverse autotrophic *Bacteria* and *Archaea*, it is autocatalytic in that it doubles the number of CO_2 acceptors with every turn, and it is central to almost all anabolic pathways (Wächtershäuser, 1994), while the latter is favored based on biogeochemical and energetic grounds (Russell and Martin, 2004).

With time, the evolution of some primitive enzymes would have extended microbial metabolic capabilities. One such enzyme would have been an iron-sulfur protein, containing either Fe_2S_2 or Fe_4S_4 clusters (Beinert et al., 1997). Perhaps derived from the FeS membranes that could have spontaneously precipitated around alkaline seepages (as in above), these catalytic proteins may have been able to catalyze the reduction of CO_2 to produce some simple organic compounds. Their chemical versatility at accepting, donating, shifting, and storing electrons could also have facilitated a primitive electron transport chain. In fact, Daniel and Danson (1995) argue that oxidized ferrodoxins (i.e., those with centers in the $(Fe_4S_4)^{2+}$ state) were the earliest biological catalysts to mediate catabolism prior to the appearance of NADPH.

The generation of H_2 as a byproduct from reaction (7.4) would have proven fortuitous because it could eventually have been used as an electron donor in subsequent metabolic reactions, gradually supplementing and then replacing the original energy source when the first primitive hydrogenase activity had evolved (Kandler, 1994). It can easily be imagined that once progenotes could bind H_2 onto a simple plasma membrane, and then by some means separate H_2 into electrons and protons, a simple proton motive force would have been created to drive a primitive PPase or ATPase (Fig. 7.6). Simultaneously, the electrons would have needed disposing of, and the elemental sulfur available around the vents (formed from the rapid cooling of volcanogenic sulfur gases) may have served this purpose nicely (reaction (7.6)). Such a reaction would have been quite easy for the progenote because the redox reaction in the Fe-S cluster is reversible, which allowed them to change their role from the oxidation of H_2S to S_2^{2-} (as in pyrite) to a role in the reduction of S^0 (via S_2^{2-}) to H_2S.

$$H_2 + S^0 \longleftrightarrow H_2S \qquad (7.6)$$

The evolution of the first cytochromes would have improved the oxidative phosphorylation capabilities so that additional pyrophosphate

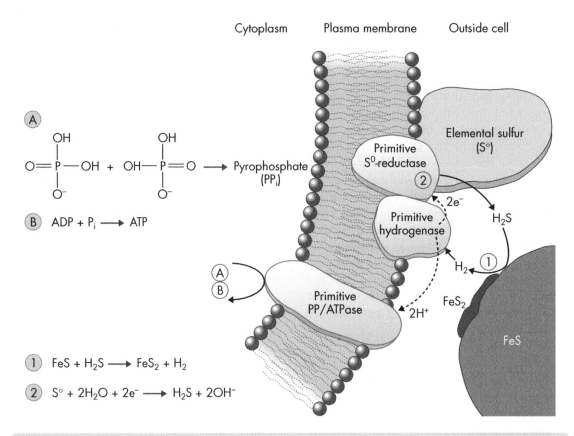

Figure 7.6 A hypothetical energy-generating metabolism for the progenotes, prior to the advent of more complex electron transport systems. Hydrogen gas generated from reaction (1) is bound to a primitive hydrogenase, where it is split into protons and electrons. The electrons then are used in the reduction of elemental sulfur (reaction 2), which might have been abundant around the hydrothermal vents. Hydrogen sulfide gas is then released to the external environment, where it reacts with more FeS. Meanwhile, the protons are used in the generation of either pyrophosphate or ATP, depending on whether the progenotes had already acquired the ability to manipulate the latter. (Modified from Madigan et al., 2003.)

or ATP could now be generated by exploiting stronger oxidants as electron acceptors, including SO_4^{2-} (reaction (7.7)) and Fe(III) (reaction (7.8)):

$$4H_2 + SO_4^{2-} \rightarrow H_2S + 2OH^- + 2H_2O \quad (7.7)$$

$$0.5H_2 + Fe(OH)_3 \rightarrow Fe^{2+} + 2OH^- + H_2O \quad (7.8)$$

Sulfate could have formed through photochemical reactions or from the reaction of volcanic SO_2 with H_2O at temperatures below 400°C, and the subsequent disproportionation to H_2S and SO_4^{2-}. Ferric hydroxide may have precipitated as a byproduct of the photochemical oxidation of Fe(II) by ultraviolet radiation at the ocean's surface (recall reaction (7.2)).

The progenotes may even have utilized a number of possible amino acid synthesis reactions (e.g., reaction (7.9)), because at 100°C there is sufficient free energy released in these reactions to overcome the energy requirements of protein synthesis, despite the fact that the above calculations were made assuming ammonium concentrations in hydrothermal solutions of $<3 \times 10^{-6}$ mol L^{-1} (Amend and Shock, 1998).

$$3H_2 + 2CO_2 + NH_4^+ \rightarrow$$
$$CH_2NH_2COOH \text{ (glycine)} + H^+ + 2H_2O \quad (7.9)$$

What is very interesting is that amino acid synthesis reactions become more exergonic as temperatures increase from 0°C to 100°C, suggesting that hyperthermophiles expend considerably less energy synthesizing proteins than do their mesophilic counterparts. Maybe this is part of the explanation for the phenomenal rates of micro- and macrofaunal biomass production around modern hydrothermal vents (Lutz et al., 1994).

Another redox reaction naturally available to the earliest H_2-oxidizing chemolithoautotrophs was methanogenesis:

$$4H_2 + CO_2 \rightarrow CH_4 + 2H_2O \quad (7.10)$$

Energy coupling in methanogenesis was more advanced than the metabolic set-up of their predecessors, involving an electron transport system that employed several unique membrane-bound enzymes to reduce CO_2 to CH_4. The methane produced, in turn, could have been used to reduce the Fe(III) in hydrothermal magnetite, some of which formed in the serpentinization reactions (recall reaction (7.3)). The methane-magnetite reaction is strongly exergonic (Amend et al., 2003):

$$4Fe_3O_4 + CH_4 + 24H^+ \rightarrow 12Fe^{2+} + 14H_2O + CO_2$$
$$\Delta G° = -94 \text{ to } -123\,kJ/e^- \quad (7.11)$$

Which metabolism was actually first used may never be known. Nevertheless, it is possible that in a short timespan mutation and/or adaptation to specific geothermal conditions may have given rise to different microbial phenotypes on the basis of each of the different chemolithoautotrophies (Nisbet, 1995). Thus, the last universal common ancestor (LUCA) may actually have been a multiphenotypical population of progenotes living under slightly different hydrothermal conditions. Concomitantly, the transfer of genetic information was probably pervasive amongst the primitive genomes, making each entity capable of acquiring any evolutionary innovation occurring within the entire microbial community (Jain et al., 1999; Vellai and Vida, 1999). Not surprisingly, a number of proteins involved in the respiratory pathways are homologous in archaeal and bacterial species, signifying that LUCA was rather advanced in terms of its metabolism and in its capability to face a wide range of environmental conditions (Castresana and Moreira, 1999). Indeed, an absence of barriers to DNA exchange could have characterized the beginning stages of cellular evolution, and it was not until genetic subsystems became more complex and less compatible with exogenous parts that lateral gene transfer was significantly restricted. This stage invariably marked the origins of cell individuality and speciation, and hence the first true prokaryotes emerged, the ancestors of *Bacteria* and *Archaea* (Woese, 1998). For an alternate view, one suggesting that *Archaea* arose rather recently (some 850 Ma (million years ago)) from *Bacteria* as an adaptation to hyperthermophily, see Cavalier-Smith (2002).

7.2.2 Deepest-branching *Bacteria* and *Archaea*

At present there are 24 recognized phyla in the *Bacteria* domain (Garrity et al., 2004). The most deeply rooted branch consists of the thermophilic to hyperthermophilic order *Aquificales*. Most species (e.g., *Aquifex pyrophilus*) grow by the oxidation of H_2, using nitrate or O_2 as oxidants, but some (e.g., *Persephonella marina*) use elemental sulfur, while others (e.g., *Sulfurihydrogenibium azorense*) reduce Fe(III) with H_2 (Aguiar et al., 2004). They all employ the reductive citric acid cycle for CO_2 fixation (e.g., Huber et al., 1992). Intriguingly if oxygen utilization by members of the *Aquificales* is actually a primitive characteristic, then this would suggest that free O_2 must have been locally available even at such early times. Possible sources include, the pyrolysis of water within submarine lava flows or disproportionation of atmospherically derived hydrogen peroxide (H_2O_2) by early catalase enzymes (reaction (7.12)). Paradoxically, phylogenetic reconstructions of molecular sequences from the cytochromes responsible for the reduction of O_2 to H_2O in aerobic respiration indicate an origin that seemingly predates the evolution of cyanobacteria (Castresana and Saraste, 1995).

$$2H_2O_2 \rightarrow 2H_2O + O_2 \qquad (7.12)$$

Alternatively, the ability of some *Aquificales* to use O_2 as a terminal electron acceptor might instead reflect a more recent metabolic capability aquired once more advanced terminal oxidases evolved. Such retrofitted acquisitions are not unique to the *Aquificales*, and it is now widely believed that organisms in all domains may have obtained key metabolic genes during evolution via lateral gene transfer from distantly related organisms (Ochman et al., 2000). The ability of the *Aquificales* to use a wide range of electron acceptors and donors, coupled with their deeply rooted position, may therefore reflect their evolutionary history in the changing atmosphere of early Earth (Reysenbach and Shock, 2002).

There are two, possibly three, archaeal phyla. Our knowledge about *Korarchaeota* is limited to a few sequences isolated from near-boiling, sulfidic hot spring effluent in Yellowstone National Park (Barns et al., 1996). Thereafter, a fundamental split appears to have taken place, leading to the divergence of the *Crenarchaeota* and the *Euryarchaeota*. All known *Crenarchaeota* are sulfur-dependent, with the deepest lineages (e.g., *Pyrodictium* and *Thermoproteus* species) being hyperthermophilic, chemolithoautotrophic anaerobes that utilize S^0 as an oxidant, and H_2 as the electron donor. Many extant, deep-branching *Crenarchaeota* are also capable of using H_2 to reduce Fe(III) to support chemolithoautotrophic growth (Vargas et al., 1998). They can even use quinone moieties as electron shuttles between solid-phase iron minerals and H_2, thereby alleviating the need for direct contact between the cell and mineral surface (Lovley et al., 2000). On the early Earth, such processes may have been the forerunners of electron transport systems where ubiquinones and membrane-bound iron-containing proteins functioned as electron shuttles. In terms of CO_2 fixation,

Crenarchaeota exhibit a mosaic of three, possibly four, autotrophic pathways (Hügler et al., 2003).

The earliest *Euryarchaeota* has proven to be a point of contention. Phylogenetic trees based on 16S rRNA data indicate that the primitive methanogen, *Methanopyrus kandleri*, a species capable of growth up to 110°C, is the deepest lineage (Burggraf et al., 1991a). This view, however, contradicts more recent alternate gene (other than 16S rRNA) and whole-genome-based phylogenetic analyses that show *Methanopyrus*, and all other methanogens, evolving late within *Euryarchaeota* (House et al., 2003). The deepest branch is now believed to be occupied by species of the hyperthermophilic and anaerobic, S^0-oxidizing genus *Pyrococcus*, with methanogenesis originating only after the divergence of the order *Thermococcales* from the other *Euryarchaeota* (Forterre et al., 2002). The complex biochemistry involved in the H_2/CO_2 couple also suggests that methanogenesis likely represents a secondary stage during the diversification of *Archaea* from a sulfur-based metabolism. Interestingly, the methanogens and the sulfur-metabolizing *Archaeoglobales* order share a number of traits: (i) both posses certain coenzymes exclusive to them; (ii) all methanogens are universally capable of utilizing the H_2/S^0 couple, while the *Archaeoglobales* are capable of producing small amounts of methane; and (iii) both use the reductive acetyl-CoA pathway for carbon synthesis (e.g., Achenbach-Richter et al., 1987).

Further uncertainty lies in trying to place the origins of sulfate reduction within the *Euryarchaeota*. The fact that species of the *Archaeoglobales* (the only archaeal order to reduce sulfate) contain genes that are unknown in other sequenced archaeal genomes, but have homologs in *Bacteria* might imply that these genes were present in LUCA, and subsequently lost in other *Archaea* (Wagner et al., 1998). Alternatively, they could have been obtained by the *Archaeoglobales* via lateral gene transfer from deep-branching, thermophilic, sulfate-reducing *Bacteria* (see below), thus clouding the issue of antiquity (e.g., Klein et al., 2001).

7.2.3 The fermenters and initial respirers

As primitive chemolithoautotrophs died, and their organic debris began to accumulate in close proximity to hydrothermal vents, some microorganisms slowly evolved the means to utilize the reduced organic carbon as an energy source. The process of fermentation appears to have developed relatively early in both domains, as suggested by the observation that extant hyperthermophilic species of deep-branching *Bacteria* (e.g., *Thermotoga* sp.) and *Archaea* (e.g., *Pyrococcus* sp., *Pyrodictium* sp.) are primarily fermenters (Madigan et al., 2003). Fermentation involves internally balanced redox reactions, whereby the oxidation of the initial substrate is coupled to the subsequent reduction of another organic compound generated from that catabolism. Typical products can include organic acids (lactic, acetic), alcohols (ethanol, methanol), ketones (acetone), and gases (H_2 and CO_2). This novel metabolism would undoubtedly have benefited the existing chemolithoautotrophs because it would have reduced their reliance on hydrothermal H_2, thus allowing the cells to expand to the peripheries of the vents, and maybe even further, wherever organic remains were scattered through dispersive processes.

During fermentation only a small portion of the potential energy tied up in the organic bonds is converted to ATP. This inefficiency would have restricted cell yields and invariably led to the advent of a biochemical process that could completely oxidize the organic matter to CO_2. This process was respiration. Its origins likely stem from reversing the pre-existing reductive citric acid cycle already employed by the chemolithoautotrophic cells; most modern H_2-oxidizing bacteria are actually facultative chemolithoautotrophs, meaning that they can use H_2 in simple organic substrates to reduce a TEA. This new-found ability would have opened up the possibility for anaerobic respiration pathways that could transfer electrons originally from organic matter, via fermentation products, to some locally available TEAs (Fig. 7.7). One potential TEA was S^0,

and many deep-branching hyperthermophilic *Crenarchaeota* (e.g., *Desulfurococcus* sp., *Thermoproteus* sp.), *Euryarchaeota* (e.g., *Thermococcus* sp.), and *Bacteria* (some *Aquificales*) grow as sulfur respirers (reaction (7.13)):

$$CH_3COO^- + 4S^0 + 4H_2O \rightarrow 4H_2S + 2HCO_3^- + H^+ \tag{7.13}$$

Similar modifications might have been made by thermophilic chemolithoautotrophic sulfate-reducers. Not only can some of the *Archaeoglobales* grow chemolithoautotrophically on the H_2/SO_4^{2-} couple, but two species can couple the reduction of SO_4^{2-} to the oxidation of simple fermentation products, such as reaction (7.14) (Burggraf et al., 1990). There are a number of thermophilic sulfate-reducing *Bacteria* as well, but two genera branch deeply within the domain, *Thermodesulfobacterium* and *Themodesulfovibrio* (Castro et al., 2000).

$$CH_3COO^- + SO_4^{2-} + H_2O \rightarrow H_2S + 2HCO_3^- + OH^- \tag{7.14}$$

What has also become a contentious issue is whether the capacity for sulfate reduction in thermophiles is yet another example of a retrofitted metabolism. Comparative analyses of the genes encoding APS (adenosine phosphosulfate) reductase show rampant lateral gene transfer across all bacterial divisions, and in some cases from mesophilic δ-*Proteobacteria* (see below) into thermophilic genera, such as *Thermodesulfobacterium* (Friedrich, 2002). Yet, similar comparisons for dissimilatory sulfite reductase, another essential enzyme in the sulfate reduction pathway, show deep rooting in thermophilic bacterial lineages and, importantly, the same genes display histories congruent with 16S rRNA phylogenies (Klein et al., 2001).

A case has similarly been made for the antiquity of Fe(III)-reducing bacteria based on the metabolic capabilities of extant hyperthermophiles (Kashefi and Lovley, 2000) and on geochemical grounds, such as the unique Fe isotopic fractionations found in 2.9 Gyr grains of magnetite that

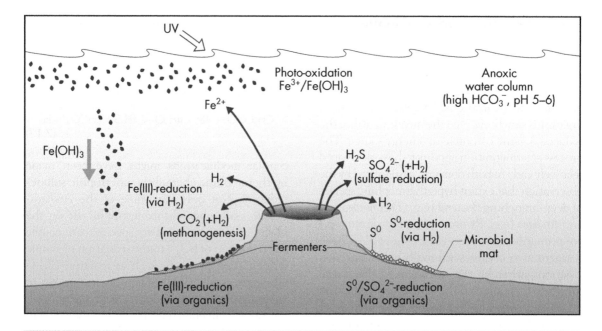

Figure 7.7 The potential metabolic diversity available at primitive hydrothermal vent systems. The most probale electron doner for chemolithoautotrophy was H_2, while terminal electron acceptors might have included S^0, SO_4^{2-}, Fe(III), and CO_2. Biomass from the lysed autotrophs would also have provided an organic substrate for the burgeoning population of fermenters and chemoheterotrophs.

closely mimic those observed during modern dissimilatory Fe(III) reduction (e.g., Yamaguchi et al., 2005). Ferric iron may have been available over relatively vast tracts of the seafloor, formed as a byproduct of surface photochemical reactions, in the absence of an effective UV screen:

$$CH_3COO^- + 8Fe(OH)_3 \rightarrow$$
$$8Fe^{2+} + 2HCO_3^- + 15OH^- + 5H_2O \qquad (7.15)$$

An active Fe cycle is certainly supported by the presence of Early Archaen banded iron formations (BIF), the laminated sedimentary rocks that contain an anomalously high content of Fe(II)/ Fe(III), with a mineralogy consisting primarily of quartz, magnetite, hematite, various carbonates, and iron-silicates. The oldest putative deposits are the Isua Greenstone Belt, which are around 3.8 Gyr. Significantly, these BIFs have also been reported as containing the earliest evidence for life on Earth (Box 7.1).

7.3 Evolution of photosynthesis

7.3.1 Early phototrophs

With life firmly entrenched aside deep-sea hydrothermal vents, an evolutionary step was needed that could facilitate microbial expansion into the sterile world around them. The first step in this process may have arisen when a rudimentary pigment, such as bacteriorhodopsin, became incorporated into the plasma membrane. Upon illumination a proton gradient could be produced by a relatively simple photochemical reaction (Deamer et al., 1994). Today, a few nonphotosynthetic halophiles (e.g., *Halobacterium halobium*) utilize this metabolic pathway when conditions are not conducive to their regular method of aerobic respiration. Recently, it has also been demonstrated that a rhodopsin-type compound

Box 7.1 Carbon isotopes

Life forms that existed during deposition of the earliest sediments on Earth left no morphological fossils that could sustain the high degrees of metamorphic re-crystallization and deformation that has affected all rocks older then 3.6 Gyr. If life had evolved prior to then, its identification will rely heavily on geochemical signatures in rocks that are clearly characteristic of biological activity. Such evidence has been reported to be in the form of graphite inclusions from two different localities in the southern region of West Greenland. Within what is known as the Akilia association, rocks tentatively dated at 3.85 Gyr, and interpreted as a banded iron formation, were formed in association with mafic rocks. Significantly, the purported iron formation contains trace amounts of apatite that house micrometer-size graphite inclusions, with a range of $\delta^{13}C$ values from -21 to -49‰, and a weighted mean of -37‰, relative to the Pee Dee Belemite (PDB) standard (Mojzsis et al., 1996). The authors believe that the host apatite protected the graphite from isotopic exchange with associated metasomatically derived carbonates, thus maintaining the primary biological signal (see below). Yet others have suggested that the graphite was instead produced through thermal decomposition of siderite, thereby negating the paleobiological significance (Van Zuilen et al., 2002). Perhaps even more scientifically damaging is the re-interpretation of the Akilia rocks as being ultramafic intrusions or volcanic rocks that were later altered by the metasomatic introduction of iron and silica (Fedo and Whitehouse, 2002), but this too has been countered on the basis of Fe isotope variations that suggest some of the original rocks (e.g., BIFs) are sedimentary in origin (Dauphas et al., 2004).

A second Greenland location comes from turbiditic and pelagic sediments of the Isua Greenstone Belt, dated at 3.7–3.8 Gyr. Considerable amounts of graphite (>0.1 wt%) have also been identified within these rocks, with isotopic compositions yielding ^{13}C-depleted values

around -19‰ (Rosing, 1999). The graphite in these samples does not coexist with Fe-bearing carbonate phases, and as such these samples may represent the best current evidence in favor of a primitive biological signature.

These negative carbon isotope values are seemingly indicative of life because the transformation of inorganic carbon (e.g., CO_2 or HCO_3^-) via autotrophic pathways into organic carbon involves the preferential incorporation of the lighter isotope, ^{12}C, into the organic phase, leaving behind a reservoir enriched in the heavier isotope, ^{13}C. This bias in favor of ^{12}C is principally due to kinetic isotope effects inherent in the diffusional transport of inorganic carbon into the cell and the first enzymatic carboxylation reaction that fixes CO_2 into the COOH (carboxyl) group of an organic acid. Consequently, organic compounds produced by autotrophic pathways display a marked preference for the light isotope, while the heavy carbon is retained in the surface reservoir of oxidized carbon, mostly as dissolved bicarbonate, and later incorporated into precipitated carbonate minerals, e.g., calcite and dolomite, or as atmospheric CO_2 (Schidlowski, 2000). Both inorganic and organic sedimentary carbon have been shown to retain the isotopic compositions of their progenitors, such that carbonate rocks preserve the isotopic composition of their parent sediments, which, in turn, is a good approximation of the isotopic composition of marine bicarbonate at the time of deposition, while kerogens (the insoluble and refractory organic compounds that represent the stable end products of buried organic matter) preserve the biological signature. Therefore, the original isotopic fractionation between reduced and oxidized carbon, as established in the ancient surficial environment, is preserved after the entry of the respective phases into the sedimentary rock record.

Unfortunately, a number of carbon fixation pathways (reductive acetyl-CoA, reductive citric acid, Calvin cycle, hydroxypropionate) have overlapping degrees of carbon fractionation.

continued

Box 7.1 *continued*

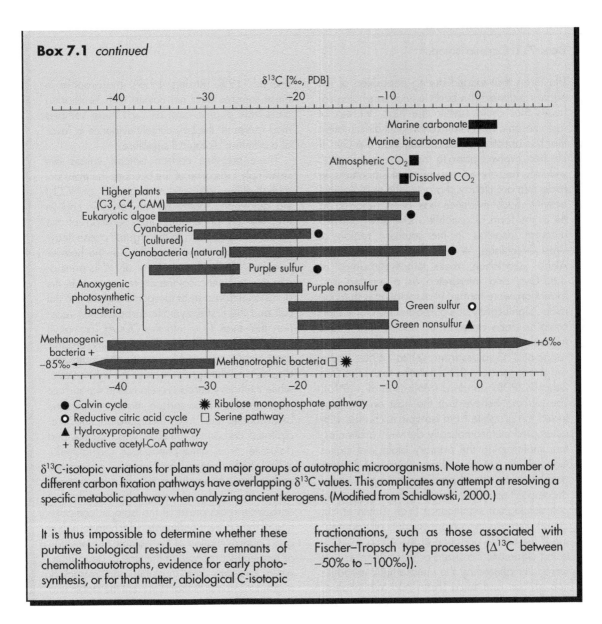

$\delta^{13}C$-isotopic variations for plants and major groups of autotrophic microorganisms. Note how a number of different carbon fixation pathways have overlapping $\delta^{13}C$ values. This complicates any attempt at resolving a specific metabolic pathway when analyzing ancient kerogens. (Modified from Schidlowski, 2000.)

It is thus impossible to determine whether these putative biological residues were remnants of chemolithoautotrophs, evidence for early photosynthesis, or for that matter, abiological C-isotopic fractionations, such as those associated with Fischer–Tropsch type processes ($\Delta^{13}C$ between −50‰ to −100‰)).

is present in extant marine *Proteobacteria*, suggesting that the legacy of a simple bacterially mediated light-driven energy generation process may yet exist in modern oceans, and thus offer a potential insight into the evolution of photosynthesis in the marine realm (Béjà et al., 2000).

A more sophisticated process would have evolved when refinement of the original porphyrins, through replacement of the iron core by magnesium and small structural changes in the ring structure, eventually led to the development of more efficient light-sensitive pigments, the bacteriochlorophylls (Jones, 1985). There are several different bacteriochlorophylls, some of which absorb light in the near infrared. Their capacity to absorb at long wavelengths led Nisbet et al. (1995) to propose that primitive chemolithoautotrophs may have enhanced their survival by being thermotactic. So, by being able to detect hot deep-sea vents by sensing infrared radiation, they could adjust their position accordingly. If such a microorganism happened to spread into a shallow, hydrothermal setting, those primary pigments may have evolved from heat sensing to the use of infrared radiation for chemical energy in photosynthesis, whilst still maintaining the ability to use electrons from H_2. Some might even have used H_2S, but whether there was as a high enough concentration of dissolved sulfide throughout the oceans, like the modern Black Sea, or whether it was confined to hydrothermal settings due to bulk ocean dilution is unknown (at that time the mesophilic sulfate reducing bacteria had not yet evolved). For an alternate view, Zavarin (1993) has suggested that photosynthesis evolved in an epicontinental soda lake, with subsequent photosynthetic life forms later moving into hypersaline lagoons.

By adapting mobile intracellular ubiquinones to form a rudimentary reaction center complex, light energy could now make use of an established electron transport system to facilitate electron flow from the excited pigments, through a series of intermediate carrier enzymes of decreasing electrode potential, back to the oxidized donor pigment (Barber and Anderson, 1994). Cyclic photophosphorylation resembles respiration and chemolithoautotrophy because electron flow establishes a proton gradient, but unlike the other two, there is no net input or consumption of electrons for ATP generation. This meant that instead of reduced inorganic substrates being the ultimate source of the reducing potential, a photochemical charge separation would have conferred an energetic advantage to the phototrophs over other cells that were competing for the same electron sources and sinks.

The nature of the first photoautotrophic microorganism may never be known, but considering the lack of photosynthetic *Archaea* (bacteriorhodopsin is adequate for supplementing but not wholly supporting a cell), it surely evolved through the *Bacteria* domain. That is, however, where the agreement ends. Based on 16S rRNA phylogenies, the earliest evolving photosynthetic lineage was originally suggested to be a green nonsulfur bacterium, similar to the modern genus *Chloroflexus* (e.g., Woese, 1987). Using heat shock proteins as markers, Gupta et al. (1999) concluded instead that heliobacteria were the earliest evolving photosynthetic lineage. Meanwhile, phylogenetic analyses of multiple photosynthesis genes has brought to light a different scenario, one that identifies the purple bacteria as the oldest photosynthesizers (Xiong et al., 2000). More recently, Raymond et al. (2002) carried out whole-genome comparisons of representative species in all five photosynthetic bacterial phyla and found that lateral gene transfer was pervasive amongst them. Furthermore, when their DNA was compared, there was a distinct lack of unanimous support for a single gene topology, indicating that photosynthetic species comprise a mosaic of genes with very different evolutionary histories. This is not overly surprising given that the earliest phototrophs probably grew in microbial mats (see below), where their close juxtaposition would strongly facilitate the exchange of genetic material.

Despite the current uncertainties, it is interesting that the bacteriochlorophyll used in purple bacteria absorbs light in the near-infrared (800 to 1040 nm, using either bchl *a* or *b*). The purple bacteria belong to the largest and most physiologically diverse of all bacterial phyla, the *Proteobacteria*. It consists of five major classes; alpha, beta, gamma, delta, and epsilon, all of which may have diversified from one ancestral phototroph (Woese, 1987). The alpha (including the purple nonsulfur bacteria), gamma (purple sulfur bacteria), and some of the beta subdivisions have members that are still phototrophic, using either H_2 (reaction (7.16)) or H_2S (reaction (7.17)) as reductants during autotrophic carbon fixation. Alternatively, some species might have utilized pre-existing organic substrates, thereby acting as photoheterotrophs.

$$12H_2 + 6CO_2 \rightarrow C_6H_{12}O_6 + 6H_2O \quad (7.16)$$

$$3H_2S + 6CO_2 + 6H_2O \rightarrow$$
$$C_6H_{12}O_6 + 3SO_4^{2-} + 6H^+ \quad (7.17)$$

It is also curious that the purple nonsulfur bacteria prefer H_2 as their electron donor for autotrophic growth, thus negating the need for reverse electron transport. Might this reflect their origins proximal to a hydrothermal vent system where H_2 was locally abundant, and later the process of reversed electron transport evolved as a means of coping with either changing environmental conditions or expansion into new habitats, where only less electronegative reductants, e.g., H_2S or Fe(II), were available?

Other genera within those subdivisions, as well those from the delta and epsilon subdivisions, are entirely nonphototrophic, having evolved later by subsequent exchange of their original photosynthetic capacity for other energy-yielding strategies, i.e., chemoheterotrophy and chemolithoautotrophy. Unlike the deeply branching lineages that preceded it, only some extant purple bacteria grow under thermophilic (40–60°C) conditions; most are mesophilic, implying that hyperthermophily was no longer an obligatory trait (Ward et al., 1989).

It is further possible that a primitive purple bacterium may have been capable of two-electron transfers utilizing hydrogen peroxide as the electron donor, generating O_2 as a by-product (recall reaction (7.12)). Although not likely abundant in the Archean oceans, H_2O_2 could have been formed, and concentrated on pyrite, around hydrothermal vents (Borda et al., 2001). Its oxidation can be carried out by a moderately oxidizing species ($E^{\circ\prime} = 0.27$ V), and it is possible that the ancient purple bacteria may have utilized a Mn-containing catalase, first as a means to detoxify the hydrogen peroxide, but later as part of the first oxygen-evolving complex (see below).

The purple bacteria were also apparently the first microorganisms to employ the Calvin cycle, which is dependent on rubisco, the enzyme capable of either carboxylase or oxygenase functions, used in photosynthesis and photorespiration, respectively. It is the most common modern protein, intimately involved in oxygenic photosynthesis. Its presence, therefore, in very ancient

photosynthetic lineages may imply that early oxygen management was needed by the primitive microbial communities. Conceivably a cell that had rubisco for photosynthesis later moved into a mildly oxygenated environment, and as the O_2:CO_2 balance shifted, its enzymes took on an oxygenase capability (Peretó et al., 1999).

Chloroflexus-like species might have evolved after the purple bacteria were already established, when the reaction center that had initially used to sense heat (e.g., bchl a) was modified with the addition of antenna complexes to harvest red light at shorter wavelengths (e.g., bchl c – 740 nm). The metabolism of extant *Chloroflexus* is relatively versatile, being able to grow photoautotrophically on H_2S, H_2, and Fe(II), photoheterotrophically, or chemoheterotrophically in the presence of O_2. Although purple photosynthetic bacteria and *Chloroflexus* share distinct similarities between their reaction centers (both type II containing pheophytin and ubiquinone molecules, Blankenship, 1992), the latter fixes carbon through the unique hydroxypropionate pathway.

Green sulfur bacteria could have come next. They have a different reaction center (type I containing Fe–S clusters) that is more electronegative upon illumination, making it possible to directly reduce $NADP^+$ without the need for reverse electron transport. They also employ the reductive citric acid pathway for CO_2 fixation. Green sulfur bacteria are tolerant of high sulfide conditions and can grow in conditions that are toxic to other nonsulfur photosynthetic bacteria. They are also strictly anaerobic, growing today in the lower layers of microbial mats along with *Chloroflexus*, commonly underlain by a basal layer of fermenters and methanogens and overlain by purple bacteria.

At some stage, these various phototrophic bacteria, and those that retained their chemotrophic origins, would have begun growing together as part of a microbial mat community, but still near a hydrothermal source. Some of those mats would have trapped and bound sediment; others may have induced mineral precipitation. In either case, the organosedimentary structures

Box 7.2 Archean stromatolites

The oldest well-preserved stromatolites have been found in sedimentary rocks from the ~3.5 Gyr Warrawoona Group of the Pilbara Supergroup in Western Australia and the ~3.3 Gyr Fig Tree Group of the Swaziland Supergroup, South Africa (e.g., see Plate 15). The Warrawoona stromatolites are found interstratified amongst mafic volcanic units, and consist of relatively simple laminated domes to columnar structures, frequently preserved in chert (Lowe, 1980; Walter et al., 1980). The fine microstructures show wavy and wrinkled laminae predominantly 50–200 μm thick, although many laminae are as thin as 20 μm. Moreover, some of the stromatolites show evidence of carbonate sediment trapping and binding, a feature common in modern intertidal mats (Van Kranendonk et al., 2003).

The Fig Tree stromatolites formed directly above silicified komatiite flows, volcaniclastic sediments,

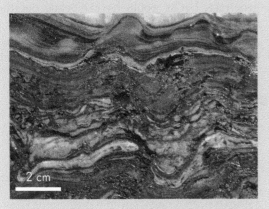

Pseudocolumnar stromatolite from the ~3.46 Gyr Dresser Formation, Warrawoona Group at North Pole, northwestern Australia. These structures occur overlying evaporative gypsum (now barite) and underlying tuffaceous sandstone with sedimentary features attesting to their syn-sedimentary origin and very shallow water environment of deposition. A biological origin is suggested by their complexity of alternating light-and-dark laminae (three size orders), and the presence of kerogen in the stromatolites and desiccation clasts. (Courtesy of Roger Buick.)

and banded ferruginous and carbonaceous chert (Byerly et al., 1986). The most common stromatolites are either stratiform or low-relief, asymmetric domes up to 3 cm wide and 3 cm high, and preserved within a chert matrix that shows laminae ranging in thickness from 50 to 100 μm.

Both the Warrawoona and Fig Tree Groups have geological features suggesting that the distribution of Early Archean mats were limited to shallow, semi-restricted evaporative basins adjacent to eroding volcanic and hydrothermal sources, characterized by rapid fluctuations in water depth and subject to burial by lava, pyroclastic debris, and a variety of detrital, evaporitic (carbonates, gypsum) and hydrothermal (silica) sediments. More stable, shallow water environments were less common.

Given that the microorganisms constructing ancient stromatolites grew in an environment subject to continuous sedimentation, and because they required nutrients in the overlying fluid, they needed strategies for coping with the dangers of sediment burial. It is widely believed that the ability of these microorganisms to move, or grow, in the direction of accretion faster than the rate of sedimentation was one means by which they survived. Sunlight would clearly have been an available stimulus for cell tropism, and it is probable that Archean stromatolites were built by microorganisms that were photoresponsive and thus could orient themselves towards the sediment–water interface (Awramik, 1992). What types of phototrophs they were remains a point of conjecture considering that both anoxygenic bacteria and cyanobacteria (see below) can form modern stromatolites with fabrics reminiscent of the ancient analogs.

One aspect does seem rather certain, that is under high sedimentation the phototrophic microorganisms would have had problems in maintaining their position at the sediment–water interface, thereby restricting their ability to construct stromatolites. Thus, the general rarity of Archean stromatolites may be due to physical, rather than biological, factors (Walter, 1994).

became lithified to form stromatolites, the first biosedimentary structures in the fossil record (Box 7.2). One of the characteristic features of stromatolites are their laminations, formed by a combination of environmental factors and behavioral responses of the microorganisms whilst alive. As such, they represent a cumulative record of an upwardly displaced microbial community over time, with their overall morphology (stratiform, domal, columnar) indicative of how the microorganisms responded to continuous accretion (Golubic, 1991).

Although the early atmosphere may have been potentially rich in sulfur gases, the early photoautotrophs would have been nonetheless subject to detrimental levels of solar radiation, particularly in the short-wavelength ultraviolet (Cockell, 2000). This is because O_2 levels were too low to produce ozone, which absorbs radiation in the wavelength range from 220 to 300 nm. Radiation with wavelengths between 190 and 280 nm (UV-C) is quite damaging to DNA, hence, primitive phototrophs would have needed to develop several different lines of defense (recall section (6.1.2)).

In addition to those employed by extant bacteria, the early phototrophs may have used their matting habit as a means of protection, whereby the topmost population of cells were killed by the radiation, while the underlying populations were effectively screened and self-shaded (Margulis et al., 1976). As long as the growth and replacement rates of the underlying cells were fast enough to exceed the death rate of the surface population, the overall community would have persisted. Other phototrophs may simply have grown under existing sediment, using the minerals as a form of sunscreen. Notably, ferric iron minerals strongly absorb in the wavelength 220–270 nm (Pierson et al., 1993). Solute-rich hydrothermal fluids would also have provided the necessary ingredients for bacteria to precipitate their own sediment cover (Phoenix et al., 2001). Amorphous silica, containing some iron (similar to modern sinters), would have allowed essential wavelengths of light to penetrate, but it would have blocked out the harmful UV (e.g., Fig. 7.8).

Some species might even have established themselves into the terrestrial subsurface, in rock

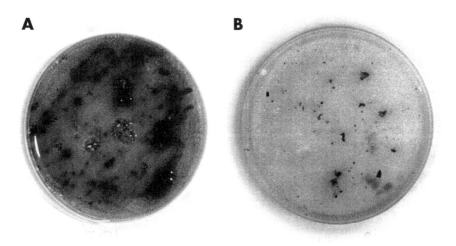

A **B**

Figure 7.8 Two agar plates with cultures of *Calothrix* sp. showing effects of UV-C radiation on biomass. (A) The cyanobacteria were initially mineralized for 20 days in a solution with 300 mg L^{-1} Si and 50 mg L^{-1} Fe(III), and then continuously irradiated for 16 days under UV-C (254 nm wavelength) at 0.35 W m^{-2}. Over 90% of the cells remained viable after irradiation for the full 16 days. (B) The cyanobacteria were not mineralized prior to irradiation. After only 4 days, less than 15% of the initial filaments were still viable. (Based on unpublished data from Phoenix et al., 2001.)

fractures and pores or under soils. Certainly, ancient soil horizons (known as paleosols) appear to have existed early in Earth's history. For instance, overlying 2.7 Gyr Archean ultramafic basement rocks in South Africa, and covered by 2.6 Gyr sediments of the Transvaal Sequence, are several weathering profiles, some up to 50 m thick, that are enriched in organic carbon (>0.1%) presumed to be of microbial origin (Martini, 1994). This premise has subsequently been supported by the occurrence of crystallographic structures, chemical signatures, and the presence of carbonaceous matter in seams (up to 1 mm thick) within a paleosol sequence from one of those locales, that seemingly indicates the previous existence of localized microbial mats that formed concurrently with the soil (Watanabe et al., 2000).

7.3.2 Photosynthetic expansion

Early photosynthetic growth was probably constrained to the peripheries of shallow hydrothermal vents, where a continuous supply of H_2 and H_2S (and possibly H_2O_2) were available. Thus, it is inevitable that they eventually experimented with other electron donors that would allow them to escape their hydrothermal confines and colonize new distal habitats. One such possibility was the abundance of dissolved Fe(II), which through its oxidation would have additionally yielded an external UV protecting shield (Pierson, 1994). The input of Fe(II) from midocean ridges was almost certainly greater during the Archean, a view supported by the Fe-rich nature of sandstones and shales of that time (Kump and Holland, 1992). Today, purple nonsulfur bacteria, green sulfur bacteria, and maybe even *Chloroflexus*, grow by photoferrotrophy (recall section 2.2.3), forming ferric hydroxide in the process (reaction (7.18)). Several such species grow today as plankton deep within the water column, and it is conceivable that their Archean predecessors may have lived in a similar manner, sheltered from UV radiation, and harvesting the light filtering down to their depths, with perhaps only their accessory carotenoid pigments (Kappler et al., 2005).

$$24Fe^{2+} + 6CO_2 + 66H_2O \rightarrow$$
$$C_6H_{12}O_6 + 24Fe(OH)_3 + 48H^+ \qquad (7.18)$$

It has also been suggested that in the CO_2-rich Archean oceans, primitive anoxygenic photosynthesizers may have taken advantage of readily available Mn-bicarbonate clusters, such as $Mn_2(HCO_3)_4$, generating O_2 as a reaction byproduct. These clusters are highly efficient precursors in the assembly of the tetramanganese water-oxidizing complex found in oxygenic photosynthesizers (Dismukes et al., 2001).

If either of these metal-based photochemical reactions were used in the Early Archean, then it is imaginable that planktonic species would eventually have diversified to colonize much more of the shallow coastal areas, and perhaps even the open ocean. Simultaneously, fermentative and respiring bacteria, exploiting their cellular debris, would have expanded beneath them on the seafloor. The anoxygenic photosynthetic production of sulfate and ferric iron would certainly have enhanced the magnitude of existing anaerobic respiratory pathways, but it is unclear by how much. Based on the presence of marine barite deposits in the Early Archean, Schidlowski (1989) has argued for the existence of a biologically mediated marine sulfate reservoir, as early as 3.5 Ga (Box 7.3). His premise has, to some extent, been supported by a recent study of evaporites from the 3.47 billion year old Dresser Formation, Western Australia. Large isotopic differences (averaging 10‰) were measured between small grains of pyrite embedded in barite and the barite itself, suggesting that this environment may have supported bacterial sulfate reduction (e.g., Shen et al., 2001). Furthermore, these deposits were originally composed of gypsum, and considering that gypsum forms at temperatures below 60°C, the findings point towards the existence of mesophilic sulfate-reducing bacteria already at that time. But, as alluded to earlier, whether this implies an earlier thermophilic origin of sulfate reduction or whether the respiratory pathway evolved in a warm evaporitic setting, like that of the Dresser Formation, remains unresolved.

Box 7.3 Sulfur isotopes

Sulfur isotopic ratios provide valuable clues regarding the presence of sulfur-based metabolic activity on the early Earth. Dissimilatory sulfate reduction is the principal process by which large-scale, low temperature sulfide formation occurs. During this reaction, microorganisms preferentially incorporate the lighter isotope, ^{32}S, into their cells, thereby enriching the external aqueous milieu in ^{34}S. With part of this H_2S escaping biological re-oxidation and ending up as sedimentary pyrite, the process is responsible for the fractionation of the sulfur pool into a light sulfide component, enriched in ^{32}S, and a heavy sulfate component, enriched in ^{34}S. Since the isotopic composition of dissolved sulfate is preserved in sulfate evaporites, the sedimentary rock record thus records the state of the sulfur cycle through geological history (Schidlowski et al., 1983).

Sulfur fractionations are directly dependent upon a number of variables, one of them being sulfate concentration. When limited sulfate is

available (<200 μmol L⁻¹), fractionations are suppressed to values near 0‰ (Habicht et al., 2002). By contrast, sizeable isotopic shifts in the negative direction, as far as −46‰, have been measured for sulfate reduction when sulfate concentrations exceed 1 mmol L⁻¹.

In addition to sulfate availability, under optimal growth conditions different SRB fractionate with different efficiencies (between −2‰ and −42‰), indicating that there is also a species-specific physiological effect (Detmers et al., 2001). Moreover, the fractionation depends on whether SRB completely oxidize the carbon source to CO_2 (inducing a large fractionation) or whether they oxidize incompletely and release acetate (inducing a smaller fractionation). Some mineral sulfides in anoxic modern sediments and euxinic waters are depleted in ^{34}S by 45–70‰ relative to seawater sulfate, a pattern that can be accounted for by a cyclic process; first reduction of sulfate to H_2S, next re-oxidation of H_2S to S^0, then the disproportionation of S^0 into sulfate and sulfide, followed again by the same cycle (Canfield and Thamdrup, 1994).

Sulfur isotopic values of sulfide (pyrite) and sulfate (barite) minerals in the Early Archean generally display a relatively narrow spread around $0 ± 3$‰ for sulfides and $4 ± 1$‰ for sulfates (Strauss, 2003). The slight fractionations between the minerals ($Δ^{34}S$ averaging 4‰) appears consistent with the abiological processes of evaporative precipitation of sulfate and sulfide from a volcanogenic sulfur source. Others, however, believe that these fractionations reflect the activity of bacterial sulfate reduction (e.g., Ohmoto et al., 1993), although the low levels of sulfate available at that time were arguably not sufficient to support SRB on a global scale. Therefore, if there is an ancient biological signal for sulfate reduction, it is only in localized, SO_4^{2-}-rich environments, where measurable isotopic differences between sulfur sources could take place.

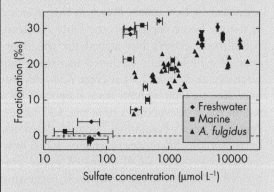

^{34}S-isotopic fractionation as a function of sulfate concentration for freshwater and marine populations of sulfate-reducing bacteria and for the hyperthermophile *Archaeoglobus fulgidus*. The horizontal bars represent the range of sulfate concentrations used in the experiments. (Reprinted with permission from Habicht et al., 2002. Copyright (2002) AAAS.)

With the evolution of photoferrotrophy, biological Fe(II) oxidation would have superseded photochemical oxidation because the bacteria could grow deeper in the water column where UV radiation would be effectively absorbed (Konhauser et al., 2002b). As long as iron and nutrients were available, the Fe(II)-utilizing phototrophs could have contributed to ferric hydroxide deposition onto vast areas of seafloor, some of which became manifest as banded iron formations (Box 7.4).

Furthermore, it has been suggested that those communities could have generated up to 1.9×10^{13} mol C yr^{-1} throughout all oceans (Kharecha et al., 2005). To put this number into context, Canfield (2005) has estimated that total net primary productivity in the late Archean-Palaeoproterozoic oceans was $1.8–5.6 \times 10^{14}$ mol C $year^{-1}$. Correspondingly, when existing heterotrophic cells eventually moved into the open ocean and pelagic sediments, the reduction of

---

Box 7.4 Banded iron formations (BIFs)

BIF deposition has occurred sporadically over 3 billion years. During most of the period from 3.5 to 2.7 Ga, BIF formed as part of greenstone successions. They are relatively thin, tectonically deformed and highly metamorphosed, and their poor preservation makes the nature of their deposition uncertain. These are the "Algoma" type BIF and their high content of iron and silica is thought to have been derived from volcanic effusive and hydrothermal sources.

The early BIFs are then followed by a period of massive iron deposition, the "Superior" type BIFs, leading to some of the most abundant chemical sedimentary deposits formed during the period 2.7–1.8 Ga, including those in the Hamersley Group, Western Australia (see Plate 16) and the

Generalized view of the 2.5 Gyr Brockman Iron Formation, Hamersley Group, Western Australia. The characteristic mesobanding of BIF, consisting of centimeter-wide alterations between iron minerals (hematite, magnetite) and chert, is shown.

Transvaal Supergroup, South Africa (Trendall, 2002). These rocks formed under tectonically stable conditions over tens of millions of years. Two characteristic features of the depositional setting were: (i) water depths below wave base, but shallow enough for carbonate precipitation during periods of non-BIF deposition; and (ii) deep ocean water was able to circulate freely into and out of them, but some form of physical barrier was nonetheless required to explain the absence of terrigenous siliciclastic sediment coarser then clay size. It is generally assumed that those conditions were best met on partially isolated, submerged platforms on the continental shelves of older cratons (e.g., Morris and Horowitz, 1983). However, recent reports have suggested instead that they were deposited in an abyssal plain setting, with the non-Fe layers forming from shelf-derived turbidity currents, and the Fe-rich layers resulting from the resuspension of ferrigenous pelagic muds that formed on the flanks of submarine volcanoes (e.g., Krapež et al., 2003). What makes these rocks particularly unique is their laminae, which can be observed on a wide range of scales, from coarse macrobands (meters in thickness) to mesobands (centimeter-thick units), to millimeter and submillimeter layers. Among the latter are the wide variety of varve-like repetitive laminae, known as microbands.

Although it is generally accepted that the iron in BIF is ultimately of hydrothermal origin, the mechanism(s) for its deposition remains unresolved (Morris, 1993). The ferric iron content of BIF minerals is ostensibly indicative of oxidative conditions during mineralization. Herein lies the

continued

Box 7.4 *continued*

dilemma. On the one hand, Fe(III) is virtually insoluble in the pH range of most oxygenated waters, so it is accepted that the iron in BIF was originally transported in the reduced oxidation state, e.g., Fe(II). This means that large portions of the deep oceans must have been anoxic. On the other hand, the vast amount of iron tied up in BIFs (the world's largest BIFs from the Late Archean contain 10^{14} tons of iron; Isley, 1995) suggests that a substantial oxidation mechanism was present. Moreover, the average oxidation state of BIF is $Fe^{2.4+}$, significantly higher than that of primary igneous rocks (Klein and Beukes, 1992).

To explain the earliest BIF deposits, experiments were developed to demonstrate how high unattenuated fluxes of UVC or UVA could have photochemically oxidized Fe(II) to Fe(III) in acidic (Cairns-Smith, 1978) or circumneutral (Braterman et al., 1983) pH waters, respectively. Subsequent models then showed that this process alone could have accounted for all the Fe(III) in BIF, prior to the onset of oxygenated conditions (François, 1986). However, the initial experiments were never performed using complex, multicomponent solutions that actually mimicked Precambrian marine waters, and therefore, this oxidative mechanism remains contentious.

Another potential precipitation mechanism centers on the activity of microorganisms. Cloud (1965) originally proposed that primitive oxygen-releasing photosynthetic bacteria (the precursors to modern cyanobacteria), that may have lacked suitably advanced oxygen-mediating enzymes, attached any free O_2 that was generated to convenient oxygen acceptors in the immediate vicinity of the cells. The abundant dissolved Fe(II) at that time would have served as such an acceptor, thereby maintaining the reducing environment necessary for their survival. Therefore, the photosynthetic microorganisms would have flourished when iron and nutrients were episodically available, allowing them to induce ferric hydroxide precipitation, but they declined in numbers to background levels (no Fe mineralization period) when these components became depleted. Similar processes are presently observed in the oceans when massive phytoplankton blooms develop following iron fertilization (e.g., Coale et al., 1996). In the absence of Fe(III) precipitation,

amorphous silica (manifest today as chert) and carbonates (siderite, ankerite, ferroan dolomite) were deposited, giving rise to the non-Fe layers (e.g., Hamade et al., 2003).

It is also possible that the bacterial surfaces simply functioned as sorption sites for Fe^{2+}, and over a short period of time the cells become completely encrusted in amorphous iron as abiological surface catalysis accelerated the rate of mineral precipitation (Konhauser, 2000). The process of transferring oxygen to nonbiological oxygen acceptors may have later evolved into a source of energy for chemolithotrophic, microaerophilic bacteria such as *Gallionella ferruginea* (Holm, 1989). Alternatively, a number of purple and green bacteria can couple Fe(II) oxidation to the reduction of CO_2 during anoxygenic photosynthesis. If indeed this was the case, then at the time in Earth's history when BIFs were being deposited, it was light rather than oxygen that coupled the Fe cycle to the C cycle (Hartman, 1984). At the heart of the argument lies the question of whether O_2 was present in the ocean waters, but in either case, it can be theoretically demonstrated, based on extant biological Fe(II) oxidation rates, that chemolithoautotrophic and phototrophic bacteria had the means to be important agents in iron deposition, in addition to the indirect role played by cyanobacteria (Konhauser et al., 2002b).

Once buried, ferric hydroxide would have transformed into: (i) siderite or magnetite through microbial Fe(III) reduction; (ii) clays, when co-precipitated silica, entrained by iron during deposition, reacted with other cationic species in high pH pore waters; or (iii) hematite, through dehydration of any residual $Fe(OH)_3$, after the respiratory processes consumed the labile organic carbon fraction (Morris, 1993). That magnetite, and possibly some siderite, were microbially catalyzed is suggested by: (i) textural features in BIF, such as magnetite overgrowths on hematite and (ii) isotopically light $\delta^{13}C$ values in BIF carbonate minerals. The latter implies that organic carbon in the bottom sediments was oxidized to bicarbonate, with the concomitant reduction of some electron acceptor, presumably ferric hydroxide (Walker, 1984). Significantly, this oxidation process explains the paucity of organic matter in BIFs (<0.5 wt%; Gole and Klein, 1981).

Fe(III), coupled to organic oxidation, would have become a favorable metabolic pathway (Konhauser et al., 2005). In support of this is the observation that Fe(III) reduction is broadly distributed amongst several known Proteobacterial genera, suggesting that this form of metabolism may be due to convergent evolution or lateral gene transfer between different species over the course of evolution (Barns and Nierzwicki-Bauer, 1997). Direct evidence comes from Fe isotopic ratios in Fe-bearing minerals from Archaen sedimentary rocks (see Box 7.10).

7.3.3 The cyanobacteria

At some time during the Archean, a significant advance was made that would irreversibly shape conditions on the Earth's surface. This was the advent of oxygenic photosynthesis and its liberation of O_2 into the euphotic zone of the oceans, and possibly the atmosphere:

$$6H_2O + 6CO_2 \rightarrow C_6H_{12}O_6 + 6O_2 \qquad (7.19)$$

The process required two major evolutionary events to have simultaneously arisen: the development of an oxidant with a sufficiently high electrode potential to remove electrons from water and a complex that could collect and store the four electrons generated during the photochemical event. The first requirement was met with some fine-tuning of the charge-separating bacteriochlorophylls, perhaps the bchl g of heliobacteria, leading to the evolution of the chlorophyll a molecule (Kobayashi et al., 1998). Chlorophyll a is able to absorb light at much shorter wavelengths (680 nm) than any of the pre-existing bacteriochlorophylls, and as a result, it could facilitate the oxidation of water. This reaction, though, leads to the intermediate formation of hydrogen peroxide, which if left unchecked within the cell can cause irreparable damage. That is where the previous development of catalase (and possibly superoxide dismutase) comes into play because it would have quickly oxidized the hydrogen peroxide to O_2. As some of the anoxygenic photo-synthesizers were already equipped with catalase, then the second innovation of charge accumulation may have arisen through the association of a purple bacterium's reaction center with a Mn-catalase enzyme that was already in place. Given that Mn-catalase has a binuclear center that is structurally equivalent to half of the tetra-manganese water oxidizing complex of extant oxygenic photosynthesizers, then subsequent developments could convert the binuclear Mn site into a tetranuclear site capable of accumulating more electrons than before (Blankenship and Hartman, 1998). The potential abundance of dissolved Mn-bicarbonate clusters may have been the source of the additional Mn^{2+} ions during the assembly of the water oxidizing complex (Baranov et al., 2000).

The microorganisms taking advantage of this new process were the cyanobacteria. They possess two reaction centers and incorporate several distinct components inherited from earlier species: the type I reaction center from green sulfur bacteria and the type II reaction center and Calvin cycle from purple bacteria. It is possible that the two photosystems may have become genetically fused when a green sulfur bacterium transferred its reaction center into a purple bacterium while the two species were growing together as part of a microbial mat community (Hartman, 1998). Interestingly, cyanobacteria have maintained the capacity to carry out anoxygenic photosynthesis in the absence of O_2 when H_2S is present. In the latter case they photosynthesize like green sulfur bacteria, using H_2S rather than water to fix CO_2. This capability may be a relict of the presymbiotic times when a green sulfur epibiont dealt with high sulfur, anoxic conditions, while the purple host handled the more oxidizing microenvironments. Once fused, the proto-cyanobacterium could protect the oxygen-sensitive green bacterium within, while continuing to emit oxygen (Nisbet and Fowler, 1999).

The benefits for cyanobacteria were immense. Their newly developed ability to strip electrons from the virtually unlimited supply of water meant that they were no longer dependent upon the availability of hydrothermally derived reductants. Instead, they could now rapidly proliferate and diversify throughout the euphotic zone of the oceans. To highlight the significance of this innovation we use modern Earth as an example. The delivery of H_2S, Fe^{2+}, Mn^{2+}, H_2, and CH_4 at hydrothermal vents is estimated to sustain at most 2×10^{12} moles of organic carbon production annually by bacteria capable of using them as their energy source (Des Marais, 2000). By comparison, modern global photosynthetic productivity is estimated at 9000×10^{12} mol C yr^{-1} (Field et al., 1998).

Just how abundant the cyanobacteria communities were in the Early Archean, however, remains debatable. On the one hand, carbonaceous black shales have been taken as evidence of their widespread occurrence in Archean open ocean waters, underlain by a deep anoxic zone that helped preserve the organic-rich sediment (Lowe, 1994). On the other hand, their populations must have been constrained to some extent by high incident fluxes of UV radiation. The optical properties of the Archean oceans are not known, but their transparency would have been affected by the absence of land-derived humic material and conversely by the vast reservoir of ferrous iron (Garcia-Pichel, 1998). There is also uncertainty regarding nutrient availability at that time. Although recent evidence points towards Late Archean crustal volume being rather similar to today (Sylvester et al., 1997), if higher seafloor spreading rates prevailed in the Archean, sea level should have been higher, thus favoring a lower estimate for land area relative to continental mass than today (Drever et al., 1988). That might have meant that the bulk of the solutes came from the deep ocean, and with that being the case, those nutrients would have first been utilized by the underlying anoxygenic photosynthetic populations in the water column.

It has also been proposed that there was a phosphate crisis, driven by phosphate adsorption onto the sedimenting ferric oxyhydroxide particles that contributed to banded iron formation (Bjerrum and Canfield, 2002). For example, at 3.2 Ga, marine phosphate concentrations may have been only 10–25% of present-day values, which should significantly have reduced the rates of photosynthesis, and hence biomass production (recall Canfield's estimate on page 321). A number of trace metals (Zn, Mo, Co, Cu, and Ni) are also essential nutrients, and any sorption reactions to sinking iron particulates could similarly have inhibited planktonic growth. Reduced primary productivity, in turn, should have led to less organic carbon burial, and recent models of the isotopic record of calcium carbonate through time support this view (Bjerrum and Canfield, 2004). This model, however, apparently contradicts the longstanding observations that the organic content of the average sedimentary rock has remained fairly uniform from Archean to Recent, oscillating around a long-term mean close to 0.5–0.6% (e.g., Schidlowski, 1993). Whether this reflects similar productivity through time based on nutrient limitation, or lower Archean productivity coupled to greater organic preservation, is unresolved, but it has been suggested that Precambrian bottom ocean waters may have accumulated higher concentrations of dissolved organic carbon (DOC) compared to present-day simply because they would have degraded more slowly under anoxic conditions (Rothman et al., 2003).

Aside from the open oceans, cyanobacteria could also have incorporated themselves into existing shallow-water mat and sediment communities. Based on the contention that microfossils within 3.5 Gyr stromatolitic cherts from Australia and South Africa are actually cyanobacteria, or their predecessors, then shallow-water oxygenic photosynthesis may already have existed in the Early Archean (Box 7.5). A firmer benchmark for the presence of cyanobacteria comes from the 2.7 Gyr stromatolitic assemblages of the Tumbiana Formation, Western Australia, that were accreted by phototrophic bacteria. Based on their habitat in a sulfate-deficient evaporative lake, they almost certainly metabolized by oxygenic photosynthesis (Buick, 1992). This view is supported by the earliest recognized fossil assemblage of colonial coccoid cells, from the

Box 7.5 Archean microfossils

Arguably the most persuasive evidence for the existence of early life comes from the examination of microfossils that have been recovered from Archean siliceous strata. There are only about 30 putatively microfossiliferous units which are known from the Archean, compared to nearly 3000 bona fide microfossils in Proterozoic strata (Schopf, 1994). The majority of the oldest microfossils have filamentous or coccoid morphologies that appear to have grown as microbial mats in semirestricted basins, where a combination of high productivity, limited water circulation, and high salinity facilitated greater cellular preservation (Horodyski et al., 1992). Rapid mineralization was also necessary to limit post mortem degradation, a feature most commonly met by primary silicification in shallow, silica-supersaturated waters (Knauth and Lowe, 2003). Additionally, the microfossils consist almost exclusively of the remains of cell sheath and wall material, while the cytoplasmic contents have either completely decayed or formed remnant structures unrelated to the original structure (e.g., Knoll et al., 1988). This is unsurprising given that the host rocks have undergone extended periods of metamorphism.

Structures resembling bacteria from 3.45 billion year old Apex cherts of the Warrawoona Group have, until recently, been deemed the oldest morphological evidence for life on Earth (Schopf, 1993). These "microfossils" were thought to occur within weathered clasts that were originally formed in an older sedimentary chert unit that, subsequent to its lithification, was eroded to produce the lithic fragments. The biological origin of the filaments was inferred from the degree of regularity of cell shape and dimensions, the apparent presence of partial septations, and by their morphological similarity to extant filamentous prokaryotes. Moreover, laser-Raman spectroscopic imagery of the filaments was used to infer the presence of kerogen in higher concentrations than the surrounding matrix (e.g., Schopf et al., 2002). Some of the

Apex specimens also exhibit features reminiscent of unbranched, partitioned trichomes, that has led to the suggestion that, not only were the Archean "microorganisms" capable of gliding and possibly phototactic motility, but that cyanobacteria may already have been in existence at that time (e.g., Awramik, 1992). Support for this conclusion came from the earlier discovery of large spheroidal, sheath-like structures (up to 20 μm in diameter), in cherts from the underlying Towers Formation, that resemble modern-day cyanobacteria (Schopf and Packer, 1987).

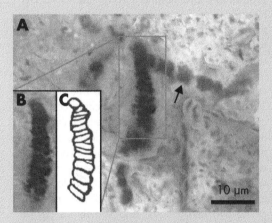

Automontage of an inferred cyanobacterial microfossil, *Archaeoscillatoriopsis disciformis*. (A) From the Apex chert. (B,C) Digital image and interpretative sketch in the style of figure 3M in Schopf (1993) that omits the side branch (arrow). (Courtesy of Owen Green.)

An alternate view was provided by Walter et al. (1972), who suggested that some Archean microfossils may instead be likened to modern filamentous, anoxygenic photosynthetic bacteria, e.g., *Chloroflexus* sp. On a general morphological level, especially when finer cytoplasmic details have been lost as a result of silicification, *Chloroflexus* could easily be mistaken for cyanobacteria: the main ultrastructural difference being that *Chloroflexus* has a smaller filament diameter

continued

Box 7.5 *continued*

and a simpler intracellular structure. It is also possible that the "microfossils" could represent primitive thermophilic chemolithoautotrophs that grow today as filamentous mats adjacent to hot spring vents (Lalonde et al., 2005).

Among the other notable Archean examples are the microfossils found within various formations from the 3.48–3.27 Gyr Onverwacht Group, Swaziland Supergroup, in the eastern Transvaal, South Africa. The Onverwacht Group is composed largely of komatiitic and basaltic volcanic rocks that are interdispersed with generally thin sedimentary layers of chert. Within the ~3.45 Gyr Hooggenoeg Formation, some cherts contain thread-like or cylindrical filamentous structures, ranging from less than 0.2 µm to 2.5 µm in diameter (Walsh and Lowe, 1985). The overlying Kromberg Formation, includes mostly mafic volcanic and volcaniclastic rocks, but the basal 150–400 meters is a unit of silicified carbonaceous and ferruginous sediment called the Buck Reef Chert. This unit has layers ranging from those deposited in shallow evaporative ponds to deep, open marine waters. The shallow water cherts contain wavy stromatolite-like laminae with thin (0.1–2.2 µm diameter), solid or hollow, kerogenous, and sometimes pyrite-encrusted microorganism-like filaments that lie parallel to subparallel to the fabric. The solid filaments are frequently broken in a regularly distributed manner, suggesting possible cellularity (Walsh and Lowe, 1985). The deeper marine sequence possess rounded, sand-sized, detrital carbonaceous matter, derived from the mats, and distributed by waves and currents to locations deeper offshore (Tice and Lowe, 2004). Based on the δ^{13} values of the carbonaceous matter (−35 to −20‰), the presence of siderite, and the lack of primary ferric iron oxides, the same authors have inferred that some form of chemical stratification existed in the water column, such that Fe^{2+} was absent from the shallow waters populated by mats of photosynthetic bacteria.

Within the Onverwacht pillow lavas, there is also putative evidence for the presence of endolithic bacteria weathering the volcanic glass soon after their crystallization (Furnes et al., 2004).

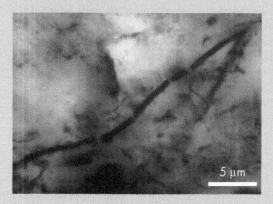

Photomicrograph of a single segmented filamentous structure within carbonaceous laminae in sedimentary chert of the Kromberg Formation, Onverwacht Group, South Africa (approximately 3.4 Ga). (Courtesy of Maud Walsh.)

Meanwhile, the discovery of pyritic, thread-like filaments in 3.2 Gyr volcanogenic massive sulfides from the Pilbara Craton of Australia seemingly provides evidence that chemolithoautotrophic thermophiles also lived in, or around, hydrothermal systems at that time (Rasmussen, 2000).

A re-examination of the Apex chert by Brasier et al. (2002) has, however, called into question the biogenicity of the filamentous structures and the sedimentary origins of the earliest "fossiliferous" deposits. They suggested that the structures are probably secondary artefacts formed by Fischer–Tropsch type reactions associated with seafloor hydrothermal systems. Indeed, Buick (1988) has maintained that the actual chert units from which the "oldest microfossils" derive are secondary hydrothermal deposits of much younger age that cross-cut the primary bedding. Buick (1991) has persuasively argued that the criteria needed to affirm the existence of a fossil (i.e., they should

continued

Box 7.5 *continued*

occur in thin sections of rocks of sedimentary origin having undergone low grade metamorphism, consist of kerogen, exceed the minimum size for independently viable cells, coexist with others of similar morphology, have hollow structures, and display cellular elaboration) are not met in the spheroids and filamentous structures in the Warrawoona samples.

The Warrawoona samples face criticisms on yet two other fronts. First, experiments have demonstrated that the mere presence of kerogen in a microfossiliferous sample does not in itself indicate biogenicity because simple organic

hydrocarbons, whose sources are abiological (e.g., formaldehyde), readily condense onto silica-carbonate inorganic filaments, and subsequently polymerize under gentle heating to yield kerogenous products (García-Ruiz et al., 2003). Second, the same kind of Raman spectral signature, as is obtained from kerogen, is also obtained from many other poorly ordered carbonaceous materials that arise from abiological processes. Thus, the presence of carbonaceous material in the Warrawoona samples is not sufficient to prove biogenicity (Pasteris and Wopenka, 2003).

2.6 Gyr Campbell Group, South Africa, which appears to include cyanobacterial genera, such as *Phormidium* or *Lyngbya*. These microorganisms contributed to the formation of stromatolitic reefs in shallow subtidal to intertidal settings (Altermann and Schopf, 1995). High concentrations of 2α-methylhopane biomarkers in the 2.6 Gyr Marra Mamba Formation, Western Australia provide another piece of evidence consistent with the appearance of cyanobacteria in the Late Archean (Box 7.6).

Despite the apparent evidence for cyanobacteria by at least 2.7 Ga, the geological rock record does not record widespread oxidation on land until around 2.4–2.2 Ga, "the Great Oxidation Event" (see below). There are several possible explanations for the temporal separation between cyanobacterial evolution and the expression of their metabolism, the most likely being the continued consumption of O_2 during the protracted and progressive oxidation of the crust, and its reaction with reduced volcanic gases in the atmosphere and dissolved solutes in the deep sea (see Kasting, 2001; Holland, 2002). Thus, the oxygen fluxes to the atmosphere did not reflect gross oxygenic photosynthesis, but instead the net balance between O_2 generation and O_2 consumption. Significantly, the presence of O_2 in the surface waters of the oceans does

not necessarily imply that the atmosphere was similarly oxygenated because the rate at which O_2 can flow between the surface ocean and atmosphere is limited by diffusion through the gas–liquid interface (Kump et al., 2004).

7.4 Metabolic diversification

7.4.1 Obligately anaerobic respirers

Cyanobacteria absorb light in regions of the visible and near infrared portions of the solar spectrum complementary to photosynthetic bacteria, and their capacity to generate O_2 would have made them highly successful competitors for available space in the surface layers of mats and sediment. This invariably led to a highly stratified ecosystem with oxygenic photosynthesis taking place in the upper levels, followed by fermenters, anoxygenic photosynthesizers, and anaerobic respirers in the underlying layers – the same sequence as occurs in extant mats. In sediments, the Fe(III)- and sulfate-reducers, methanogens, and acetogens would have made use of available nutrients in minerals and pore waters, as well as the fermentable substrates

Box 7.6 Biomarkers

Unlike mineralized protists, such as coccolithophores, foraminifera, diatoms, and radiolarians, most planktonic bacteria do not leave behind microscopically identifiable remains in the sedimentary record. Instead, we are forced to rely on the preservation of specific biomarkers, the organic compounds derived from more complex precursors by either reductive or oxidative geochemical reactions, to provide information on species abundance in ancient environments. Generally, lipids, pigments, and other membranous materials are better preserved than the more labile components, such as amino acids, sugars, etc. What makes biomarkers particularly useful is that they retain some resemblance to the original biological molecules, even after a long history of decomposition and alteration that accompanies burial and diagenesis. A biomarker must also be diagnostic, and fortuitously *Bacteria*, *Archaea*, and *Eukarya* each have signature membrane lipids with recalcitrant carbon skeletons that resist degradation. Therefore, detailed characterization of the biomarker compositions allows for an assessment of the major contributing species to depositional processes in modern environments, and crucially, in the sedimentary record.

The extraction of biomarkers from Archean rocks has had the most profound affect in our understanding of early O_2 availability. In kerogenous shales from the 2.6 Gyr Marra Mamba Iron Formation and the 2.5 Gyr Mt McRae Shale, both of the Hamersley Group, Western Australia,

there are abundant amounts of 2α-methylhopanes, the derivatives of prominent lipids in cyanobacteria (methyl-bacteriohopanepolyols) that serve to make the cell membrane rigid (Brocks et al., 1999; Summons et al., 1999). These lipids have not yet been found in quantitatively significant amounts in any other groups of organisms, and therefore, they would appear diagnostic of cyanobacteria at that time of their inclusion into the sediments. With that said, our confidence in cyanobacterial markers will be tested because a recent study of genomic databases indicates that several facultative and obligate anaerobes possess the appropriate genes for hopanoids biosynthesis, and *Geobacter sulfurreducens* actually produces a wide variety of complex hopanoids under strictly anaerobic conditions in pure culture (Fischer et al., 2005).

A second important suite of biomarkers are specific steranes (28- to 30-carbon isomers) from the underlying 2.7 Gyr shales of the Jeerinah Formation, Hamersley Group. They are presently believed to represent unique alteration products of the sterols used in eukaryotic cell membranes, and the only prokaryotes known to synthesize sterols have biosynthetic pathways leading to different structural isomers. Importantly, O_2 is required for the biosynthesis of sterols, hence their extraction form Archean rocks suggests that at least some dissolved oxygen (>0.002 ml O_2 L^{-1}) must have been present in surface waters at that time (Runnegar, 1991).

---

derived form the lysed cyanobacteria above, and moved deeper into the aphotic zone (below the level of effective light penetration) of unoccupied sediments, away from the presence of free O_2. This stage may have marked the appearance of the obligate anaerobes, microorganisms that were already in existence, but then became limited to anoxic niches away from the toxic effects of oxygen. Accordingly, once segregation of bacteria into distinct niches became the norm, evolutionary divergences would have become inevitable, ultimately leading to the extreme

diversity of heterotrophs encountered in modern environments (Nisbet, 1995).

Two such examples of this evolutionary divergence are mesophilic sulfate-reducing bacteria and methanogens. The SRB likely increased in number and variety, leading to the heterogeneous groups of bacteria represented today in the delta and epsilon classes of *Proteobacteria* (e.g., the genera *Desulfovibrio*, *Desulfotomaculum*, *Desulfomonas*, and *Desulfobacter*). Despite their diversification into new niches, the sulfate reducers were limited in actual number because oceanic sulfate levels

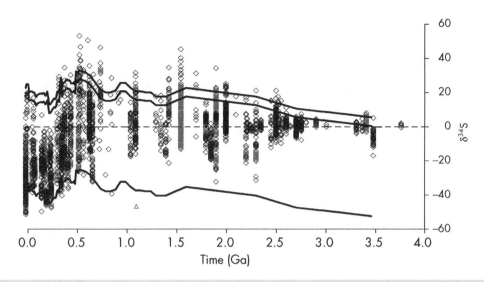

Figure 7.9 Compilation of the isotopic composition of sedimentary sulfides (diamonds) over geological time. The double line at the top of the figure is a reconstruction of the isotopic composition of seawater sulfate, while the bottom single line is the isotopic composition of seawater sulfate offset by 55‰, representing the maximum fractionation between sulfate and sulfide through the past 600 million years. The bottom line has been extrapolated back in time to account for the paucity of isotopic compositions before 1.7 Ga. (From Canfield, 2005. Reprinted with permission from the *Annual Review of Earth and Planetary Sciences*. Volume 33 © 2005 by Annual Reviews. www.annualreviews.org)

remained low (<1% of present-day levels, which are 28 mmol L^{-1}) until some time in the Late Archean (Fig. 7.9). Indeed, the 2.75 Gyr sedimentary sulfides of the Michipicoten Iron Formation, with $\delta^{34}S$ values as low as −10.5‰, are possibly the first presumptive evidence for widespread bacterial sulfate reduction (Goodwin et al., 1976); larger scale fractionations that seem unambiguously due to bacterial sulfate reduction are observed in the 2.35 Gyr Timeball Hill Formation, Transvaal Supergroup, South Africa (Cameron, 1982).

With sulfate reduction repressed, respiration prior to 2.7 Ga may have been dominated by methanogenesis (Kral et al., 1998). Some methanogens of the order *Methanobacteriales* expanded their metabolic diversity by disproportionating methyl-containing compounds, such as methanol, into methane and CO_2 (reaction (7.20)) (e.g., *Methanosphaera* sp.), while others from the order *Methanosarcinales* (e.g. *Methanosaeta* sp.) oxidized the carboxyl group in acetate to bicarbonate and reduced the methyl carbon to methane (reaction (7.21)):

$$4CH_3OH \rightarrow 3CH_4 + CO_2 + 2H_2O \quad (7.20)$$

$$CH_3COO^- + H_2O \rightarrow CH_4 + HCO_3^- \quad (7.21)$$

In addition, if fermentation-generated H_2 was abundantly available in those early mats, as proposed by Hoehler et al. (2001), then production of methane through the H_2/CO_2 couple could have yielded fluxes in excess of geothermal sources by several orders of magnitude. Moreover, upward diffusion of H_2 to the atmosphere, and its eventual loss to space, combined with UV photolysis of methane in the upper atmosphere (to H_2 and CO_2), would have contributed to the irreversible oxidation of the Earth's crust, and the inevitable build up of O_2 in the atmosphere hundreds of millions of years thereafter (Catling et al., 2001).

Other *Methanosarcinales* (e.g., *Methanosarcina* sp.) may have developed syntrophic relationships with H_2-oxidizing sulfate reducers (recall section 2.5.3).

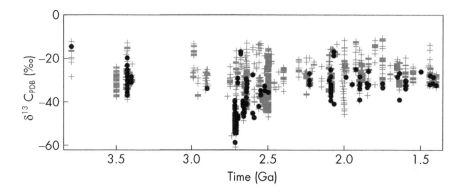

Figure 7.10 Compilation of published carbon isotope data (δ^{13}C versus PDB standard) for Archean to Mesoproterozoic kerogen carbon (solid circles) and total organic carbon (crosses). Lowest values are between 2.8 and 2.6 Gyr. (From Pavlov et al., 2001. Reproduced with permission from the Geological Society of America.)

A few methanogens diverged considerably to display a number of unusual physiological modifications. For example, *Methanomicrobiales* that moved into hypersaline environments (e.g., *Methanospirillum hungatei*) adapted to the high salt concentrations by using salt pumps to prevent osmotic dehydration. Today those species comprise the halophilic lineage, the *Halobacteriales*, an order of *Euryarchaeota* that have lost their ability to produce methane but instead have acquired both oxygen tolerance and an innovative respiratory enzyme system (Burggraf et al., 1991b). Both rRNA and ribosomal protein-based phylogenetics show that the *Methanomicrobiales* and *Halobacteriales* are sister groups (Forterre et al., 2002).

The increased role of methanogenesis had global implications for early atmospheric chemistry. It has been estimated that prebiotic methane flux rates were quite small ($<10^{10}$ molecules cm^{-2} s^{-1}). Then, once the methanogens became widely established (beginning first with those that utilized the H_2/CO_2 couple, and then the heterotrophic varieties), they may have generated methane fluxes approaching 10^{12} molecules cm^{-2} s^{-1} (Kasting et al., 1983). If the primitive methanogens produced methane at the same rate as today, the atmospheric methane concentrations could have exceeded 1000 ppm, more than 600 times its present atmospheric level. But did this take place, and if so, when? The δ^{13}C values, of Precambrian kerogens are generally within the range $-25‰$ to $-40‰$, with the lower value characteristic of methanogenic activity (Fig. 7.10). Therefore, the isotopically light values at 3.5 Ga may already represent active methanogenesis. This would not be too surprising given the recent views that a greenhouse gas was needed to explain why the young Earth avoided a deep freeze, as CO_2 concentrations were dropping due to carbonatization of the oceanic crust, tectonic loss to the mantle, and continental weathering (e.g., Sleep and Zahnle, 2001).

Although few scientists argue against the presence of early methanogens, the level of their impact is controversial. Pavlov et al. (2001) have argued for a prominent role by the Late Archean, with high methane levels leading to a greenhouse-like climate that helped offset the lower solar luminosity at that time. Such conditions, however, would inevitably have prompted a negative feedback loop because individual methane molecules would have polymerized to form complex hydrocarbons that condensed into dust-like particles. Then, if the $CH_4:CO_2$ ratio approached, or exceeded, unity, a high altitude organic-rich haze would have resulted, eventually offsetting the greenhouse effect by absorbing incoming solar radiation and re-radiating it back to space. The haze would have led to an anti-greenhouse effect, much like that on Saturn's moon, Titan. But, an organic haze would also have had a number of beneficiary effects, including minimizing the penetration of ultraviolet light into the surface oceans; and shielding ammonia from photodissociation (Sagan and Chyba, 1997). This, in turn, may have facilitated the spread of the phototrophic microorganisms into shallow-water environments, and afforded the burgeoning populations with an usable source of nitrogen. It is interesting though that no abnormally ^{13}C-depleted carbon-rich deposits, as might be anti-cipated from organic haze deposition, have been identified in pre-3.0 Ga rock records, despite such negative excursions occurring in the interval from 2.8 to 2.6 Ga (see Box 7.8). Moreover, the presence of nahcolite ($NaHCO_3$) as a primary evaporite in 3.4 Gyr rocks of the Barberton greenstone belt, and evidence for intensive weathering of first-cycle clastic sedimentary rocks, argues in favor of a $CH_4:CO_2$ ratio of less than 1 (Lowe and Tice, 2004).

7.4.2 Continental platforms as habitats

The end of the Archean witnessed significant environmental changes, beginning with the formation of several large blocks of continental crust in Australia (Pilbara craton) and South Africa (Kaapvaal craton) about 3.3–3.1 Ga, increased physical and chemical weathering, higher concentrations of Ca^{2+} and HCO_3^- transported to the oceans, and the resultant formation of stable carbonate continental plat-forms (Lowe and Tice, 2004). This facilitated the expansion of microbial mats throughout the shallow seas, which both consumed more CO_2 through photosynthesis and released O_2 into the euphotic zone (Fig. 7.11). This resulted in two important features: (i) an ocean that was either stratified, with a vast reservoir of deep, metal-rich bottom waters overlain by a con-tinuous layer of oxygenated water, or where the

presence of oxygen was limited to "oasis", in otherwise anoxic waters (e.g., Klein and Beukes, 1989); and (ii) highly oversaturated oceans with respect to calcium carbonate, which led to the indiscriminate sedimentation of massive amounts of calcite or aragonite, depending on the Mg/Ca mole ratio in seawater at that time (Hardie, 2003). Prior to then, occurrences of well-preserved carbonates are less abundant: a few are known from the Warrawoona and Onverwacht Groups, but they are all extremely thin, consisting of dis-continuous limestone and dolomite units that are often extensively replaced by chert (e.g., Buick and Dunlop, 1990). Then, by the Late Archean calcium carbonates precipitated *in situ* on the seafloor as decimeter- to meter-thick beds that extended over thousands of square kilometers. Such was the state of supersaturation that the cyanobacterial mats were likely incidental to the accretion of the stromatolites forming then (Grotzinger and Knoll, 1999).

Another important consequence of CO_2 re-moval from the atmosphere, and its incorporation into marine sediment as carbonate minerals, was the eventual initiation of Earth's first glacia-tion events at around 2.9 Ga, the glaciogenic deposits in the Mozaan Group, South Africa (Young et al., 1998). Supporting geological evid-ence for reduced atmospheric CO_2 comes from the occurrence of Fe(II)-rich carbonate as weather-ing rinds on pebbles in 3.2 Gyr river gravels that provide a lower limit pCO_2 of $10^{-1.9}$ atm at surface temperatures of 50°C (Hessler et al., 2004), while paleosol measurements at 2.8 Ga indicate an upper limit pCO_2 of $10^{-1.4}$ atm – about 100 times today's levels (Rye et al., 1995): for a dissenting view, see Ohmoto et al. (2004) who suggest that CO_2 levels must have been higher to take into account the presence of massive siderite deposits still forming at 1.8 Ga. If CO_2 levels continued to drop, as believed by most, then methane could have taken on greater relative importance (as discussed above). With time and continued carbonate deposition on the newly formed plat-forms, the degree of supersaturation decreased, and by the early Paleoproterozoic precipitation

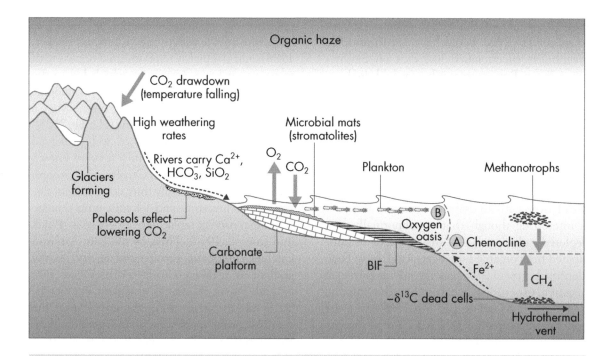

Figure 7.11 Model of the Late Archean (between 3.0 and 2.5 Ga), when large continental blocks were being weathered, causing CO_2 transfer from the atmosphere to the seafloor as stable carbonate platforms. Correspondingly, lower atmospheric CO_2 is reflected in the composition of the paleosols and Earth's first glaciation. As CO_2 levels continued to diminish, CH_4 took on greater prominence, maybe even exceeding CO_2 as the greenhouse gas. With new broad shelves, cyanobacterial mats flourished, and through the various calcification processes, they were preserved as stromatolites. Different planktonic varieties would also have evolved. Whether they collectively produced enough O_2 to create a permanently stratified ocean (A), with a continuous oxic layer overlying a deep anoxic water column, or whether they only produced enough for an oxygen oasis (B), will be largely dependent upon nutrient availability, amongst other factors. By 2.7 Ga, Superior-type banded iron formations also began to form on the continental shelves, with the Fe^{2+} derived from hydrothermal sources and silica from continental runoff. As long as Fe^{2+} and nutrients were available, microbial activity induced the precipitation of ferric hydroxide.

became limited to intertidal settings and the cementation of previously deposited sediments (Grotzinger, 1994).

Simultaneously, cyanobacteria began to take on a greater role in stromatolite formation as their filaments trapped and bound existing carbonate sediments, while the underlying heterotrophic communities created conditions conducive to mat calcification (recall section 6.1.4(b)). This period of time is reflected by the emergence of increased stromatolite diversity (Box 7.7), yet surprisingly, the geological rock record at that time shows a distinct scarcity of calcified cyanobacteria (Riding, 1994). One explanation is that the appearance of cyanobacterial calcification arose from the evolution of new species amenable to mineralization, perhaps not until the Neoproterozoic, or even later into the Cambrian (Défarge et al., 1994). Alternatively, the relative amount of alkalinization induced via photosynthesis in the Precambrian oceans was not sufficient to induce supersaturation when compared to the high dissolved inorganic carbon pool at that time (Arp et al., 2001).

Box 7.7 Proterozoic stromatolites

By the Proterozoic, stromatolites had reached their maximum abundance and multiplicity in form. Some had vertical relief from microscopic scales to reefs hundreds of meters high, and extended laterally from fringing reefs and atolls to huge barrier reefs that stretched hundreds of kilometers. Unlike the Archean, where stromatolites were limited to shallow-marine evaporitic basins, by the Proterozoic, stromatolites were also developing in nearshore and intertidal settings in the form of siliciclastic deposits (Schieber, 1999) and in epicontinental alkaline lakes and rivers, where

Composite stratiform and branching bulbous stromatolite from the 2.72 Gyr Tumbiana Formation, Fortescue Group at Billadunna, northwestern Australia. These structures contain white fenestrae, calcite-filled voids that formerly contained gases derived from respiratory process, and at the microscopic scale, microbial filaments aggregated at the crests of flexures, indicative of photoresponsive behavior. Biomarker, geochemical relationships, and carbon isotopes all show that the microbial community was complex, containing cyanobacteria that lived by oxygenic photosynthesis, methanogens that degraded dead organic matter, and methanotrophs that oxidized the resulting methane. (Courtesy of Roger Buick.)

the mats were dolomitized and silicified (Buck, 1980). Morphologies ranged from domes and simple columns to elaborately branched and bulbous structures.

There were two major periods of diversification, the first during the Paleoproterozoic (2500–1650 Ma), and the second during the latter stages of the Mesoproterozoic (1350–1000 Ma). From the first period of diversification some 180 different forms of stromatolites have been described (McNamara and Awramik, 1992). The appearance of new stromatolites at this time may have resulted from a combination of the development of new habitable spaces on the broad continental shelves and the evolution of new types of cyanobacteria with greater motility and faster growth rates that could better compensate for continuous accretion (Knoll, 1984). The stromatolites contributed to the construction of continental shelves similar to those later built by corals, calcareous algae, and other skeletal organisms. In the second period of diversification over 340 stromatolite types have been described. Although more enigmatic, the reasons for that diversification may stem from the expansion of eukaryotic algae which in some way promoted the development of new morphological architectures.

By the late Neoproterozoic and early Palaeozoic, there was a sharp decline in stromatolite diversity. The causes for this remain speculative, but grazing and burrowing of mats by metazoans, competition for nutrients, and a reduction in the carbonate saturation state of the oceans (that decreased the efficiency by which mat layers were accreted), may all have been contributory factors (Grotzinger, 1994). Coincident with this decline was a change in biota. Cyanobacterial-dominated stromatolites were gradually replaced by reefs constructed of crustose red algae, calcareous sponges, and/ or coelenterates (Golubic, 1994). Also possibly related to the stromatolite decline was the evolution of thrombolites. Unlike stromatolites

continued

7.4.3 Aerobic respiratory pathways

As oxygen concentrations increased, some microorganisms began to elaborate on their electron transport chain to include special terminal oxidase enzymes, i.e. cytochrome aa_3 (but see Castresana and Saraste, 1995 who suggest aerobic respirers predate cyanobacteria). This now made it possible to pass electrons directly onto O_2. The benefit for those cells was that they could now harness more energy from the inorganic and organic substrates they oxidized. The biochemical reduction of O_2, however, leads to the formation of several reactive intermediate products, including the superoxide anion (O_2^-), hydrogen peroxide, and the hydroxyl radical ($OH\bullet$). Consequently, the aerobic microorganisms required antioxidant molecules, such as superoxide dismutase and catalase. As discussed earlier, these enzymes appear to have ancient origins, a view supported by the fact that they are widely distributed in extant microorganisms, including obligately anaerobic bacteria (Hewitt and Morris, 1975).

The increased levels of available oxygen would have triggered a diversification of chemolithoautotrophic pathways, inevitably leading to the expansion of the proteobacterial lineages. No longer was chemolithoautotrophic activity limited to H_2-oxidation, but now reduced compounds with higher electrode potentials, e.g., CH_4, reduced sulfur compounds, Fe(II), Mn(II), NH_4^+, and NO_2^- were usable substrates. The net effect of this diversification was that microbial

activity began to play a wider role in global elemental cycling, with the products of the various chemolithoautotrophic metabolisms now providing TEAs for chemoheterotrophic activity, and vice versa.

One group of microorganisms that had an impact on the early Earth, at least by 2.8 Ga, were the methanotrophic bacteria. Their presence in the Late Archean has been inferred from the extreme levels of ^{13}C depletion in kerogens (Box 7.8) and the presence of 3β-methylhopane biomarkers in 2.5 Gyr shales of the Hamersley Group (Brocks et al., 2003). These species represented a new link between the production and utilization of biogenic methane, which importantly triggered the loss of methane from the atmosphere and the concomitant cooling of Earth's surface temperatures. This scenario could help explain the onset of the first widespread, and possibly global glaciation at about 2.3–2.2 Ga, known as the Huronian glaciation (Fig. 7.12).

Primitive terrestrial hyperthermophiles, such as *Sulfolobus* sp., may have developed oxygen tolerance quite early in archaeal evolution, using it to oxidize sulfide (in the form of FeS_2) as a means of attaining energy (Segerer et al., 1986). However, it was not until the advent of the colorless, nonphotosynthetic sulfur bacteria (all of which are *Proteobacteria*) that biological sulfide oxidation became more globally important. Two broad ecological classes exist, namely those living at neutral pH under microaerophilic conditions (e.g., *Beggiatoa* sp.) and those living in acid environments where they are involved in

Box 7.8 Kerogens

The carbon isotopic compositions in Precambrian kerogens have yielded some interesting findings, the most prominent being that they are isotopically light (i.e., more ^{12}C-enriched) compared to carbonate carbon. For example, kerogens of the Warrawoona and Fig Tree Groups have average $\delta^{13}C$ values varying from $-32‰$ to $-28‰$, respectively, which are indicative of biological carbon fixation (Schidlowski, 1988). Between 2.8 and 2.6 Ga, the $\delta^{13}C$ fall even further, to values between $-40‰$ and $-60‰$ (recall Fig. 7.10). Those extremely light isotopic values are from marine sediments, but some of the lowest come from paleosols (e.g., Rye and Holland, 2000). To become so isotopically light requires an exceptionally ^{12}C-rich source, and it is likely that those kerogens are the result of methanogenic ^{12}C-rich gas production, the incorporation of this methane into the biomass of methanotrophic bacteria, and inevitably the preservation of ^{12}C-enriched organic matter (Hayes, 1983). The methanotrophs could have either been aerobes (certainly at least for those kerogens in soils) or, alternatively, they could have been sulfate-reducing bacteria that anaerobically oxidized methane in oceanic environments (recall section 6.2.5(c)). Irrespective of which type of microbial population, their metabolism would have been inextricably linked to oxygen availability; the first use it directly as a TEA, while the others require it indirectly to generate sulfate. After the onset of an oxygenated atmosphere, methanogens no longer played a significant role in the global carbon cycle because more widespread oxic conditions prompted aerobic respiration and competition with SRB due to increased sulfate levels. Only at times of exceptional methane release, i.e., through gas hydrate dissolution, did methane once again make a strong reduced carbon contribution to the atmosphere. As a result, the extremely light $\delta^{13}C$ values were never seen again on a global scale.

The elemental composition of ancient kerogens have also been altered over time. In particular,

one of the characteristic features of organic matter as it becomes buried and metamorphically altered is that its hydrogen component is progressively released, in part as methane, and the kerogenous residue gradually approaches the composition of graphite. Not surprisingly, kerogens isolated from Precambrian sedimentary rocks exhibit a much lower hydrogen to carbon ratio than similar rocks in the Phanerozoic, with for example, kerogen in Archean sediments tending to have H/C ratios less than 0.3. By contrast, kerogen from Cambrian sediments have H/C ratios over 1.0 (Hayes et al., 1983).

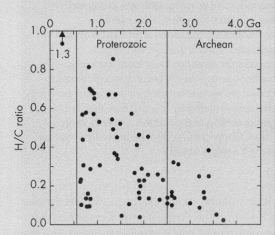

H/C ratios of kerogen isolated from Precambrian sediments of various ages. Note how the kerogens from the Archean have lower H/C ratios than those from the Neoproterozoic. (Modified from Schopf, 1994.)

Isotopic changes also occur due to metamorphism. When organic matter is heated to temperatures between 275 and 300°C, mobilization and loss of isotopically light lipids and labile functional groups, as well as preferential breakage of ^{12}C–^{12}C bonds, begins to occur and the gases released (e.g., CO_2, CH_4) tend to be enriched in the ^{12}C isotope.

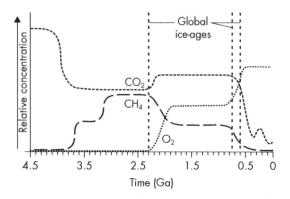

Figure 7.12 The relative concentrations of major atmospheric gases may help explain why global ice ages (vertical dotted lines) occurred in Earth's history. As long as methanogens flourished, temperatures were balanced between greenhouse warming and an antigreenhouse prompted by organic haze formation when methane concentrations approached, or maybe even exceeded, carbon dioxide. As oxygen concentrations began to increase markedly, methanogens suddenly became marginalized to anoxic environments, along with an accompanying decrease in methane. This decrease then led to global cooling and glaciation. (From Kasting, 2004. Copyright © (2004) by Scientific American, Inc. All rights reserved.)

sulfide mineral oxidation (e.g., *Acidithiobacillus* sp.). During the Archean, the former may have been limited to mat communities, oxidizing the hydrogen sulfide generated by the underlying SRB (reaction (7.22)). They were not likely important in the open oceans because the hydrogen sulfide generated in anoxic deep waters was oxidized by anoxygenic photoautotrophs and effectively consumed before it came into contact with the oxygenated surface waters (i.e., whatever chemocline existed in the Archean, it would have been positioned within the euphotic zone). The acidophiles would have been limited to terrestrial environments, where they grew in association with exposed sulfide minerals, like their hyperthermophilic counterparts.

$$H_2S + 0.5O_2 \rightarrow S^0 + H_2O \qquad (7.22)$$

As long as the O_2 partial pressures in ocean waters were extremely low, bacteria would not have been able to gain energy from aerobic oxidation of Fe(II) to form Fe(III). Energy can be gained from reaction (7.23) at O_2 partial pressures of approximately 10^{-59} bars, yet the stability of detrital pyrite in 3.0–2.5 Gyr Witwatersrand sediments, South Africa (see below), suggests that O_2 partial pressures at that time were likely below 10^{-70} bars. As a consequence, Archean bacterial communities would not have attained sufficient energy from Fe(II) oxidation (Nesbitt, 1997).

$$2Fe^{2+} + 0.5O_2 + 2H^+ \rightarrow 2Fe^{3+} + H_2O \qquad (7.23)$$

As O_2 partial pressures eventually increased beyond the threshold concentration, perhaps by the early Paleoproterozoic, Fe(II) oxidation by microaerophilic, chemolithoautotrophic bacteria such as *Gallionella* (beta subdivision of *Proteobacteria*) may have become possible, and likely even favored compared to abiological or photosynthetic reactions (Holm, 1989).

Manganese (II) oxidation requires much more energy than Fe(II), and in modern environments it proceeds almost exclusively via bacterial mediation:

$$Mn^{2+} + 0.5O_2 + H_2O \rightarrow MnO_2 + 2H^+ \qquad (7.24)$$

Mn(II)-oxidizing bacteria are phylogenetically diverse, falling within all the *Proteobacteria* classes (Tebo et al., 1997). Major deposits of Mn associated with BIFs occur only twice in the known geological record, and both follow directly after inferred snowball Earth events; global glaciations in which the world's oceans are hypothesized to have been almost completely covered by a continuous carapace of sea ice that formed a barrier between the oceans and atmosphere, resulting in severely diminished biological productivity. The first of these events occurred between 2.4 and 2.1 Ga (e.g., the Huronian event). When surface temperatures returned to above freezing, as a result of continued volcanic activity, rapid melting of the ice ensued and the surviving microorganisms would have proliferated in the metalliferous, reducing oceans. Oxygen released by cyanobacterial blooms could have reacted not only with Fe^{2+} to form BIF, but also with Mn^{2+} to form extensive deposits such as the 2.4 Gyr Kalahari Manganese member of the Hotazel Formation, South Africa (Kirschvink et al., 2000). The oxidation of so much manganese (and iron) certainly points towards cyanobacterial photosynthesis being a significant source of surface water oxygenation at that time.

Nitrifying bacteria may also have evolved in the Late Archean–Early Paleoproterozoic, with NH_4^+ initially oxidized to NO_2^- by ammonium-oxidizing *Proteobacteria* (e.g., species of the beta subdivision, *Nitrosomonas*) (reaction (7.25)), and NO_2^- then oxidized to NO_3^- by nitrite-oxidizers (e.g., the alpha subdivision, *Nitrobacter*) (reaction (7.26)):

$$NH_4^+ + 1.5O_2 \rightarrow NO_2^- + H_2O + 2H^+ \quad (7.25)$$

$$NO_2^- + 0.5O_2 \rightarrow NO_3^- \quad (7.26)$$

Combined, these processes would have contributed to significant ammonium losses, with an attendant increase in dissolved nitrate (Falkowski, 1997). The loss of ammonium would initially have had a dramatic effect on the early biosphere, and necessitated primitive cells to develop new methods of nitrogen acquisition. One method employed by some species was the evolution of nitrogen fixation, the energy expensive process whereby one mole of atmospheric N_2 is initially reduced to two moles of ammonia via the nitrogenase enzymes, and then converted into an organic form:

$$N_2 + 3H_2 \rightarrow 2NH_3 + 2R\text{-}OH \rightarrow 2R\text{-}NH_2 + 2H_2O \quad (7.27)$$

The timing for the evolution of nitrogen fixation is very controversial, and is invariably based on when the demand for nitrogen by the burgeoning microbial communities exceeded the supply from abiological sources. Navarro-González et al. (2001) have calculated that sufficient inorganic nitrogen was available throughout the Archean and Paleoproterozoic. When CO_2 was the dominant atmospheric gas in the Hadean and Early Archean, nitric oxides would have formed during lightning discharges via the production of oxygen atoms obtained by the splitting of CO_2 and H_2O. As discussed above, this is likely to have been the case before 3.5 Ga, but then as CO_2 levels dropped, biogenic methane levels increased and became a significant greenhouse gas (Kasting and Siefert, 2002). At this stage, reaction of methane with N_2 could instead have led to HCN formation (which is hydrolyzed in solution to yield NH_4^+ cations) when the CH_4 mixing ratios exceed 10^{-5} for an atmosphere with 100 times the present atmospheric levels of CO_2 (Zahnle, 1986). It was only when sufficient O_2 became available to oxidize methane (via methanotrophy) and to generate enough dissolved sulfate for SRB to outcompete methanogens for reduced substrates that these abiological processes would have diminished in importance. At this stage, a nitrogen crisis may have developed, and N_2-fixation became necessary. In modern oceans, N_2-fixation can fuel as much as half of the new N production, but N_2-fixers make up a relatively small amount of the total phytoplankton biomass because their high energy demands means that they can only outcompete non-N_2-fixing bacteria when nutrient N concentrations are low compared to nutrient P (Tyrrell, 1999).

The main argument in favor of an early origin for N_2-fixation comes from genetic studies that point to the antiquity of nitrogenase (Young, 1992). However, nitrogenase is not present in all representatives of any particular lineage, and even if it is ancient, it is possible that nitrogenase was developed for some other metabolic purpose. Perhaps the ancestral counterpart of nitrogenase may originally have been used to protect the bacteria against high levels of HCN, and only after nitrogen became limiting did the detoxyase evolve into nitrogenase (Fani et al., 2000). More-over, nitrogenases may have been a retrofitted late acquisition that was spread and utilized throughout the microbial communities by lateral gene transfer after the benefits to the cells that evolved them became apparent (Kasting and Siefert, 2001).

The impact of oxygenic photosynthesis would have extended beyond chemolithoautotrophic metabolic pathways. The net production of O_2 could have swept the oceans clear of iron and other reductants in thousands of years (Towe, 1990). With a shortage of geological sinks, the only mechanism that could halt an oxygen runaway would have been another biological process, that being aerobic respiration:

$$C_6H_{12}O_6 + 6O_2 \rightarrow 6CO_2 + 6H_2O \qquad (7.28)$$

The initial O_2-utilizing chemoheterotrophs were probably amphiaerobic, meaning that they retained the capability to live via fermentation or anaerobic respiration when necessary, i.e., when O_2 concentrations were temporarily below the so-called Pasteur Point, and aerobic respiration could not supplant anaerobic fermentation as the principal source of energy. Some present-day amphiaerobes, also known as facultative anaerobes, include those that use NO_3^- and Mn(IV).

The assimilation of nitrate was of paramount importance as it obviated the need for some species to utilize the metabolically expensive process of nitrogen fixation. For other species, denitrification (reaction (7.29)) presented them with a new form of metabolism:

$$2.5C_6H_{12}O_6 + 12NO_3^- \rightarrow$$
$$6N_2 + 15CO_2 + 12OH^- + 9H_2O \qquad (7.29)$$

The probable origins of denitrification can be traced back to the early Paleoproterozoic when sufficient O_2 became available to form usable amounts of dissolved nitrate. This view appears to be supported by both nitrogen isotopic data (Box 7.9) and the distribution and phylogeny of nitrate reductase gene sequences in various denitrifying bacteria that imply an ancient lineage (Petri and Imhoff, 2000).

Evidence for the origins of bacterial Mn(IV) reduction comes from isotopic analyses of manganese carbonates in the same ~2.4 Gyr Kalahari Manganese deposits discussed above. They show depleted $\delta^{13}C$ values, suggesting that they were, in part, formed during diagenetic MnO_2 reduction (Kirschvink et al., 2000):

$$CH_3COO^- + 4MnO_2 + 3H_2O \rightarrow$$
$$4Mn^{2+} + 2HCO_3^- + 7OH^- \qquad (7.30)$$

With so many oxygen sinks, there was probably no net accumulation in the atmosphere, and levels may have remained below 10^{-13} present atmosphere levels (PAL) (Kasting, 1987). However, some time between 2.4 and 2.2 Ga, oxygen accumulated to sufficiently high levels to exceed the rate of supply of reductants. At this stage O_2 levels would have increased abruptly (to values between 10^{-5} and 10^{-2} PAL), thereby altering the global atmosphere from predominantly neutral to oxidizing. The timing of this transition can be constrained by a number of features in the rock record. Evidence for low oxygen in the Archean comes from the following:

1 Easily oxidized detrital uraninite, pyrite, and siderite have been recovered in fluvial siliciclastic sediments of the 3.2–2.7 Gyr Pilbara craton (Rasmussen and Buick, 1999). They are not abundant in fluvial systems younger than 2.3 Gyr.

2 The herringbone textures of some Archean carbonates result from the presence of Fe^{2+} in shallow, surface waters (Sumner, 1997). Along similar lines, a number of Archean carbonates exhibit high Fe(II) and Mn(II) contents (e.g., Veizer et al., 1989). As Fe^{2+} concentrations decreased, due to oxygenation, the style of carbonate precipitation also changed from seafloor cements to water column micrites.

3 Low levels of marine sulfate suggest minimal oxidative weathering of sulfide minerals and subsequent delivery of sulfate to the oceans (Canfield et al., 2000).

Box 7.9 Nitrogen isotopes

The primary source of organic nitrogen today is through the fixation of atmospheric N_2 by certain microorganisms. After the death of these organisms, nitrogen from their nucleic and amino acids undergoes several transformations. Some organic nitrogen remains as part of the residual organic fraction, ultimately becoming incorporated as kerogen into the rock record, while the bulk is mineralized into ammonium cations. The fate of the ammonium includes: uptake by other microorganisms or higher plants; interlayer fixation by clays such as vermiculite; loss as ammonia gas; or transformation into nitrite and nitrate through nitrification. The nitrate is then assimilated by microorganisms and higher plants, reduced by ammonification or denitrification, the latter process releasing N_2 back into the atmosphere, or it is left as a dissolved anion.

In terms of nitrogen isotopic fractionations, biological N_2 fixation discriminates against the heavy isotope ^{15}N, and the resulting biomass is observed to be isotopically light in comparison to atmospheric N_2, with $\delta^{15}N$ values of -2 to $-4‰$ (e.g., Macko et al., 1987). Similarly, incorporation of reduced forms of nitrogen (e.g., NH_4^+) leads to depleted $\delta^{15}N$ values (-4 to $-27‰$) for biomass, with the largest negative fractionations occurring when external ammonium concentrations are sufficiently low that the cells are forced to invoke active transport mechanisms (Hoch et al., 1992). During nitrification, significant fractionation in the other direction can take place, with the residual ammonium being enriched in ^{15}N and the nitrate formed depleted in ^{15}N (Wada and Hattori, 1978). Denitrification then preferentially returns ^{14}N to the atmosphere, leaving the remaining nitrate with $\delta^{15}N$ values up to $+20‰$, but with a mean value of $+6‰$, as measured in modern marine waters (Wada et al., 1975). As NO_3^- is the main assimilable nitrogen species, marine organisms record its isotopic signature, and consequently, marine sedimentary organic matter preserves a positive enrichment.

Kerogen samples from a number of Precambrian cherts display a range of $\delta^{15}N$ values, with the oldest samples (between 3.5 and 3.4 Ga) being isotopically light down to $-6.2‰$. Between 2.7 and 2.5 Ga, values range from -6 to $+12‰$. It was not until the Paleoproterozoic (at around 2.0 Ga)

that negative $\delta^{15}N$ values disappear altogether and kerogens become exclusively positive, with $\delta^{15}N$ values ranging from $0.3‰$ to $10.1‰$, similar to Phanerozoic samples (Beaumont and Robert, 1999; Pinti et al., 2001). These results would appear to suggest that nitrification, and hence reduction by denitrification, did not become important metabolic processes until some time during the early Paleoproterozoic.

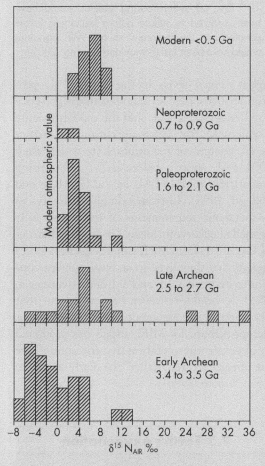

Histograms showing the $\delta^{15}N$-isotopic composition of organic matter through geological time. Note the general depletion in ^{15}N of organic matter from Early Archean cherts, relative to modern marine sediments. The transition from negative to positive values appears to coincide with the onset of atmospheric oxygenation. (Reprinted from Beaumont and Robert, 1999 with permission from Elsevier.)

4 The discovery of mass-independent fractionation (MIF) of sulfur isotopes (^{33}S and ^{36}S) in sulfide and sulfate-containing rocks deposited prior to 2.4 Ga, but not after 2.1 Ga. This indicates that the sulfur cycle changed from one governed by gas-phase photochemical reactions in an O_2-free atmosphere to one dominated by oxidative weathering, where the different sulfur species lost their MIF signal (Farquhar et al., 2000; Mojzsis et al., 2003). In order to preserve the MIF in Archean and early Paleoproterozoic sediments, the atmospheric oxygen concentration must have been less than 10^{-5} PAL. By contrast, in atmospheres with O_2 concentrations greater than 10^{-5} PAL, all sulfur-bearing species would have been oxidized to sulfate before becoming incorporated into the sediment, so any MIF signature would have been lost (Pavlov and Kasting, 2002).

Higher oxygen levels in the middle Paleoproterozoic atmosphere are manifest in the presence of red-beds after 2.2 Ga and the oxidation state of paleosols; the latter appearing to restrict the O_2 content of the atmosphere to less than 0.004 PAL shortly before 2.2–2.3 Ga, rising dramatically to 0.15 PAL by 2.0 Ga (Rye and Holland, 1998). Correspondingly, estimates of O_2 metabolic requirements of the oldest fully organelled eukaryotic algae, *Grypania spiralis*, in 2.1 Gyr cherts from Michigan (see section 7.5.2) suggests that they lived in an environment containing between 0.01 and 0.1 PAL (Runnegar, 1991). It should, however, be pointed out that each of the above arguments is not unambiguous (except perhaps for MIF), which has prompted the alternate interpretation that the atmosphere was already oxygenated in the Early Archean (e.g., Ohmoto, 1996).

7.5 Earth's oxygenation

7.5.1 The changing Proterozoic environment

One of the most significant outcomes of an increase in atmospheric O_2 is that it would have been accompanied by an increase in strato-spheric ozone. As photochemical models show, the thickness of the ozone layer increases non-linearly with atmospheric O_2, such that even small amounts of O_2 (0.001 PAL) could have produced a relatively effective UV screen, perhaps as early as 2.3 Ga (Fig. 7.13). This development had three biological consequences:

1 It permitted the increased occupation of the euphotic zone of the open oceans by planktonic microorganisms. More phytoplanktonic biomass, in turn, led to increased O_2 production and higher rates of organic sedimentation (Des Marais et al., 1992). Evidence for such a sequence of events appears in the large positive $\delta^{13}C$ excursion in carbonate sediments between approximately 2.20 and 2.06 Ga, suggesting that the fraction of carbon gases reduced to organic carbon during this period was much larger than before (Karhu and Holland, 1996).

2 Biofilms would have begun to colonize vast tracts of terrestrial rock surfaces, no longer requiring shelter in soil pore spaces or rock fractures for protection. Thus began an accelerated role for biology in mediating weathering processes on land (e.g., Schwartzman and Volk, 1991). Evidence of a terrestrial biosphere in the Paleoproterozoic comes from 2.2–2.0 Gyr leaching patterns of laterite profiles, indicative of the presence of a subaerial microbial cover (Gutzmer and Beukes, 1998). Similarly, as the O_2 content in the atmosphere rose, biological pyrite oxidation would have accelerated the production of H_2SO_4. This would have increased the overall rate of chemical weathering, thereby delivering more nutrients (e.g., $H_2PO_4^-$) to the oceans to support the diversified phytoplankton community, while the increased sulfate flux permitted the underlying SRB to maintain deepwater anoxia.

3 Increased atmospheric O_2 paved the way for the diversification of the *Eukarya* (see below).

After the atmosphere became oxidizing, the final stages of banded iron deposition took place by around 1.8 Ga. It was generally believed that this sedimentation process terminated when oxic bottom waters developed, thus removing the supply of deep-water Fe(II) before its transport onto the shelf. Canfield (1998), however, has

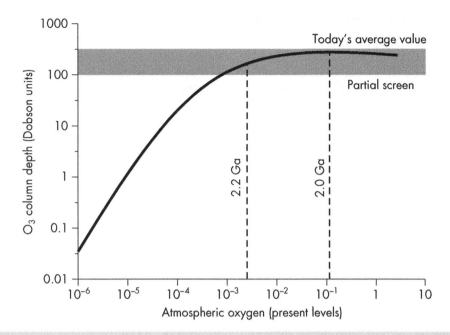

Figure 7.13 Photochemical models have shown that most of the harmful UV radiation is absorbed when the "O_3 column depth" exceeds about 100 Dobson units. As the figure demonstrates, this column depth is reached at an O_2 level of only 0.001 present atmospheric levels (PAL). Dashed lines are time line estimates. (Modified from Kump et al., 2004.)

convincingly argued that anoxic waters persisted until well after banded iron formations ceased, and that it was HS^-, rather than O_2, that was responsible for removing Fe(II) from the deep ocean. Based on comparisons of Fe speciation and S-isotope profiles in the 1.88 Gyr Gunflint Iron Formation in Ontario, with the overlying shales of the Rove Formation, Poulton et al. (2004b) seemed to have pinpointed the actual transition from Fe-rich to sulfidic bottom waters (resembling modern euxinic basins such as the Black Sea), at around 1.84 Ga. The recent measurement of iron isotopes on a variety of Precambrian pyrite grains also corroborates the sulfidic ocean hypothesis (Box 7.10). The source of the sulfide was the increased level of bacterial sulfate reduction taking place in the oceans, which until around 800 Ma saw the $\Delta^{34}S$ between sedimentary sulfides and sulfates reach values ~40‰, near the maximum associated with SRB activity. In the geological record, this

supplementary sulfur flux is also evident in the large amounts of pyritized iron in Paleoproterozoic black shales (Shen et al., 2003) and the first appearance of similarly aged massive bedded gypsum deposits (Lowe, 1983).

During the Mesoproterozoic (1.6–1.0 Ga) a number of unusual features are observed in the rock record. In terms of the carbon isotope record, there is strikingly little variation in $\delta^{13}C$ values of carbonates (the average values are $0 \pm 2‰$), suggesting perhaps unique long-term stability in the carbon cycle. This feature may reflect a fairly constant rate of organic carbon burial throughout that time interval (Buick et al., 1995), which, in turn, might be due to either a period of high productivity but more effective microbial recycling through enhanced respiration (given both O_2 and SO_4^{2-} availability), or lower overall productivity, and hence diminished organic burial. In the case of the latter, a potential cause for arguing lower productivity may lie with the deep sulfidic waters

Box 7.10 Iron isotopes

Iron has a number of isotopes, of which mass 54 and 56 are most abundant. The $^{56}Fe/^{54}Fe$ ratio in igneous rocks is nearly constant ($\delta^{56}Fe = 0.0 \pm 0.05‰$), regardless of the emplacement time, location, or tectonic setting. By contrast, chemically precipitated iron minerals in sedimentary environments undergoing active diagenesis tend to have variable δ^{56} Fe values (Beard et al., 1999). For example, Late Archean to Early Proterozoic BIFs from the Transvaal Supergroup, South Africa, have isotopic fractionations that span the range from as low as −2.5 in pyrites to positive values in hematite and magnetite, as high as +0.60‰ to +1.1‰, respectively (Johnson et al., 2003).

Due to the small variation in the relative masses of the Fe isotopes, their isotopic ratios are generally less affected by abiological fractionation processes, such as temperature changes,

although there can be marked kinetic and equilibrium effects (see Anbar, 2004 for details). Fe isotopic fractionation is also susceptible to biologically mediated oxidation and reduction reactions. For instance, laboratory experiments have shown that single-step bacterial Fe(III) reduction leads to isotopically light ($\delta^{56}Fe < -1.2‰$) aqueous Fe(II) relative to the initial ferric oxide substrate (Icopini et al., 2004). Even larger fractionations can be generated through multiple steps of Fe(II) oxidation and Fe(III) reduction, or at high reduction rates, where rapid formation and sorption of Fe(II) to ferric oxides leads to fractionations as large as -2.3‰ (Johnson et al., 2005). Meanwhile photosynthetic Fe(II)-oxidizing bacteria can form ferric hydroxide enriched in the heavy isotope by $1.5 \pm 0.2‰$ relative to Fe(II) (Croal et al., 2004).

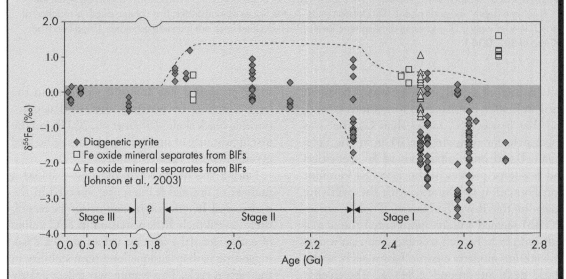

Plot of $\delta^{56}Fe$ versus sample age for Fe-sulfides from black shales and Fe-oxides from BIFs. On the basis of the $\delta^{56}Fe$ values, the ocean cycle can be divided into three stages: stage 1 is from >2.8 to 2.3 Ga; stage II is from 2.3 to 1.7 Ga; stage 3 is less than1.7 Ga. The gray diamonds correspond to Fe isotope compositions of pyrite, and open squares and triangles denote Fe isotope compositions of magnetite- and hematite-rich BIF samples, respectively. The gray area corresponds to $\delta^{56}Fe$ values of Fe derived from igneous rocks (at 0.1‰) and hydrothermal sources (at −0.5‰). Dashed lines contour maximum and minimum Fe isotope compositions of sedimentary sulfides used to define the three stages. (From Rouxel et al., 2005. Copyright (2005) AAAS. Reprinted with permission.)

continued

Box 7.10 *continued*

Fe isotopic variations can be used to ascertain the potential biological role in iron cycling in past and present environments. For example, analyses of carbon- and magnetite-rich rocks, from the 2.9 Gyr Rietkuil Formation, Witwatersrand Supergroup, South Africa, revealed $\delta^{56}Fe$ values as low as $-2.3‰$. These negative fractionations have been interpreted as being indicative of the presence of dissimilatory Fe(III)-reducing bacteria in the Late Archean (Yamaguchi et al., 2005). Rouxel et al. (2005) have similarly conducted a detailed survey of the variation in $\delta^{56}Fe$ in Fe-sulfide and Fe-oxide minerals during the Precambrian. Highly variable, but negative, values in pyrite from black shales ($0.5‰$ to $-3.5‰$) between 2.8 and 2.3 Ga suggest that Archean oceans were globally Fe rich and that Fe concentrations and isotopic compositions were affected by deposition of Fe-oxides (e.g., BIFs), that preferentially removed isotopically heavy ^{56}Fe, driving the ocean waters to negative $\delta^{56}Fe$ values, as recorded in pyrite. A global change in $\delta^{56}Fe$ at 2.3 Ga is observed at approximately the same time that atmospheric oxygen is thought to have increased. Between 2.3 and 1.7 Ga, positive iron isotope ratios of pyrite then reflect an increase in the precipitation of iron sulfides relative to iron oxides, supporting the premise that the deep oceans remained anoxic and were slightly sulfidic; the sulfate being derived from increased oxygenation of the Earth's land masses. Subsequent sulfidic conditions in the deep sea then ensured that dissolved Fe(II) did not accumulate in the ocean but rather it was removed as an insoluble component of the sediment, thereby retaining the isotopic composition of the source hydrothermal Fe input. This is observed in the lack of substantial $\delta^{56}Fe$ value variations in pyrites from black shales younger than 1.5 Gyr.

because Fe and Mo, both of which are important for biological nitrogen fixation and nitrate reduction, are removed from solution in sulfide-bearing waters (Anbar and Knoll, 2002). Therefore, as long as the deep oceans contained excess sulfide, they would have undergone a period of exceptional nitrogen stress. This is not too dissimilar from modern oceanic conditions where low concentrations of iron limit N_2-fixation, making nitrogen the primary major limiting nutrient for phytoplankton growth (Sunda, 2000).

By contrast to the Mesoproterozoic, the beginning of the Neoproterozoic marks some of the most dramatic events in Earth's history. It began with the formation of the first supercontinent, known as Rodinia, sometime around 1.0 Ga, and is thought to have persisted for nearly 250 million years (Windley, 1984). There were also major changes in atmospheric and oceanic chemistry. Atmospheric oxygen levels may have accumulated to nearly 18% PAL, and along with the concomitant increase in ocean oxygenation, there would have been an attendant deepening of the oxic–anoxic interface, decreased sulfide levels owing to increased aerobic respiration (hence less sulfate reduction), and the direct reaction of O_2 with HS^-, producing more complex sulfide intermediates (Canfield and Teske, 1996). At this stage, $\Delta^{34}S$ values approached 70‰, indicating that wide-scale initiation of the oxidative sulfur cycle, as possibly driven by a combination of newly evolved nonphotosynthetic sulfide-oxidizing and sulfur-disproportionating bacteria, had taken place. Others have suggested that sulfur disproportionation may have been active for much of the Proterozoic (and thus unlinked to sulfide-oxidizing bacteria), but the $\Delta^{34}S$ values remained lower than 46‰ as a consequence of low sulfate concentrations and efficient burial of HS^- in the form of pyrite (Hurtgen et al., 2005). Increased levels of O_2 could even have facilitated the emergence of metazoans, primitive soft-bodied, multicellular animals, on the shallow seafloor, as early as

1.0 Ga. Evidence in support of their evolution include the decline in stromatolite diversity and the presence of trace fossils in rocks of this age (Conway-Morris, 1993).

By the middle of the Neoproterozoic, the most extensive glaciations for over 1.5 billion years took place, with ice maybe reaching as far as the equator during the second Snowball Earth episode (Harland, 1964). At least two main phases of ice advance occurred, the first between 760 and 700 Ma, called the Sturtian glaciation, and the second between 620 and 590 Ma, called the Varanger glaciation. It has been speculated that the "global" glaciations arose due to a pre-ponderance of land masses, after the break-up of Rodinia, being positioned in the middle and low latitudes that raised planetary albedo, a situation that has not been encountered at any subsequent time in Earth's history (Kirschvink, 1992). The ice would further have increased Earth's albedo by lowering sealevel, exposing continental shelves and inland seas. In addition, the absence of pelagic calicifiers at that time (e.g., coccolithophores and foraminifera had not yet evolved) meant that carbonate deposition was limited to shallow water shelf environments. Therefore, as sealevels began to drop during the onset of glaciation, calcium carbonate precipitation decreased, high carbonate anion concentrations were maintained, and this might have had the effect of inducing lower atmospheric CO_2 levels due to the aqueous carbonate equilibrium ($CO_2 + CO_3^{2-} + H_2O \longleftrightarrow 2HCO_3^-$) being shifted to the right (Ridgwell et al., 2003).

One other explanation for the Neoproterozoic glaciations lies with atmospheric methane concentrations. It is striking that there is such a lengthy time interval between global glaciations, and it has been suggested that this could be attributed to residual methane levels, perhaps as high as 100 mg L^{-1}, still present in the atmosphere (Kasting, 2004). It was only when O_2 levels had become a major atmospheric constituent that methane concentrations were diminished to present-day values and any greenhouse effect it exerted would be nullified (Pavlov et al., 2003).

Figure 7.14 A cap carbonate sequence in Namibia, deposited at the termination of the Sturtian glaciation. The white highlighting delineates the cap carbonates from the underlying glacial deposits. (Courtesy of Paul Hoffman.)

The snowball events were, however, short-lived because volcanoes would have continued to release CO_2 into the atmosphere (and ocean), which would eventually have warmed surface temperatures and induced melting a few million years later. Then, as the ice melted and ocean circulation became re-established, the metals that had accumulated in the stagnant waters became oxidized and formed a suite of BIF (albeit on a smaller scale than the Superior-type) in the oxic zone of upwelling areas. Major deposits include the 700 Myr Rapitan Group of the Mackenzie Mountains in Canada, which was deposited close to the equator and is overlain by hundreds of meters of glacial outwash. Simultaneously, a transient but intense greenhouse climate ensued, leading to enhanced weathering of the glacially eroded landscape, increased alkalinity and carbonate precipitation (Hoffman and Schrag,

2002). This resulted in the rapid deposition of unusually ^{13}C-depleted, finely laminated dolomicrites, known as cap carbonates, directly on top of the glacial debris (Fig. 7.14).

What effect global glaciation had on the microbiota is unclear. Certainly, chemotrophic and anaerobically respiring heterotrophic prokaryotes could have survived burial under ice, but the phototrophic bacteria, eukaryotes, and the primitive animals must have fared worse. To some extent, this is indicated by the $\delta^{13}C$ ratios of the cap carbonates (and the rocks which immediately pre-date them), that seemingly point towards a severe diminishment of oceanic photosynthesis during the snowball event (Hoffman et al., 1998). Additionally, the fossil record tells us that there was a marked decline in microfaunal diversity at the time of the glaciations (e.g., Vidal and Knoll, 1982), but the existence of extant photosynthetic species, known from preglacial times, indicates that some found refuge somewhere, possibly in pockets of open water in the circumequatorial ocean or around shallow hot springs associated with volcanic islands (Hyde et al., 2000). In fact, the recent report of biomarkers from 740–700 Myr black shales confirms that a diverse photosynthetic community was sustained, at least locally, during times of glaciation (Olcott et al., 2005).

Following the snowball event, greater primary productivity by phytoplankton facilitated increased ocean water oxygenation. Concomitantly, their lysed cells were extensively reworked (by zooplankton) and packaged as rapidly sinking particulate organic carbon, in the form of fecal pellets. This meant that organic matter no longer remained in suspension, or as bottom water DOC, but instead it accumulated at the seafloor (Logan et al., 1995). More efficient transport of carbon to the seafloor also had important ramifications for phosphogenesis. Before fecal pellets, organic carbon reaching the seafloor was rarely rich in phosphate, but thereafter, higher levels were delivered to the sediment surface, where the now prevailing oxic conditions favored their retention as phosphorite deposits (Brasier, 1992).

With increased ventilation, the organic reservoir at the seafloor was short-lived. Then, once the deep ocean waters no longer remained an immediate sink for oxygen, atmospheric O_2 finally began to climb towards present values. Crucially, with bottom waters now oxic, large, metabolically active animals could diversify from shallow waters to pelagic sediments. This initially gave way to a group of soft-bodied marine animals (e.g., annelids, coelenterates), the so-called "Ediacaran faunas", that were preserved worldwide as impressions, casts, and molds in rocks as old, or older, than 575 Ma (e.g., Narbonne, 2004). They, in turn, were eclipsed near the Precambrian–Cambrian boundary (545 Ma) by more complex animal phyla, many of them skeletal with modern body plans and displaying a higher degree of behavioral sophistication, in what is known as the "Cambrian radiation" (Knoll and Carroll, 1999). The cause for the decline in Ediacarans remains speculative, but it probably involved a combination of factors, such as a rise in predation, increased competition for nutrients, more active bioturbation, or a mass extinction event prompted by an environmental perturbation. Their disappearance from the rock record may also simply be due to lack of their preservation relative to organisms that evolved mineralizing capabilities. Indeed, Brennan et al. (2004) have suggested that a surge in calcium concentrations during the early Cambrian spurred the onset of calcium carbonate biomineralization, which led to a number of marine biota developing calcium carbonate shells, in addition to the advent of calcifying cyanobacteria.

7.5.2 Eukaryote evolution

As briefly mentioned above, the appearance of eukaryotes marks the last major innovation in single cell evolution. These organisms include protozoans, algae, and fungi. Eukaryotes differed from the prokaryotes by having a membrane-bound nucleus that stores chromosomes, a cytoskeleton, a cell membrane stiffened by sterols, the presence of organelles (e.g., mitochondria and chloroplasts), and a larger genome size. Their origin, however, remains enigmatic. Several decades ago, Margulis (1970) suggested that the first eukaryotic cells originated when the cytoplasm of an autotrophic microorganism, incapable of respiration, was invaded by one that could (the process of endosymbiosis). The respiring bacterium provided its host with energy (as ATP) in exchange for a stable, protected environment and a readily available source of reduced carbon and nutrients. Eventually, the relationship of the host cell with the symbiont became one of absolute interdependence, with the symbiont becoming the mitochondrion, and thus losing its capacity for an independent existence. In a similar fashion, the symbiotic uptake of a cyanobacterium may have led to the forerunner of the chloroplast. The *Prochlorococcus* sp., an important component of today's marine phytoplankton, may be the living ancestor of the cyanobacterium involved in this event (Lewin, 1976). For several years thereafter, this then-radical view was challenged, mainly on the basis that some of the most deeply divergent *Eukarya* were amitochondriate, meaning they lack mitochondria. This seemed to suggest that *Eukarya* evolution required two independent stages. In one theory, the first eukaryote initially arose from the stem of the phylogenetic tree by evolving a nucleus, endomembrane, and cytoskeleton, and much later on, some obtained a mitochondrion by the endocytosis of a bacterium, while others remained amitochondriate (Cavalier-Smith, 1987). The alternate "chimera hypothesis" proposed that a fusion event occurred between ancestral *Archaea* and *Bacteria*, with the resulting cell eventually evolving eukaryotic structures and later mitochondria as above (Sogin, 1991). Recent molecular studies, however, now advocate that modern amitochondriates did indeed once possess mitochondria, perhaps even earlier than the nucleus, but they were either lost or converted to hydrogenosomes (H_2-producing mitochondria) (Bui et al., 1996).

A two-stage hypothesis is still favored today, but instead the initial phase involved the gradual cellular merger between some ancient anaerobic archaeal host and a respiring bacterium, leading to the first complex cell. After an initial period of close symbiotic association, the symbiotic bacterium eventually developed into a mitochondrion with a lipid bilayer, thereby reorganizing the host's original metabolism via compartmentalization, while the host cell developed a nuclear membrane around their expanded genetic material (Vellai and Vida, 1999). The intracellular incorporation of the symbiont also led to an increase in size and complexity of the host, providing the genetic basis for eventual evolution into real nucleus-bearing eukaryotic cells with secondary cytological features like the cytoskeleton (Gray et al., 1999). In fact, the increased genetic potential even in a single cell could now encode for endocytosis, a eukaryotic invention. Such levels of complexity in all probability could not have happened until after the acquisition of an efficient energy-conserving organelle, such as the mitochondrion.

So what were the factors responsible for this symbiosis to occur in the first place? According to the "hydrogen hypothesis" anaerobic *Archaea* (possibly methanogens), that were strictly autotrophic and H_2 dependent, became displaced into an environment where H_2 was limiting. Under those inhibiting conditions, they were forced to establish tight physical associations with nearby fermentative bacteria that generated H_2 as waste (Fig. 7.15). In order to maximize contact, some *Archaea* may have evolved cell shapes of large surface area that could surround the symbionts, but not endocytose them. Initially, the advantage of this cellular merger was the utilization by the host of H_2 excreted by the symbionts. In time, the symbiont was engulfed and the transfer of genes (such as the symbiont's carbon importers) into the archaeal cell may have enabled the host to bring in small organic molecules and break them down into simple substrates for both itself and the symbiont. As the host could now meet both its own carbon and energy needs heterotrophically, methanogenesis was no longer needed. The end result was

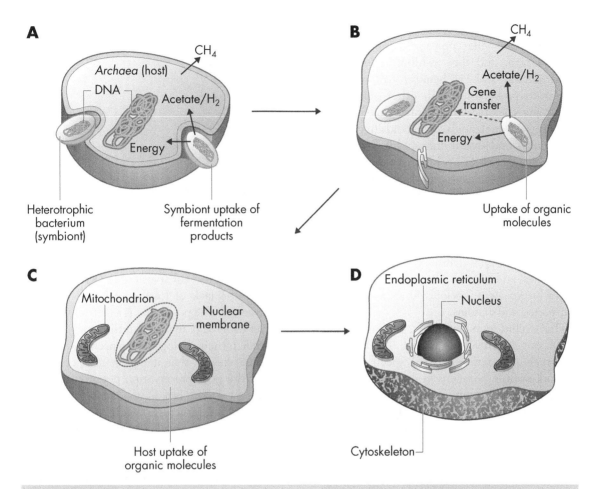

Figure 7.15 Endosymbiosis as proposed for the origin of eukaryotic cells. (A) An archaeal host cell, incapable of respiration, and likely methanogenic, becomes dependent on the metabolic waste (H_2, acetate) of a fermentative bacterium. (B) Engulfment by an unknown mechanism yields a symbiont bacterium that continues to degrade organic material anaerobically in the cytoplasm of the host, contributing additional metabolites and allowing for the enlargement of the host cell. (C) The synergistic relationship allows for the compartmentalization of the DNA, as illustrated by the evolution of the nuclear membrane. Meanwhile, the original symbiont becomes the mitochondrion of the host cell. (D) The resulting eukaryotic cell is now dependent on respiration facilitated by a mitochondrion that possesses bacterial characteristics. Further evolution provides the eukaryote with features such as a cytoskeleton and intracellular organelles. (Adapted from Vellai et al., 1998.)

that the host evolved into a chemoheterotroph, while the symbiont was transformed into either a mitochondrion or a hydrogenosome (Martin and Müller, 1998). An alternate view is that the anaerobic *Archaea* might have found it advantageous to reduce the increasingly high levels of free O_2 by incorporating oxygen-respiring bacteria (Vellai et al., 1998).

The age of *Eukarya* remains uncertain, but 18S rRNA-based molecular phylogenies suggest that they evolved as a distinct domain early in the history of life (recall Fig. 1.1 for *Eukarya* 18S rRNA universal phylogenetic tree). The most phylogenetically ancient eukaryotes are the parasitic diplomonads, microsporidia, and the trichomonads (e.g., Sogin et al., 1989). These aerotolerant anaerobes lack mitochondria and many other organelles compared to more recently evolved taxa, and they exhibit a number of ultrastructural and biochemical features more similar to prokaryotes than other eukaryotes. Next came several seemingly unrelated organisms, including the euglenoids, amoeboids (a group of protozoans), and several slime molds. Thereafter, a subsequent burst in evolutionary diversity led to the remaining eukaryotes that comprise the "crown" of the eukaryotic subtree – alveolates, stramenopiles, red algae, fungi, green algae/plants, and animals (Sogin, 1994). The alveolates are a complex evolutionary assemblage that includes various protozoans, such as the flagellates, sporozoans, and ciliates. The stramenopiles consist of brown algae, crysophytes, xanthophytes, oomycetes, and diatoms.

This evolutionary pattern is, to some extent, borne out in the rock record. As discussed earlier, the extraction of some specific sterane biomarkers from bitumens in the 2.7 Gyr shales of the Jeerinah Formation appears to set a new benchmark for their existence (Brocks et al., 1999). Despite this apparently early evolution, there is little physical evidence for their existence until the oldest recognizable eukaryote fossils at 2.1 Ga, consisting of the remains of the *Grypania spiralis*, a stringy, corkscrew-shaped organism that grew on the seafloor and reached a maximum size of about 0.6 m in length and 2 mm in diameter (Han and Runnegar, 1992). *Grypania* has no certain living relatives, but is regarded as a likely eukaryotic alga because of its complexity, structural rigidity, and large size (Fig. 7.16). The oldest acritarchs (organic-walled microfossils of unknown biological affinity, but with features interpreted as eukaryotic), consisting of spheroidal organic vesicles up to 160 μm in diameter, have been recovered from the 1.5 Gyr Roper Group in northern Australia (Javaux et al., 2001), while the first remains of multicellular red algae occur in the 1.2 Gyr Huntington Formation in Canada (Butterfield et al., 1990). Yet, even then algal fossils continued to be of low diversity and limited abundance until the late Neoproterozoic, when

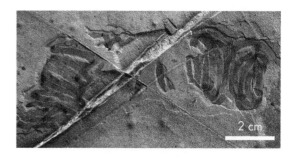

Figure 7.16 Photograph of *Grypania spiralis*, from the Negaunee Iron Formation in Michigan. (Courtesy of Bruce Runnegar.)

they underwent a massive expansion to form the crown of the eukaryotic tree (Knoll, 1992). So why the lack of fossils until well into the Proterozoic? One possibility is that that they had not yet evolved (Cavalier-Smith, 2002), but this contradicts their phylogenetic placement. A more accepted view is that it represents a combination of the poor preservation potential of early eukaryotes (that lacked degradation-resistant or mineralized structures) with an insufficient sedimentary rock record older than 2 Gyr.

By the Late Neoproterozoic–Early Cambrian, the first silica-precipitating protists appear. The radiolaria form skeletons of amorphous silica, and extensive accumulations of their siliceous remains, now in the form of chert, are found throughout Palaeozoic strata. Along with the advent of sponges, their expansion had the significant effect of causing seawater to slowly become undersaturated with respect to amorphous silica. In the Cambrian, the first important carbonate-precipitating protists evolved, the foraminifera. They were followed in the upper Triassic (~200 Ma) by the coccolithophores. These two microorganisms invaded the open oceans, shifting calcium carbonate production away from the shallow seas, thereby increasing the levels of carbonate deposition world-wide (Falkowski and Raven, 1997). Both protists form shells that are relatively resistant to decay, and hence their deposition into sediment leads to easily identifiable fossils in the rock record. Diatoms also appeared in the late Triassic–early Jurassic, however they did not become quantitatively important constituents of the marine phytoplanktonic community until the middle Cretaceous, some 100 million years ago. Their proliferation reinforced the role of radiolarians in marine Si cycling by shifting the main locus of silica deposition from the shallow shelf (recall as in BIF formation) to the open ocean (Maliva et al., 2005).

A major bolide collision at the Cretaceous/Tertiary boundary (65 Ma) eliminated a major proportion of the marine phytoplankton diversity that had developed over the preceding hundred million years, particularly amongst the coccolithophores. By contrast, the diatoms increased in diversity throughout the Cenozoic and they soon displaced most of the other eukaryotic species in the oceans (Falkowski et al., 2004).

7.6 Summary

Life on Earth appears to have already existed at 3.8 Ga, when isotopic evidence suggests that primitive autotrophic microorganisms were growing in an environment subject to reducing hydrothermal fluids, continuous burial by volcaniclastic material and chemical precipitates, and under a greenhouse atmosphere fundamentally different in composition than that of today. Whether those microorganisms were representative of the original progenotes, or whether they were the survivors of evolutionary bottlenecks, is unclear. What does seem likely, however, is that life diversified near hydrothermal vents, where essential biochemical elements were available and large redox gradients (e.g., H_2/S^0) were established by the continuous discharge of reductants into a weakly oxidizing ocean. Indeed, the deepest lineages on the universal phylogenetic tree correspond to a hydrothermal existence. The earliest cells were also autotrophic, but with time the dead cells would accumulate, creating a potential niche for new species with the capability of recycling the reduced organic matter using various electron acceptors available to them. Accordingly, the primitive biosphere developed an active carbon cycle and a remarkable level of metabolic sophistication. With some evolutionary adaptation, photosynthesis took root, inevitably facilitating the diversification of new species throughout the planet's surface. Their biological innovations led to lasting environmental change, but in turn, geology fed back on biology, creating new opportunities for evolutionary innovation. A well-illustrated example of this type of feedback is the development of oxygenic photosynthesis, which may have its roots even before 3.5 Ga. As eloquently pointed out by Knoll (2003), this biological process modified the redox history of the atmosphere and oceans, making it possible for the evolution of aerobically respiring prokaryotes and more complex eukaryotic species, including plants and animals. Yet, photosynthetic species were affected by the availability of nutrients supplied by the weathering of landmasses and on having shallow continental shelves that permitted their proliferation and diversification. On some occasions in Earth's history, physical processes exceeded the environmental buffering capacity conferred by the existing microbial communities (e.g., Snowball events), that if truly global, must have been detrimental to many existing life forms, thereby inducing significant biological change. Thus, there is a historical, and ongoing, interaction between life and the Earth that fortuitously is manifest today as a habitable planet.

References

Absolom, D.R., Lamberti, F.V., Policova, Z., et al. 1983. Surface thermodynamics of bacterial adhesion. Applied and Environmental Microbiology, 46:90–97.

Achenbach-Richter, L., Stetter, K.O., and Woese, C.R., 1987. A possible biochemical missing link among archaebacteria. Nature, 327:348–340.

Adams, L.F. and Ghiorse, W.C., 1986. Physiology and ultrastructure of *Leptothrix discophora* SS-1. Archives of Microbiology, 145:126–135.

Aguiar, P., Beveridge, T.J., and Reysenbach, A.-L., 2004. *Sulfurihydrogenibium azorense*, sp. nov., a thermophilic hydrogen-oxidizing microaerophile from terrestrial hot springs in the Azores. International Journal of Systematic and Evolutionary Microbiology, 54:33–39.

Aharon, P. and Fu, B., 2003. Sulfur and oxygen isotopes of coeval sulfate-sulfide in pore fluids of cold seep sediments with sharp redox gradients. Chemical Geology, 195:201–218.

Ahimou, F., Paquot, M., Jacques, P., Thonart, P., and Rouxhet, P.G., 2001. Influence of electrical properties on the evaluation of the surface hydrophobicity of *Bacillus subtilis*. Journal of Microbiological Methods, 45:119–126.

Ahmann, D., Roberts, A.L., Krumholz, L.R., and Morel, F.M.M., 1994. Microbe grows by reducing arsenic. Nature, 371:750.

Aislabie, J. and Loutit, M.W., 1986. Accumulation of Cr(III) by bacteria isolated from polluted sediment. Marine Environmental Research, 20:221–232.

Akthar, M.N., Sastry, K.S., and Mohan, P.M., 1996. Mechanism of metal ion biosorption by fungal biomass. BioMetals, 9:21–28.

Al-Agha, M.R., Burley, S.D., Curtis, C.D., and Esson, J., 1995. Complex cementation textures and authigenic mineral assemblages in Recent concretions from the Lincolnshire Wash (east coast, UK) driven by Fe(0) to Fe(II) oxidation. Journal of the Geological Society of London, 152:157–171.

Allègre, C.J., Staudacher, T., and Sarda, P., 1987. Rare gas systematics: formation of the atmosphere, evolution and structure of the Earth's mantle. Earth and Planetary Science Letters, 81:127–150.

Aller, R.C., 1980. Quantifying solute distributions in the bioturbated zone of marine sediments by defining an average microenvironment. Geochimica et Cosmochimica Acta, 44:1955–1965.

Aller, R.C., 1990. Bioturbation and manganese cycling in hemipelagic sediments. Philosophical Transactions of the Royal Society of London, 331:51–68.

Aller, R.C. and Rude, P.D., 1988. Complete oxidation of solid phase sulfides by manganese and bacteria in anoxic marine sediments. Geochimica et Cosmochimica Acta, 52:751–765.

Aller, R.C., Macklin, J.E., and Cox, R.T., Jr, 1986. Diagenesis of Fe and S in Amazon inner shelf muds: apparent dominance of Fe reduction and implications for genesis of ironstones. Continental Shelf Research, 6:263–289.

Alt, J.C., 1988. Hydrothermal oxide and nontronite deposits on seamounts in the Eastern Pacific. Marine Geology, 81:227–239.

Altermann, W. and Schopf, J.W., 1995. Microfossils from the Neoarchean Campbell Group, Griqualand West Sequence of the Transvaal Supergroup, and their palaeoenvironmental and evolutionary implications. Precambrian Research, 75:65–90.

Amann, R.I., Ludwig, W., and Schleifer, K.-H., 1995. Phylogenetic identification and in situ detection of individual microbial cells without cultivation. Microbiological Reviews, 59:143–169.

Amend, J.P. and Shock, E.L., 1998. Energetics of amino acid synthesis in hydrothermal ecosystems. Science, 281:1659–1662.

Amend, J.P., Rogers, K.L., Shock, E.L., Gurrieri, S., and Inguaggiato, S., 2003. Energetics of chemolithoautotrophy in the hydrothermal system of Vulcano Island, southern Italy. Geobiology, 1:37–58.

Aminuddin, M., 1980. Substrate level versus oxidative phosphorylation in the generation of ATP in *Thiobacillus denitrificans*. Archives of Microbiology, 128:19–25.

Aminuddin, M. and Nicholas, D.J.D., 1973. Sulfide oxidation linked to the reduction of nitrate and nitrite in *Thiobacillus denitrificans*. Biochimica et Biophysica Acta, 325:81–93.

Amonette, J.E., Workman, D.J., Kennedy, D.W., Fruchter, J.S., and Gorby, Y.A., 2000. Dechlorination of carbon tetrachloride by Fe(II) associated with goethite. Environmental Science and Technology, 34:4606–4613.

Amouric, M. and Parron, C., 1985. Structure and growth mechanism of glauconite as seen by high-resolution transmission electron microscopy. Clays and Clay Minerals, 33:473–482.

Anbar, A.D., 2004. Iron stable isotopes: beyond biosignatures. Earth and Planetary Science Letters, 217:223–236.

Anbar, A.D. and Knoll, A.H., 2002. Proterozoic ocean chemistry and evolution: a bioinorganic bridge? Science, 297:1137–1142.

Anderson, R.T. and Lovley, D.R., 2002. Microbial redox interactions with uranium: an environmental perspective. In: Keith-Roach, M.J. and Livens, F.R. (eds), Interactions of Microorganisms with Radionuclides. Elsevier, Amsterdam, pp. 205–223.

Anderson, R.T., Chapelle, F.H., and Lovley, D.R., 1998. Evidence against hydrogen-based microbial ecosystems in basalt aquifers. Science, 281:976–977.

Anderson, R.T., Vrionis, H.A., Ortiz-Bernad, I. et al. 2003. Stimulating the in situ activity of *Geobacter* species to remove uranium from the groundwater of a uranium-contaminated aquifer. Applied and Environmental Microbiology, 69:5884–5891.

Anthony, C., 1986. Bacterial oxidation of methane and methanol. Advances in Microbial Physiology, 27:113–210.

Andrews, G.R., 1988. The selective adsorption of thiobacilli to dislocation sites on pyrite surfaces. Biotechnology and Bioengineering, 31:378–381.

Andrews, J.E., Brimblecomb, P., Jickells, T.D., Liss, P.S., and Reid, B.J., 2004. An Introduction to Environmental Chemistry, 2nd edn. Blackwell Science, Oxford.

Archer, D.E., Emerson, S., and Reimers, C., 1989. Dissolution of calcite in deep-sea sediments: pH and O_2 microelectrode results. Geochimica et Cosmochimica Acta, 53:2831–2845.

Armstrong, H.A. and Brasier, M.D., 2005. Microfossils. Blackwell Publishing, Oxford.

Arp, G., Reimer, A., and Reitner, J., 2001. Photosynthesis-induced biofilm calcification and calcium concentrations in Phanerozoic oceans. Science, 292:1701–1704.

Arrhenius, S., 1908. Worlds in the Making. Harper Brothers, New York.

Atlas, R.M., 1981. Microbial degradation of petroleum hydrocarbons: an environmental perspective. Microbiology Reviews, 45:180–209.

Awramik, S.M., 1992. The oldest records of photosynthesis. Photosynthesis Research, 33:75–89.

Azam, F., Fenchel, T., Field, J.G., Gray, J.S., Meyer-Reil, L.A., and Thingstad, F., 1983. The ecological role of water-column microbes in the Sea. Marine Ecology Progress Series, 10:257–263.

Bagdigian, R.M. and Myerson, A.S., 1986. The adsorption of *Thiobacillus ferrooxidans* on coal surfaces. Biotechnology and Bioengineering, 28:467–479.

Bak, F. and Cypionka, H., 1987. A novel type of energy metabolism involving fermentation of inorganic sulfur compounds. Nature, 326:891–892.

Baker, B.J. and Banfield, J.F., 2003. Microbial communities in acid mine drainage. FEMS Microbiology Ecology, 44:139–152.

Baker, B.J., Hugenholtz, P., Dawson, S.C., and Banfield, J.F., 2003. Extremely acidophilic protists from acid mine drainage host *Rickettsiales*-lineage endosymbionts that have intervening sequences in their 16S rRNA genes. Applied and Environmental Microbiology, 69:5512–5518.

Baldan, A., 2002. Progress in Ostwald ripening theories and their applications to nickel-base superalloys. Journal of Materials Science, 37:2171–2202.

Baltscheffsky, H. and Baltscheffsky, M., 1994. Molecular origin and evolution of early biological energy conversion. In: Bengtson, S. (ed.), Early Life on Earth. Columbia University Press, New York, pp. 81–90.

Banfield, J.F. and Hamers, R.J., 1997. Processes at minerals and surfaces with relevance to microorganisms and prebiotic synthesiz. In: Banfield, J.F. and Nealson, K.H. (eds), Geomicrobiology: interactions between microbes and minerals, vol. 35. Mineralogical Society of America, Washington, DC, pp. 81–122.

Banfield, J.F., Welch, S.A., Zhang, H., Ebert, T.T., and Penn, R.L., 2000. Aggregation-based crystal growth and microstructure development in natural iron oxyhydroxides biomineralization products. Science, 289:751–754.

Baranov, S.V., Ananyez, G.M., Klimov, V.V., and Dismukes, G.C., 2000. Bicarbonate accelerates assembly of the inorganic core of the water oxidizing complex in manganese-depleted Photosystem II: a proposed biogeochemical role for atmospheric carbon dioxide in oxygenic photosynthesis. Biochemistry, 39:6060–6065.

Barber, J. and Anderson, B., 1994. Revealing the blueprint of photosynthesis. Nature, 370:31–34.

Barker, W.W. and Banfield, J.F., 1996. Biologically versus inorganically-mediated weathering reactions: relationships between minerals and extracellular microbial polymers in lithobiontic communities. Chemical Geology, 132:55–69.

Barker, W.W., Welch, S.A., and Banfield, J.F., 1997. Biogeochemical weathering of silicate minerals. In: Banfield, J.F. and Nealson, K.H. (eds), Geomicrobiology: interactions between microbes and minerals, vol. 35. Mineralogical Society of America, Washington, DC, pp. 391–428.

Barns, S.M. and Nierzwicki-Bauer, S.A., 1997. Microbial diversity in ocean, surface and subsurface environments. In: Banfield, J.F. and Nealson, K.H. (eds), Geomicrobiology: interactions between microbes and minerals, vol. 35. Mineralogical Society of America, Washington, DC, pp. 35–79.

Barns, S.M., Delwiche, C.F., Palmer, J.D., and Pace, N.R., 1996. Perspectives on archaeal diversity, thermophily and monophyly from environmental rRNA sequences. Proceedings of the National Academy of Sciences, USA, 93:9188–9193.

Bar-Nun, A., Bar-Nun, N., Bauer, S.H., and Sagan, C., 1970. Shock synthesis of amino acids in simulated primitive environments. Science, 168:470–473.

Baross, J.A., Lilley, M.D., and Gordon, L.I., 1982. Is the CH_4, H_2 and CO venting from submarine hydrothermal systems produced by thermophilic bacteria? Nature, 298:366–368.

Bartley, J.K., 1996. Actualistic taphonomy of cyanobacteria: implications for the Precambrian fossil record. Palaios, 11:571–586.

Battista, J.R., 1997. Against all odds: the survival strategies of *Deinococcus radiodurans*. Annual Review of Microbiology, 51:203–224.

Bazylinski, D.A., 1996. Controlled biomineralization of magnetic minerals by magnetotactic bacteria. Chemical Geology, 132:191–198.

Bazylinski, D.A. and Blakemore, R.P., 1983. Denitrification and assimilatory nitrate reduction in *Aquaspirillum magnetotacticum*. Applied and Environmental Microbiology, 46:1118–1124.

Bazylinski, D.A. and Frankel, R.B., 2000. Biologically controlled mineralization of magnetic iron minerals by magnetotactic bacteria. In: Lovley, D.R. (ed.), Environmental Microbe–Metal Interactions. ASM Press, Washington, DC, pp. 109–143.

Bazylinski, D.A., Heywood, B.R., Mann, S., and Frankel, R.B., 1993. Fe_3O_4 and Fe_3S_4 in a bacterium. Nature, 366:218.

Bazylinski, D.A., Frankel, R.B., Heywood, B.R., et al. 1995. Controlled biomineralization of magnetite (Fe_3O_4) and greigite (Fe_3S_4) in a magnetotactic bacterium. Applied and Environmental Microbiology, 61:3232–3239.

Bazylinski, D.A., Schlezinger, D.R., Howes, B.H., Frankel, R.B., and Epstein, S.S., 2000. Occurrence and distribution of diverse populations of magnetic protists in a chemically stratified coastal salt pond. Chemical Geology, 169:319–328.

Beales, N., 2003. Adaptation of microorganisms to cold temperatures, weak acid preservatives, low pH and osmotic stress: a review. Comprehensive Reviews in Food Science and Food Safety, 3:1–20.

Beard, B.L., Johnson, C.M., Cox, L., Sun, H., Nealson, K.H., and Aguilar, C., 1999. Iron isotope biosignatures. Science, 285:1889–1892.

Beaumont, V. and Robert, F., 1999. Nitrogen isotope ratios of kerogens in Precambrian cherts: a record of the evolution of atmosphere chemistry? Precambrian Research, 96:63–82.

Bebié, J. and Schoonen, M.A.A., 1999. Pyrite and phosphate in anoxia and an origin-of-life hypothesis. Earth and Planetary Science Letters, 171:1–5.

Bebié, J., Schoonen, M.A.A., Fuhrman, M., and Strongin, D.R., 1998. Surface charge development on transition metal sulfides: an electrokinetic study. Geochimica et Cosmochimica Acta, 62:633–642.

Bebout, B., Paerl, H., Crocker, K., and Prufert, L., 1987. Diel interactions of oxygenic photosynthesis and N₂ fixation (acetylene reduction) in a marine microbial mat community. Applied and Environmental Microbiology, 53:2353–2362.

Bebout, B.M., Hoehler, T.M., Thamdrup, B., et al. 2004. Methane production by microbial mats under low sulphate concentrations. Geobiology, 2:87–96.

Beech, I.B. and Gaylarde, C.C., 1991. Microbial polysaccharides and corrosion. International Biodeterioration, 27:95–107.

Beinert, H., Holm, R.H., and Münck, E., 1997. Iron-sulfur clusters: nature's modular, multipurpose structures. Science, 277:653–659.

Béjà, O., Aravind, L., Koonin, E.V., et al. 2000. Bacterial rhodopsin: evidence for a new type of phototrophy in the sea. Science, 289:1902–1906.

Bender, B. and Heggie, D.T., 1984. Fate of organic carbon reaching the deep sea: a status report. Geochimica et Cosmochimica Acta, 48:977–986.

Benner, R., Maccubbin, A.E., and Hodson, R.E., 1984. Anaerobic biodegradation of the lignin and polysaccharide components of lignocellulose and synthetic lignin by sediment microflora. Applied and Environmental Microbiology, 47:998–1004.

Bennett, P. and Siegel, D.I., 1987. Increased solubility of quartz in water due to complexing by organic compounds. Nature, 326:684–687.

Bennett, P.C., Hiebert, F.K., and Choi, W.J., 1996. Microbial colonization and weathering of silicates in a petroleum-contaminated groundwater. Chemical Geology, 132:45–53.

Bennett, P.C., Melcer, M.E., Siegel, D.I., and Hassett, J.P., 1988. The dissolution of quartz in dilute aqueous solutions of organic acids at 25°C. Geochimica et Cosmochimica Acta, 52:1521–1530.

Benning, L.G., Wilkin, R.T., and Barnes, H.L., 2000. Reaction pathways in the Fe-S system below 100°C. Chemical Geology, 167:25–51.

Benning, L.G., Phoenix, V., Yee, N., and Konhauser, K.O., 2003. The dynamics of cyanobacterial silicification: an infrared micro-spectroscopic investigation. Geochimica et Cosmochimica Acta, 68:743–757.

Bergman, B., Gallon, J.R., Rai, A.N., and Stal., L.J., 1997. N₂ fixation by non-heterocystous cyanobacteria. FEMS Microbiology Reviews, 19:139–185.

Berner, R.A., 1964. An idealized model of dissolved sulfate distribution in recent sediments. Geochimica et Cosmochimica Acta, 28:1497–1503.

Berner, R.A., 1978. Rate control of mineral dissolution under Earth surface conditions. American Journal of Science, 278:1235–1252.

Berner, R.A., 1980. Early Diagenesis: A theoretical approach. Princeton University Press, Princeton, New Jersey.

Berner, R.A., 1982. Burial of organic carbon and pyrite sulfur in the modern ocean: its geochemical and environmental significance. American Journal of Science, 282:451–473.

Berner, R.A., 1984. Sedimentary pyrite formation: an update. Geochimica et Cosmochimica Acta, 48:605–615.

Berner, R.A., 1995. Sedimentary organic matter preservation: an assessment and speculative synthesis. Marine Chemistry, 49:121–122.

Berner, R.A. and Morse, J.A., 1974. Dissolution kinetics of calcium carbonate in sea water: IV. Theory of calcite dissolution. American Journal of Science, 274:108–134.

Berner, R.A. and Raiswell, R., 1983. Burial of organic carbon and pyrite sulfur in sediments over Phanerozoic time: a new theory. Geochimica et Cosmochimica Acta, 47:855–862.

Berner, R.A. and Westrich, J.T., 1985. Bioturbation and the early diagenesis of carbon and sulfur. American Journal of Science, 285:193–206.

Berner, R.A., Lasaga, A.C., and Garrels, R.M., 1983. The carbon-silicate geochemical cycle and its effect on atmospheric carbon dioxide over the past 100 million years. American Journal of Science, 283:641–683.

Bernstein, M.P., Sandford, S.A., Allamandola, L.J., Gillette, J.S., Clemett, S.J., and Zare, R.N., 1999. UV irradiation of polycyclic aromatic hydrocarbons in ices: production of alcohols, quinones, and ethers. Science, 283:1135–1138.

Berthelin, J., 1983. Microbial weathering processes. In: Krumbein, W.E. (ed.), Microbial Geochemistry. Blackwell, Oxford, pp. 223–262.

Bethke, C.M. and Brady, P.V., 2000. How the Kd approach undermines ground water cleanup. Ground Water, 38:435–443.

Beveridge, T.J., 1981. Ultrastructure, chemistry and function of the bacterial wall. International Review of Cytology, 72:229–317.

Beveridge, T.J., 1984. Mechanisms of the binding of metallic ions to bacterial walls and the possible impact on microbial ecology. In: Reddy, C.A. and Klug, M.J. (eds), Current Perspectives in Microbial Ecology. American Society for Microbiology, Washington, DC, pp. 601–607.

Beveridge, T.J., 1988. The bacterial cell surface: general considerations towards design and function. Canadian Journal of Microbiology, 34:363–372.

Beveridge, T.J., 1989a. The structure of bacteria. In: Poindexter, J.S. and Leadbetter, E.R. (eds), Bacteria in Nature. Plenum Press, New York, pp. 1–65.

Beveridge, T.J., 1989b. Interactions of metal ions with components of bacterial cell walls and their biomineralization. In: Poole, R.K. and Gadd, G.M. (eds), Metal–Microbe Interactions. Special Publications of the Society for General Microbiology, vol. 26. IRL Press, Oxford, pp. 65–83.

Beveridge, T.J. and Davies, J.A., 1983. Cellular responses of *Bacillus subtilis* and *Escherichia coli* to the Gram stain. Journal of Bacteriology, 156:846–858.

Beveridge, T.J. and Graham, L.L., 1991. Surface layers of bacteria. Microbiological Reviews, 55:684–705.

Beveridge, T.J. and Koval, S.F., 1981. Binding of metals to cell envelopes of *Escherichia coli* K-12. Applied and Environmental Microbiology, 42:325–335.

Beveridge, T.J. and Murray, R.G.E., 1976. Uptake and retention of metals by cell walls of *Bacillus subtilis*. Journal of Bacteriology, 127:1502–1518.

Beveridge, T.J. and Murray, R.G.E., 1979. How thick is the *Bacillus subtilis* cell wall? Current Microbiology, 2:1–4.

Beveridge, T.J. and Murray, R.G.E., 1980. Sites of metal deposition in the cell wall of *Bacillus subtilis*. Journal of Bacteriology, 141:876–887.

Beveridge, T.J. and Schultze-Lam, S., 1996. The response of selected members of the Archaea to the Gram stain. Microbiology, 142:2887–2895.

Beveridge, T.J., Forsberg, C.W. and Doyle, R.J., 1982. Major sites of metal binding in *Bacillus licheniformis* walls. Journal of Bacteriology, 150:1438–1448.

Bidle, K.D. and Azam, F., 1999. Accelerated dissolution of diatom silica by marine bacterial assemblages. Nature, 397:508–512.

Biebl, H. and Pfennig, N., 1977. Growth of sulfate-reducing with sulfur as electron acceptor. Archives of Microbiology, 112:115–117.

Biebl, H. and Pfennig, N., 1978. Growth yields of green sulfur bacteria in mixed cultures with sulfur and sulfate reducing bacteria. Archives of Microbiology, 117:9–16.

Bigham, J.M., Schwertmann, U., and Pfab, G., 1996. Influence of pH on mineral speciation in a bioreactor simulating acid mine drainage. Applied Geochemistry, 11:845–849.

Birch, L. and Bachofen, R., 1990. Complexing agents from microorganisms. Experientia, 46:827–834.

Bjerrum, C.J. and Canfield, D.E., 2002. Ocean productivity before about 1.9 Gyr limited by phosphorous adsorption onto iron oxides. Nature, 417:159–162.

Bjerrum, C.J. and Canfield, D.E., 2004. New insights into the burial history of organic carbon on the early Earth. Geochemistry, Geophysics, Geosystems, 5(8):Q08001,9pp.

Blackburn, T.H., 1980. Seasonal variation in the rate of organic-N mineralization in anoxic sediments. Colloque International CNRS (Marseilles), no 293,173–183.

Blackburn, T.H. and Blackburn, N.D., 1993. Coupling of cycles and global significance of sediment diagenesis. Marine Geology, 113:101–110.

Blake, R. and Johnson, D.B., 2000. Phylogenetic and biochemical diversity among acidophilic bacteria that respire on iron. In: Lovley, D.R. (ed.), Environmental Microbe–Metal Interactions. ASM Press, Washington, DC, pp. 53–78.

Blake, R.C., Shute, E.A., and Howard, G.T., 1994. Solubilization of minerals by bacteria: electrophoretic mobility of *Thiobacillus ferrooxidans* in the presence of iron, pyrite, and sulfur. Applied and Environmental Microbiology, 60:3349–3357.

Blakemore, R., 1975. Magnetotactic bacteria. Science, 190:377–379.

Blakemore, R.P. and Blakemore, N.A., 1990. Magnetotactic magnetogens. In: Frankel, R.B. and Blakemore, R.P. (eds), Iron Biominerals. Plenum Press, New York, pp. 51–67.

Blankenship, R.E., 1992. Origin and early evolution of photosynthesis. Photosynthesis Research, 33:91–111.

Blankenship, R.E. and Hartman, H., 1998. The origin and evolution of oxygenic photosynthesis. Trends in Biochemical Science, 23:94–97.

Blowes, D.W., Al, T., Lortie, L., Gould, W.D., and Jambor, J.L., 1995. Microbiological, chemical, and mineralogical characterization of the Kidd Creek mine tailings impoundment, Timmins area, Ontario. Geomicrobiology Journal, 13:13–31.

Blum, A. and Lasaga, A., 1988. Role of surface speciation in the low-temperature dissolution of minerals. Nature, 331:431–433.

Boetius, A., Ravenschlag, K., Schubert, C.J., et al. 2000. A marine microbial consortium apparently mediating anaerobic oxidation of methane. Nature, 407:623–626.

Bond, P.L., Smriga, S.P., and Banfield, J.F., 2000. Phylogeny of microorganisms populating a thick, subaerial, pre-dominantly lithotrophic biofilm at an extreme acid mine drainage site. Applied and Environmental Microbiology, 66:3842–3849.

Boogerd, R.C. and de Vrind, J.P.M., 1987. Manganese oxidation by *Leptothrix discophora*. Journal of Bacteriology, 169:489–494.

Boopathy, R. and Daniels, L., 1991. Effect of pH on anaerobic mild steel corrosion by methanogenic bacteria. Applied and Environmental Microbiology, 57:2104–2108.

Booth, I.R., 1988. Bacterial transport: energetics and mechanisms. In: Anthony, C. (ed.), Bacterial Energy Transduction. Academic Press, London, pp. 377–428.

Borda, M.J., Elsetinow, A.R., Schoonen, M.A., and Strongin, D.R., 2001. Pyrite-induced hydrogen peroxide formation as a driving force in the evolution of photosynthetic organisms on an early Earth. Astrobiology, 1:283–288.

Borowitzka, M.A., 1989. Carbonate calcification in algae – initiation and control. In: Mann, S., Webb, J., and Williams, R.J.P. (eds), Biomineralization: chemical and biochemical perspectives. VCH Verlagsgesellschaft, Weinheim, Germany, pp. 63–94.

Boschker, H.T.S., Nold, S.C., Wellsbury, P., et al. 1998. Direct linking of microbial populations to specific biogeochemical processes by ^{13}C-labeling of biomarkers. Nature, 392:801–805.

Boudreau, B.P., 1991. Modelling the sulfide-oxygen reaction and associated pH gradients in porewaters. Geochimica et Cosmochimica Acta, 55:145–159.

Boudreau, B.P. and Canfield, D.E., 1993. A comparison of closed-and-open system models of porewater pH and calcite-saturation state. Geochimica et Cosmochimica Acta, 57:317–334.

Bowring, S.A., Kinh, J.E., Housh, T.B., Isachsen, C.E., and Podosek, F.A., 1989. Neodymium and lead isotope evidence for the enriched early Archaean crust in North America. Nature, 340:222–225.

Bradley, J.P., Harvey, R.P., and McSween, Jr H.Y., 1997. No "nanofossils" in Martian meteorite. Nature, 390:454.

Brady, N.C., 2002. The Nature and Properties of Soils, 13th edn. Prentice Hall. Upper Saddle River, New Jersey.

Brady, P.V. and Walther, J.V., 1989. Controls on silicate dissolution rates in neutral and basic pH solutions at 25°C. Geochimica et Cosmochimica Acta, 40:41–49.

Brady, P.V. and Walther, J.V., 1990. Kinetics of quartz dissolution at low temperatures. Chemical Geology, 82:253–264.

Bragg, J.R., Prince, R.C., Harner, E.J., and Atlas, R.M., 1994. Effectiveness of bioremediation for the *Exxon Valdez* oil spill. Nature, 368:413–418.

Braissant, O., Cailleau, G., Duprez, C., and Verrecchia, E.P., 2003. Bacterially induced mineralization of calcium carbonate in terrestrial environments. The role of exopolysaccharides and amino acids. Journal of Sedimentary Research, 73:485–490.

Braissant, O., Cailleau, G., Aragno, M., and Verrecchia, E.P., 2004. Biologically induced mineralization in the tree *Milicia excelsa* (Moraceae): its causes and consequences to the environment. Geobiology, 2:59–66.

Brake, S.S., Hasiotis, S.T., Dannelly, H.K., and Connors, K.A., 2002. Eukaryotic stromatolite builders in acid mine drainage: implications for Precambrian iron formations and oxygenation of the atmosphere? Geology, 30:599–602.

Brandes, J.A., Boctor, N.Z., Cody, G.D., Cooper, B.A., Hazen, R.M., and Yoder, H.S., 1998. Abiotic nitrogen reduction on the early Earth. Nature, 395:365–367.

Brandt, U., 1996. Energy conservation by bifurcated electron-transfer in the cytochrome-bc_1 complex. Biochimica et Biophysica Acta, 1275:41–46.

Brasier, M.D., 1992. Nutrient-enriched waters and the early skeletal fossil record. Journal of the Geological Society of London, 149:621–629.

Brasier, M.D., Green, O.R., Jephcoat, A.P., et al. 2002. Questioning the evidence for Earth's oldest fossils. Nature, 416:76–81.

Braterman, P.S., Cairns-Smith, A.G., and Sloper, R.W., 1983. Photo-oxidation of hydrated Fe^{2+} – significance for banded iron formations. Nature, 303:163–164.

Bremner, J.M., 1980. Concretionary phosphorite from SW Africa. Journal of the Geological Society of London, 137:773–786.

Brennan, S.T., Lowenstein, T.K., and Horita, J., 2004. Seawater chemistry and the advent of biocalcification. Geology, 32:473–476.

Brierley, C.L., Brierley, J.A., and Davidson, M.S., 1989. Applied microbial processes for recovery and removal from wastewater. In: Beveridge, T.J. and Doyle, R.J. (eds), Metal Ions and Bacteria. John Wiley, New York, pp. 359–382.

Briggs, D.E.G. and Kear, A.J., 1993. Fossilization of soft tissue in the laboratory. Science, 259:1439–1442.

Brocks, J.J., Logan, G.A., Buick, R., and Summons, R.E., 1999. Archean molecular fossils and the early rise of eukaryotes. Science, 285:1033–1036.

Brocks, J.J., Buick, R., Summons, R.E., and Logan, G.A., 2003. A reconstruction of Archean biological diversity based on molecular fossils from the 2.78 to 2.45 billion-year-old Mount Bruce Supergroup, Hamersley Basin, Western Australia. Geochimica et Cosmochimica Acta, 67:4321–4335.

Brownlow, A.H., 1996. Geochemistry, 2nd edn. Prentice-Hall, Upper Saddle River, New Jersey.

Brune, D.C., 1989. Sulfur oxidation by phototrophic bacteria. Biochimica et Biophysica Acta, 975:189–221.

Bruneel, O., Personné, J.-C., Casio, C., et al. 2003. Mediation of arsenic oxidation by *Thiomonas* sp. in acid-mine drainage (Carnoulès, France). Journal of Applied Microbiology, 95:492–499.

Bryant, R.D. and Laishley, E.J., 1989. The role of hydrogenase in anaerobic biocorrosion. Canadian Journal of Microbiology, 36:259–264.

Buck, S.G., 1980. Stromatolite and ooid deposits within the fluvial and lacustrine sediments of the Precambrian Ventersdorp Supergroup of South Africa. Precambrian Research, 12:311–330.

Büdel, B., Weber, B., Kühl, M., Pfanz, H., Sültemeyer, D., and Wessels, D., 2005. Reshaping of sandstone surfaces by cryptoendolithic cyanobacteria: bioalkalization causes chemical weathering in arid landscapes. Geobiology, 2:261–268.

Buffle, J., 1990. Complexation Reactions in Aquatic Systems: an analytical approach, 2nd edn. Ellis Horwood, London.

Bui, E.T.N., Bradley, P.J., and Johnson, P.J., 1996. A common evolutionary origin for mitochondria and hydrogenosomes. Proceedings of the National Academy of Sciences, USA, 93:9651–9656.

Buick, R., 1988. Carbonaceous filaments from North Pole, western Australia – are they fossil bacteria in Archean stromatolites. A reply. Precambrian Research, 39:311–317.

Buick, R., 1991. Microfossil recognition in Archean rocks: an appraisal of spheroids and filaments from a 3500 M.y. old chert-barite unit at North Pole, Western Australia. Palaios, 5:441–459.

Buick, R., 1992. The antiquity of oxygenic photosynthesis: evidence for stromatolites in sulfate-deficient Archaean lakes. Science, 255:74–77.

Buick, R. and Dunlop, J.S.R., 1990. Evaporitic sediments of Early Archean age from the Warrawoona Group, North Pole, Western Australia. Sedimentology, 37:247–277.

Buick, R., Des Marais, D.J., and Knoll, A.H., 1995. Stable isotopic compositions of carbonates from the Mesoproterozoic Bangemall Group, northwestern Australia. Chemical Geology, 123:153–171.

Burdige, D.J., 1993. The biogeochemistry of manganese and iron reduction in marine sediments. Earth-Science Reviews, 35:249–284.

Burdige, D.J. and Gieskes, J.M., 1983. A pore water/solid phase diagenetic model for manganese in marine sediments. American Journal of Science, 283:29–47.

Burdige, D.J. and Nealson, K.H., 1986. Chemical and microbiological studies of sulfide-mediated manganese reduction. Geomicrobiology Journal, 4:361–387.

Burdige, D.J., Dhakar, S.P., and Nealson, K.H., 1992. Effects of manganese oxide mineralogy on microbial and chemical manganese reduction. Geomicrobiology Journal, 10:27–48.

Burggraf, S., Jannasch, H.W., Nicolaus, B., and Stetter, K.O., 1990. *Archaeglobus profundus* sp. nov., represents a new species within the sulfate-reducing Archaebacteria. Systematic and Applied Microbiology, 13:24–28.

Burggraf, S., Stetter, K.O., Rouviere, P., and Woese, C.R., 1991a. *Methanopyrus kandleri*: an archaeal methanogen unrelated to all other known methanogens. Systematic and Applied Microbiology, 14:346–351.

Burggraf, S., Ching, A., Stetter, K.O., and Woese, C.R., 1991b. The sequence of *Methanospririllum hungatei* 23S rRNA confirms the specific relationship between the extreme halophiles and the *Methanomicrobiales*. Systematic and Applied Microbiology, 14:358–363.

Burne, R.V. and Moore, L.S., 1987. Microbialites: organosedimentary deposits of benthic microbial communities. Palaios, 2:241–254.

Burnett, B.R. and Nealson, K.H., 1981. Organic films and microorganisms associated with manganese nodules. Deep-Sea Research, 28A:637–645.

Burnett, W.C., 1977. Geochemistry and origin of phosphorite deposits from off Peru and Chile. Geological Society of America Bulletin, 88:813–823.

Butterfield, N.J., Knoll, A.H., and Swett, K., 1990. A bangiophyte red alga from the Proterozoic of arctic Canada. Science, 250:104–107.

Byerly, G.R., Lowe, D.R., and Walsh, M.M., 1986. Stromatolites from the 3,300–3,500-Myr Swaziland Supergroup, Barberton Mountain Land, South Africa. Nature, 319:489–491.

Caccavo, F., Jr, Schamberger, P.C., Keiding, K., and Nielsen, P.H., 1997. Role of hydrophobicity in adhesion of the dissimilatory Fe(III)-reducing bacterium *Shewanella alga* to amorphous Fe(III) oxide. Applied and Environmental Microbiology, 63:3837–3843.

Cady, S.L. and Farmer, J.D. 1996. Fossilization processes in siliceous thermal springs: trends in preservation along thermal gradients. In: Brock, G.R. and Goode, J.A. (eds), Evolution of Hydrothermal Ecosystems on Earth (and Mars?). Wiley, Chichester, pp. 150–173.

Caesar-Tonthat, T-C., van Ommen Kloeke, F., Geesey, G.G., and Henson, J.M., 1995. Melanin production by a filamentous soil fungus in response to copper and localization of copper sulfide by sulfide-silver staining. Applied and Environmental Microbiology, 61:1968–1975.

Cail, T.L. and Hochella, M.F., Jr, 2005. The effects of solution chemistry on the sticking efficiencies of viable *Enterococcus faecalis*: an atomic force microscopy and modeling study. Geochimica et Cosmochimica Acta, 69:2959–2969.

Cairns-Smith, A.G., 1978. Precambrian solution photochemistry, inverse segregation, and banded iron formations. Nature, 276:807–808.

Cairns-Smith, A.G., 1982. Genetic Takeover and the Mineral Origins of Life. Cambridge University Press, Cambridge.

Cairns-Smith, A.G., Hall, A.J., and Russell, M.J., 1992. Mineral theories of the origin of life and an iron sulfide example. Origins of Life and Evolution of the Biosphere, 22:161–180.

Calamita, H.G., Ehringer, W.D., Koch, A.L., and Doyle, R.J., 2001. Evidence that the cell wall of *Bacillus subtilis* is protonated during respiration. Proceedings of the National Academy of Sciences, USA, 98:15260–15263.

Caldwell, M.M., Teramura, A.H., and Tevini, M., 1989. The changing solar ultraviolet climate and the ecological consequences for higher plants. Trends in Ecology and Evolution, 4:363–367.

Calvert, S.E., Karlin, R.E., Toolin, L.J., Donahue, D.J., Southon, J.R., and Vogel, J.S., 1991. Low organic carbon accumulation rates in Black Sea sediments. Nature, 350:692–695.

Camacho, A., Vicente, E., and Miracle, M.R., 2000. Spatio-temporal distribution and growth dynamics of phototrophic sulfur bacteria populations in the sulfide-rich Lake Arcas. Aquatic Sciences, 62:334–349.

Cameron, E.M., 1982. Sulfate and sulfate reduction in early Precambrian oceans. Nature, 296:145–148.

Canfield, D.E., 1989. Reactive iron in marine sediments. Geochimica et Cosmochimica Acta, 53:619–632.

Canfield, D.E., 1993. Organic matter oxidation in marine sediments. In: Wollast, R., Chou, L., and Mackenzie, F. (eds), Interactions of C, N, P and S Biogeochemical Cycles and Global Change. Springer-Verlag, Berlin, pp. 333–363.

Canfield, D.E., 1994. Factors influencing organic carbon preservation in marine sediments. Chemical Geology, 114:315–329.

Canfield, D.E., 1998. A new model for Proterozoic ocean chemistry. Nature, 396:450–453.

Canfield, D.E., 2005. The early history of atmospheric oxygen: homage to Robert M. Garrels. Annual Reviews in Earth and Planetary Sciences, 33:17.1–17.36

Canfield, D.E. and Des Marais, 1991. Aerobic sulfate reduction in microbial mats. Science, 251:1471–1473.

Canfield, D.E. and Des Marais, D.J., 1993. Biogeochemical cycles of carbon, sulfur, and free oxygen in a microbial mat. Geochimica et Cosmochimica Acta, 57:3971–3984.

Canfield, D.E. and Raiswell, R., 1991. Pyrite formation and fossil preservation. In: Allison, P.A. and Briggs, D.E.G. (eds), Taphonomy. Releasing the data locked in the fossil record. Plenum Press, New York, pp. 337–387.

Canfield, D.E. and Teske, A., 1996. Late Proterozoic rise in atmospheric oxygen concentration inferred from phylogenies and sulfur-isotope studies. Nature, 382:127–132.

Canfield, D.E. and Thamdrup, B., 1994. The production of ^{34}S-depleted sulfide during bacterial disproportionation of elemental sulfur. Science, 266:1973–1975.

Canfield, D.E., Raiswell, R., and Bottrell, S., 1992. The reactivity of sedimentary iron minerals toward sulfide. American Journal of Science, 292:659–683.

Canfield, D.E., Thamdrup, B., and Hansen, J.W., 1993. The anaerobic degradation of organic matter in Danish coastal sediments: iron reduction, manganese reduction, and sulfate reduction. Geochimica et Cosmochimica Acta, 57:3867–3883.

Canfield, D.E., Habicht, K.S., and Thamdrup, B., 2000. The Archean sulfur cycle and the early history of atmospheric oxygen. Science, 288:658–661.

Canfield, D.E., Sørensen, K.B., and Oren, A., 2004. Biogeochemistry of a gypsum-encrusted microbial ecosystem. Geobiology, 2:133–150.

Cano, R.J. and Borucki, M., 1995. Revival and identification of bacterial spores in 25 to 40 million year old Dominican amber. Science, 268:1060–1064.

Canuto, V.M., Levine, J.S., Augustsson, T.R., and Imhoff, C.L., 1982. UV radiation from the young Sun and oxygen and ozone levels in the prebiological palaeoatmosphere. Nature, 296:816–820.

Carlile, M.J., 1980. Positioning mechanisms. The role of motility, taxis and tropism in the life of microorganisms. In: Ellwood, D.C., Hedger, J.N., Latham, M.J., Lynch, J.M., and Slater, J.H. (eds), Contemporary Microbial Ecology. Academic Press, London, pp. 55–74.

Carlson, C.A. and Ingraham, J.L., 1983. Comparison of denitrification by *Pseudomonas stutzeri*, *Pseudomonas aeruginosa*, and *Paracoccus denitrificans*. Applied and Environmental Microbiology, 45:1247–1253.

Carpenter, E.J., Lin, S.J., and Capone, D.G., 2000. Bacterial activity in South Pole snow. Applied and Environmental Microbiology, 66:4514–4517.

Carstensen, E.L. and Marquis, R.E., 1968. Passive electrical properties of microorganisms. III. Conductivity of isolated bacterial cell walls. Biophysics Journal, 8:536–548.

Carstensen, E.L., Cox, H.A. Jr, Mercer, W.B., and Natale, L.A., 1965. Passive electrical properties of microorganisms. 1. Conductivity of *Escherichia coli* and *Micrococcus lysodeikticus*. Biophysical Journal, 5:289.

Casey, W.H. and Bunker, B., 1990. Leaching of mineral and glass surfaces during dissolution. In: Hochella, M.F. and White, A.F. (eds), Mineral–Water Interface Geochemistry, vol 23. Mineralogical Society of America, Washington, DC, pp. 397–426.

Casey, W.H. and Ludwig, C., 1995. Silicate mineral dissolution as a ligand-exchange reaction. In: White, A.F. and Brantley, S.L. (eds), Chemical Weathering Rates of Silicate Minerals. Reviews in Mineralogy, 31:87–114.

Castanier, S., Le Métayer-Levrel, G., and Perthuisot, J.-P., 2000. Bacterial roles in the precipitation of carbonate minerals. In: Riding, R.E. and Awramik, S.M. (eds), Microbial Sediments. Springer-Verlag, Berlin, pp. 32–39.

Castenholz, R.W., 1973. The possible photosynthetic use of sulfide by the filamentous phototrophic bacteria of hot springs. Limnology and Oceanography, 18:863–876.

Castenholz, R.W., 1988. The green sulfur and nonsulfur bacteria of hot springs. In: Olson, J.M., Ormerod, J.G., Amesz, J., Stackebrandt, E., and Trüper, H.G. (eds), Green Photosynthetic Bacteria. Plenum, New York, pp. 243–255.

Castenholz, R.W. and Waterbury, J.B., 1989. Group I. Cyanobacteria. In: Stanley, J.T., Bryant, M.P., Pfennig, N., and Holt, J.G. (eds), Bergey's Manual of Systematic Bacteriology, vol. 3. Williams and Williams, Baltimore, pp. 1710–1727.

Castenholz, R.W., Jørgensen, B.B., D'Amelio, E., and Bauld, J., 1991. Photosynthetic and behavioural versatility of the cyanobacterium *Oscillatoria boryana* in a sulfide-rich microbial mat. FEMS Microbial Ecology, 86:43–58.

Castresana, J. and Moreira, D., 1999. Respiratory chains in the last common ancestor of living organisms. Journal of Molecular Evolution, 49:453–460.

Castresana, J. and Saraste, M., 1995. Evolution of energetic metabolism: the respiration-early hypothesis. Trends in Biochemical Sciences, 20:443–448.

Castro, H.F., Williams, N.H., and Ogram, A., 2000. Phylogeny of sulfate-reducing bacteria. FEMS Microbiology Ecology, 31:1–9.

Catling, D.C., Zahnle, K.J., and McKay, C.P., 2001. Biogenic methane, hydrogen escape, and the irreversible oxidation of early Earth. Science, 293:839–843.

Cavalier-Smith, T., 1987. Eukaryotes with no mitochondria. Nature, 326:332–333.

Cavalier-Smith, T., 2002. The neomuran origin of archaebacteria, the negibacterial root of the universal tree and bacterial megaclassification. International Journal of Systematic and Evolutionary Microbiology, 52:7–76.

Cavicchioli, R., 2002. Extremophiles and the search for extraterrestrial life. Astrobiology, 2:281–292.

Cervantes, C., Ji, G., Ramírez, J.L., and Silver, S., 1994. Resistance to arsenic compounds in microorganisms. FEMS Microbiology Reviews, 15:355–367.

Chafetz, H.S., 1986. Marine peloids: a product of bacterially induced precipitation of calcite. Journal of Sedimentary Petrology, 56:812–817.

Chafetz, H.S. and Buczynski, C., 1992. Bacterially induced lithification of microbial mats. Palaios, 7:277–293.

Chafetz, H.S., Utech, N.M., and Fitzmaurice, S.P., 1991. Differences in the $\delta^{18}O$ and $\delta^{13}C$ signatures of seasonal laminae comprising travertine stromatolites. Journal of Sedimentary Petrology, 61:1015–1028.

Chameides, W.L. and Walker, J.C.G., 1981. Rates of fixation by lightning of carbon and nitrogen in possible primitive atmospheres. Origins of Life and Evolution of the Biosphere, 11:291–302.

Chang, S., 1994. The planetary setting of prebiotic evolution. In: Bengtson, S. (ed.), Early Life on Earth. Columbia University Press, New York, pp. 10–23.

Chang, S.-B.R. and Kirschvink, J.L., 1989. Magnetofossils, the magnetization of sediments, and the evolution of magnetite biomineralization. Annual Reviews in Earth and Planetary Sciences, 17:169–195.

Chang, J.-S., Law, R., and Chang, C.-C., 1997. Biosorption of lead, copper and cadmium by biomass of *Pseudomonas aeruginosa* PU21. Water Research, 31:1651–1658.

Chapelle, F.H. and Lovley, D.R., 1992. Competitive exclusion of sulfate reduction by Fe(III)-reducing bacteria: a mechanism for producing discrete zones of high-iron ground water. Ground Water, 30:29–36.

Chapelle, F.H., O'Neill, K., Bradley, P.M., et al. 2002. A hydrogen-based subsurface microbial community dominated by methanogens. Nature, 415:312–315.

Characklis, W.G., 1973. Attached microbial growths. 1. Attachment and growth. Water Research, 7:1113–1127.

Characklis, W.G., 1981. Bioengineering report/fouling biofilm development: a process analysis. Biotechnology and Bioengineering, 23:1923–1960.

Charles, A.M. and Suzuki, I., 1966. Purification and properties of sulfite: cytochrome *c* oxidoreductase from *Thiobacillus novellus*. Biochimica et Biophysica Acta, 128:522–534.

Chaudhuri, S.K., Lack, J.G., and Coates, J.D., 2001. Biogenic magnetite formation through anaerobic biooxidation of Fe(II). Applied and Environmental Microbiology, 67:2844–2848.

Chelikani, P., Fita, I., and Loewen, P.C., 2004. Diversity of structures and properties among catalases. Cellular and Molecular Life Sciences, 61:192–208.

Chester, R., 2003. Marine Geochemistry, 2nd edn. Blackwell Publishing, Oxford.

Childers, S.E., Ciufo, S., and Lovley, D.R., 2002. *Geobacter metallireducens* accesses insoluble Fe(III) oxide by chemotaxis. Nature, 416:767–769.

Christensen, B.E. and Characklis, W.G., 1990. Physical and chemical properties of biofilms. In: Characklis, W.G. and Marshall, K.C. (eds), Biofilms. John Wiley, New York, pp. 93–130.

Christensen, J.P., Murray, J.W., Devol, A.H., and Codispoti, L.A., 1987. Denitrification in continental shelf sediments has major impact on the oceanic nitrogen budget. Global Biogeochemical Cycles, 1:97–116.

Chukhrov, F.V., Zvyagin, B.B., Gorshkov, A.I., Yermilova, L.P., and Balachova, V.V., 1973. Ferrihydrite. International Geology Reviews, 16:1131–1143.

Chyba, C.F., 1987. The cometary contribution to the oceans of primitive Earth. Nature, 330:632–635.

Chyba, C.F. and Phillips, C.B., 2001. Possible ecosystems and the search for life on Europa. Proceedings of the National Academy of Sciences, USA, 98:801–804.

Chyba, C. and Sagan, C., 1992. Endogenous production, exogenous delivery and impact-shock synthesis of organic molecules: an inventory for the origins of life. Nature, 355:125–131.

Chyba, C.F., Thomas, P.J., Brookshaw, L., and Sagan, C., 1990. Cometary delivery of organic molecules to the early Earth. Science, 249:366–373.

Clarke, W., Konhauser, K.O., Thomas, J., and Bottrell, S.H., 1997. Ferric hydroxide and ferric hydroxysulfate precipitation by bacteria in an acid mine drainage lagoon. FEMS Microbiology Reviews, 20:351–361.

Clemett, S.J., Dulay, M.T., Gillette, J.S., Chillier, X.D.F., Mahajan, T.B., and Zare, R.N., 1998. Evidence for the extraterrestrial origin of polycyclic aromatic hydrocarbons in the Martian meteorite ALH84001. Faraday Discussions, 109:417–436.

Cloud, P.E., Jr, 1965. Significance of the Gunflint (Precambrian) microflora. Science, 148:27–35.

Coale, K.H. and Bruland, K.W., 1988. Copper complexation in the Northeast Pacific. Limnology and Oceanography, 33:1084–1101.

Coale, K.H., Johnson, K.S., Fitzwater, S.E., et al. 1996. A massive phytoplankton bloom induced by an ecosystem-scale iron fertilization experiment in the equatorial Pacific Ocean. Nature, 383:495–501.

Cockell, C.S., 2000. Ultraviolet radiation and the photobiology of Earth's early oceans. Origins of Life and Evolution of the Biosphere, 30:467–499.

Cocozza, C., Tsao, C.C.G., Cheah, S.-F., et al. 2002. Temperature dependence of goethite dissolution promoted by trihydroxamate siderophores. Geochimica et Cosmochimica Acta, 66:431–438.

Cody, G.D., 2004. Transition metal sulfides and the origins of metabolism. Annual Reviews in Earth and Planetary Sciences, 32:569–599.

Cody, G.D., Boctor, N.Z., Filley, T.R., et al. 2000. Primordial carbonylated iron-sulfur compounds and the synthesis of pyruvate. Science, 289:1337–1340.

Cohen, Y., Padan, E., and Shilo, M., 1975. Facultative anoxygenic photosynthesis in the cyanobacterium *Oscillatoria limnetica*. Journal of Bacteriology, 123:855–861.

Coleman, M.L., 1985. Geochemistry of diagenetic nonsilicate minerals. Kinetic considerations. Philosophical Transactions of the Royal Society of London, 315:39–56.

Coleman, M.L., Hedrick, D.B., Lovley, D.R., White, D.C., and Pye, K., 1993. Reduction of Fe(III) in sediments by sulfate-reducing bacteria. Nature, 361:436–438.

Collins, Y.E. and Stotzky, G., 1992. Heavy metals alter the electrokinetic properties of bacteria, yeasts, and clay minerals. Applied and Environmental Microbiology, 58:1592–1600.

Conway-Morris, S., 1993. The fossil record and the early evolution of the Metazoa. Nature, 361:219–225.

Cooper, D.C., Picardal, F., Rivera, J., and Talbot, C., 2000. Zinc immobilization and magnetite formation via ferric oxide reduction by *Shewanella putrefaciens* 200. Environmental Science and Technology, 34:100–106.

Copping, A.E. and Lorenzen, C.J., 1980. Carbon budget of a marine phytoplankton herbivore–system with carbon-14 as a tracer. Limnology and Oceanography, 25:873–882.

Cord-Ruwisch, R., 2000. Microbially influenced corrosion of steel. In: Lovley, D.R. (ed.), Environmental Microbe–Metal Interactions. ASM Press, Washington, DC, pp. 159–173.

Cord-Ruwisch, R. and Widdel, F., 1986. Corroding iron as hydrogen source for sulfate reduction in growing cultures of sulfate-reducing bacteria. Applied Microbiology and Biotechnology, 25:169–174.

Cord-Ruwisch, R., Kleinitz, W., and Widdel, F., 1987. Sulfate-reducing bacteria and their activities in oil production. Journal of Petroleum Technology, 1:97–106.

Corliss, J.B., Dymond, J., Gordon, L.I., et al. 1979. Submarine thermal springs on the Galápagos Rift. Science, 203:1073–1083.

Costerton, J.W., Irvin, R.T., and Cheng, K-J., 1981. The role of bacterial surface structures in pathogenesis. Critical Reviews in Microbiology, 9:303–338.

Costerton, J.W., Cheng, K-J., Geesey, G.G., et al. 1987. Bacterial biofilms in nature and disease. Annual Reviews of Microbiology, 41:435–464.

Costerton, J.W., Lewandowski, Z., DeBeer, D., Caldwell, D., Korber, D., and James, G., 1994. Biofilms, the customized microniche. Journal of Bacteriology, 176:2137–2142.

Costerton, J.W., Lewandowski, Z., Caldwell, D.E., Korber, D.R., and Lappin-Scott, H.M., 1995. Microbial biofilms. Annual Reviews of Microbiology, 49:711–745.

Coupland, K. and Johnson, D.B., 2004. Geochemistry and microbiology of an impounded subterranean acidic water body at Mynydd Parys, Anglesey, Wales. Geobiology, 2:77–86.

Cowen, J.P., Massoth, G.J., and Baker, E.T., 1986. Bacterial scavenging of Mn and Fe in a mid- to far-field hydrothermal particle plume. Nature, 322:169–171.

Cox, J.C. and Brand, M.D., 1984. Iron oxidation and energy conservation in the chemoautotroph *Thiobacillus ferrooxidans*. In: Sprott, G.D. and Jarrell, K.F. (eds), Electrochemical Potential and Membrane Properties of Methanogenic Bacteria. Ohio State Press, Colombus, Ohio, pp. 31–46.

Cox, J.S., Smith, D.S., Warren, L.A., and Ferris, F.G., 1999. Characterizing heterogeneous bacterial surface functional groups using discrete affinity spectra for proton binding. Environmental Science and Technology, 33:4514–4521.

Crist, R.H., Oberholser, K., Schank, N., and Nguyen, M., 1981. Nature of bonding between metallic ions and algal cell walls. Environmental Science and Technology, 15:1212–1217.

Crist, R.H., Oberholser, K., McGarrity, J., Crist, D.R., Johnson, J.K., and Brittsan, J.M., 1992. Interaction of metals and protons with algae. 3. Marine algae, with emphasis on lead and aluminum. Environmental Science and Technology, 26:496–502.

Croal, L.R., Johnson, C.M., Beard, B.L., and Newman, D.K., 2004. Iron isotope fractionation by Fe(II)-oxidizing photo-autotrophic bacteria. Geochimica et Cosmochimica Acta, 68:1227–1242.

Cronin, J.R., 1989. Origin of organic compounds in carbonaceous meteorites. Advances in Space Research, 9:54–64.

Curtis, C.D., Coleman, M.L., and Love, L.G., 1986. Pore water evolution during sediment burial from isotopic and mineral chemistry of calcite, dolomite and siderite concretions. Geochimica et Cosmochimica Acta, 50:2321–2334.

D'Hondt, S., Rutherford, S., and Spivack, A.J., 2002. Metabolic activity of subsurface life in deep-sea sediments. Science, 295:2067–2070.

da Silva, J.J.R.F. and Williams, R.J.P., 2001. The Biological Chemistry of the Elements: the inorganic chemistry of life. Oxford University Press, Oxford.

Daniel, R.M. and Danson, M.J., 1995. Did primitive microorganisms use nonheme iron proteins in place of NAD/P? Journal of Molecular Evolution, 40:559–563.

Dannenberg, S., Kroder, M., Dilling, W., and Cypionka, H., 1992. Oxidation of H_2, organic compounds and inorganic sulfur compounds coupled to reduction of O_2 or nitrate by sulfate-reducing bacteria. Archives of Microbiology, 158:93–99.

Darnall, D.W., Greene, B., Henzl, M.T., et al. 1986. Selective recovery of gold and other ions from an algal biomass. Environmental Science and Technology, 20:206–208.

Daughney, C.J. and Fein, J.B., 1998. The effect of ionic strength on the adsorption of H^+, Cd^{2+}, Pb^{2+}, and Cu^{2+} by *Bacillus subtilis* and *Bacillus licheniformis*: a surface complexation model. Journal of Colloid and Interface Science, 198:53–77.

Daughney, C.J., Fein, J.B., and Yee, N., 1998. A comparison of the thermodynamics of metal adsorption onto two common bacteria. Chemical Geology, 144:161–176.

Daughney, C.J., Fowle, D.A., and Fortin, D., 2001. The effect of growth phase on proton and metal adsorption by *Bacillus subtilis*. Geochimica et Cosmochimica Acta, 65:1025–1035.

Dauphas, N., van Zuilen, M., Wadhwa, M., Davis, A.M., Marty, B., and Janney, P.E., 2004. Clues from Fe isotope variations on the origin of early Archean BIFs from Greenland. Science, 306:2077–2080.

Davaud, E. and Girardclos, S., 2001. Recent freshwater ooids and oncoids from western Lake Geneva (Switzerland): indications of a common organically mediated origin. Journal of Sedimentary Research, 71:423–429.

Davies, D.G. and Geesey, G.G., 1995. Regulation of the alginate biosynthesis gene *algC* in *Pseudomonas aeruginosa* during biofilm development in continuous culture. Applied and Environmental Microbiology, 61:860–867.

Davis, C.C., Chen, H-W., and Edwards, M., 2002. Modelling silica sorption to iron hydroxide. Environmental Science and Technology, 36:582–587.

Davis, J.A. and Kent, D.B., 1990. Surface complexation modeling in aqueous geochemistry. In: Hochella, M.F. and White, A.F. (eds), Reviews in Mineralogy, Mineralogical Society of America, 23:177–248.

Davis, J.B. and Yarborough, H.E., 1966. Anaerobic oxidation of hydrocarbons by *Desulfovibrio desulfuricans*. Chemical Geology, 1:137–144.

Davison, W., Grime, G.W., Morgan, J.A.W., and Clarke, K., 1991. Distribution of dissolved iron in sediment pore waters at submillimeter resolution. Nature, 352:323–324.

Dawson, M.P., Humphrey, B.A., and Marshall, K.C., 1981. Adhesion: a tactic in the survival strategy of a marine vibrio during starvation. Current Microbiology, 6:195–199.

De Leeuw, J.W. and Largeau, C., 1993. A review of macromolecular organic compounds that comprise living organisms and their role in kerogen, coal, and petroleum formation. In: Engel, M.H. and Macko, S.A. (eds), Organic Geo-chemistry. Principles and applications. Plenum Press, New York, pp. 23–72.

De Vrind-de Jong, E.W. and De Vrind, J.P.M., 1997. Algal deposition of carbonates and silicates. In: Banfield, J.F. and Nealson, K.H. (eds), Geomicrobiology: interactions between microbes and minerals, vol. 35. Mineralogical Society of America, Washington, DC, pp. 267–307.

De Wit, R. and van Gemerden, H., 1990. Growth of the phototrophic purple sulfur bacterium *Thiocapsa roseopersicina* under oxic/anoxic regimes in the light. FEMS Microbial Ecology, 73:69–76.

De Wit, R. and van Gemerden, H., 1987. Chemolithoautotrophic growth of the phototrophic sulfur bacterium *Thiocapsa roseopersicina*. FEMS Microbial Ecology, 45:117–126.

Deamer, D.W., 1985. Boundary structures are formed by organic components of the Murchison carbonaceous chondrite. Nature, 317:792–794.

Deamer, D.W., Mahon, E.H., and Bosco, G., 1994. Self-assembly and function of primitive membrane structures. In: Bengtson, S. (ed.), Early Life on Earth. Columbia University Press, New York, pp. 107–123.

Dean, W.E. and Greeson, P.E., 1979. Influences of algae on the formation of freshwater ferromanganese nodules, Oneida Lake, New York. Archives of Hydrobiology, 86:181–192.

Dean, W.E., Moore, W.S., and Nealson, K.H., 1981. Manganese cycles and the origin of manganese nodules, Oneida Lake, New York, USA. Chemical Geology, 34:53–64.

Decho, A.W., 2000. Exopolymer microdomains as a structuring agent for heterogeneity within microbial biofilms. In: Riding, R.E. and Awramik, S.M. (eds), Microbial Sediments. Springer-Verlag, Berlin, pp. 9–15.

Défarge, C., Trichet, J., and Couté, A., 1994. On the appearance of cyanobacteria calcification in modern stromatolites. Sedimentary Geology, 94:11–19.

DeFlaun, M.F. and Mayer, L.M., 1983. Relationships between bacteria and grain surfaces in intertidal sediments. Limnology and Oceanography, 28:873–881.

Delano, J.W., 2001. Redox history of the Earth's interior since ~3900 Ma: implications for prebiotic molecules. Origins of Life and Evolution of the Biosphere, 31:311–341.

DeLong, E.F., 2000. Resolving a methane mystery. Nature, 407:577–579.

DeLong, E.F., Wickham, G.S., and Pace, N.R., 1989. Phylogenetic stains: ribosomal RNA-based probes for the identification of single cells. Science, 243:1360–1362.

DeLong, E.F., Frankel, R.B., and Bazylinski, D.A., 1993. Multiple evolutionary origins of magnetotaxis in bacteria. Science, 259:803–806.

DeMaster, D.J., 1981. The supply and accumulation of silica in the marine environment. Geochimica et Cosmochimica Acta, 45:1715–1732.

Deming, J.W., 2002. Psychrophiles and polar regions. Current Opinion in Microbiology, 5:301–309.

Des Marais, D.J., 1997. Long-term evolution of the biogeochemical carbon cycle. In: Banfield, J.F. and Nealson, K.H. (eds), Geomicrobiology: interactions between microbes and minerals, vol. 35. Mineralogical Society of America, Washington, DC, pp. 429–448.

Des Marais, D.J., 2000. When did photosynthesis emerge on Earth? Science, 289:1703–1705.

Des Marais, D.J., Strauss, H., Summons, R.E., and Hayes, J.M., 1992. Carbon isotope evidence for the stepwise oxidation of the Proterozoic environment. Nature, 359:605–609.

Detmers, J., Brüchert, V., Habicht, K.S., and Kuever, J., 2001. Diversity of sulfur isotope fractionations by sulfate-reducing prokaryotes. Applied and Environmental Microbiology, 67:888–894.

Devereux, R., Kane, M.D., Winfrey, J., and Stahl, D.A., 1992. Genus- and group-specific hybridization probes for determinative and environmental studies of sulfate-reducing bacteria. Systematic and Applied Microbiology, 15:601–609.

Devol, A.H., 1991. Direct measurement of nitrogen gas fluxes from continental shelf sediments. Nature, 349:319–321.

Dew, D.W., Lawson, E.N., and Broadhurst, J.L., 1997. The BIOX® Process for Biooxidation of gold-bearing ores or concentrates. In: Rawlings, D.E. (ed.), Biomining. Theory, microbes and industrial processes. Springer-Verlag, Berlin, pp. 45–80.

DiRienzo, J.M., Nakamura, K., and Inouye, M., 1978. The outer membrane proteins of gram-negative bacteria: biosynthesis, assembly, and functions. Annual Review of Biochemistry, 47:481–532.

Dichristina, T.J., 1992. Effects of nitrate and nitrite on dissimilatory iron reduction by *Shewanella putrefaciens* 200. Journal of Bacteriology, 174:1891–1896.

Dickens, G.R., 2003. Rethinking the global carbon cycle with a large, dynamic and microbially mediated gas hydrate capacitor. Earth and Planetary Science Letters, 213:169–183.

Dill, R.F., Shinn, E.A., Jones, A.T., Kelly, K., and Steinen, R.P., 1986. Giant subtidal stromatolites forming in normal salinity waters. Nature, 324:55–58.

Dillig, W. and Cypionka, H., 1990. Aerobic respiration in sulfate-reducing bacteria. FEMS Microbial Letters, 71:123–128.

Dismukes, G.C., Klimov, V.V., Baranov, S.V., Kozlov, Y.N., and Tyryshkin, A., 2001. The origin of atmospheric oxygen on Earth: innovation of oxygenic photosynthesis. Proceedings of the National Academy of Sciences, USA, 98:2170–2175.

DiSpirito, A.A., Taaffe, L.R., and Hooper, A.B., 1985. Localization and concentration of hydroxylamine oxidoreductase and cytochromes c-552, c-554, c_m-553, c_m-552 and a in *Nitrosomonas europaea*. Biochimica et Biophysica Acta, 806:320–330.

Doemel, W.N. and Brock, T.D., 1977. Structure, growth, and decomposition of laminated algal-bacterial mats in alkaline hot springs. Applied and Environmental Microbiology, 34:433–452.

Dojka, M.A., Harris, J.K., and Pace, N.R., 2000. Expanding the known diversity and environmental distribution of an uncultured phylogenetic division of bacteria. Applied and Environmental Microbiology, 66:1617–1621.

Doolittle, W.F., 1999. Phylogenetic classification and the universal tree. Science, 284:2124–2127.

Doolittle, W.F. and Brown, J.R., 1994. Tempo, mode, the progenote and the universal root. Proceedings of the National Academy of Sciences, USA, 91:6721–6728.

Dorn, R.I. and Oberlander, T.M., 1981. Microbial origin of desert varnish. Science, 213:1245–1247.

Dorobantu, L.S., Yeung, A.K.C., Fought, J.M., and Gray, M.R., 2004. Stabilization of oil-water emulsions by hydrophobic bacteria. Applied and Environmental Microbiology, 70:6333–6336.

dos Santos Afonso, M. and Stumm, W., 1992. Reductive dissolution of iron (III) (hydr)oxides by hydrogen sulfide. Langmuir, 8:1671–1675.

Dowdle, P.R., Laverman, A.M., and Oremland, R.S., 1996. Bacterial dissimilatory reduction of arsenic(V) to arsenic(III) in anoxic sediments. Applied and Environmental Microbiology, 62:1664–1669.

Doyle, R.J., 1989. How cell walls of gram-positive bacteria interact with metal ions. In: Beveridge, T.J. and Doyle, R.J. (eds), Metal Ions and Bacteria. John Wiley, New York, pp. 275–293.

Doyle, R.J. and Koch, A.L., 1987. The functions of autolysins in the growth and division of Bacillus subtilis. Critical Reviews in Microbiology, 15:169–222.

Doyle, R.J., McDannel, M.L., Streips, U.N., Birdsell, D.C., and Young, F.E., 1974. Polyelectrolyte nature of bacterial teichoic acids. Journal of Bacteriology, 118:606–615.

Doyle, R.J., McDannel, M.L., Helman, J.R., and Streips, U.N., 1975. Distribution of teichoic acid in the cell wall of Bacillus subtilis. Journal of Bacteriology, 122:152–158.

Doyle, R.J., Matthews, T.H., and Streips, U.N., 1980. Chemical basis for selectivity of metal ions by the Bacillus subtilis cell wall. Journal of Bacteriology, 143:471–480.

Drake, H.L., 1994. Acetogenesis. Chapman and Hall, New York.

Drever, J.I., 1988. The Geochemistry of Natural Waters, 2nd edn. Prentice Hall, Englewood Cliffs, New Jersey.

Drever, J.I., Li, Y.-H., and Maynard, J.B., 1988. Geochemical cycles: the continental crust and the oceans. In: Gregor, C.B., Garrels, R.M., Mackenzie, F.T., and Maynard, J.B. (eds), Chemical Cycles in the Evolution of the Earth. Wiley, New York, pp. 17–53.

Dreybrodt, W., 1980. Deposition of calcite from thin films of natural calcareous solutions and the growth of speleothems. Chemical Geology, 29:89–105.

Drobner, E., Huber, H., Wächtershäuser, G., Rose, D., and Stetter, K.O., 1990. Pyrite formation linked with hydrogen evolution under anaerobic conditions. Nature, 346:742–744.

Druschel, G.K., Hamers, R.J., and Banfield, J.F., 2003. Kinetics and mechanism of polythionate oxidation to sulfate at low pH by O_2 and Fe^{3+}. Geochimica et Cosmochimica Acta, 67:4457–4469.

Dubina, G.A., 1981. The role of microorganisms in the formation of the recent iron-manganese lacustrine ores. Microbiology, 47:471–478.

Dymond, J., Lyle, M., Finney, B., et al. 1984. Ferromanganese nodules from MANOP sites H, S, and R – control of mineralogical and chemical composition by multiple accretionary processes. Geochimica et Cosmochimica Acta, 48:931–949.

Dzombak, D.A. and Morel, F.M.M., 1990. Surface Complexation Modeling. Hydrous ferric oxide. John Wiley, New York.

Easton, R.M., 1997. Lichen–rock–mineral interactions: an overview. In: McIntosh, J.M. and Groat, L.A. (eds), BiologicalMineralogical Interactions. Mineralogical Association of Canada Short Course Series, vol. 25, Ottawa, pp. 209–239.

Eccles, H., 1995. Removal of heavy metals from effluent streams. Why select a biological process. International Biodeterioration and Biodegradation, 35:5–16.

Edwards, C., 1990. Microbiology of Extreme Environments. Open University Press, Milton Keynes, UK.

Edwards, H.G.M., Edwards, K.A.E., Farwell, D.W., Lewis, I.R., and Seaward, M.R.D., 1994. An approach to stone and fresco lichen biodegradation through Fourier-transform Raman microscopic investigation of thallus substratum encrustations. Journal of Raman Spectroscopy, 25:99–103.

Edwards, K.J., Goebel, B.M., Rodgers, T.M., et al. 1999. Geomicrobiology of pyrite (FeS_2) dissolution: case study at Iron Mountain, California. Geomicrobiology Journal, 16:155–179.

Edwards, K.J., Bond, P.L., Gihring, T.M., and Banfield, J.F., 2000a. An archaeal iron-oxidizing extreme acidophile important in acid mine drainage. Science, 287:1796–1799.

Edwards, K.J., Bond, P.L., Druschel, G.K., McGuire, M.M., Hamers, R.J., and Banfield, J.F., 2000b. Geochemical and biological aspects of sulfide mineral dissolution: lessons from iron Mountain, California. Chemical Geology, 169:383–397.

Edwards, K.J., Hu, B., Hamers, R.J., and Banfield, J.F., 2001. A new look at microbial leaching patterns on sulfide minerals. FEMS Microbial Ecology, 34:197–206.

Edwards, K.J., McCollom, T.M., Konishi, H., and Buseck, P.R., 2003. Seafloor bioalteration of sulfide minerals: results from in situ studies. Geochimica et Cosmochimica Acta, 67:2843–2856.

Ehrenfreund, P. and Charnley, S.B., 2000. Organic molecules in the interstellar medium, comets and meteorites: a voyage from dark clouds to the early Earth. Annual Review of Astronomy and Astrophysics, 38:427–483.

Ehling-Schulz, M., Bilger, W., and Scherer, S., 1997. UV-B-induced synthesis of photoprotective pigments and extracellular polysmers in the terrestrial cyanobacterium *Nostoc commune*. Journal of Bacteriology, 179:1940–1945.

Ehrenreich, A. and Widdel, F., 1994. Anaerobic oxidation of ferrous iron by purple bacteria, a new type of phototrophic metabolism. Applied and Environmental Microbiology, 60:4517–4526.

Ehrlich, H.L., 2002. Geomicrobiology, 4th edn. Marcel Dekker, New York.

Ehrlich, H.L. and Salerno, J.C., 1990. Energy coupling in Mn^{2+} oxidation by a marine bacterium. Archives of Microbiology, 154:12–17.

Ellwood, D.C. and Tempest, D.W., 1972. Effects of environment on bacterial wall content and composition. Advances in Microbial Physiology, 7:83–117.

Emeis, K.-C., Richnow, H.-H., and Kempe, S., 1987. Travertine formation in Plitvice National Park, Yugoslavia: chemical versus biological control. Sedimentology, 34:595–609.

Emerson, D., 2000. Microbial oxidation of Fe(II) and Mn(II) at circumneutral pH. In: Lovley, D.R. (ed.), Environmental Microbe–Metal Interactions. ASM Press, Washington, DC, pp. 31–52.

Emerson, D. and Moyer, C.L., 2002. Neutrophilic Fe-oxidizing bacteria are abundant at the Loihi Seamount hydrothermal vents and play a major role in Fe oxide deposition. Applied and Environmental Microbiology, 68:3085–3093.

Emerson, D., and Revsbech, N.P., 1994a. Investigation of an iron-oxidizing microbial mat community located near Aarhus, Denmark. Field studies. Applied and Environmental Microbiology, 60:4022–4031.

Emerson, D., and Revsbech, N.P., 1994b. Investigation of an iron-oxidizing microbial mat community located near Aarhus, Denmark: laboratory studies. Applied and Environmental Microbiology, 60:4032–4038.

Emerson, S. and Bender, M., 1981. Carbon fluxes at the sediment–water interface of the deep-sea: calcium carbonate preservation. Journal of Marine Research, 39:139–162.

Emerson, S. and Hedges, J.I., 1988. Processes controlling the organic carbon content of open ocean sediments. Paleooceanography, 3:621–634.

Falkowski, P.G., 1997. Evolution of the nitrogen cycle and its influence on the biological sequestration of CO_2 in the ocean. Nature, 387:272–275.

Falkowski, P.G. and Raven, J.A., 1997. Aquatic Photosynthesis. Blackwell, Malden, Massachusetts.

Falkowski, P.G., Katz, M.E., Knoll, A.H., et al. 2004. The evolution of modern eukaryotic phytoplankton. Science, 305:354–360.

Fani, R., Gallo, R., and Lio, P., 2000. Molecular evolution of nitrogen fixation: the evolutionary history of the *nifD*, *nifK*, and *nifN* genes. Journal of Molecular Evolution, 51:1–11.

Farmer, V.C., Krishnamurti, G.S.R., and Huang, P.M., 1991. Synthetic allophane and layer-silicate formation in SiO_2-Al_2O_3-FeO-Fe_2O_3-MgO-H_2O systems at 23°C and 89°C in a calcareous environment. Clays and Clay Minerals, 39:561–570.

Farquhar, J., Bao, H., and Thiemans, M., 2000. Atmospheric influence of Earth's earliest sulfur cycle. Science, 289:756–758.

Fattom, A. and Shilo, M., 1984. Hydrophobicity as an adhesion mechanism of benthic cyanobacteria. Applied and Environmental Microbiology, 47:135–143.

Faure, G., 1998. Principles and Applications of Geochemistry, 2nd edn. Prentice Hall, Upper Saddle River, New Jersey.

Fedo, C.M. and Whitehouse, M.J., 2002. Metasomatic origin of quartz-pyroxene rock, Akilia Greenland, and implications for Earth's earliest life. Science, 296:1448–1452.

Fegley, B., Prinn, R.G., Hartman, H., and Watkins, G.H., 1986. Chemical effects of large impacts on the Earth's primitive atmosphere. Nature, 319:305–308.

Fein, J.B., 2000. Quantifying the effects of bacteria on adsorption reactions in water–rock systems. Chemical Geology, 169:265–280.

Fein, J.B., Martin, A.M., and Wightman, P.G., 2001. Metal adsorption onto bacterial surfaces: development of a predictive model. Geochimica et Cosmochimica Acta, 65:4267–4273.

Fein, J.B., Scott, S., and Rivera, N., 2002. The effect of Fe on Si adsorption by *Bacillus subtilis* cell walls: insights into nonmetabolic bacterial precipitation of silicate minerals. Chemical Geology, 182:265–273.

Fein, J.M., Daughney, C.J., Yee, N., and Davis, T.A., 1997. A chemical equilibrium model for metal adsorption onto bacterial surfaces. Geochimica et Cosmochimica Acta, 61:3319–3328.

Feldmann, M. and McKenzie, J., 1998. Stromatolite–thrombolite associations in a modern environment, Lee Stocking Island, Bahamas. Palaios, 13:201–212.

Feller, G. and Gerday, C., 2003. Psychrophilic enzymes: hot topics in cold adaptation. Nature Reviews, 1:200–208.

Fenchel, T., 1994. Motility and chemosensory behaviour of the sulfur bacterium *Thiovulum majus*. Microbiology, 140:3109–3116.

Fenchel, T. and Bernard, C., 1995. Mats of colorless sulfur bacteria. I. Major microbial processes. Marine Ecology Progress Series, 128:161–170.

Fenchel, T. and Straarup, B.J., 1971. Vertical distribution of photosynthetic pigments and the penetration of light in marine sediments. Oikos, 22:172–182.

Fenchel, T., King, G.M., and Blackburn, T.H., 2000. Bacterial Biogeochemistry. The ecophysiology of mineral cycling, 2nd edn. Academic Press, San Diego.

Ferguson, S.J., 1988. Periplasmic electron transport reactions. In: Anthony, C. (ed.), Bacterial Energy Transduction. Academic Press, London, pp. 151–182.

Ferreira, K.N., Iverson, T.M., Maghlaoui, K., Barber, J., and Iwata, S., 2004. Architecture of the photosynthetic pxygen-evolving center. Science, 303:1831–1838.

Ferris, F.G., 1989. Metallic ion interactions with the outer membrane of gram-negative bacteria. In: Beveridge, T.J. and Doyle, R.J. (eds), Metal Ions and Bacteria. John Wiley, New York, pp. 295–323.

Ferris, F.G., 1997. Formation of authigenic minerals by bacteria. In: Biological–Mineralogical Interactions. Mineralogical Association of Canada Short Course Series, vol. 25. Ottawa, pp. 187–208.

Ferris, F.G. and Beveridge, T.J., 1984. Binding of a paramagnetic metal cation to *Escherichia coli* K-12 outer-membrane vesicles. FEMS Microbiology Letters, 24:43–46.

Ferris, F.G. and Beveridge, T.J., 1986a. Site specificity of metallic ion binding in *Escherichia coli* K-12 lipopolysaccharide. Canadian Journal of Microbiology, 32:52–55.

Ferris, F.G. and Beveridge, T.J., 1986b. Physiochemical roles of soluble metal cations in the outer membrane of *Escherichia coli* K-12. Canadian Journal of Microbiology, 32: 594–601.

Ferris, F.G. and Lowson, E.A., 1997. Ultrastructure and geochemistry of endolithic microorganisms in limestone of the Niagara Escarpment. Canadian Journal of Microbiology, 43:211–219.

Ferris, F.G., Fyfe, W.S., and Beveridge, T.J., 1987. Bacteria as nucleation sites for authigenic minerals in a metal-contaminated lake sediment. Chemical Geology, 63:225–232.

Ferris, F.G., Fyfe, W.S., and Beveridge, T.J., 1988. Metallic ion binding by *Bacillus subtilis*: implications for the fossilization of microorganisms. Geology, 16:149–152.

Ferris, F.G., Fyfe, W.S., Witten, T., Schultze, S., and Beveridge, T.J., 1989. Effect of mineral substrate hardness on the population density of epilithic microorganisms in two Ontario rivers. Canadian Journal of Microbiology, 35:744–747.

Ferris, F.G., Fratton, C.M., Gerits, J.P., Schultze-Lam, S., and Sherwood Lollar, B., 1995. Microbial precipitation of a strontium calcite phase at a groundwater discharge zone near Rock Creek, British Columbia, Canada. Geomicrobiology Journal, 13:57–67.

Ferris, F.G., Konhauser, K.O., Lyvén, B., and Pedersen, K., 1999. Accumulation of metals by bacteriogenic iron oxides in a subterranean environment. Geomicrobiology Journal, 16:181–192.

Ferris, J.P., 1992. Chemical markers of prebiotic chemistry in hydrothermal systems. Origins of Life and Evolution of the Biosphere, 22:109–134.

Ferris, J.P., Hill, A.R., Rihe, L., and Orgel, L.E., 1996. Synthesis of long prebiotic oligomers on mineral surfaces. Nature, 318:59–61.

Ferry, J.G., 1993. Methanogenesis: ecology, physiology, biochemistry, and genetics. Chapman and Hall, New York.

Field, C.B., Behrenfeld, M.J., Randerson, J.T., and Falkowski, P., 1998. Primary production of the biosphere: integrating terrestrial and oceanic components. Science, 281:237–240.

Finlay, B.J., 2002. Global dispersal of free-living microbial eukaryote species. Science, 296:1061–1063.

Fischer, W.W., Summons, R.E., and Pearson, A., 2005. Targeted genomic detection of biosynthetic pathways: anaerobic production of hopanoid biomarkers by a common sedimentary microbe. Geobiology, 3:33–40.

Fisher, Q.J., Raiswell, R., and Marshall, J.D., 1998. Siderite concretions from nonmarine shales (Westphalian A) of the Pennines, England: controls on their growth and composition. Journal of Sedimentary Research, 68:1034–1045.

Fitz-Gibbon, S.T. and House, C.H., 1999. Whole genome-based phylogenetic analysis of free-living microorganisms. Nucleic Acids Research, 27:4218–4222.

Flemming, C.A., Ferris, F.G., Beveridge, T.J., and Bailey, G.W., 1990. Remobilization of toxic heavy metals adsorbed to bacterial wall-clay composites. Applied and Environmental Microbiology, 56:3191–3203.

Fletcher, M., 1977. The effects of culture concentration and age, time, and temperature on bacterial attachment to polystyrene. Canadian Journal of Microbiology, 23:1–6.

Fletcher, M., 1985. Effect of solid surfaces on the activity of attached bacteria. In: Savage, D. and Fletcher, M.M. (eds), Bacterial Adhesion. Mechanisms and physiological significance. Plenum Press, New York, pp. 339–361.

Fontes, D.E., Mills, A.L., Hornberger, G.M., and Herman, J.S., 1991. Physical and chemical factors influencing transport of microorganisms through porous media. Applied and Environmental Microbiology, 57:2473–2481.

Force, E.R. and Cannon, W.F., 1988. Depositional model for shallow-marine manganese deposits around black shale basins. Economic Geology, 83:93–117.

Forterre, P., Brochier, C., and Philippe, H., 2002. Evolution of the Archaea. Theoretical Population Biology, 61:409–422.

Fortin, D. and Beveridge, T.J., 1997. Microbial sulfate reduction within sulfidic mine tailings: formation of authigenic Fe-sulfides. Geomicrobiology Journal, 15:1–21.

Fortin, D. and Ferris, F.G., 1998. Precipitation of iron, silica, and sulfate on bacterial cell surfaces. Geomicrobiology Journal, 15:309–324.

Fossing, H., Nielsen, H.P., Schultz, H., et al. 1995. Concentration and transport of nitrate by the mat-forming sulfur bacterium *Thioploca*. Nature, 374:713–715.

Fouke, B.W., Bonheyo, G.T., Sanzenbacher, B., and Frias-Lopez, J., 2003. Partitioning of bacterial communities between travertine depositional facies at Mammoth Hot Springs, Yellowstone National Park. Canadian Journal of Earth Sciences, 40:1531–1548.

Fournier, R.O., 1985. The behavior of silica in hydrothermal solutions. In: Berger, B.R. and Bethke, P.M. (eds), Geology and Geochemistry of Epithermal Systems. Reviews in Economic Geology, 2:45–61.

Fowle, D.A. and Fein, J.B., 1999. Competitive adsorption of metals onto two gram positive bacteria: testing the chemical equilibrium model. Geochimica et Cosmochimica Acta, 63:3059–3067.

Fowle, D.A. and Fein, J.B., 2000. Experimental measurements of the reversibility of metal–bacteria adsorption reactions. Chemical Geology, 168:27–36.

Fowle, D.A. and Fein, J.B., 2001. Quantifying the effects of *Bacillus subtilis* cell walls on the precipitation of copper hydroxide from aqueous solution. Geomicrobiology Journal, 18:77–91.

Fowle, D.A., Fein, J.B., and Martin, A.M., 2000. Experimental study of uranyl adsorption onto *Bacillus subtilis*. Environmental Science and Technology, 34:3737–3741.

Fox, G.E., Stackebrandt, R.B., Hespell, J., et al. 1980. The phylogeny of prokaryotes. Science, 209:457–463.

Francis, A.J., Dodge, C.J., and Gillow, J.B., 1992. Biodegradation of metal citrate complexes and implications for toxic-metal mobility. Nature, 356:140–142.

François, L.M., 1986. Extensive deposition of banded iron formations was possible without photosynthesis. Nature, 320:352–354.

Frank, H.A. and Cogdell, R.J., 1996. Carotenoids in photosynthesis. Photochemistry and Photobiology, 63:257–264.

Frankel, R.B., 1987. Anaerobes pumping iron. Nature, 330:208.

Frankel, R.B. and Blakemore, R.P., 1989. Magnetite and magnetotaxis in microorganisms. Bioelectromagnetics, 10:223–237.

Frankel, R.B., Papaefthymiou, G.C., Blakemore, R.P. and O'Brien, W., 1983. Fe_3O_4 precipitation in magnetotactic bacteria. Biochimica et Biophysica Acta, 763:147–159.

Frankel, R.B., Bazylinski, D.A., Johnson, M.S., and Taylor, B.L., 1997. Magneto-aerotaxis in marine coccoid bacteria. Biophysical Journal, 73:994–1000.

Frea, J.I., 1984. Methanogenesis: its role in the carbon cycle. In: Sprott, G.D. and Jarrell, K.F. (eds), Electrochemical Potential and Membrane Properties of Methanogenic Bacteria. Ohio State Press, Colombus, Ohio, pp. 229–253.

Friedberg, E.C., 1985. DNA Repair. W.H. Freeman, New York.

Friedmann, E.I., 1982. Endolithic microorganisms in the Antarctic cold desert. Science, 215:1045–1053.

Friedmann, E.I. and Weed, R., 1987. Microbial trace-fossil formation, biogenous, and abiotic weathering in the Antarctic cold desert. Science, 236:703–705.

Friedrich, M.W., 2002. Phylogenetic analysis reveals multiple lateral transfers of adenosine-5′-phosphosulfate reductase genes among sulfate-reducing microorganisms. Journal of Bacteriology, 184:278–289.

Froelich, P.N., Klinkhammer, G.P., Bender, M.L., et al. 1979. Early oxidation of organic matter in pelagic sediments of the eastern equatorial Atlantic: suboxic diagenesis. Geochimica et Cosmochimica Acta, 43:1075–1090.

Froelich, P.N., Arthur, M.A., Burnett, W.C., et al. 1988. Early diagenesis of organic matter in Peru continental margin sediments: phosphorite precipitation. Marine Geology, 80:309–343.

Fry, N.K., Fredrickson, J.K., Fishbain, S., Wagner, M., and Stahl, D.A., 1997. Population structure of microbial communities associated with two deep, anaerobic, alkaline aquifers. Applied and Environmental Microbiology, 63:1498–1504.

Furnes, H., Staudigel, I.H., Thorseth, T., Torsvik, T., Muehlenbachs, K., and O. Tumyr, 2001. Bioalteration of basaltic glass in the oceanic crust. Geochemistry, Geophysics and Geosystems, 2:2000GC000150, 30 pp.

Furnes, H., Banerjee, N.R., Muehlenbachs, K., Staudigel, H., and de Wit, M., 2004. Early life recorded in Archean pillow lava. Science, 304:578–581.

Gächter, R., Meyer, J.S., and Mares, A., 1988. Contribution of bacteria to release and fixation of phosphorous in lake sediments. Limnology and Oceanography, 33:1542–1558.

Gadd, G.M., 1993. Interactions of fungi with toxic metals. New Phytologist, 124:25–60.

Gadd, G.M., 1999. Fungal production of citric and oxalic acid: importance in metal speciation, physiology and biogeochemical processes. Advanced Microbial Physiology, 41:47–92.

Gadd, G.M., 2002. Microbial interactions with metals/radionuclides: the basis of bioremediation. In: Keith-Roach, M.J. and Livens, F.R. (eds), Interactions of Microorganisms with Radionuclides. Elsevier, Amsterdam, pp. 179–203.

Gadd, G.M. and Sayer, J.A., 2000. Influence of fungi on the environmental mobility of metals and metalloids. In: Lovley, D.R. (ed.), Environmental Microbe–Metal Interactions. ASM Press, Washington, DC, pp. 237–256.

Galtier, N., Tourasse, N., and Gouy, M., 1999. A nonhyperthermophilic common ancestor to extant life forms. Science, 283:220–221.

Galun, M., Keller, P., Malki, D., et al. 1983. Removal of uranium (VI) from solution by fungal biomass and fungal wall-related biopolymers. Science, 219:285–286.

Ganesh, R., Robinson, K.G., Reed, G.D., and Sayler, G.S., 1997. Reduction of hexavalent uranium from organic complexes by sulfate- and iron-reducing bacteria. Applied and Environmental Microbiology, 63:4385–4391.

Gannon, J.T., Manilal, V.B., and Alexander, M., 1991. Relationship between cell surface properties and transport of bacteria through soil. Applied and Environmental Microbiology, 57:190–193.

Garcia-Pichel, F., 1994. A model for internal self-shading in planktonic organisms and its implications for the usefulness of ultraviolet sunscreens. Limnology and Oceanography, 39:1704–1717.

Garcia-Pichel, F., 1998. Solar ultraviolet and the evolutionary history of cyanobacteria. Origins of Life and Evolution of the Biosphere, 28:321–347.

Garcia-Pichel, F., Mechling, M., and Castenholz, R.W., 1994. Diel migrations of microorganisms within a benthic, hypersaline mat community. Applied and Environmental Microbiology, 60:1500–1511.

García-Ruiz, J.M., Hyde, S.T., Carnerup, A.M., Christy, A.G., van Kranendonk, M.J., and Welham, N.J., 2003. Self-assembled silica-carbonate structures and detection of ancient microfossils. Science, 302:1194–1197.

Garrels, R.M. and Mackenzie, F.T., 1971. Evolution of Sedimentary Rocks. W.W. Norton, New York.

Garrity, G.M., Bell, J., and Lilburn, T.G., 2004. Taxonomic Outline of the Prokaryotes. Bergey's Manual of Systematic Bacteriology, 2nd edn, Release 5.0, NY DOI: 10.1007/bergeysoutline, Springer-Verlag, New York.

Gautier, D.L., 1982. Siderite concretions: indicators of early diagenesis in the Gammon shale (Cretaceous). Journal of Sedimentary Petrology, 52:859–871.

Gautret, P. and Trichet, J., 2005. Automicrites in modern cyanobacterial deposits of Rangiroa, Tuamotu Archipelago, French Polynesia: biochemical parameters underlying their formation. Sedimentary Geology, 178:55–73.

Gedulin, B. and Arrhenius, G., 1994. Sources and geochemical evolution of RNA precursor molecules: the role of phosphate. In: Bengtson, S. (ed.), Early Life on Earth. Columbia University Press, New York, pp. 91–106.

Geesey, G.G. and Jang, L., 1989. Interactions between metal ions and capsular polymers. In: Beveridge, T.J. and Doyle, R.J. (eds), Metal Ions and Bacteria. John Wiley, New York, pp. 325–357.

Gerdes, G., Krumbein, W.E., and Noffke, N., 2000. Evaporite microbial sediments. In: Riding, R.E. and Awramik, S.M. (eds), Microbial Sediments. Springer-Verlag, Berlin, pp. 196–208.

Ghiorse, W.C., 1984. Biology of iron- and manganese-depositing bacteria. Annual Reviews in Microbiology, 38:515–550.

Ghiorse, W.C. and Ehrlich, H.L., 1992. Microbial biomineralization of iron and manganese. In: Skinner, H.C.W. and Fitzpatrick, R.W. (eds), Biomineralization Processes. Catena Supplement. Cremlingen, Germany, pp. 75–99.

Gibson, E.K. Jr, McKay, D.S., Thomas-Keprta, K.L., et al. 2001. Life on Mars: evaluation of the evidence within Martian meteorites ALH84001, Nakhla and Shergotty. Precambrian Research, 106:15–34.

Gilbert, P., Evans, D.J, Evans, E., Duguid, I.G., and Brown, M.R.W., 1991. Surface characteristics and adhesion of *Escherichia coli* and *Staphylococcus epidermidis*. Journal of Applied Bacteriology, 71:72–77.

Gilichinsky, D.A., Soina, V.S., and Petrova, M.A., 1993. Cryoprotective properties of water in the Earth cryolithosphere and its role in exobiology. Origins of Life and Evolution of the Biosphere, 23:65–75.

Gilichinsky, D.A., Wagener, S., and Vishnevetskaya, T.A., 1995. Permafrost microbiology. Permafrost and Periglacial Processes, 6:281–291.

Gilvarry, J.J. and Hochstim, A.R., 1963. Possible role of meteorites in the origin of life. Nature, 197:624–625.

Glasauer, S., Langley, S., and Beveridge, T.J., 2001. Sorption of Fe hydr(oxides) to the surface of *Shewanella putrefaciens*: cell-bound fine-grained minerals are not always formed de novo. Applied and Environmental Microbiology, 67:5544–5550.

Glasauer, S., Langley, S., and Beveridge, T.J., 2002. Intracellular iron minerals in a dissimilatory iron-reducing bacterium. Science, 295:117–119.

Glasauer, S., Weidler, P.G., Langley, S., and Beveridge, T.J., 2003. Controls on Fe reduction and mineral formation by a subsurface bacterium. Geochimica et Cosmochimica Acta, 67:1277–1288.

Glazer, A.N., 1985. Light harvesting by phycobilisomes. Annual Review of Biophysics and Biophysical Chemistry, 14:47–77.

Glenn, C.R. and Arthur, M.A., 1988. Petrology and major element geochemistry of Peru margin phosphorites and associated diagenetic minerals: authigenesis in modern organic-rich sediments. Marine Geology, 80:231–276.

Gogarten-Boekels M., Hilario, H., and Gogarten, J.P., 1995. The effects of heavy meteorite bombardment on early evolution – the emergence of the three domains of life. Origins of Life and Evolution of the Biosphere, 25:251–264.

Goldhaber, M.B., 1983. Experimental study of metastable sulfur oxyanion formation during pyrite oxidation at pH 6–9 at 30°C. American Journal of Science, 283:193–217.

Gole, M.J. and Klein, C., 1981. Banded iron-formations through much of Precambrian time. Journal of Geology, 89:169–183.

Golubic, S., 1976. Organisms that build stromatolites. In: Walter, M.R. (ed.), Stromatolites: developments in sedimentology. Elsevier, Amsterdam, pp. 113–126.

Golubic, S., 1991. Modern stromatolites: a review. In: Riding, R. (ed.), Calcareous Algae and Stromatolites. Springer-Verlag, Berlin, pp. 541–561.

Golubic, S., 1994. The continuing importance of cyanobacteria. In: Bengtson, S. (ed.), Early Life on Earth. Columbia University Press, New York, pp. 335–340.

Golubic, S., Brent, G., and Le Campion-Alsumard, T., 1970. Scanning electron microscopy of endolithic algae and fungi using a multipurpose casting-embedding technique. Lethai, 3:203–209.

Golubic, S. and Seong-Joo, L., 1999. Early cyanobacteria fossil record: preservation, palaeoenvironments and identification. European Journal of Phycology, 34:339–348.

Gonçalves, M.L.S., Sigg, L., Reulinger, M., and Stumm, W., 1987. Metal ion binding by biological surfaces: voltammetric assessment in the presence of bacteria. Science of the Total Environment, 60:105–119.

González-Munoz, M.T., Fernández-Luque, B., Martínez-Ruiz, F., et al. 2003. Precipitation of barite by *Myxococcus xanthus*: possible implications for the biogeochemical cycle of barium. Applied and Environmental Microbiology, 69:5722–5725.

Goodwin, A.M., Monster, J., and Thode, H.G., 1976. Carbon and sulfur isotope abundances in Archean iron-formations and early Precambrian life. Economic Geology, 71:870–891.

Gorby, Y.A. and Lovley, D.R., 1992. Enzymatic uranium precipitation. Environmental Science and Technology, 26:205–207.

Gorby, Y.A., Beveridge, T.J., and Blakemore, R.P., 1988. Characterization of the bacterial magnetosome membrane. Journal of Bacteriology, 170:834–841.

Gordon, R. and Drum, R.W., 1994. The chemical basis of diatom morphogenesis. International Review of Cytology, 150: 243–372.

Gottschal, J.C. and Kuenen, J.G., 1980. Mixotrophic growth of *Thiobacillus* A2 on acetate and thiosulfate as growth limiting substrates in the chemostat. Archives of Microbiology, 126:33–42.

Gounot, A.-M., 1986. Psychrophilic and psychrotrophic microorganisms. Experientia, 42:1192–1197.

Grant, W.D. and Tindall, B.J., 1986. The alkaline, saline environment. In: Herbert, R.A. and Codd, G.A. (eds), Microbes in Extreme Environments. Academic Press, London, pp. 22–54.

Graustein, W.C., Cromack, K. Jr, and Sollins, P., 1977. Calcium oxalate: occurrence in soils and effect on nutrient and geochemical cycles. Science, 198:1252–1254.

Gray, M.W., Burger, G., and Lang, B.F., 1999. Mitochondrial evolution. Science, 283:1476–1481.

Greene, B., Hosea, M., McPherson, R., Henzl, M., Alexander, M.D., and Darnall, D.W., 1986. Interaction of gold(I) and gold(III) complexes with algal biomass. Environmental Science and Technology, 20:627–632.

Grote, G. and Krumbein, W.E., 1992. Microbial precipitation of manganese by bacteria and fungi from desert rock and rock varnish. Geomicrobiology Journal, 10:49–57.

Grotzinger, J.P., 1994. Trends in Precambrian carbonate sediments and their implications for understanding evolution. In: Bengtson, S. (ed.), Early Life on Earth. Columbia University Press, New York, pp. 245–258.

Grotzinger, J.P. and Kasting, J.F., 1993. New constraints on Precambrian ocean composition. Journal of Geology, 101:235–243.

Grotzinger, J.P. and Knoll, A.H., 1999. Stromatolites in Precambrian carbonates: evolutionary mileposts or environmental dipsticks. Annual Reviews in Earth and Planetary Sciences, 27:313–358.

Gulbrandsen, R.A., 1969. Physical and chemical factors in the formation of marine apatite. Economic Geology, 64:365–382.

Gundersen, J.K., Jørgensen, B.B., Larsen, E., and Jannasch, H.W., 1992. Mats of giant sulfur bacteria on deep-sea sediments due to fluctuating hydrothermal flow. Nature, 360:454–456.

Gupta, R.S., Mukhtar, T., and Singh, B., 1999. Evolutionary relationships among photosynthetic prokaryotes (*Heliobacterium chlorum*, *Chloroflexus aurantiacus*, cyanobacteria, *Chlorobium tepidum*, and Proteobacteria): implications regarding the origin of photosynthesis. Molecular Microbiology, 32:893–906.

Gutzmer, J. and Beukes, N.J., 1998. Earliest laterites and possible evidence for terrestrial vegetation in the Early Proterozoic. Geology, 26:263–266.

Haas, J.R., Dichristina, T.J., and Wade, R., Jr, 2001. Thermodynamics of U(VI) sorption onto *Shewanella putrefaciens*. Chemical Geology, 180:33–54.

Habicht, K.S., Gade, M., Thamdrup, B., Berg, P., and Canfield, D.E., 2002. Calibration of sulfate levels in the Archean ocean. Science, 298:2372–2374.

Hackl, R.P., 1997. Commercial applications of bacteria–mineral interactions. In: McIntosh, J.M. and Groat, L.A. (eds), Biological–Mineralogical Interactions. Mineralogical Association of Canada Short Course Series, vol. 25. Ottawa, pp. 143–167.

Hafenbrandl, D., Keller, M., Dirmeier, R., et al. 1996. *Ferroglobus placidus* gen. nov., sp. nov. a novel hyperthermophilic archaeum that oxidizes Fe^{2+} at neutral pH under anoxic conditions. Archives of Microbiology, 166:308–314.

Hallbeck, L. and Pederson, K., 1990. Culture parameters regulating stalk formation and growth rates of *Gallionella ferruginea*. Journal of General Microbiology, 136:1675–1680.

Hallbeck, L. and Pederson, K., 1991. Autotrophic and mixotrophic growth of *Gallionella ferruginea*. Journal of General Microbiology, 137:2657–2661.

Hamade, T., Konhauser, K.O., Raiswell, R., Morris, R.C., and Goldsmith, S., 2003. Using Ge:Si ratios to decouple iron and silica fluxes in Precambrian banded iron formations. Geology, 31:35–38.

Hamilton, W.A., 1988. Energy transduction in anaerobic bacteria. In: Anthony, C. (ed.), Bacterial Energy Transduction. Academic Press, London, pp. 83–149.

Hamm, C.E., Merkel, R., Springer, O., et al. 2003. Architecture and material properties of diatom shells provide effective mechanical protection. Nature, 421:841–843.

Hammond, D.E., Fuller, C., Harmon, D., et al. 1985. Benthic fluxes in San Francisco Bay. Hydrobiologia, 129:69–90.

Han, T.M. and Runnegar, B., 1992. Megascopic eukaryotic algae from the 2.1-billion-year-old Negaunee Iron-Formation, Michigan. Science, 257:232–235.

Hancock, R.E.W., 1987. Role of porins in outer membrane permeability. Journal of Bacteriology, 169:929–933.

Hanert, H.H., 1992. The genus *Gallionella*. In: Balows, A. (ed.), The Prokaryotes. Springer, Berlin, pp. 4082–4088.

Hansen, J.I., Henriksen, K., and Blackburn, T.H., 1981. Seasonal distribution of nitrifying bacteria and rates of nitrification in coastal marine sediments. Microbial Ecology, 7:297–304.

Hansen, T.A. and van Gemerden, H., 1972. Sulfide utilization by purple nonsulfur bacteria. Archives of Microbiology, 86:49–56.

Harden, V.P. and Harris, J.O., 1953. The isoelectric point of bacterial cells. Journal of Bacteriology, 65:198–202.

Hardie, L.A., 2003. Secular variations in Precambrian seawater chemistry and the timing of Precambrian aragonite seas and calcite seas. Geology, 31:785–788.

Harland, W.B., 1964. Critical evidence for a great infra-Cambrian glaciation. Geologische Rundschau, 54:45–61.

Harrison, A.P. Jr, 1978. Microbial succession and mineral leaching in an artificial coal spoil. Applied and Environmental Microbiology, 36:861–869.

Hart, B.T., 1982. Uptake of trace metals by sediments and suspended particulates: a review. Hydrobiologia, 91:299–313.

Hart, W.M., Batarseh, K., Swaney, G.P., and Stiller, A.H., 1991. A rigorous model to predict the AMD production rate of mine waste rock. In: Proceedings of the Second International Conference on the Abatement of Acidic Drainage, Montreal, Canada, vol. 2, pp. 257–270.

Hartman, H., 1984. The evolution of photosynthesis and microbial mats: a speculation on the banded iron formations. In: Cohen, Y., Castenholz, R.W., and Halvorson, H.O. (eds), Microbial Mats: stromatolites. Alan Liss, New York, pp. 449–453.

Hartman, H., 1998. Photosynthesis and the origin of life. Origins of Life and Evolution of the Biosphere, 28:515–521.

Harvey, R.W. and Leckie, J.O., 1985. Sorption of lead onto two gram-negative marine bacteria in seawater. Marine Chemistry, 15:333–344.

Harvey, R.W., George, L.H., Smith, R.L., and LeBlanc, D.R., 1989. Transport of microspheres and indigenous bacteria through a sandy aquifer: results of natural- and-forced gradient tracer experiments. Environmental Science and Technology, 23:51–56.

Harwood, D.M. and Nikolaev, V.A., 1995. Cretaceous diatoms, morphology, taxonomy, biostratigraphy. In: Blome, C.D. (ed.), Siliceous Microfossils. Paleontological Society Short Course in Paleontology, no. 8. The Paleontological Society, Pittsburgh, Pennsylvania, pp. 81–106.

Hayatsu, R. and Anders, E., 1981. Organic compounds in meteorites and their origins. Topics in Current Chemistry. 99:1–37.

Hayes, J.M., 1983. Geochemical evidence bearing on the origin of aerobiosis, a speculative hypothesis. In: Schopf, J.W. (ed.), Earth's Earliest Biosphere, Its Origin and Evolution. Princeton University Press, Princeton, pp. 291–301.

Hayes, J.M., Kaplan, I.R., and Wedeking, K.W., 1983. Precambrian organic geochemistry; preservation of the record. In: Schopf, J.W. (ed.), Earth's Earliest Biosphere, Its Origin and Evolution. Princeton University Press, Princeton, pp. 93–134.

He, L.M. and Tebo, B.M., 1998. Surface charge properties and Cu(II) adsorption by spores of the marine *Bacillus* sp. strain SG-1. Applied and Environmental Microbiology, 64:1123–1129.

Heath, G.R., 1974. Dissolved silica and deep-sea sediments. In: Hay, W.W. (ed.), Studies in Paleo-oceanography. Society of Economic Palaeontologists and Mineralogists, Special Publication, 20:77–93.

Hedges, J.I. and Keil, R.G., 1995. Sedimentary organic matter preservation: an assessment and speculative synthesis. Marine Chemistry, 49:81–115.

Hedges, J.I., Clark, W.A., and Cowie, G.L., 1988. Fluxes and reactivities of organic matter in a coastal marine bay. Limnology and Oceanography, 33:1137–1152.

Hein, J.R., Scholl, D.W., Barron, J.A., Jones, M.G., and Miller, J., 1978. Diagenesis of late Cenozoic diatomaceous deposits and formation of bottom simulating reflector in the southern Bering Sea. Sedimentology, 25:155–181.

Heinen, W. and Lauwers, A.M., 1996. Organic sulfur compounds resulting from the interaction of iron sulfide, hydrogen sulfide, and carbon dioxide in an anaerobic aqueous environment. Origins of Life and Evolution of the Biosphere, 26:131–150.

Heising, S. and Schink, B., 1998. Phototrophic oxidation of ferrous iron by a *Rhodomicrobium vannielii* strain. Microbiology, 144:2263–2269.

Heising, S., Richter, L., Ludwig, W., and Schink, B., 1999. *Chlorobium ferrooxidans* sp. nov., a phototrophic green sulfur bacterium that oxidizes ferrous iron in coculture with a "*Geospirillum*" sp. strain. Archives of Microbiology, 172:116–124.

Helgeson, H.C., Murphy, W.M., and Aagaard, P., 1984. Thermodynamic and kinetic constraints on reaction rates among minerals and aqueous solution: II. Rate constants, effective surface area and the hydrolysis of feldspars. Geochimica et Cosmochimica Acta, 48:2405–2432.

Hem, J.D. and Lind, C.J., 1983. Nonequilibrium models for predicting forms of precipitated manganese oxides. Geochimica et Cosmochimica Acta, 47:2037–2046.

Hennet, R.J.C., Holm, N.G., and Engel, M.H., 1992. Abiotic synthesis of amino acids under hydrothermal conditions and the origin of life: a perpetual phenomenon? Naturwissenschaften, 79:361–365.

Henrichs, S.M., 1992. Early diagenesis of organic matter in marine sediments: progress and perplexity. Marine Chemistry, 39:119–149.

Henrichs, S.M. and Reeburgh, W.S., 1987. Anaerobic mineralization of marine sediment organic matter: rates and the role of anaerobic processes in the ocean carbon economy. Geomicrobiology Journal, 5:191–237.

Hering, J.G. and Stumm, W., 1990. Oxidative and reductive dissolution of minerals. In: Hochella, M.F. and White, A.F. (eds), Mineral–Water Interface Geochemistry, vol. 23. Mineralogical Society of America, pp. 427–465.

Hernlem, B.J., Vane, L.M., and Sayles, G.D., 1996. Stability constants for complexes of the siderophore desferrioxamine B with selected heavy metal cations. Inorganica Chimica Acta, 244:179–184.

Herring, C., 1951. Some theorems on the free energies of crystal surfaces. Physical Review, 82:87–93.

Hersman, L.E., 2000. The role of siderophores in iron oxide dissolution. In: Lovley, D.R. (ed.), Environmental Microbe–Metal Interactions. ASM Press, Washington, DC, pp. 145–157.

Hersman, L.E., Lloyd, T., and Sposito, G., 1995. Siderophore-promoted dissolution of hematite. Geochimica et Cosmochimica Acta, 59:3327–3330.

Hersman, L.E., Huang, A., Maurice, P.A., and Forsythe, J.H., 2000. Siderophore production and iron reduction by *Pseudomonas mendocina* in response to iron deprivation. Geomicrobiology Journal, 17:261–273.

Hesse, R., 1990. Origin of chert: diagenesis of biogenic siliceous sediments. In: McIlreath, I.A. and Morrow, D.W. (eds), Diagenesis. Geoscience Canada Reprint Series 4. Geological Society of Canada, St John's, pp. 227–251.

Hessler, A.M., Lowe, D.R., Jones, R.L., and Bird, D.K., 2004. A lower limit for atmospheric carbon dioxide levels 3.2 billion years ago. Nature, 428:736–738.

Hewitt, J. and Morris, J.G., 1975. Superoxide dismutase in some obligately anaerobic bacteria. FEBS Letters, 50:315–318.

Heywood, B.R., Bazylinski, D.A., Garratt-Reed, A., Mann, S., and Frankel, R.B., 1990. Controlled biosynthesis of greigite (Fe_3S_4) in magnetotactic bacteria. Naturwissenschaften, 77:536–538.

Hider, R.C., 1984. Siderophore mediated absorption of iron. Structure and Bonding, 58:25–87.

Hiebert, F.K. and Bennett, P.C., 1992. Microbial control of silicate weathering in organic-rich ground water. Science, 258:278–281.

Higgins, I.J., Best, D.J., Hammond, R.C., and Scott, D., 1981. Methane-oxidizing microorganisms. Microbiology Reviews, 45:556–590.

Hildebrand, M., Volcani, B.E., Gassmann, W., and Schroeder, J.I., 1997. A gene family of silicon transporters. Nature, 385:688–689.

Hill, C., O'Driscoll, B., and Booth, I., 1995. Acid adaptation and food poisoning microorganisms. International Journal of Food Microbiology, 28:245–254.

Hinkle, P.C. and McCarty, R.E., 1978. How cells make ATP. Scientific American, 238:104–123.

Hinrichs, K-U., Hayes, J.M., Sylva, S.P., Brewer, P.G., and DeLong, E.F., 1999. Methane-consuming archaebacteria in marine sediments. Nature, 398:802–805.

Hirschler, A., Luca, J., and Hubert, J.-C., 1990. Apatite genesis: a biologically induced or biologically controlled mineral formation process? Geomicrobiology Journal, 8:47–57.

Hobot, J.A., Carlemalm, E., Villiger, W., and Kellenberger, E., 1984. Periplasmic gel: new concept resulting from the reinvestigation of bacterial cell envelope ultrastructure by new methods. Journal of Bacteriology, 160:143–152.

Hoch, M.P., Fogel, M.L., and Kirchman, D.L., 1992. Isotope fractionation associated with ammonium uptake by a marine bacterium. Limnology and Oceanography, 37:1447–1459.

Hoehler, T.M., Alperin, M.J., Albert, D.B., and Martens, C.S., 1994. Field and laboratory studies of methane oxidation in an anoxic marine sediment: evidence for a methanogen-sulfate reducer consortium. Global Biogeochemical Cycles, 9:451–463.

Hoehler, T.M., Bebout, B.M., and Des Marais, D.J., 2001. The role of microbial mats in the production of reduced gases on the early Earth. Nature, 412:324–327.

Hoffman, P.F. and Schrag, D.P., 2002. The snowball Earth hypothesis: testing the limits of global change. Terra Nova, 14:129–155.

Hoffman, P.F., Kaufman, A.J., Halverson, G.P., and Schrag, D.P., 1998. A Neoproterozoic snowball Earth. Science, 281:1342–1346.

Hoiczyk, E., 1998. Structural and biochemical analysis of the sheath of *Phormidium uncinatum*. Journal of Bacteriology, 180:3923–3932.

Holland, H.D., 1984. The Chemical Evolution of the Atmosphere and Oceans. Princeton University Press, Princeton, New Jersey.

Holland, H.D., 2002. Volcanic gases, black smokers, and the Great Oxidation Event. Geochimica et Cosmochimica Acta, 66:3811–3826.

Holligan, P.M., Viollier, M., Harbour, D.S., Camus, P., and Chanpagne-Philippe, M., 1983. Satellite and ship studies of coccolithophore production along a continental shelf edge. Nature, 304:339–342.

Holm, N.G., 1987. Possible biological origin of banded iron-formations from hydrothermal solutions. Origin of Life, 17:229–250.

Holm, N.G., 1989. The $^{13}C/^{12}C$ ratios of siderite and organic matter of a modern metalliferous hydrothermal sediment and their implications for banded iron formations. Chemical Geology, 77:41–45.

Holmén, B.A. and Casey, W.H., 1996. Hydroxamate ligands, surface chemistry, and the mechanism of ligand-promoted dissolution of goethite [α-FeOOH(s)]. Geochimica et Cosmochimica Acta, 60:4403–4416.

Hooper, A.B., 1984. Ammonia oxidation and energy transduction in the nitrifying bacteria. In: Strohl, W.R. and Tuovinen, O.H. (eds), Microbial Chemoautotrophy. Ohio State University Press, Columbus, Ohio, pp. 133–167.

Horneck, G. and Brack, A., 1992. Study of the origin, evolution and distribution of life with emphasis on exobiology experiments in Earth's orbit. Advances in Space Biology and Medicine, 2:229–262.

Horneck, G., Bücker, H., and Reitz, G., 1994. Long-term survival of bacterial spores in space. Advances in Space Research, 14:41–45.

Horodyski, R.J., Bauld, J., Lipps, J.H., and Mendelson, C.V., 1992. Preservation of prokaryotes and organic-walled and calcareous and siliceous protists. In: Schopf, J.W. and Klein, C. (eds), The Proterozoic Biosphere. Cambridge University Press, Cambridge, pp. 185–193.

Horvath, R.S., 1972. Microbial co-metabolism and the degradation of organic compounds in nature. Bacteriological Reviews, 36:146–155.

Hosea, M., Greene, B., McPherson, R., Henzl, M., Alexander, M.D., and Darnall, D.W., 1986. Accumulation of gold on the alga *Chlorella vulgaris*. Inorganica Chimica Acta, 123:161–165.

Hou, C.T., 1984. Microbiology and biochemistry of methylotrophic bacteria. In: Hou, C.T. (ed.), Methylotrophs: microbiology, biochemistry, and genetics. CRC Press, Boca Raton, Florida, pp. 2–53.

House, C.H., Runnegar, B., and Fitz-Gibbon, S.T., 2003. Geobiological analysis using whole genome-based tree building applied to the Bacteria, Archaea, and Eukarya. Geobiology, 1:15–26.

Howarth, R.W., 1979. Pyrite: its rapid formation in a salt marsh and its importance to ecosystem metabolism. Science, 203:49–51.

Hoyle, B. and Beveridge, T.J., 1983. Binding of metallic ions to the outer membrane of *Escherichia coli*. Applied and Environmental Microbiology, 46:749–752.

Hoyle, B. and Beveridge, T.J., 1984. Metal binding by the peptidoglycan sacculus of *Escherichia coli* K-12. Canadian Journal of Microbiology, 30:204–211.

Hsieh, K.M., Lion, L.W., and Schuler, M.L., 1985. Bioreactor for the study of defined interactions of toxic metals and biofilms. Applied and Environmental Microbiology, 50:1155–1161.

Hu, M.Z.-C., Norman, J.M., Faison, B.D., and Reeves, M.E., 1996. Biosorption of uranium by *Pseudomonas aeruginosa* strain CSU: characterization and comparison studies. Biotechnology and Bioengineering, 51:237–247.

Huber, C. and Wächtershäuser, G., 1997. Activated acetic acid by carbon fixation on (Fe, Ni)S under primordial conditions. Science, 276:245–247.

Huber, C. and Wächtershäuser, G., 1998. Peptides by activation of amino acids with CO on (Ni,Fe)S surfaces: implications for the origin of life. Science, 281:670–672.

Huber, H., Hohn, M.J., Rachel, R., Fuchs, T, Wimmer V.C., and Stetter, K.O., 2002. A new phylum of Archaea represented by a nanosized hyperthermophilic symbiont. Nature, 417:63–67.

Huber, R., Wilharm, T., Huber, D., et al. 1992. *Aquifex pyrophilus* gen. nov. sp. nov., represents a novel group of marine hyperthermophilic hydrogen-oxidizing bacteria. Systematic and Applied Microbiology, 15:340–351.

Hughes, K.A. and Lawley, B., 2003. A novel Antarctic microbial endolithic community within gypsum crusts. Environmental Microbiology, 5:555–565.

Hughes, M.N. and Poole, R.K., 1989. Metal mimicry and metal limitation in studies of metal–microbe interactions. In: Poole, R.K. and Gadd, G.M. (eds), Metal–Microbe Interactions. IRL Press, Oxford, pp. 1–17.

Hügler, M., Huber, H., Stetter, K.O., and Fuchs, G., 2003. Autotrophic CO_2 fixation pathways in archaea (Crenarchaeota). Archives of Microbiology, 179:160–173.

Hulth, S., Aller, R.C., and Gilbert, F., 1999. Coupled anoxic nitrification/manganese reduction in marine sediments. Geochimica et Cosmochimica Acta, 63:49–66.

Hunt, S., 1986. Diversity of biopolymer structure and its potential for ion-binding applications. In: Eccles, H. and Hunt, S. (eds), Immobilization of Ions by Bio-Sorption. Ellis Harwood, Chichester, UK, pp. 15–46.

Hunter, R.J., 2001. Foundations of Colloid Science. Oxford University Press, New York.

Hurd, D.C., 1983. Physical and chemical properties of siliceous skeletons. In: Aston, S. (ed.), Silicon Geochemistry and Biogeochemistry. Academic Press, New York, pp. 187–244.

Hurtgen, M.T., Arthur, M.A., and Halverson, G.P., 2005. Neoproterozoic sulfur isotopes, the evolution of microbial sulfur species, and the burial efficiency of sulfide as sedimentary pyrite. Geology, 33:41–44.

Hutchins, D.A. and Bruland, K.W., 1998. Iron-limited diatom growth and Si:N uptake ratios in a coastal upwelling regime. Nature, 393:561–564.

Hyde, W.T., Crowley, T.J., Baum, S.K., and Peltier, W.R., 2000. Neoproterozoic "snowball Earth" simulations with a coupled climate/ice-sheet model. Nature, 405:425–429.

Icopini, G.A., Anbar, A.D., Ruebush, S.S., Tien, M., and Brantley, S.L., 2004. Iron isotope fractionation during microbial reduction of iron: the importance of adsorption. Geology, 32:205–208.

Iler, R.K., 1979. Chemistry of Silica. Wiley-Interscience, New York.

Imhoff, J.F., 1992. Taxonomy, phylogeny, and general ecology of anoxygenic phototrophic bacteria. In: Mann, N.H. and Carr, N.G. (eds), Photosynthetic Prokaryotes. Plenum Press, New York, pp. 53–92.

Ingall, E. and Jahnke, R., 1997. Influence of water-column anoxia on the elemental fractionation of carbon and phosphorus during sediment diagenesis. Marine Geology, 139:219–229.

Ingledew, W.J., 1982. *Thiobacillus ferrooxidans*. The bioenergetics of an acidophilic chemolithotroph. Biochimica et Biophysica Acta, 683:89–117.

Ingledew, W.J., Cox, J.C., and Halling, P.J., 1977. A proposed mechanism for energy conservation during Fe^{2+} oxidation by *Thiobacillus ferro-oxidans*: chemiosmotic coupling to net H^+ influx. FEMS Microbiology Letters, 2:193–197.

Irwin, H., Curtis, C., and Coleman, M., 1977. Isotopic evidence for source of diagenetic carbonates formed during burial of organic-rich sediments. Nature, 269:209–213.

Islam, F.S., Gault, A.G., Boothman, C., et al. 2004. Role of metal-reducing bacteria in arsenic release from Bengal delta sediments. Nature, 430:68–71.

Isley, A.E., 1995. Hydrothermal plumes and the delivery of iron to banded iron formation. Journal of Geology, 103:169–185.

Issacs, C.M., 1981. Porosity reduction during diagenesis of the Monterey Formation, Santa Barbara coastal area, California. In: Garrison, R.E. (ed.), Pacific Section. SEPM Special Publication, 15: 257–271.

Ivarson, K.C., 1973. Microbiological formation of basic ferric sulfates. Canadian Journal of Soil Science, 53:315–323.

Ivarson, K.C. and Sojak, M., 1978. Microorganisms and ochre deposits in field drains of Ontario. Canadian Journal of Soil Science, 58:1–17.

Iverson, W.P., 1987. Microbial corrosion of metals. Advances in Applied Microbiology, 32:1–36.

Iwabe, N., Kuma, K.-I., Hasegawa, M., Osawa, S., and Miyata, T., 1989. Evolutionary relationship of archaebacteria, eubacteria and eukaryotes inferred from phylogenetic trees of duplicated genes. Proceedings of the National Academy of Sciences, USA, 86:9355–9359.

Jackson, B.E. and McInerney, M.J., 2002. Anaerobic microbial metabolism can proceed close to thermodynamic limits. Nature, 415:454–456.

Jackson, J.B., 1988. Bacterial photosynthesis. In: Anthony, C. (ed.), Bacterial Energy Transduction. Academic Press, London, pp. 318–375.

Jackson, T.A. and Keller, W.D., 1970. A comparative study of the role of lichens and inorganic processes in the chemical weathering of recent Hawaiian lava flows. American Journal of Science, 269:446–466.

Jahnke, R.A., 1984. The synthesis and solubility of carbonate fluorapatite. American Journal of Science, 284:58–78.

Jain, R., Rivera, M.C., and Lake, J.A., 1999. Horizontal gene transfer among genomes: the complexity hypothesis. Proceedings of the National Academy of Sciences, USA, 96:3801–3806.

Jakosky, B.M. and Shock, E.L., 1998. The biological potential of Mars, the early Earth and Europa. Journal of Geophysical Research, 103:19359–19364.

Jang, L.-K., Chang, P.W., Findley, J.E., and Yen, T.F., 1983. Selection of bacteria with favorable transport properties through porous rock for the application of microbial-enhanced oil recovery. Applied and Environmental Microbiology, 46:1066–1072.

Jannasch, H.W. and Mottl, M.J., 1985. Geomicrobiology of deep-sea hydrothermal vents. Science, 229:717–725.

Javaux, E.J., Knoll, A.H., and Walter, M.R., 2001. Morphological and ecological complexity in early eukaryotic eco-systems. Nature, 412:66–69.

Jewett, D.J., Hilbert, T.A., Logan, B.E., Arnold, R.G., and Bales, R.C., 1995. Bacterial transport in laboratory columns and filters: influence of ionic strength and pH on collision efficiency. Water Research, 29:1673–1680.

Jiang, G., Kennedy, M.J., and Christie-Blick, N., 2003. Stable isotopic evidence for methane seeps in Neoproterozoic postglacial cap carbonates. Nature, 426:822–826.

Jiao, Y., Kappler, A., Croal, L.R., and Newman, D.K., 2005. Isolation and characterization of a genetically tractable photoautotrophic Fe(II)-oxidizing bacterium, *Rhodopseudomonas palustris* strain TIE-1. Applied and Environmental Microbiology, 71:4487–4496.

Johnson, C.M., Beard, B.L., Beukes, N.J., Klein, C., and O'Leary, J.M., 2003. Ancient geochemical cycling in the Earth as inferred from Fe isotope studies of banded iron formations from the Transvaal craton. Contributions to Mineralogy and Petrology, 144:523–547.

Johnson, C.M., Roden, E.E., Welch, S.A., and Beard, B.L., 2005. Experimental constraints on Fe isotope fractionation during magnetite and Fe carbonate formation coupled to dissimilatory hydrous ferric oxide reduction. Geochimica et Cosmochimica Acta, 69:963–993.

Johnson, D.B. and Roberto, F.F., 1997. Heterotrophic acidophiles and their roles in the bioleaching of sulfide minerals. In: Rawlings, D.E. (ed.), Biomining. Theory, microbes and industrial processes. Springer-Verlag, Berlin, pp. 259–279.

Johnson, W.P. and Logan, B.E., 1996. Enhanced transport of bacteria in porous media by sediment-phase and aqueous-phase natural organic matter. Water Research, 30:923–931.

Jollife, L.K., Doyle, R.J., and Streips, U.N., 1981. The energized membrane and cellular autolysis in *Bacillus subtilis*. Cell, 25:753–763.

Jones, B., 1989. The role of microorganisms in phytokarst development on dolostones and limestones, Grand Cayman, British West Indies. Canadian Journal of Earth Sciences, 26:2204–2213.

Jones, B. and Renaut, R.W., 1996. Influence of thermophilic bacteria on calcite and silica precipitation in hot springs with water temperatures above 90°C: evidence from Kenya and New Zealand. Canadian Journal of Earth Sciences, 33:72–83.

Jones, B., Renaut, R.W., and Rosen, M.R., 1997. Biogenicity of silica precipitation around geysers and hot-spring vents, North Island, New Zealand. Journal of Sedimentary Research, 67:88–104.

Jones, B., Renaut, R.W., and Rosen, M.R., 1998. Microbial biofacies in hot-spring sinters: a model based on Ohaaki Pool, North Island, New Zealand. Journal of Sedimentary Research, 68:413–434.

Jones, B., Renaut, R.W., and Rosen, M.R., 2001. Taphonomy of silicified filamentous microbes in modern geothermal sinters – implications for identification. Palaios, 16:580–592.

Jones, B., Renaut, R.W., Rosen, M.R., and Ansdell, K.M., 2002. Coniform stromatolites from geothermal systems, North Island, New Zealand. Palaios, 17:84–103.

Jones, B., Renaut, R.W., and Rosen, M.R., 2003. Silicified microbes in a geyser mound: the enigma of low-temperature cyanobacteria in a high-temperature setting. Palaios, 18:87–109.

Jones, B., Konhauser, K., Renaut, R.W., and Wheeler, R., 2004. Microbial silicification in Iodine Pool, Waimangu geothermal area, North Island, New Zealand: implications for recognition and identification of ancient silicified microbes. Geological Society of London, 161:983–993.

Jones, B.E., Grant, W.D., Duckworth, A.W., and Owenson, G.G., 1998. Microbial diversity of soda lakes. Extremophiles, 2:191–200.

Jones, C.A., Langner, H.W., Anderson, K., McDermott, T.R., and Inskeep, W.P., 2000. Rates of microbially mediated arsenate reduction and solubilization. Soil Science Society of America Journal, 64:600–608.

Jones, C.W., 1985. The evolution of bacterial respiration. In: Schliefer, K.H. and Stackebrandt, E. (eds), Evolution of Prokaryotes. Academic Press, London, pp. 175–204.

Jones, C.W., 1988. Membrane-associated energy conservation in bacteria: a general introduction. In: Anthony, C. (ed.), Bacterial Energy Transduction. Academic Press, London, pp. 1–82.

Jones, D., Wilson, M.J., and Tait, J.M., 1980. Weathering of a basalt by *Pertusaria corallina*. Lichenologist, 12:277–289.

Jones, D., Wilson, M.J., and McHardy, W.J., 1981. Lichen weathering of rock-forming minerals: application of scanning electron microscopy and microprobe analysis. Journal of Microscopy, 124:95–104.

Jones, D.L. and Kochian, L.V., 1996. Aluminum–organic acid interactions in acid soils. Plant and Soil, 18:221–228.

Jones, J.G., 1985. Microbes and microbial processes in sediments. Philosophical Transactions of the Royal Society of London, 315:3–17.

Jones, R.A., Koval, S.F., and Nesbitt, H.W., 2003. Surface alteration of arsenopyrite (FeAsS) by *Thiobacillus ferrooxidans*. Geochimica et Cosmochimica Acta, 67:955–965.

Jones, R.D. and Morita, R.Y., 1983. Methane oxidation by *Nitrosococcus oceanus* and *Nitrosomonas europaea*. Applied and Environmental Microbiology, 45:401–410.

Jørgensen, B.B., 1977. The sulfur cycle of a coastal marine sediment (Limfjorden, Denmark). Limnology and Oceanography, 22:814–832.

Jørgensen, B.B., 1980. Mineralization and the bacterial cycling of carbon, nitrogen and sulphur in marine sediments. In: Ellwood, D.C., Hedger, J.N., Latham, M.J., Lynch, J.M., and Slater, J.H. (eds), Contemporary Microbial Ecology. Academic Press, London, pp. 239–252.

Jørgensen, B.B., 1982a. Ecology of the bacteria of the sulfur cycle with special reference to anoxic–oxic interface environments. Philosophical Transactions of the Royal Society of London, 298:543–561.

Jørgensen, B.B., 1982b. Mineralization of organic matter in the seabed – the role of sulfate reduction. Nature, 296:643–645.

Jørgensen, B.B., 1983. Processes at the sediment–water interface. In: Bolin, B.C. (ed.), The Major Biogeochemical Cycles and their Interactions. SCOPE, Wiley, New York, pp. 477–509.

Jørgensen, B.B., 1989. Light penetration, absorption, and action spectra in cyanobacterial mats. In: Cohen, Y. and Rosenberg, E. (eds), Microbial Mats: physiological ecology of benthic microbial communities. American Society for Microbiology, Washington, DC, pp. 123–137.

Jørgensen, B.B., 1990. A thiosulfate shunt in the sulfur cycle of marine sediments. Science, 249:152–154.

Jørgensen, B.B. and Bak, F., 1991. Pathways and microbiology of thiosulfate transformations and sulfate reduction in a marine sediment (Kattegat, Denmark). Applied and Environmental Microbiology, 57:847–856.

Jørgensen, B.B. and Des Marais, D.J., 1986. Competition for sulfide among colorless and purple sulfur bacteria in cyanobacterial mats. FEMS Microbial Ecology, 38:179–186.

Jørgensen, B.B. and Revsbech, N.P., 1983. Colorless sulfur bacteria. *Beggiatoa* spp. and *Thiovulum* spp. in O_2 and H_2S microgradients. Applied and Environmental Microbiology, 45:1261–1270.

Jørgensen, B.B. and Revsbech, N.P., 1985. Diffusive boundary layers and the oxygen uptake of sediments and detritus. Limnology and Oceanography, 30:111–122.

Jørgensen, B.B. and Sørensen, J., 1985. Seasonal cycles of O_2, NO_3^- and SO_4^{2-} reduction in estuarine sediments: the significance of an NO_3^- reduction maximum in the spring. Marine Ecology Progress Series, 24:65–74.

Jørgensen, B.B., Revsbech, N.P., and Cohen, Y., 1983. Photosynthesis and structure of benthic microbial mats: microelectrode and SEM studies of four cyanobacterial communities. Limnology and Oceanography, 28:1075–1093.

Joshi, H.M. and Tabita, F.R., 1996. A global two component signal transduction system that integrates the control of photosynthesis, carbon dioxide assimilation, and nitrogen fixation. Proceedings of the National Academy of Sciences, USA, 93:14515–14520.

Joye, S.B., Boetius, A., Orcutt, B.N., et al. 2004. The anaerobic oxidation of methane and sulfate reduction in sediments from Gulf of Mexico cold seeps. Chemical Geology, 205:219–238.

Jull, A.J.T., Courtney, C., Jeffrey, D.A., and Beck, J.W., 1998. Isotopic evidence for a terrestrial source of organic compounds found in Martian meteorites Allan Hills 84001 and Elephant Moraine 79001. Science, 279:366–369.

Junge, K., Krembs, C., Deming, J.W., Stierle, A., and Eicken, H., 2001. A microscopic approach to investigate bacteria under in situ conditions in sea-ice samples. Annals of Glaciology, 33:304–310.

Junghans, K. and Straube, G., 1991. Biosorption of copper by yeasts. Biology of Metals, 4:233–237.

Juniper, S.K. and Fouquet, Y., 1988. Filamentous iron-silica deposits from modern and ancient hydrothermal sites. Canadian Mineralogist, 26:859–869.

Juniper, S.K. and Tebo, B.M., 1995. Microbe–metal interactions and mineral deposition at hydrothermal vents. In: Karl, D.M. (ed.), The Microbiology of Deep-Sea Hydrothermal Vents. CRC Press, Boca Raton, pp. 219–253.

Kalin, M., Wheeler, W.N., and Meinrath, G., 2005. The removal of uranium from mining waste water using algal/microbial biomass. Journal of Environmental Radioactivity, 78:151–177.

Kalinowski, B.E., Liermann, L.J., Givens, S., and Brantley, S.L., 2000a. Rates of bacteria-promoted solubilization of Fe from minerals: a review of problems and approaches. Chemical Geology, 169:357–370.

Kalinowski, B.E., Liermann, L.J., Brantley, S.L., and Pantano, C.G., 2000b. X-ray electron evidence for bacteria-enhanced dissolution of hornblende. Geochimica et Cosmochimica Acta, 64:1331–1343.

Kandler, O., 1994. The early diversification of life. In: Bengtson, S. (ed.), Early Life on Earth, Columbia University Press, New York, pp. 152–160.

Kappler, A. and Newman, D.K., 2004. Formation of Fe(III)-minerals by Fe(II)-oxidizing photoautotrophic bacteria. Geochimica et Cosmochimica Acta, 68:1217–1226.

Kappler, A., Pasquero, C., Konhauser, K.O., and Newman, D.K., 2005. Deposition of banded iron formations by phototrophic Fe(II)-oxidizing bacteria. Geology, 33:865–868.

Karhu, J.A. and Holland, H.D., 1996. Carbon isotopes and the rise of atmospheric oxygen. Geology, 24:867–870.

Karl, D., Letelier, R., Tupas, L., Dore, J., Christian, J., and Hebel, D., 1997. The role of nitrogen fixation in biogeochemical cycling in the subtropical North Pacific Ocean. Nature, 388:533–538.

Karl, D., Michaels, A., Bergman, B., et al., 2002. Dinitrogen fixation in the world's oceans. Biogeochemistry, 57/58:47–98.

Karl, D.M., Knauer, G.A., Martin, J.H., and Ward, B.B., 1984. Bacterial chemolithotrophy in the ocean is associated with sinking particles. Nature, 309:54–56.

Karl, D.M., McMurtry, G.M., Malahoff, A., and Garcia, M.O., 1988. Loihi Seamount, Hawaii: a mid-plate volcano with a distinctive hydrothermal system. Nature, 335:532–535.

Karl, D.M., Bird, D.F., Björkman, K., Houlihan, T., Shackelford, R., and Tupas, L., 1999. Microorganisms in the accreted ice of Lake Vostok, Antarctica. Science, 286:2144–2147.

Karlin, R., Lyle, M., and Heath, G.R., 1987. Authigenic magnetite formation in suboxic marine sediments. Nature, 326:490–493.

Karner, M.B., DeLong, E.F., and Karl, D.M., 2001. Archaeal dominance in the mesopelagic zone of the Pacific Ocean. Nature, 409:507–510.

Kashefi, K. and Lovley, D.R., 2000. Reduction of Fe(III), Mn(IV), and toxic metals at 100°C by *Pyrobaculum islandicum*. Applied and Environmental Microbiology, 66:1050–1056.

Kashefi, K. and Lovley, D.R., 2003. Extending the upper temperature limit for life. Science, 301:934.

Kashefi, K., Tor, J.M., Nevin, K.P., and Lovley, D.R., 2001. Reductive precipitation of gold by dissimilatory Fe(III)-reducing bacteria and Archaea. Applied and Environmental Microbiology, 67:3275–3279.

Kashket, E.R., 1985. The proton motive force in bacteria: a critical assessment of methods. Annual Reviews of Microbiology, 39:219–242.

Kasting, J.F., 1982. Stability of ammonia in the primitive terrestrial atmosphere. Journal of Geophysical Research, 87:3091–3098.

Kasting, J.F., 1987. Theoretical constraints on oxygen and carbon dioxide concentrations in the Precambrian atmosphere. Precambrian Research, 34:205–229.

Kasting, J.F., 1990. Bolide impacts and the oxidation state of carbon in the Earth's early atmosphere. Origins of Life and Evolution of the Biosphere, 20:199–231.

Kasting, J.F., 2001. The rise of atmospheric oxygen. Science, 293:819–820.

Kasting, J.F., 2004. When methane made climate. Scientific American, 291:78–85.

Kasting, J.F. and Ackerman, T.P., 1986. Climatic consequences of very high carbon dioxide levels in the Earth's early atmosphere. Science, 234:1383–1385.

Kasting, J.F. and Siefert, J.L., 2001. The nitrogen fix. Nature, 412:26–27.

Kasting, J.F. and Siefert, J.L., 2002. Life and the evolution of Earth's atmosphere. Science, 296:1066–1068.

Kasting, J.F., Zahnle, K.J., and Walker, J.C.G., 1983. Photochemistry of methane in the Earth's early atmosphere. Precambrian Research, 20:121–148.

Kasting, J.F., Zahnle, K.J., Pinto, J.P., and Young, A.T., 1989. Sulfur, ultraviolet radiation, and the early evolution of life. Origins of Life and Evolution of the Biosphere, 19:95–108.

Kasting, J.F., Eggler, D.H., and Raeburn, S.P., 1993. Mantle redox evolution and the oxidation state of the Archean atmosphere. Journal of Geology, 101:245–257.

Kastner, M., Keene, J.B., and Gieskes, J.M., 1977. Diagenesis of siliceous oozes. 1. Chemical controls on the rate of opal-A to opal-CT transformation – an experimental study. Geochimica et Cosmochimica Acta, 41:1041–1059.

Kato, C., Li, L., Nogi, Y., Nakamura, Y., Tamaoka, J., and Horikoshi, K., 1998. Extremely barophilic bacteria isolated from the Mariana Trench, Challenger Deep, at a depth of 11,000 meters. Applied and Environmental Microbiology, 64:1510–1513.

Kaufman, E.N., Little, M.H., and Selvaraj, P.T., 1996. Recycling of FGD gypsum to calcium carbonate and elemental sulfur using mixed sulfate-reducing bacteria with sewage digest as a carbon source. Journal of Chemical Technology and Biotechnology, 66:365–374.

Kazumi, J., Haggblom, M.M., Young, L.Y., 1995. Degradation of monochlorinated and nonchlorinated aromatic compounds under iron-reducing conditions. Applied and Environmental Microbiology, 61:4069–4073.

Ke, B., 2001. Photosynthesis: photobiochemistry and photobiophysics. Kluwer, Dordrecht.

Kelley, D.S., Karson, J.A., Blackman, D.K., et al. 2001. An off-axis hydrothermal vent field near the Mid-Atlantic Ridge at 30°N. Nature, 412:145–149.

Kelly, D.P., 1974. Growth and metabolism of the obligate photolithotroph *Chlorobium thiosulfatophilum* in the presence of added organic nutrients. Archives of Microbiology, 100:163–178.

Kempe, S., Kazmierczak, J., Landmann, G., Konuk, T., Reimer, A., and Lipp, A., 1991. Largest known microbialites discovered in Lake Van, Turkey. Nature, 349:605–608.

Kennard, J.M. and James, N.P., 1986. Thrombolites and stromatolites: two distinct types of microbial structures. Palaios, 1:492–503.

Kennedy, C.B., Martinez, R.E., Scott, S.D., and Ferris, F.G., 2003. Surface chemistry and reactivity of bacteriogenic iron oxides from Axial Volcano, Juan de Fuca Ridge, north-east Pacific Ocean. Geobiology, 1:59–70.

Kennedy, M.J., Christie-Blick, N., and Sohl, L.E., 2001. Are Proterozoic cap carbonates and isotopic excursions a record of gas hydrate destabilization following Earth's coldest intervals? Geology, 29:443–446.

Kennett, J.P., Cannariato, K.G., Hendy, I.L., and Behl, R.J., 2000. Carbon isotopic evidence for methane hydrate instability during Quaternary interstadials. Science, 288:128–133.

Keswick, B.H., Wang, D.S., and Gerba, C.P., 1982. The use of microorganisms as ground-water tracers: a review. Ground Water, 20:142–149.

Khanna, S. and Nicholas, D.J.D., 1982. Utilization of tetrathionate and ^{35}S-labelled thiosulfate by washed cells of *Chlorobium vibrioforme* f. sp. *thiosulfatophilum*. Journal of General Microbiology, 128:1027–1034.

Khummongkol, D., Canterford, G.S., and Fryer, C., 1982. Accumulation of heavy metals in unicellular algae. Biotechnology and Bioengineering, 24:2643–2660.

Kim, D-S., Thomas, S., and Fogler, H.S., 2000. Effects of pH and trace minerals on long-term starvation of *Leuconostoc mesenteroides*. Applied and Environmental Microbiology, 66:976–981.

Kirk, T.K. and Farrell, R.L., 1987. Enzymatic "combustion": the microbial degradation of lignin. Annual Reviews in Microbiology, 41:465–505.

Kirschvink, J.L., 1982. Paleomagnetic evidence for fossil biogenic magnetite in western Crete. Earth and Planetary Science Letters, 59:388–392.

Kirschvink, J.L., 1992. Late Proterozoic low-latitude global glaciation: the Snowball Earth. In: Schopf, J.W. and Klein, C. (eds), The Proterozoic Biosphere. Cambridge University Press, Cambridge, pp. 51–52.

Kirschvink, J.L. and Chang, S-B.R., 1984. Ultrafine-grained magnetite in deep-sea sediments: possible bacterial magnetofossils. Geology, 12:559–562.

Kirschvink, J.L. and Weiss, B.P., 2002. Mars, Panspermia and the origin of life: where did it all begin? Palaeontologia Electronica, 25 January, 1–8.

Kirschvink, J.L., Gaidos, E.J., Bertani, L.E., et al. 2000. Paleoproterozoic snowball Earth: extreme climatic and geochemical global change and its biological consequences. Proceedings of the National Academy of Sciences, USA, 97:1400–1405.

Kjelleberg, S. and Hermansson, M., 1984. Starvation-induced effects on bacterial surface characteristics. Applied and Environmental Microbiology, 48:497–503.

Klein, C. and Beukes, N.J., 1989. Geochemistry and sedimentology of a facies transition from limestone to iron-formation deposition in the Early Proterozoic Transvaal Supergroup, South Africa. Economic Geology, 84:1733–1774.

Klein, C. and Beukes, N.J., 1992. Time distribution, stratigraphy, and sedimentologic setting, and geochemistry of Precambrian iron-formations. In: Schopf, J.W. and Klein, C. (eds), The Proterozoic Biosphere: a multidisciplinary study. Cambridge University Press, Cambridge, pp. 139–146.

Klein, M., Friedrich, M., Roger, A.J., et al. 2001. Multiple lateral transfers of dissimilatory sulfite reductase genes between major lineages of sulfate-reducing prokaryotes. Journal of Bacteriology, 183:6028–6035.

Kleinmann, R.L.P. and Crerar, D.A., 1979. *Thiobacillus ferrooxidans* and the formation of acidity in simulated coal mine environments. Geomicrobiology Journal, 1:373–388.

Kleinmann, R.L.P., Crerar, D.A., and Pacelli, R.R., 1981. Biogeochemistry of acid mine drainage and a method to control acid formation. Mining Engineering, 1981:300–305.

Knauer, G.A., Martin, J.H., and Bruland, K.W., 1979. Fluxes of particulate carbon, nitrogen and phosphorus in the upper water column of the northeast Pacific. Deep-Sea Research, 26:97–108.

Knauth, L.P. and Lowe, D.R., 2003. High Archean climatic temperatures inferred from oxygen isotope geochemistry of cherts in the 3.5 Ga Swaziland Supergroup, South Africa. Geological Society of America Bulletin, 155:566–580.

Knoll, A.H., 1984. The Archean/Proterozoic transition: a sedimentary and paleobiological perspective. In: Holland, H.D. and Trendall, A.F. (eds), Patterns of Change in Earth Evolution. Springer-Verlag, Berlin, pp. 221–242.

Knoll, A.H., 1985. Exceptional preservation of photosynthetic organisms in silicified carbonates and silicified peats. Philosophical Transactions of the Royal Society of London, 311B:111–122.

Knoll, A.H., 1992. The early evolution of eukaryotes: a geological perspective. Science, 256:622–627.

Knoll, A.H., 2003. The geological consequences of evolution. Geobiology, 1:3–14.

Knoll, A.H. and Carroll, S.B., 1999. Early animal evolution: emerging views from comparative biology and geology. Science, 284:2129–2137.

Knoll, A.H., Strother, P.K., and Rossi, S., 1988. Distribution and diagenesis of microfossils from the Lower Proterozoic Duck Creek Dolomite, Western Australia. Precambrian Research, 38:257–279.

Knowles, R., 1982. Denitrification. Microbial Reviews, 46:43–70.

Knowles, R., 1985. Microbial transformation as sources and sinks of nitrogen oxides. In: Caldwell, D.E., Brierley, J.A., and Brierley, C.L. (eds), Planetary Ecology. Van Nostrand Reinhold, New York, pp. 411–426.

Kobayashi, M., Hamano, T., Akiyama, M., et al. 1998. Light-independent isomerization of bacteriochlorophyll g to chlorophyll a catalyzed by weak acid in vitro. Analytica Chimica Acta, 365:199–203.

Koch, A.L., 1983. The surface stress theory of microbial morphogenesis. Advances in Microbial Physiology, 24:301–367.

Koch, A.L., 1986. The pH in the neighbourhood of membranes generating a protonmotive force. Journal of Theoretical Biology, 120:73–84.

Köhler, B., Singer, A., and Stoffers, P., 1994. Biogenic nontronite from marine white smoker chimneys. Clays and Clay Minerals, 42:689–701.

Komeili, A., Vali, H., Beveridge, T.J., and Newman, D.K., 2004. Magnetosome vesicles are present before magnetite formation, and MamA is required for their activation. Proceedings of the National Academy of Sciences, USA, 101:3839–3844.

Konhauser, K.O., 1998. Diversity of bacterial iron mineralization. Earth-Science Reviews, 43:91–121.

Konhauser, K.O., 2000. Hydrothermal bacterial biomineralization: potential modern-day analogues for Precambrian banded iron formation. In: Glenn, C.R., Lucas, J., and Prévôt, L. (eds), Marine Authigenesis: from global to microbial. SEPM Special Publication no. 66, pp. 133–145.

Konhauser, K.O. and Ferris, F.G., 1996. Diversity of iron and silica precipitation by microbial biofilms in hydrothermal waters, Iceland: implications for Precambrian Iron Formations. Geology, 24:323–326.

Konhauser, K.O. and Fyfe, W.S., 1993. Biogeochemical cycling of metals in freshwater algae from Manaus and Carajás, Brazil. Energy Sources, 15:3–16.

Konhauser, K.O. and Urrutia, M.M., 1999. Bacterial clay authigenesis: a common biogeochemical process. Chemical Geology, 161:399–413.

Konhauser, K.O., Mann, H., and Fyfe, W.S., 1992. Prolific SiO_2 precipitation in a solute-deficient river: Rio Negro, Brazil. Geology, 20:227–230.

Konhauser, K.O., Fyfe, W.S., Ferris, F.G., and Beveridge, T.J., 1993. Metal sorption and mineral precipitation by bacteria in two Amazonian river systems: Rio Solimões and Rio Negro. Geology, 21:1103–1106.

Konhauser, K.O., Fyfe, W.S., Schultze-Lam, S., Ferris, F.G., and Beveridge, T.J., 1994. Iron phosphate precipitation by epilithic microbial biofilms in Arctic Canada. Canadian Journal of Earth Sciences, 31:1320–1324.

Konhauser, K.O., Fisher, Q.J., Fyfe, W.S., Longstaffe, F.J., and Powell, M.A., 1998. Authigenic mineralization and detrital clay binding by freshwater biofilms: the Brahmani River, India. Geomicrobiology Journal, 15:209–222.

Konhauser, K.O., Phoenix, V.R., Bottrell, S.H., Adams, D.G., and Head, I.M., 2001. Microbial–silica interactions in Icelandic hot spring sinter: possible analogues for Precambrian siliceous stromatolites. Sedimentology, 48:415–433.

Konhauser, K.O., Schiffman, P., and Fisher, Q.J., 2002a. Microbial mediation of authigenic clays during hydrothermal alteration of basaltic tephra, Kilauea Volcano. Geochemistry, Geophysics, and Geosystems, 3:2002GC000317, 13 pp.

Konhauser, K.O., Hamade, T., Morris, R.C., et al. 2002b. Did bacteria form Precambrian banded iron formations? Geology, 30:1079–1082.

Konhauser, K.O., Jones, B., Reysenbach, A.-L., and Renaut, R.W., 2003. Hot spring sinters: keys to understanding Earth's earliest life forms. Canadian Journal of Earth Sciences, 40:1713–1724.

Konhauser, K.O., Jones, B., Phoenix, V.R., Ferris, G., and Renaut, R.W., 2004. The microbial role in hot spring silicification. Ambio, 33:552–558.

Konhauser, K.O., Newman, D.K., and Kappler, A., 2005. The potential significance of microbial Fe(III)-reduction during Precambrian banded iron formations. Geobiology, 3:167–177.

König, H., 1988. Archaeobacterial cell envelopes. Canadian Journal of Microbiology, 34:395–406.

Koonin, E.V., 2003. Comparative genomics, minimal gene-sets and the last universal common ancestor. Nature Reviews, 1:127–136.

Korber, D.R., Lawrence, J.R., Hendry, M.J., and Caldwell, D.E., 1993. Analysis of spatial variability within Mot+ and Mot− *Pseudomonas fluorescens* biofilms using representative elements. Biofouling, 7:339–358.

Kostka, J.E., Dalton, D.D., Skelton, H., Dollhopf, S., and Stucki, J.W., 2002. Growth of Iron(III)-reducing bacteria on clay minerals as the sole electron acceptor and comparison of growth yields on a variety of oxidized iron forms. Applied and Environmental Microbiology, 68:6256–6262.

Koval, S.F., 1988. Paracrystalline protein surface arrays on bacteria. Canadian Journal of Bacteriology, 34:407–414.

Krajewski, K.P., Van Cappellen, P., Trichet, J., et al. 1994. Biological processes and apatite formation in sedimentary environments. Eclogae Geologicae Helvetiae, 87:701–745.

Kral, T.A., Brink, K.M., Miller, S.L., and McKay, C.P., 1998. Hydrogen consumption by methanogens on the early Earth. Origins of Life and Evolution of the Biosphere, 28:311–319.

Krapež, B., Barley, M.E., and Pickard, A.L., 2003. Hydrothermal and resedimented origins of the precursor sediments to banded iron formations: sedimentological evidence from the early Palaeoproterzoic Brockman Supersequence of Western Australia. Sedimentology, 50:979–1011.

Krauskopf, K.B. and Bird, D.K., 1995. Introduction to Geochemistry, 3rd edn. McGraw-Hill, New York.

Kröger, N., Deutzmann, R., and Sumper, M., 1999. Polycationic peptides from diatom biosilica that direct silica nanosphere formation. Science, 286:1129–1132.

Krom, M.D. and Berner, R.A., 1980. Adsorption of phosphate in anoxic marine sediments. Limnology and Oceanography, 25:797–806.

Krom, M.D. and Berner, R.A., 1981. The diagenesis of phosphorous in nearshore marine sediment. Geochimica et Cosmochimica Acta, 45:207–216.

Krulwich, T.A., Ito, M., Hicks, D.B., Gilmour, R., and Guffanti, A.A., 1997. Mechanisms of cytoplasmic pH regulation in alkaliphilic strains of *Bacillus*. Extremophiles, 1:163–169.

Krumbein, W.E. and Jens, K., 1981. Biogenic rock varnishes of the Negev Desert (Israel): an ecological study of iron and manganese transformation by cyanobacteria and fungi. Oecologia, 50:25–38.

Krumholz, L.R., McKinley, J.P., Ulrich, G.A., and Suflita, J.M., 1997. Confined subsurface microbial communities in Cretaceous rock. Nature, 386:64–66.

Kuhn, W.R. and Atreya, S.K., 1979. Ammonia photolysis and the greenhouse effect in the primordial atmosphere of the Earth. Icarus, 37:207–213.

Kump, L.R. and Garrels, R.M., 1986. Modeling atmospheric O_2 in the global sedimentary redox cycle. American Journal of Science, 286:337–360.

Kump, L.R. and Holland, H.D., 1992. Iron in Precambrian rocks: implications for the global oxygen budget of the ancient Earth. Geochimica et Cosmochimica Acta, 56:3217–3223.

Kump, L.R., Brantley, S.L., and Arthur, M.A., 2000. Chemical weathering, atmospheric CO_2, and climate. Annual Reviews of Earth and Planetary Science, 28:611–667.

Kump, L.R., Kasting, J.F., and Crane, R.G., 2004. The Earth System, 2nd edn. Pearson Education, Upper Saddle River, New Jersey.

Kuypers, M.M.M., Sliekers, A.O., Lavik, G., et al. 2003. Anaerobic ammonium oxidation by anammox bacteria in the Black Sea. Nature, 422:608–611.

Kuyucak, N. and Volesky, B., 1989a. Accumulation of gold by algal biosorbent. Biorecovery, 1:189–204.

Kuyucak, N. and Volesky, B., 1989b. The elution of gold on a natural biosorbent. Biorecovery, 1:205–218.

Kvenvolden, K.A., 1988. Methane hydrate – a major reservoir of carbon in the shallow geosphere? Chemical Geology, 71:41–51.

Kvenvolden, K.A., 1999. Potential effects of gas hydrate on human welfare. Proceedings of the National Academy of Sciences, USA, 96:3420–3426.

L'Haridon, S., Reysenbach, A.-L., Glénat, P., Prieur, D., and Jeanthon, C., 1995. Hot subterranean biosphere in a continental oil reservoir. Nature, 377:223–224.

La Roche, J., Boyd, P.W., McKay, R.M.L., and Geider, R.J., 1996. Flavodoxin as *in situ* marker for iron stress in phytoplankton. Nature, 382:802–805.

Labrenz, M., Druschel, G.K., Thomsen-Ebert et al. 2000. Formation of sphalerite (ZnS) deposits in natural biofilms of sulfate-reducing bacteria. Science, 290:1744–1747.

Lacey, D.T. and Lawson, F., 1970. Kinetics of the liquid-phase oxidation of acid ferrous sulfate by the bacterium *Thiobacillus ferrooxidans*. Biotechnology and Bioengineering, 12:29–50.

Lahav, N. and Nir, S., 1997. Emergence of template-and-sequence-directed (TSD) syntheses: 1. A biogeochemical model. Origins of Life and Evolution of the Biosphere, 27:377–395.

Lalonde, S.V., Konhauser, K.O., Reysenbach, A.-L., and Ferris, F.G., 2005. Thermophilic silicification: the role of Aquificales in hot spring sinter formation. Geobiology, 3:41–52.

Lange, C.C., Wackett, L.P., Minton, K.W., and Daly, M.J., 1998. Engineering a recombinant *Deinococcus radiodurans* for organopollutant degradation in radioactive mixed waste environments. Nature Biotechnology, 16:929–933.

Langmuir, D., 1997. Aqueous Environmental Geochemistry. Prentice Hall, New Jersey.

Larrson, L., Gunnel, O., Holst, O., and Karlsson, H.T., 1993. Oxidation of pyrite by *Acidianus brierleyi*: importance of close contact between pyrite and the microorganism. Biotechnology Letters, 15:99–104.

Larsen, I., Little, B., Nealson, K.H., Ray, R., Stone, A., and Tian, J., 1998. Manganite reduction by *Shewanella putrefaciens* MR-4. American Mineralogist, 83:1564–1572.

Lasaga, A.C. and Luttge, A., 2001. Variation of crystal dissolution rate based on a dissolution stepwave model. Science, 291:2400–2404.

Laverman, A.M., Blum, J.S., Schaefer, J.K., Phillips, E.J.P., Lovley, D.R., and Oremland, R.S., 1995. Growth of strain SES-3 with arsenate and other diverse electron acceptors. Applied and Environmental Microbiology, 61:3556–3561.

Lawrence, J.R. and Hendry, M.J., 1996. Transport of bacteria through geologic media. Canadian Journal of Microbiology, 42:410–422.

Lawrence, J.R., Kwong, Y.T.J., and Swerhone, G.D.W., 1997. Colonization and weathering of natural sulfide mineral assemblages by *Thiobacillus ferrooxidans*. Canadian Journal of Microbiology, 43:178–188.

Lazaroff, N., Sigal, W. and Wasserman, A., 1982. Iron oxidation and precipitation of ferric hydroxysulfates by resting *Thiobacillus ferrooxidans* cells. Applied and Environmental Microbiology, 43:924–938.

Ledin, M., 1999. Accumulation of metals by microorganisms. Processes and importance for soil systems. Earth-Science Reviews, 51:1–31.

Ledin, M. and Pedersen, K., 1996. The environmental impact of mine wastes roles of microorganisms and their significance in treatment of mine wastes. Earth-Science Reviews, 41:67–108.

Ledin, M., Pedersen, K., and Allard, B., 1997. Effects of pH and ionic strength on the adsorption of Cs, Sr, Eu, Zn, Cd and Hg by *Pseudomonas putida*. Water, Air and Soil Pollution, 93:367–381.

Ledin, M., Krantz-Rülcker, C., and Allard, B., 1999. Microorganisms as metal sorbents: comparison with other soil constituents in multi-compartment systems. Soil Biology and Biochemistry, 31:1639–1648.

Lee, C., 1992. Controls on organic carbon preservation: the use of stratified water bodies to compare intrinsic rates of decomposition in oxic and anoxic systems. Geochimica et Cosmochimica Acta, 56:3323–3335.

Lee, W., Lewandowski, Z., Nielsen, P.H., and Hamilton, W.A., 1995. Role of sulfate-reducing bacteria in corrosion of mild steel: a review. Biofouling, 8:165–194.

Lerman, L., 1986. Potential role of bubbles and droplets in primordial and planetary chemistry: exploration of the liquid–gas interface as a reaction zone for condensation process. Origins of Life and Evolution of the Biosphere, 16:201–202.

Lewin, R.A., 1976. Prochlorophyta as a proposed new division of algae. Nature, 261:697–698.

Liang, L., McNabb, J.A., Paulk, J.M., Gu, B., and McCarthy, J.F., 1993. Kinetics of Fe(II) oxygenation at low partial pressure of oxygen in the presence of natural organic matter. Environmental Science and Technology, 27:1864–1870.

Liermann, L.J., Kalinowski, B.E., Brantley, S.L., and Ferry, J.G., 2000. Role of bacterial siderophores in dissolution of hornblende. Geochimica et Cosmochimica Acta, 64:587–602.

Lilley, M.D., Butterfield, D.A., Olson, E.J., Lupton, J.E., Macko, S.A., and McDuff, R.E., 1993. Anomalous CH_4 and NH_4^+ concentrations at an unsedimented mid-ocean-ridge hydrothermal system. Nature, 364:45–47.

Lindqvist, R. and Enfield, C.G., 1992. Biosorption of dichlorodiphenyltrichloroethane and hexachlorobenzene in groundwater and its implications for facilitated transport. Applied and Environmental Microbiology, 58:2211–2218.

Lindström, E.B., Gunneriusson, E., and Tuovinen, O.H., 1992. Bacterial oxidation of refractory sulfide ores for gold recovery. Critical Reviews in Biotechnology, 12:133–155.

Lineweaver, C.H. and Davis, T.M., 2002. Does the rapid appearance of life on Earth suggest that life is common in the universe? Astrobiology, 2:293–304.

Little, B.J., Wagner, P.A., and Lewandowski, Z., 1997. Spatial relationships between bacteria and mineral surfaces. In: Banfield, J.F. and Nealson, K.H. (eds), Geomicrobiology: interactions between microbes and minerals, vol. 35. Mineralogical Society of America, Washington, DC, 123–159.

Lizama, H.M. and Suzuki, I., 1988. Bacterial leaching of a sulfide ore by *Thiobacillus ferrooxidans*: 1. Shake flask studies. Biotechnology and Bioengineering, 32:110–116.

Lizama, H.M. and Suzuki, I., 1989. Bacterial leaching of a sulfide ore by *Thiobacillus ferrooxidans* and *Thiobacillus thiooxidans*. Part II: Column leaching studies. Hydrometallurgy, 22:301–310.

Lloyd, J.R. and Macaskie, L.E., 2000. Bioremediation of radionuclide-containing wastewaters. In: Lovley, D.R. (ed.), Environmental Microbe–Metal Interactions. ASM Press, Washington, DC, pp. 277–327.

Lockhart, D.J. and Winzeler, E.A., 2000. Genomics, gene expression and DNA arrays. Nature, 405:827–835.

Logan, G.A., Hayes, J.M., Hieshima, G.B., and Summons, R.E., 1995. Terminal Proterozoic reorganization of biogeochemical cycles. Nature, 376:53–56.

Lovelock, J.E., 2000. Gaia: a new look at life on Earth. Oxford University Press, Oxford.

Lovelock, J.E. and Whitfield, M., 1982. Life span of the biosphere. Nature, 296:561–563.

Lovley, D.R., 1990. Magnetite formation during microbial dissimilatory iron reduction. In: Frankel, R.B. and Blakemore, R.P. (eds), Iron Biominerals. Plenum Press, New York, pp. 151–166.

Lovley, D.R., 1991. Dissimilatory Fe(III) and Mn(IV) reduction. Microbiology Reviews, 55:259–287.

Lovley, D.R., 1995. Bioremediation of organic and metal contaminants with dissimilatory metal reductions. Journal of Industrial Microbiology, 14:85–93.

Lovley, D.R., 2003. Cleaning up with genomics: applying molecular biology to bioremediation. Nature Reviews, 1:35–44.

Lovley, D.R. and Chapelle, F.H., 1995. Deep subsurface microbial processes. Reviews of Geophysics, 33:365–381.

Lovley, D.R. and Coates, J.D., 1997. Bioremediation of metal contamination. Current Opinion in Biotechnology, 8:285–289.

Lovley, D.R. and Goodwin, S., 1988. Hydrogen concentrations as an indicator of the predominant terminal electron-accepting reactions in aquatic sediments. Geochimica et Cosmochimica Acta, 52: 2993–3003.

Lovley, D.R. and Klug, M.J., 1983. Methanogenesis from methanol and methylamines and acetogenesis from hydrogen and carbon dioxide in the sediments of a eutrophic lake. Applied and Environmental Microbiology, 45:1310–1315.

Lovley, D.R. and Klug, M.J., 1986. Model for the distribution of sulfate reduction and methanogenesis in freshwater sediments. Geochimica et Cosmochimica Acta, 50:11–18.

Lovley, D.R. and Lonergan, D.J., 1990. Anaerobic oxidation of toluene, phenol, and p-cresol by the dissimilatory iron-reducing organism, GS-15. Applied and Environmental Microbiology, 56:1858–1864.

Lovley, D.R. and Phillips, E.J.P., 1986. Organic matter mineralization with reduction of ferric iron in anaerobic sediments. Applied and Environmental Microbiology, 51:683–689.

Lovley, D.R. and Phillips, E.J.P., 1987a. Competitive mechanisms for inhibition of sulfate reduction and methane production in the zone of ferric iron reduction in sediments. Applied and Environmental Microbiology, 53:2636–2641.

Lovley, D.R. and Phillips, E.J.P., 1987b. Rapid assay for microbial reducible ferric iron in aquatic sediments. Applied and Environmental Microbiology, 53:1536–1540.

Lovley, D.R. and Phillips, E.J.P., 1988a. Manganese inhibition of microbial iron reduction in anaerobic sediments. Geomicrobiology Journal, 6:145–155.

Lovley, D.R. and Phillips, E.J.P., 1988b. Novel mode of microbial energy metabolism: organic carbon oxidation coupled to dissimilatory reduction of iron and manganese. Applied and Environmental Microbiology, 54:1472–1480.

Lovley, D.R. and Phillips, E.J.P., 1992. Bioremediation of uranium contamination with enzymatic uranium reduction. Environmental Science and Technology, 26:2228–2234.

Lovley, D.R. and Phillips, E.J.P., 1994. Novel processes for anaerobic sulfate production from elemental sulfur by sulfate-reducing bacteria. Applied and Environmental Microbiology, 60:2394–2399.

Lovley, D.R., Dwyer, D.F., and Klug, M.J., 1982. Kinetic analysis of competition between sulfate reducers and methanogens for hydrogen in sediments. Applied and Environmental Microbiology, 43:1373–1379.

Lovley, D.R., Stolz, J.F., Nord, G.L. Jr, and Phillips, E.J.P., 1987. Anaerobic production of magnetite by a dissimilatory iron-reducing microorganism. Nature, 330:252–254.

Lovley, D.R., Chapelle, F.H., and Phillips, E.J.P., 1990. Fe(III)-reducing bacteria in deeply buried sediments of the Atlantic Coastal Plain. Geology, 18:954–957.

Lovley, D.R., Phillips, E.J.P., Gorby, Y.A., and Landa, E.R., 1991. Microbial reduction of uranium. Nature, 350:413–416.

Lovley, D.R., Woodward. J.C., and Chapelle, F.H., 1994. Stimulated anoxic degradation of aromatic hydrocarbons using Fe(III) ligands. Nature, 370:128–131.

Lovley, D.R., Coates, J.D., Blunt-Harris, E.L., Phillips, E.J.P., and Woodward, J.C., 1996a. Humic substances as electron acceptors for microbial respiration. Nature, 382:445–448.

Lovley, D.R., Woodward, J.C., and Chapelle, F.H., 1996b. Rapid anaerobic benzene oxidation with a variety of chelated Fe(III) forms. Applied and Environmental Microbiology, 62:288–291.

Lovley, D.R., Kashefi, K., Vargas, M., Tor, J.M., and Blunt-Harris, E.L., 2000. Reduction of humic substances and Fe(III) by hyperthermophilic microorganisms. Chemical Geology, 169:289–298.

Lovley, D.R., Holmes, D.E. and Nevin, K.P., 2004. Dissimilatory Fe(III) and Mn(IV) reduction. Advances in Microbial Physiology, 49:219–286.

Lowe, D.R., 1980. Stromatolites 3,400-Myr old from the Archean of Western Australia. Nature, 284:441–443.

Lowe, D.R., 1983. Restricted shallow-water sedimentation of early Archean stromatolitic and evaporitic strata of the Strelley Pool Chert, Pilbara Block, Western Australia. Precambrian Research, 19:239–283.

Lowe, D.R., 1994. Early environments: constraints and opportunities for early evolution. In: Bengtson, S. (ed.), Early Life on Earth. Columbia University Press, New York, pp. 24–35.

Lowe, D.R. and Tice, M.M., 2004. Geologic evidence for Archean atmospheric and climatic evolution: fluctuating levels of CO_2, CH_4 and O_2 with an overriding tectonic control. Geology, 32:493–496.

Lowenstam, H.A., 1981. Minerals formed by organisms. Science, 211:1126–1131.

Lowenstam, H.A. and Weiner, S., 1989. On Biomineralization. Oxford University Press, New York.

Lower, S.K., Hochella, M.F. Jr, and Beveridge, T.J., 2001. Bacterial recognition of mineral surfaces: nanoscale interactions between Shewanella and –FeOOH. Science, 292:1360–1363.

Lucas, W.J., 1983. Photosynthetic assimilation of exogenous HCO_3 by aquatic plants. Annual Review of Plant Physiology, 34:71–104.

Luther, G.W. III., 1987. Pyrite oxidation and reduction: molecular orbital theory considerations. Geochimica et Cosmochimica Acta, 51:3193–3199.

Luther, G.W., III., 1991. Pyrite synthesis via polysulfide compounds. Geochimica et Cosmochimica Acta, 55:2839–2849.

Luther, G.W., III, Sundby, B., Lewis, B.L., Brendel, P.J., and Silverberg, N., 1997. Interactions of manganese with the nitrogen cycle. Alternative pathways to dinitrogen. Geochimica et Cosmochimica Acta, 61:4043–4052.

Lütters-Czekalla, S., 1990. Lithoautotrophic growth of the iron bacterium Gallionella ferruginea with thiosulfate or sulfide as energy source. Archives of Microbiology, 154:417–421.

Lutz, R.A., Shank, T.M., Fornari, D.J., et al. 1994. Rapid growth at deep-sea vents. Nature, 371:663–664.

Macaskie, L.E., 1991. The application of biotechnology to the treatment of wastes produced from the nuclear fuel cycle: biodegradation and bioaccumulation as a means of treating radionuclide-containing streams. Critical Reviews in Biotechnology, 11:41–112.

Macaskie, L.E. and Lloyd, J.R., 2002. Microbial interactions with radioactive wastes and potential applications. In: Keith-Roach, M.J. and Livens, F.R. (eds), Interactions of Microorganisms with Radionuclides. Elsevier, Amsterdam, pp. 343–380.

Macaskie, L.E., Empson, R.M., Cheetham, A.K., Grey, C.P., and Skarnulis, A.J., 1992. Uranium bioaccumulation by a Citrobacter sp. as a result of enzymatically mediated growth of polycrystalline HUO_2PO_4. Science, 257:782–784.

Macaskie, L.E., Lloyd, J.R., Thomas, R.A.P., and Tolley, M.R., 1996. The use of microorganisms for the remediation of solutions contaminated with actinide elements, other radionuclides, and organic contaminants generated by nuclear fuel cycle activities. Nuclear Energy, 35:257–271.

Macaskie, L.E., Bonthrone, K.M., Yong, P., and Goddard, D.T., 2000. Enzymatically mediated bioprecipitation of uranium by a Citrobacter sp.: a concerted role of exocellular lipopolysaccharide and associated phosphatase in biomineral formation. Microbiology, 146:1855–1867.

Machel, H.G., 2001. Bacterial and thermochemical sulfate reduction in diagenetic settings – old and new insights. Sedimentary Geology, 140:143–175.

Machel, H.G. and Foght, J., 2000. Products and depth limits of microbial activity in petroliferous subsurface settings. In: Riding, R.E. and Awramik, S.M. (eds), Microbial Sediments. Springer-Verlag, Berlin, pp. 105–120.

Machemer, S.D. and Wildeman, T.R., 1992. Adsorption compared with sulfide precipitation as metal removal processes from acid mine drainage in a constructed wetland. Journal of Contaminant Hydrology, 9:115–131.

Macintyre, I.G., Prufert-Bebout, L., and Reid, R.P., 2000. The role of endolithic cyanobacteria in the formation of lithified laminae in Bahamian stromatolites. Sedimentology, 47:915–921.

Mack, E.E. and Pierson, B.K., 1988. Preliminary characterization of a temperate marine member of the Chloroflexaceae. In: Olsen, J.M., Ormerod, J.G., Amesz, J., Stackebrandt, E., and Trüper, H.G. (eds), Green Photosynthetic Bacteria. Plenum, New York, pp. 237–241.

Macko, S.A., Fogel, M.L., Hare, P.E., and Hoering, T.C., 1987. Isotopic fractionation of nitrogen and carbon in the synthesis of amino acids by microorganisms. Chemical Geology, 65:79–92.

Macko, S.A., Engel, M.H., and Parker, P.L., 1993. Early diagenesis of organic matter in sediments. In: Engel, M.H. and Macko, S.A. (eds), Organic Geochemistry. Plenum Press, New York, pp. 211–224.

MacRae, I.C. and Edwards, J.F., 1972. Adsorption of colloidal iron by bacteria. Applied Microbiology, 24:819–823.

Madigan, M.T. and Ormerod, J.G., 1996. Taxonomy, physiology and ecology of heliobacteria. In: Blankenship, R., Madigan, M.T., and Bauer, C. (eds), Anoxygenic Photosynthetic Bacteria. Kluwer, Dordrecht, pp. 17–30.

Madigan, M.T., Martinko, J.M., and Parker, J., 2003. Brock. Biology of Microorganisms, 10th edn. Prentice Hall, Upper Saddle River, New Jersey.

Mah, R.A., Hungate, R.E., and Ohwaki, K., 1976. Acetate, a key intermediate in methanogenesis. In: Schlegel, H.G. and Barnea, J. (eds), Microbial Energy Conservation. Erich Glotz KG, Göttingen, pp. 97–106.

Majidi, V., Laude, D.A., and Holcombe, J.A., 1990. Investigation of the metal-algae binding site with [113]Cd nuclear magnetic resonance. Environmental Science and Technology, 24:1309–1312.

Malard, F., Reygrobellet, J.-L., and Soulié, M., 1994. Transport and retention of fecal bacteria at sewage-polluted fractured rock sites. Journal of Environmental Quality, 23:1352–1363.

Malasarn, D., Saltikov, C.W., Campbell, K.J.M., Santini, J.M., Hering, J.G., and Newman, D.K., 2004. *arr*A is a reliable marker for As(V) respiration. Science, 306:455.

Maliva, R.G., Knoll, A.H., and Simonson, B.M., 2005. Secular change in the Precambrian silica cycle: insights from chert petrology. GSA Bulletin, 117:835–845.

Mandernack, K.W. and Tebo, B.M., 1993. Manganese scavenging and oxidation at hydrothermal vents and vent plumes. Geochimica et Cosmochimica Acta, 57:3907–3923.

Mandernack, K.W., Post, J., and Tebo, B.M., 1995. Manganese mineral formation by bacterial spores of the marine *Bacillus*, strain SG-1: evidence for the direct oxidation of Mn(II) to Mn(IV). Geochimica et Cosmochimica Acta, 59:4393–4408.

Mann, S., 1988. Molecular recognition in biomineralization. Nature, 332:119–124.

Mann, S., Frankel, R.B., and Blakemore, R.P., 1984. Structure, morphology and crystal growth of bacterial magnetite. Nature, 310:405–407.

Mann, S., Sparks, N.H.C., Frankel, R.B., Bazylinski, D.A., and Jannasch, H.W., 1990. Biomineralization of ferrimagnetic greigite (Fe_3S_4) and iron pyrite (FeS_2) in a magnetotactic bacterium. Nature, 343:258–261.

Mann, S., Archibald, D.D., Didymus, J.M., et al. 1993. Crystallization at inorganic–organic interfaces: biominerals and biomimetic synthesis. Science, 261:1286–1292.

Margulis, L., 1970. Origin of Eukaryotic Cells. Yale University Press, New Haven.

Margulis, L., Walker, J.C.G., and Rambler, M., 1976. Reassessment of roles of oxygen and ultraviolet light in Precambrian evolution. Nature, 264:620–623.

Marquis, R.E., 1968. Salt-induced contraction of bacterial cell walls. Journal of Bacteriology, 95:775–781.

Marquis, R.E., Mayzel, K., and Carstensen, E.L., 1976. Cation exchange in cell walls of gram-positive bacteria. Canadian Journal of Microbiology, 22:975–982.

Marshall, K.C., Stout, R., and Mitchell, R., 1971. Mechanism of the initial events in the sorption of marine bacteria to surfaces. Journal of General Microbiology, 68:337–348.

Martell, A.E. and Smith, R.M., 1977. Critical Stability Constants, III. Other organic ligands. Plenum Press, New York.

Martens, C.S. and Berner, R.A., 1974. Methane production in the interstitial waters of sulfate-depleted sediments. Science, 185:1167–1169.

Martens, C.S. and Klump, J.V., 1984. Biogeochemical cycling in an organic-rich coastal marine basin. 4. An organic carbon budget for sediments dominated by sulfate reduction and methanogenesis. Geochimica et Cosmochimica Acta, 48:1987–2004.

Martill, D.M., 1988. Preservation of fish in the Cretaceous Santana Formation of Brazil. Palaeontology, 31:1–18.

Martin, W. and Müller, M., 1998. The hydrogen hypothesis for the first eukaryote. Nature, 392:37–41.

Martini, J.E.J., 1994. A late Archean-Paleoproterozoic (2.6 Ga) paleosol on ultramafics in the eastern Transvaal, South Africa. Precambrian Research, 67:159–180.

Mastrapa, R.M.E., Glanzberg, H., Head, J.N., Melosh, H.J., and Nicholson, W.L., 2001. Survival of bacteria exposed to extreme acceleration: implications for Panspermia. Earth and Planetary Science Letters, 189:1–8.

Mather, T.A., Pyle, D.M., and Allen, A.G., 2004. Volcanic source for fixed nitrogen in the early Earth's atmosphere. Geology, 32:905–908.

Matis, K.A., Zouboulis, A.I., Grigoriadou, A.A., Lazaridis, N.K., and Ekateriniadou, L.V., 1996. Metal biosorption flotation. Application to cadmium removal. Applied Microbiology and Biotechnology, 45:569–573.

Matte-Tailliez, O., Brochier, C., Forterre, P., and Philippe, H., 2002. Archaeal phylogeny based on ribosomal proteins. Molecular Biology and Evolution, 19:631–639.

Mattey, M., 1992. The production of organic acids. Critical Reviews in Biotechnology, 12:87–132.

Matthews, T.H., Doyle, R.J., and Streips, U.N., 1979. Contribution of peptidoglycan to the binding of metal ions by the cell wall of Bacillus subtilis. Current Microbiology, 3:51–53.

Maurette, M., 1998. Carbonaceous micrometeorites and the origin of life. Origins of Life and Evolution of the Biosphere, 28:385–412.

Mayer, L.M., 1993. Organic matter at the sediment–water interface. In: Engel, M.H. and Macko, S.A. (eds), Organic Geochemistry. Plenum Press, New York, pp. 171–184.

McCabe, C., Sassen, R., and Saffer, B., 1987. Occurrence of secondary magnetite within biodegraded oil. Geology, 15:7–10.

McCollom, T.M., 2000. Geochemical constraints on primary productivity in submarine hydrothermal vent plumes. Deep-Sea Research, 47:85–101.

McGuire, M.M., Edwards, K.J., Banfield, J.F., and Hamers, R.J., 2001. Kinetics, surface chemistry, and structural evolution of microbially mediated sulfide mineral dissolution. Geochimica et Cosmochimica Acta, 65:1243–1258.

McIntosh, J.M., Silver, M., and Groat, L.A., 1997. Bacteria and the breakdown of sulfide minerals. In: McIntosh, J.M. and Groat, L.A. (eds), Biological–Mineralogical Interactions. Mineralogical Association of Canada Short Course Series, vol. 25. Ottawa, pp. 63–92.

McKay, D.S., Gibson, E.K. Jr, Thomas-Keprta, K., et al. 1996. Search for past life on Mars: possible relict biogenic activity in Martian meteorite ALH84001. Science, 273:924–930.

McKibben, M.A. and Barnes, H.L., 1986. Oxidation of pyrite in low temperature acidic solutions: rate laws and surface textures. Geochimica et Cosmochimica Acta, 50:1509–1520.

McMahon, P.B. and Chapelle, F.H., 1991. Microbial production of organic acids in aquitard sediments and its role in aquifer geochemistry. Nature, 349:233–235.

McNamara, K.J. and Awramik, S.M., 1992. Stromatolites: a key to understanding the early evolution of life. Science in Progress, 76:345–364.

Meixner, O., Mischack, H., Kubicek, C.P., and Rohr, M., 1985. Effect of manganese deficiency on plasma-membrane lipid composition and glucose uptake in Aspergillus niger. FEMS Microbiology Letters, 26:271–274.

Melosh, H.J., 1988. The rocky road to Panspermia. Nature, 332:687–688.

Merchant, S. and Sawaya, M.R., 2005. The light reactions: a guide to recent acquisitions for the picture gallery. The Plant Cell, 17:648–663.

Mero, J.L., 1962. Ocean-floor manganese nodules. Economic Geology, 57:747–767.

Merz-Preiß, M., 2000. Calcification in cyanobacteria. In: Riding, R.E. and Awramik, S.M. (eds), Microbial Sediments. Springer-Verlag, Berlin, pp. 51–56.

Meybeck, M., 1979. Concentrations des eaux fluviales en éléments majeurs et apports en solution aux océans. Revue de Géologie Dynamique de Géographie Physique, 21:215–246.

Michaelis, W., Seifert, R., Nauhaus, K., et al. 2002. Microbial reefs in the Black Sea fueled by anaerobic oxidation of methane. Science, 297:1013–1015.

Michalopoulos, P. and Aller, R.C., 1995. Rapid clay mineral formation in Amazon Delta sediments: reverse weathering and oceanic elemental cycles. Science, 270:614–617.

Middleburg, J.J., de Lange, G.J., and van der Weijden, C.H., 1987. Manganese solubility control in marine pore waters. Geochimica et Cosmochimica Acta, 51:759–763.

Middleburg, J.J., Vlug, T., Jaco, F., and van der Nat, W.A., 1993. Organic matter mineralization in marine systems. Global and Planetary Change, 8:47–58.

Mielke, R.E., Pace, D.L., Porter, T., and Southam, G., 2003. A critical stage in the formation of acid mine drainage: colonization of pyrite by *Acidithiobacillus ferrooxidans* under pH-neutral conditions. Geobiology, 1:81–90.

Mileikowsky, C., Cucinotta, F.A., Wilson, J.W., et al. 2000. Natural transfer of viable microbes in space – 1. From Mars to Earth and Earth to Mars. Icarus, 145:391–427.

Miller, P.C., 1997. The design and operating practice of bacterial oxidation plant using moderate thermophiles (The BacTech Process). In: Rawlings, D.E. (ed.), Biomining. Theory, microbes and industrial processes. Springer-Verlag, Berlin, pp. 81–102.

Miller, S.L., 1953. A production of amino acids under possible primitive Earth conditions. Science, 117:528–529.

Miller, S.L., 1957. The mechanism of synthesis of amino acids by electric discharges. Biochimica et Biophysica Acta, 23:480–487.

Miller, S.L. and Bada, J.L., 1988. Submarine hot springs and the origin of life. Nature, 334:609–611.

Miller, S.L. and Orgel, L.E., 1974. The Origins of Life on Earth. Prentice-Hall, Englewood Cliffs, New Jersey.

Milligan, A.J. and Morel, F.M.M., 2002. A proton buffering role for silica in diatoms. Science, 297:1848–1850.

Mills, A.L., Herman, J.S., Hornberger, G.M., and DeJesús, T.H., 1994. Effect of solution ionic strength and iron coatings on mineral grains on the sorption of bacterial cells to quartz sand. Applied and Environmental Microbiology, 60:3300–3306.

Mitchell, J.G., Martinez-Alonso, M., Lalucat, J., Esteve, I., and Brown, S., 1991. Velocity changes, long runs, and reversals in the *Chromatium minus* swimming response. Journal of Bacteriology, 173:997–1003.

Mitchell, P., 1961. Coupling of phosphorylation to electron and hydrogen transfer by a chemi-osmotic type of mechanism. Nature, 191:144–148.

Mitchell, P., 1966. Chemiosmotic coupling in oxidative and photosynthetic phosphorylation. Biology Reviews of the Cambridge Philosophical Society, 41:445–502.

Moat, A.G., Foster, J.W., and Spector, M.P., 2002. Microbial Physiology, 4th edn. Wiley-Liss Inc., New York.

Moffett, J.W. and Brand, L.E., 1996. Production of strong, extracellular Cu chelators by marine cyanobacteria in response to Cu stress. Limnology and Oceanography, 41:388–395.

Mohagheghi, A., Updegraff, D.M., and Goldhaber, M.B., 1985. The role of sulfate-reducing bacteria in the deposition of sedimentary uranium ores. Geomicrobiology Journal, 4:153–173.

Mojzsis, S.J., Arrhenius, G., McKeegan, K.D., Harrison, T.M., Nutman, A.P., and Friend, C.R.L., 1996. Evidence for life on Earth before 3,800 million years ago. Nature, 384:55–59.

Mojzsis, S.J., Coath, C.D., Greenwood, J.P., McKeegan, K.D., and Harrison, T.M., 2003. Mass-independent isotope effects in Archean (2.5 to 3.8 Ga) sedimentary sulfides determined by ion microprobe analysis. Geochimica et Cosmochimica Acta, 67:1635–1658.

Monty, C.L.V., 1976. The origin and development of cryptalgal fabrics. In: Walter, M.R. (ed.), Stromatolites, Developments in Sedimentology. Elsevier, Amsterdam, pp. 193–250.

Moore-Landecker, E., 1996. Fundamentals of the Fungi. Prentice Hall, Upper Saddle River, New Jersey.

Morales, J., Esparza, P., González, S., Salvarezza, R., and Arévalo, M.P., 1993. The role of *Pseudomonas aeruginosa* on the localized corrosion of 304 stainless steel. Corrosion Science, 34:1531–1540.

Moreira, D. and López-García, P., 2002. The molecular ecology of microbial eukaryotes unveils a hidden world. Trends in Microbiology, 10:31–38.

Moreno-Vivián, C., Cabello, P., Martínez-Luque, M., Blasco, R., and Castillo, F., 1999. Prokaryotic nitrate reduction: molecular properties and functional distinction among nitrate reductases. Journal of Bacteriology, 181:6573–6584.

Morris, R.C., 1993. Genetic modelling for banded iron-formation of the Hamersley Group, Pilbara Craton, Western Australia. Precambrian Research, 60:243–286.

Morris, R.C. and Horwitz, R.C., 1983. The origin of the iron-formation-rich Hamersley Group of Western Australia – deposition on a platform. Precambrian Research, 21:273–297.

Morse, J.W., 1983. The kinetics of calcium carbonate dissolution and precipitation. In: Reeder, R.J. (ed.), Carbonates: mineralogy and chemistry reviews, vol. 11. Mineralogical Society of America, Washington, DC, pp. 227–264.

Morse, J.W. and Arvidson, R.S., 2002. The dissolution kinetics of major sedimentary carbonate minerals. Earth-Science Reviews, 58:51–84.

Morse, J.W. and Casey, W.H., 1988. Ostwald processes and mineral paragenesis in sediments. American Journal of Science, 288:537–560.

Mortimer, R.J.G., Coleman, M.L., and Rae, J.E., 1997. Effect of bacteria on the elemental composition of early diagenetic siderite: implications for palaeoenvironmental interpretations. Sedimentology, 44:759–765.

Mortimer, R.J.G., Davey, J.T., Krom, M.D., Watson, P.G., Frickers, P.E., and Clifton, R.J., 1999. The effect of macrofauna on pore water profiles and nutrient fluxes in the intertidal zone of the Humber Estuary. Estuarine, Coastal and Shelf Science, 48:683–699.

Moser, D.P. and Nealson, K.H., 1996. Growth of the facultative anaerobe *Shewanella putrefaciens* by elemental sulfur reduction. Applied and Environmental Microbiology, 62:2100–2105.

Moses, C.O. and Herman, J.S., 1991. Pyrite oxidation at circumneutral pH. Geochimica et Cosmochimica Acta, 55:471–482.

Moses, C.O., Nordstrom, D.K., Herman, J.S., and Mills, A.L., 1987. Aqueous pyrite oxidation by dissolved oxygen and by ferric iron. Geochimica et Cosmochimica Acta, 51:1561–1571.

Moskowitz, B.M., Frankel, R.B., Bazylinski, D.A., Jannasch, H.W., and Lovley, D.R., 1989. A comparison of magnetite particles produced anaerobically by magnetotactic and dissimilatory iron-reducing bacteria. Geophysical Research Letters, 16:665–668.

Moule, A.L. and Wilkinson, S.G., 1989. Composition of lipopolysaccharides from *Altermonas putrefaciens* (*Shewanella putrefaciens*). Journal of General Microbiology, 135:163–173.

Mozes, N., Léonard, A.J., and Rouxhet, P.G., 1988. On the relations between elemental surface composition of yeasts and bacteria and their charge and hydrophobicity. Biochimica et Biophysica Acta, 945:324–334.

Mukhin, L.M., Gerasimov, M.V., and Safonova, I.N., 1989. Origin of precursors of organic molecules during evaporation of meteorites and rocks. Advances in Space Research, 9:95–97.

Mullen, M.D., Wolf, D.C., Ferris, F.G., Beveridge, T.J., Flemming, C.A., and Bailey, G.W., 1989. Bacterial sorption of heavy metals. Applied and Environmental Microbiology, 55: 3143–3149.

Müller, G. and Raymond, K.N., 1984. Specificity and mechanism of ferrioxamine-mediated iron transport in *Streptomyces pilosus*. Journal of Bacteriology, 160:304–312.

Murray, J.W., 1991. Ecology and Palaeoecology of Benthic Foraminifera. Longman, Essex, UK.

Murray, J.W. and Grundmanis, V., 1980. Oxygen consumption in pelagic marine sediments. Science, 209:1527–1530.

Mustin, C., Berthelin, J., Marion, P., and de Donato, P., 1992. Corrosion and electrochemical oxidation of pyrite by *Thiobacillus ferrooxidans*. Applied and Environmental Microbiology, 58:1175–1182.

Myers, K.H. and Nealson, K.H., 1988a. Bacterial manganese reduction and growth with manganese oxide as the sole electron donor. Science, 240:1319–1321.

Myers, K.H. and Nealson, K.H., 1988b. Microbial reduction of manganese oxides: interactions with iron and sulfur. Geochimica et Cosmochimica Acta, 52:2727–2732.

Nakajima, A., 2003. Accumulation of gold by microorganisms. World Journal of Microbiology and Biotechnology, 19:369–374.

Narbonne, G.M., 2004. Modular construction of early Ediacaran complex life forms. Science, 305:1141–1144.

Nauhaus, K., Boetius, A., Krüger, M., and Widdel, F., 2002. *In vitro* demonstration of anaerobic oxidation of methane coupled to sulfate reduction in sediment from a marine gas hydrate area. Environmental Microbiology, 4:296–305.

Navarro-González, R., McKay, C.P., and Mvondo, D.N., 2001. A possible nitrogen crisis for Archaean life due to reduced nitrogen fixation by lightening. Nature, 412:61–64.

Nealson, K.H., 1982. Microbiological oxidation and reduction of iron. In: Holland, H.D. and Schidlowski, M. (eds), Mineral Deposits and the Evolution of the Biosphere. Springer-Verlag, New York, pp. 51–66.

Nealson, K.H. and Saffarini, D., 1994. Iron and manganese in anaerobic respiration: environmental significance, physiology and regulation. Annual Reviews of Microbiology, 48:311–343.

Nealson, K.H. and Stahl, D.A., 1997. Microorganisms and biogeochemical cycles: what can we learn from layered microbial communities. In: Banfield, J.F. and Nealson, K.H. (eds), Geomicrobiology: interactions between microorganisms and minerals, vol. 35. Mineralogical Society of America, Washington, DC, pp. 5–34.

Nealson, K.H., Tebo, B.M., and Rosson, R.A., 1988. Occurrence and mechanisms of microbial oxidation of manganese. Advances in Applied Microbiology, 33:279–318.

Nealson, K.H., Rosson, R.A., and Myers, C.R., 1989. Mechanisms of oxidation and reduction of manganese. In: Beveridge, T.J. and Doyle, R.J. (eds), Metal Ions and Bacteria. John Wiley, New York, pp. 383–411.

Neihof, R.A. and Loeb, G.I., 1972. The surface charge of particulate matter in seawater. Limnology and Oceanography, 17:7–16.

Neilands, J.B., 1989. Siderophore systems of bacteria and fungi. In: Poindexter, J.S. and Leadbetter, E.R. (eds), Bacteria in Nature. Plenum Press, New York, pp. 141–163.

Nelson, D.C. and Castenholz, R.W., 1981. Use of reduced sulfur compounds by Beggiatoa sp. Journal of Bacteriology, 147:140–154.

Nelson, D.C. and Fisher, C.R., 1995. Chemoautotrophic and methanotrophic endosymbiotic bacteria at deep-sea vents and seeps. In: Karl, D.M. (ed.), The Microbiology of Deep-Sea Hydrothermal Vents. CRC Press, Boca Raton, pp. 125–167.

Nelson, D.C., Jørgensen, B.B., and Revsbech, N.P., 1986. Growth pattern and yield of a chemoautotrophic Beggiatoa sp. in oxygen-sulfide microgradients. Applied and Environmental Microbiology, 52:225–233.

Nelson, D.C., Wirsen, C.O., and Jannasch, H.W., 1989. Characterization of large autotrophic Beggiatoa spp. abundant at hydrothermal vents of the Guaymas Basin. Applied and Environmental Microbiology, 55:2909–2917.

Nelson, D.L. and Cox, M.M., 2005. Lehninger: principles of biochemistry, 4th edn. W.H. Freeman, New York.

Nelson, N. and Ben-Shem, A., 2004. The complex architecture of oxygenic photosynthesis. Nature Reviews, 5:1–12.

Nesbitt, H.W., 1997. Bacterial and inorganic weathering processes and weathering of crystalline rocks. In: McIntosh, J.M. and Groat, L.A. (eds), Biological–Mineralogical Interactions. Mineralogical Association of Canada Short Course Series, vol. 25, Ottawa, pp. 113–142.

Nesbitt, H.W. and Muir, I.J., 1994. X-ray electron spectroscopic study of a pristine pyrite surface reacted with water vapour and air. Geochimica et Cosmochimica Acta, 58:4667–4679.

Neubauer, H. and Götz, F., 1996. Physiology and interaction of nitrate and nitrite reduction in Staphylococcus carnosus. Journal of Bacteriology, 178:2005–2009.

Neumann, T., Heiser, U., Leosson, M.A., and Kersten, M., 2002. Early diagenetic processes during Mn-carbonate formation: evidence from isotopic composition of authigenic Ca-rhodochrosites of the Baltic Sea. Geochimica et Cosmochimica Acta, 66:867–879.

Nevin, K.P. and Lovley, D.R., 2000. Potential for nonenzymatic reduction of Fe(III) via electron shuttling in subsurface sediments. Environmental Science and Technology, 34:2472–2478.

Newman, D.K. and Banfield, J.F., 2002. Geomicrobiology: how molecular-scale interactions underpin biogeochemical systems. Science, 296:1071–1077.

Newman, D.K. and Kolter, R., 2000. A role for excreted quinones in extracellular electron transfer. Nature, 405:94–97.

Newman, D.K., Beveridge, T.J., and Morel, F.M.M., 1997. Precipitation of arsenic trisulfide by Desulfotomaculum auripigmentum. Applied and Environmental Microbiology, 63:2022–2028.

Newton, W.E. and Burgess, B.K., 1983. Nitrogen fixation: its scope and importance. In: Mueller, A. and Newton, W.E. (eds), Nitrogen Fixation. The chemical–biochemical–genetic interface. Plenum Press, New York, pp. 1–19.

Nielsen, A.M. and Beck, J.V., 1972. Chalcocite oxidation and coupled carbon dioxide fixation by Thiobacillus ferrooxidans. Science, 175:1124–1126.

Nielson, A.E. and Söhnel, O., 1971. Interfacial tensions electrolyte crystal-aqueous solutions, from nucleation data. Journal of Crystal Growth, 11:233–242.

Nielson, P.H., Jahn, A., and Palmgren, R., 1997. Conceptual model for production and composition of exopolymers in biofilms. Water Science and Technology, 36:11–19.

Nies, D.H., 2000. Heavy metal-resistant bacteria as extremophiles: molecular physiology and biotechnological use of Ralstonia sp. CH34. Extremophiles, 4:77–82.

Niewöhner, C., Hensen, C., Kasten, S., Zabel, M., and Schulz, H.D., 1998. Deep sulfate reduction completely mediated by anaerobic methane oxidation in sediments of the upwelling area off Namibia. Geochimica et Cosmochimica Acta, 62:455–464.

Nilsen, R.K., Beeder, J., Thorstenson, T., and Torsvik, T., 1996. Distribution of thermophilic marine sulfate reducers in North Sea oil field waters and oil reservoirs. Applied and Environmental Microbiology, 62:1793–1798.

Nisbet, E.G., 1995. Archaean ecology: a review of evidence for the early development of bacterial biomes, and speculations on the development of a global-scale biosphere. In: Coward, M.P. and Ries, A.C. (eds), Early Precambrian Processes. Geological Society Special Publication no. 95, pp. 27–51.

Nisbet, E.G. and Fowler, C.M.R., 1999. Archaean metabolic evolution of microbial mats. Proceedings of the Royal Society of London, 266:2375–2382.

Nisbet, E.G. and Sleep, N.H., 2001. The habitat and nature of early life. Nature, 409:1083–1091.

Nisbet, E.G., Cann, J.R., and van Dover, C.L., 1995. Origin of photosynthesis. Nature, 373:479–480.

Nitschke, W., Sétif, P., Liebl, U., Feiler, U., and Rutherford, A.W., 1990. Reaction center photochemistry of *Heliobacterium chlorum*. Biochemistry, 29:11079–11088.

Nordstrom, D.K., 1982. Aqueous pyrite oxidation and the consequent formation of secondary iron minerals. In: Kittrick, J.A., Fanning, D.S., and Hosner, L.R. (eds), Acid Sulfate Weathering, Soil Science Society of America, Madison, Wisconsin, pp. 37–56.

Nordstrom, D.K. and Southam, G., 1997. Geomicrobiology of sulfide mineral oxidation. In: Banfield, J.F. and Nealson, K.H. (eds), Geomicrobiology: interactions between microbes and minerals, vol. 35. Mineralogical Society of America, Washington, DC, pp. 361–390.

Norman, P.F. and Snyman, C.P., 1988. The biological and chemical leaching of an auriferous pyrite/arsenopyrite flotation concentrate: a microscopic examination. Geomicrobiology Journal, 6:1–10.

Norton, C.F. and Grant, W.D., 1988. Survival of halobacteria within fluid inclusions in salt crystals. Journal of General Microbiology, 134:1365–1373.

Novelli, P.C., Michelson, A.R., Scranton, M.I., Banta, G.T., Hobbie, J.E., and Howarth, R.W., 1988. Hydrogen and acetate in two sulfate-reducing sediments: Buzzards Bay and Town Cove, Mass. Geochimica et Cosmochimica Acta, 52:2477–2486.

Nübel, U., Bateson, M.M., Vandieken, V., Wieland, A., Kühl, M., and Ward, D.M., 2002. Microscopic examination of distribution and phenotypic properties of phylogenetically diverse *Chloroflexaceae*-related bacteria in hot spring microbial mats. Applied and Environmental Microbiology, 68:4593–4603.

Nutman, A.P., Mojzsis, S.J., and Friend, C.R.L., 1997. Recognition of > 3850 Ma water-lain sediments in West Greenland and their significance for the early Archaean Earth. Geochimica et Cosmochimica Acta, 61:2475–2484.

Ochman, H., Lawrence, J.G., and Groisman, E.A., 2000. Lateral gene transfer and the nature of bacterial innovation. Nature, 405:299–304.

Oehler, J.H., 1976. Experimental studies in Precambrian paleontology: structural and chemical changes in blue-green algae during simulated fossilization in synthetic chert. Geological Society of America Bulletin, 87:117–129.

Ohmoto, H., 1996. Evidence in pre-2.2 Ga paleosols for the early evolution of atmospheric oxygen and terrestrial biota. Geology, 24:1135–1138.

Ohmoto, H., Kakegawa, T., and Lowe, D.R., 1993. 3.4-billion-year-old biogenic pyrites from Barberton, South Africa: sulfur isotope evidence. Science, 262:555–557.

Ohmoto, H., Watanabe, Y., and Kumazawa, K., 2004. Evidence from massive siderite beds for a CO_2-rich atmosphere before ~1.8 billion years ago. Nature, 429:395–399.

Ollivier, B., Caumette, P., Garcia, J.-L., and Mah, R.A., 1994. Anaerobic bacteria from hypersaline environments. Microbiological Reviews, 58:27–38.

Olsen, G.J. and Woese, C.R., 1996. Lessons from an archaeal genome: what are we learning from *Methanococcus jannaschii*? Trends in Genetics, 12:377–379.

Olsen, G.J., Lane, D.J., Giovannoni, S.J., Pace, N.R., and Stahl, D.A., 1986. Microbial ecology and evolution: a ribosomal RNA approach. Annual Review of Microbiology, 40:337–365.

Olson, G.J., McFeters, G.A., and Temple, K.L., 1981. Occurrence and activity of iron oxidizing and sulfur oxidizing microorganisms in alkaline coal strip mine spoils. Microbial Ecology, 7:39–50.

O'Neill, J.G. and Wilkinson, J.F., 1977. Oxidation of ammonia by methane-oxidizing bacteria and the effects of ammonia on methane oxidation. Journal of General Microbiology, 100:407–412.

Orcutt, B.N., Boetius, A., Lugo, S.K., MacDonald, I.R., Samarkin, V.A., and Joye, S.B., 2004. Life at the edge of methane ice: microbial cycling of carbon and sulfur in Gulf of Mexico gas hydrates. Chemical Geology, 205:239–251.

Oremland, R.S. and Stolz, J.F., 2003. The ecology of arsenic. Science, 300:939–944.

Oremland, R.S., Marsh, L.M., and Polcin, S., 1982. Marine production and simultaneous sulfate reduction in anoxic, salt marsh sediments. Nature, 296:143–145.

Oren, A. and Shilo, M., 1979. Anaerobic heterotrophic dark metabolism in the cyanobacterium *Oscillatoria limnetica*: sulfur respiration and lactate fermentation. Archives of Microbiology, 122:77–84.

Orgel, L.E., 1998. Polymerization on the rocks: theoretical introduction. Origins of Life and Evolution of the Biosphere, 28:227–234.

Ormerod, J., 1992. Physiology of the photosynthetic prokaryotes. In: Mann, N.H. and Carr, N.G. (eds), Photosynthetic Prokaryotes. Plenum Press, New York, pp. 93–120.

Ormerod, J., 2003. "Every dogma has its day": a personal look at carbon metabolism in photosynthetic bacteria. Photosynthesis Research, 76:135–143.

Oró, J., 1961. Comets and the formation of biochemical compounds on the primitive Earth. Nature, 190:389–390.

Oró, J., 1994. Early chemical stages in the origin of life. In: Bengtson, S. (ed.), Early Life on Earth, Nobel Symposium No. 84. Columbia University Press, New York, pp. 48–59.

Orphan, V.J., House, C.H., Hinrichs, K.-U., McKeegan, K.D., and DeLong, E.F., 2001. Methane-consuming Archaea revealed by directly coupled isotopic and phylogenetic analysis. Science, 293:484–487.

Pace, N.R., Stahl, D.A., Lane, D.J., and Olsen, G.J., 1986. The analysis of natural microbial populations by ribosomal RNA sequences. Advances in Microbial Ecology, 9:1–55.

Padan, E., Zilberstein, D., and Schuldiner, S., 1981. pH homeostasis in bacteria. Biochimica et Biophysica Acta, 650:151–166.

Paerl, H.W. and Pinkney, J.L., 1996. A mini-review of microbial consortia: their roles in aquatic production and biogeochemical cycling. Microbial Ecology, 31:225–247.

Paerl, H.W., Steppe, T.F., and Reid, R.P., 2001. Bacterially mediated precipitation in marine stromatolites. Environmental Microbiology, 3:123–130.

Page, W.J., 1993. Growth conditions for the demonstration of siderophores and iron-repressible outer membrane proteins in soil bacteria, with an emphasis on free-living diazotrophs. In: Barton, L.L. and Hemming, B.C. (eds), Iron Chelation in Plants and Soil Microorganisms. Academic Press, New York, pp. 75–110.

Page, W.J. and Huyer, M., 1984. Derepression of the *Azotobacter vinelandii* siderophore system, using iron-containing minerals to limit iron repletion. Journal of Bacteriology, 158:496–502.

Pandey, A. and Mann, M., 2000. Proteomics to study genes and genomes. Nature, 405:837–846.

Paradis, S., Jonasson, I.R., Le Cheminant, G.M., and Watkinson, D.H., 1988. Two zinc-rich chimneys from the plume site, Southern Juan de Fuca. Canadian Mineralogist, 26:637–654.

Parkes, R.J., Cragg, B.A., Bale, S.J., et al. 1994. Deep bacterial biosphere in Pacific Ocean sediments. Nature, 371:410–413.

Parsons, I., Lee, M.R., and Smith, J.V., 1998. Biochemical evolution II: origin of life in tubular microstructures on weathered feldspar surfaces. Proceedings of the National Academy of Sciences, USA, 95:15173–15176.

Pasteris, J.D. and Wopenka, B., 2003. Necessary, but not sufficient: Raman identification of disordered carbon as a signature of ancient life. Astrobiology, 3:727–738.

Patel, G.B., Sprott, G.D., Humphreys, R.W., and Beveridge, T.J., 1986. Comparative analyses of the sheath structures of *Methanothrix concilii* GP6 and *Methanospirillum hungatei* strains GP1 and JF1. Canadian Journal of Microbiology, 32:623–631.

Pavlov, A.A. and Kasting, J.F., 2002. Mass-independent fractionation of sulfur isotopes in Archean sediments: strong evidence for an anoxic Archean atmosphere. Astrobiology, 2:27–41.

Pavlov, A.A., Kasting, J.F., Eigenbrode, J.L., and Freeman, K.H., 2001. Organic haze in Earth's early atmosphere: source of low-^{13}C late Archean kerogens? Geology, 29:1003–1006.

Pavlov, A.A., Hurtgen, M.T., Kasting, J.F., and Arthur, M.A., 2003. Methane-rich Proterozoic atmosphere? Geology, 31:87–90.

Peck, H.D., 1993. Bioenergetic strategies of the sulfate-reducing bacteria. In: Odom, J.M. and Singleton, R. (eds), Sulfate-Reducing Bacteria: contemporary perspectives. Springer-Verlag, New York, pp. 41–76.

Peckmann, J. and Thiel, V., 2004. Carbon cycling at ancient methane-seeps. Chemical Geology, 205:443–467.

Pedersen, K., 1993. The deep subterranean biosphere. Earth-Science Reviews, 34:243–260.

Pedersen, T.F. and Calvert, S.E., 1990. Anoxia vs. productivity: what controls the formation of organic-rich sediments and sedimentary rocks? American Association for Petroleum Geology Bulletin, 74:454–466.

Pentecost, A., 1978. Blue-green algae and freshwater carbonate deposits. Proceedings of the Royal Society of London, 200:43–61.

Pentecost, A., 1985. Association of cyanobacteria with tufa deposits: identity, enumeration, and nature of the sheath material revealed by histochemistry. Geomicrobiology Journal, 4:285–298.

Pentecost, A., 1992. Growth and distribution of endolithic algae in some North Yorkshire streams (UK). British Journal of Phycology, 27:145–151.

Pentecost, A. and Riding, R., 1986. Calcification in cyanobacteria. In: Leadbeater, B.S.C. and Riding, R. (eds), Biomineralization in Lower Plants and Animals. Systematic Association Special Volume, 30:73–90.

Perdue, E.M., 1978. Solution thermochemistry of humic substances 1. Acid base equilibria of humic acid. Geochimica et Cosmochimica Acta, 42:1351–1358.

Peretó, J.G., Velasco, A.M., Becerra, A., and Lazcano, A., 1999. Comparative biochemistry of CO_2 fixation and the evolution of autotrophy. International Microbiology, 2:3–10.

Perfettini, J.V., Revertegat, E., and Langomazino, N., 1991. Evaluation of cement degradation by the metabolic products of two fungal strains. Experientia, 47:527–533.

Perry, K., 1995. Sulfate-reducing bacteria and immobilization of metals. Marine Georesources and Geotechnology, 13:33–39.

Petersen, N., von Dobeneck, T., and Vali, H., 1986. Fossil bacterial magnetite in deep-sea sediments from the South Atlantic Ocean. Nature, 320:611–615.

Petri, R. and Imhoff, J.F., 2000. The relationship of nitrate reducing bacteria on the basis of narH gene sequences and comparison of narH and 16S rRNA based phylogeny. Systematic and Applied Microbiology, 23:47–57.

Pfennig, N., 1978. General physiology and ecology of photosynthetic bacteria. In: Clayton, R.K. and Sistrom, W.R. (eds), The Photosynthetic Bacteria. Plenum, New York, pp. 3–18.

Philippe, H. and Laurent, J., 1998. How good are deep phylogenetic trees? Current Opinion in Genetics and Development, 8:616–623.

Phoenix, V.R., Konhauser, K.O., and Adams, D.G., 1999. Photosynthetic controls on the silicification of cyanobacteria. In: Ármannsson, H. (ed.), Geochemistry of the Earth's Surface. Balkema, Rotterdam, pp. 275–278.

Phoenix, V.R., Adams, D.G., and Konhauser, K.O., 2000. Cyanobacterial viability during hydrothermal biomineralization. Chemical Geology, 169:329–338.

Phoenix, V.R., Konhauser, K.O., Adams, D.G., and Bottrell, S.H., 2001. The role of biomineralization as an ultraviolet shield: implications for Archean life. Geology, 29:823–826.

Phoenix, V.R., Martinez R.E., Konhauser, K.O., and Ferris F.G., 2002. Characterization and implications of the cell surface reactivity of *Calothrix* sp. Strain KC97. Applied and Environmental Microbiology, 68:4827–4834.

Phoenix, V.R., Konhauser, K.O., and Ferris, F.G., 2003. Experimental study of iron and silica immobilization by bacteria in mixed Fe–Si systems: implications for microbial silicification in hot-springs. Canadian Journal of Earth Sciences, 40:1669–1678.

Pierson, B.K., 1994. The emergence, diversification, and role of photosynthetic eubacteria. In: Bengtson, S. (ed.), Early Life on Earth. Columbia University Press, New York, pp. 161–180.

Pierson, B.K. and Parenteau, M.N., 2000. Phototrophs in high iron microbial mats: microstructure of mats in iron-depositing hot springs. FEMS Microbiology Ecology, 32:181–196.

Pierson, B.K., Oesterle, A., and Murphy, G.L., 1987. Pigments, light penetration, and photosynthetic activity in the multilayered microbial mats of Great Sippewisset Salt Marsh, Massachusetts. FEMS Microbial Ecology, 45:365–376.

Pierson, B.K., Mitchell, H.K., and Ruff-Roberts, A.L., 1993. *Chloroflexus aurantiacus* and ultraviolet radiation: implications for Archean shallow-water stromatolites. Origins of Life and Evolution of the Biosphere, 23:243–260.

Pike, J. and Kemp, A.E.S., 1999. Diatom mats in Gulf of California sediments: implications for the palaeoenvironmental interpretation of laminated sediments and silica burial. Geology, 27:311–314.

Pinti, D.L., Hashizume, K., and Matsuda, J.-I., 2001. Nitrogen and argon signatures in 3.8 to 2.8 Ga metasediments: clues on the chemical state of the Archean ocean and the deep biosphere. Geochimica et Cosmochimica Acta, 65:2301–2315.

Pitsch, S., Eschenmoser, A., Gedulin, B., Hui, S., and Arrhenius, G., 1995. Mineral induced formation of sugar phosphates. Origins of Life and Evolution of the Biosphere, 25:297–334.

Plette, A.C.C., Riemsdijk, W.H., Benedetti, M.F., and van der Wal, A., 1995. pH dependent charging behavior of isolated walls of a gram-positive soil bacterium. Journal of Colloid and Interface Science, 173:354–363.

Plette, A.C.C., Benedetti, M.F., and van Riemsdijk, W.H., 1996. Competitive binding of protons, calcium, cadmium, and zinc to isolated walls of a gram-positive soil bacterium. Environmental Science and Technology, 30:1902–1910.

Pollock, D.E., 1997. The role of diatoms, dissolved silicate and Antarctic glaciation in glacial/interglacial climatic change: a hypothesis. Global and Planetary Change, 14:113–125.

Popa, R., Kinkle, B.K., and Badescu, A., 2004. Pyrite framboids as biomarkers for iron–sulfur systems. Geomicrobiology Journal, 21:193–206.

Pósfai, M., Buseck, P.R., Bazylinski, D.A., and Frankel, R.B., 1998. Iron sulfides from magnetotactic bacteria: structure, composition, and phase transitions. American Mineralogist, 83:1469–1481.

Postma, D., 1982. Pyrite and siderite formation in brackish and freshwater swamp sediments. American Journal of Science, 282:1151–1183.

Postma, D. and Jakobsen, R., 1996. Redox zonation: equilibrium constraints on the Fe(III)/SO$_4^-$ reduction interface. Geochimica et Cosmochimica Acta, 60:3169–3175.

Potts, M., 1994. Dessication tolerance of prokaryotes. Microbiological Reviews, 58:755–805.

Poulton, S.W., Krom, M.D., and Raiswell, R., 2004a. A revised scheme for the reactivity of iron (oxyhydr)oxide minerals towards dissolved sulfide. Geochimica et Cosmochimica Acta, 68:3703–3715.

Poulton, S.W., Fralick, P.W., and Canfield, D.E., 2004b. The transition to a sulfidic ocean ~1.84 billion years ago. Nature, 431:173–177.

Pratt, A.R., Nesbitt, H.W., and Muir, I.J., 1994. Generation of acids from mine waste: oxidative leaching of pyrrhotite in dilute H$_2$SO$_4$ solutions at pH 3.0. Geochimica et Cosmochimica Acta, 58:5147–5159.

Prévôt, L., El Faleh, E.M., and Lucas, J., 1989. Details on synthetic apatites formed through bacterial mediation mineralogy and chemistry products. Sciences Géologiques Bulletin, 42:237–254.

Price, P.B., 2000. A habitat for psychrophiles in deep Antarctic ice. Proceedings of the National Academy of Sciences, USA, 97:1247–1251.

Priscu, J.C., Adams, E.E., Lyons, W.B., et al., 1999. Geomicrobiology of subglacial ice above Lake Vostok, Antarctica. Science, 286:2141–2143.

Pronk, J.T., de Bruyn, J.C., Bos, P., and Kuenen, J.G., 1992. Anaerobic growth of *Thiobacillus ferrooxidans*. Applied and Environmental Microbiology, 58:2227–2230.

Pulford, I.D., 1991. A review of methods to control acid mine drainage generation in pyritic coal mine wastes. In: Davies, M.C.R. (ed.), Land Reclamation. Elsevier, London, pp. 269–278.

Purves, W.K. and Orians, G.H., 1983. Life: The Science of Biology. Sinauer, Sunderland, Massachusetts.

Pye, K., Dickson, J.A.D., Schiavon, N., Coleman, M.L., and Cox, M., 1990. Formation of siderite-Mg calcite-iron sulfide concretions in intertidal marsh and sandflat sediments, north Norfolk, England. Sedimentology, 37:325–343.

Quesada, A. and Vincent, W.F., 1997. Strategies of adaptation by Antarctic cyanobacteria to ultraviolet radiation. European Journal of Phycology, 32:335–342.

Racki, G. and Cordey, F., 2000. Radiolarian palaeoecology and radiolarites: is the present the key to the past? Earth-Science Reviews, 52:83–120.

Radajewski, S., Ineson, P., Parekh, N.R., and Murrell, J.C., 2000. Stable-isotope probing as a tool in microbial ecology, 403:646–649.

Ragsdale, S.W., 1991. Enzymology of the acetyl-CoA pathway of CO$_2$ fixation. Critical Reviews in Biochemistry and Molecular Biology, 26:261–300.

Raiswell, R., 1997. A geochemical framework for the application of stable sulfur isotopes to fossil pyritization. Journal of the Geological Society of London, 154:343–356.

Raiswell, R. and Berner, R., 1985. Pyrite formation in euxinic and semi-euxinic sediments. American Journal of Science, 285:710–724.

Raiswell, R. and Berner, R.A., 1986. Pyrite and organic matter in Phanerozoic normal marine shales. Geochimica et Cosmochimica Acta, 50:1967–1976.

Raiswell, R., Buckley, F., Berner, R.A., and Anderson, T.F., 1988. Degree of pyritization of iron as a palaeoenvironmental indicator of bottom-water oxygenation. Journal of Sedimentary Petrology, 58:812–819.

Ramirez, A.J. and Rose, A.W., 1992. Analytical geochemistry of organic phosphorus and its correlation with organic carbon in marine and fluvial sediments and soils. American Journal of Science, 292:421–454.

Ramsing, N.B., Ferris, M.J., and Ward, D.M., 2000. Highly ordered vertical structure of *Synechococcus* populations within the one-millimeter-thick photic zone of a hot spring cyanobacterial mat. Applied and Environmental Microbiology, 66:1038–1049.

Rasmussen, B., 2000. Filamentous microfossils in a 3,235-million-year-old volcanogenic massive sulfide deposit. Nature, 405:676–679.

Rasmussen, B. and Buick, R., 1999. Redox state of the Archean atmosphere: evidence from detrital heavy minerals in ca. 3250–2750 Ma sandstones from the Pilbara Craton, Australia. Geology, 27:115–118.

Raven, J.A., 1983. The transport and function of silicon in plants. Biological Reviews of the Cambridge Philosophical Society, 58:179–207.

Raymond, J., Zhaxybayeva, O., Gogarten, J.P., Gerdes, S.Y., and Blankenship, R.E., 2002. Whole-genome analysis of photosynthetic prokaryotes. Science, 298:1616–1620.

Raymond, J.A. and Fritsen, C.H., 2001. Semipurification and ice recrystallization inhibition activity of ice-active substances associated with Antarctic photosynthetic organisms. Cryobiology, 43:63–70.

Reeburgh, W.S., 1980. Anaerobic methane oxidation: rate depth distributions in Skan Bay sediments. Earth and Planetary Science Letters, 47:345–352.

Rees, H.C., Grant, W.D., Jones, B.E., and Heaphy, S., 2004. Diversity of Kenyan soda lake alkaliphiles assessed by molecular methods. Extremophiles, 8:63–71.

Reid, R.P., Visscher, P.T., Decho, A.W., et al. 2000. The role of microbes in accretion, lamination and early lithification of modern marine stromatolites. Nature, 406:989–992.

Reis, M.A.M., Almeida, J.S., Lemos, P.C., and Carrondo, M.J.T., 1992. Effect of hydrogen sulfide on growth of sulfate-reducing bacteria. Biotechnology and Bioengineering, 40:593–600.

Renaut, R.W. and Jones, B., 1997. Controls on aragonite and calcite precipitation in hot spring travertines at Chemurkeu, Lake Bogoria, Kenya. Canadian Journal of Earth Sciences, 34:801–818.

Renaut, R.W. and Jones, B., 2000. Microbial precipitates around continental hot springs and geysers. In: Riding, R.E. and Awramik, S.M. (eds), Microbial Sediments. Springer-Verlag, Berlin, pp. 187–195.

Renaut, R.W., Jones, B., and Tiercelin, J.J., 1998. Rapid in situ silicification of microbes at Loburu hot springs, Lake Bogoria, Kenya Rift Valley. Sedimentology, 45:1083–1103.

Rettberg, P., Eschweiler, U., Strauch, K., et al. 2002. Survival of microorganisms in space protected by meteorite material: results of the experiment "Exobiologie" of the Perseus Mission. Advances in Space Research, 30:1539–1545.

Reynolds, P.J., Sharma, P., Jenneman, G.E., and McInerney, M.J., 1989. Mechanisms of microbial movement in subsurface materials. Applied and Environmental Microbiology, 55:2280–2286.

Reysenbach, A.-L. and Shock, E., 2002. Merging genomes with geochemistry in hydrothermal ecosystems. Science. 296:1077–1082.

Reysenbach, A.-L., Götz, D., and Yernool, D., 2002. Microbial diversity of marine and terrestrial thermal springs. In: Staley, J.T. and Reysenbach, A.-L. (eds), Biodiversity of Microbial Life. Wiley Publishers, New York, pp. 345–421.

Ribbons, D.W., Harrison, J.E., and Wadinski, A.M., 1970. Metabolism of single carbon compounds. Annual Review of Microbiology, 24:135–158.

Rice, G., Stedman, K., Snyder, J., et al. 2001. Viruses from extreme thermal environments. Proceedings of the National Academy of Sciences, USA, 98:13341–13345.

Richardson, L.L., Aguilar, C., and Nealson, K.H., 1988. Manganese oxidation in pH and O_2 microenvironments produced by phytoplankton. Limnology and Oceanography, 33:352–363.

Rickard, D.T., 1969. The microbiological formation of iron sulfides. Stockholm Contributions in Geology, 20:49–66.

Rickard, D.T., 1997. Kinetics of pyrite formation by H_2S oxidation of iron (II) monosulfide in aqueous solutions between 25°C and 125°C: the rate equation. Geochimica et Cosmochimica Acta, 61:115–134.

Ridgway, H.F., Safarik, J., Phipps, D., and Clark, D., 1990. Identification and catabolic activity of well-derived gasoline-degrading bacteria from a contaminated aquifer. Applied and Environmental Microbiology, 56:3565–3575.

Ridgwell, A.J., Kennedy, M.J., and Caldeira, K., 2003. Carbonate deposition, climate stability, and Neoproterozoic ice ages. Science, 302:859–862.

Riding, R., 1994. Evolution of algal and cyanobacterial calcification. In: Bengtson, S. (ed.), Early Life on Earth, Nobel Symposium No. 84. Columbia University Press, New York, pp. 426–438.

Riding, R., 2000. Microbial carbonates: the geological record of calcified bacterial-algal mats and biofilms. Sedimentology, 47:179–214.

Riech, V. and von Rad, U., 1979. Silica diagenesis in the Atlantic Ocean: diagenetic potential and transformations. In: Talwani, M., Hay, W., and Ryan, W.B.F. (eds), Deep Drilling Results in the Atlantic Ocean: continental margins and paleoenvironment. American Geophysical Union, Maurice Ewing Series 3, Washington, DC, pp. 315–340.

Rippka, R., Deruelles, J., Waterbury, J.B., Herdman, M., and Stanier, R.Y., 1979. Generic assignments, strain histories and properties of pure cultures of cyanobacteria. Journal of General Microbiology, 111:1–61.

Rivkina, E.M., Friedmann, E.I., McKay, C.P., and Gilichinsky, D.A., 2000. Metabolic activity of permafrost bacteria below the freezing point. Applied and Environmental Microbiology, 66:3230–3233.

Robbins, E.I., LaBerge, G.L., and Schmidt, R.G., 1987. A model for the biological precipitation of Precambrian iron formations – B: morphological evidence and modern analogs. In: Appel, P.W.U. and LaBerge, G.L. (eds), Precambrian Iron-formations. Theophrastus, Athens, pp. 97–139.

Robbins, L.L. and Blackwelder, P.L., 1992. Biochemical and ultrastructural evidence for the origin of whitings: a biologically induced calcium carbonate precipitation mechanism. Geology, 20:464–468.

Robbins, L.L, Tao, Y., and Evans, C.A., 1997. Temporal and spatial distribution of whitings on Great Bahama Bank and a new lime mud budget. Geology, 25:947–950.

Roberts, J.A., Bennett, P.C., González, L.A., Macpherson, G.L., and Milliken, K.L., 2004. Microbial precipitation of dolomite in methanogenic groundwater. Geology, 32:277–280.

Robertson, L.A. and Kuenen, J.G., 1984. Aerobic denitrification: a controversy revived. Archives of Microbiology, 139:351–354.

Robertson, M.P. and Miller, S.L., 1995. An efficient prebiotic synthesis of cytosine and uracil. Nature, 375:772–774.

Roden, E.E., 2003. Fe(III) oxide reactivity toward biological versus chemical reduction. Environmental Science and Technology, 37:1319–1324.

Roden, E.E. and Lovley, D.R., 1993. Dissimilatory Fe(III) reduction by the marine microorganism *Desulfuromonas acetoxidans*. Applied and Environmental Microbiology, 59:734–742.

Roden, E.E. and Zachara, J.M., 1996. Microbial reduction of crystalline iron (III) oxides: influence of oxide surface area and potential for cell growth. Environmental Science and Technology, 30:1618–1628.

Rojas, J., Giersig, M., and Tributsch, H., 1995. Sulfur colloids as temporary energy reservoirs for *Thiobacillus ferrooxidans* during pyrite oxidation. Archives of Microbiology, 163:352–356.

Rojas-Chapana, J.A. and Tributsch, H., 2004. Interfacial activity and leaching patterns of *Leptospirillum ferrooxidans* on pyrite. FEMS Microbiology Ecology, 47:19–29.

Romanek, C.S., Grady, M.M., Wright, I.P., et al. 1994. Record of fluid–rock interactions on Mars from the meteorite ALH84001. Nature, 372:655–657.

Rönner, U., Husmark, U., and Henriksson, A., 1990. Adhesion of bacillus spores in relation to hydrophobicity. Journal of Applied Bacteriology, 69:550–556.

Rosing, M.T., 1999. ^{13}C-depleted carbon microparticles in >3700-Ma sea-floor sedimentary rocks from West Greenland. Science, 283:674–676.

Rossi, G., 1990. Biohydrometallurgy. McGraw-Hill, Hamburg, Germany.

Rothman, D.H., Hayes, J.M., and Summons, R.E., 2003. Dynamics of the Neoproterozoic carbon cycle. Proceedings of the National Academy of Sciences, USA, 100:8124–8129.

Rothschild, L.J. and Mancinelli, R.L., 2001. Life in extreme environments. Nature, 409:1092–1101.

Rouchy, J.M. and Monty, C., 2000. Gypsum microbial sediments: Neogene and modern examples. In: Riding, R.E. and Awramik, S.M. (eds), Microbial Sediments. Springer-Verlag, Berlin, pp. 209–216.

Rouxel, O.J., Bekker, A., and Edwards, K.J., 2005. Iron isotope constraints on the Archean and Paleoproterozoic ocean redox state. Science, 307:1088–1091.

Rudd, T., Sterritt, R.M., and Lester, J.N., 1983. Mass balance of heavy metal uptake by encapsulated cultures of *Klebsiella aerogenes*. Microbial Ecology, 9:261–272.

Runnegar, B., 1991. Precambrian oxygen levels estimated from the biochemistry and physiology of early eukaryotes. Palaeogeography, Palaeoclimatology, Palaeoecology, 97:97–111.

Russell, M.J., 2003. The importance of being alkaline. Science, 302:580–581.

Russell, M.J. and Hall, A.J., 1997. The emergence of life from iron monosulfide bubbles at a submarine hydrothermal redox and pH front. Journal of the Geological Society of London, 154:377–402.

Russell, M.J. and Martin, W., 2004. The rocky roots of the acetyl-CoA pathway. Trends in Biochemical Sciences, 29:358–363.

Russell, M.J., Daniel, R.M., and Hall, A.J., 1993. On the emergence of life via catalytic iron sulfide membranes. Terra Nova, 5:343–347.

Russell, M.J., Hall, A.J., Boyce, A.J., and Fallick, A.E., 2005. On hydrothermal convection systems and the emergence of life. Economic Geology, 100:419–438.

Russell, R.S., 1977. Plant Root Systems. McGraw-Hill, London.

Rye, R. and Holland, H.D., 1998. Paleosols and the evolution of atmospheric oxygen: a critical review. American Journal of Science, 298:621–672.

Rye, R. and Holland, H.D., 2000. Life associated with a 2.76 Ga ephemeral pond? Evidence from Mount Roe #2 paleosol. Geology, 28:483–486.

Rye, R., Kuo, P.H., and Holland, H.D., 1995. Atmospheric carbon dioxide concentrations before 2.2-billion years ago. Nature, 378:603–605.

Sagan, C. and Chyba, C., 1997. The early faint sun paradox: organic shielding of ultraviolet-labile greenhouse gases. Science, 276:1217–1221.

Sageman, J., Bale, S.J., Briggs, D.E.G., and Parkes, R.J., 1999. Controls on the formation of authigenic minerals in association with decaying organic matter: an experimental approach. Geochimica et Cosmochimica Acta, 63:1083–1095.

Saier, M.H., Jr, 1987. Enzymes in Metabolic Pathways. A comparative study of mechanism, structure, evolution and control. Harper and Row Publishers, New York.

Saiz-Jimenez, C. and Shafizadeh, F., 1984. Iron and copper binding by fungal phenolic polymers: an electron spin resonance study. Current Microbiology, 10:281–285.

Sakaguchi, T., Burgess, J.G., and Matsunaga, T., 1993. Magnetite formation by a sulfate-reducing bacterium. Nature, 365:47–49.

Sánchez-Navas, A., Martín-Algarra, A., and Nieto, F., 1998. Bacterially-mediated authigenesis of clays in phosphate stromatolites. Sedimentology, 45:519–533.

Sand, W. and Bock, E., 1991. Biodeterioration of mineral materials by microorganisms. Biogenic sulfuric and nitric acid corrosion of concrete and natural stone. Geomicrobiology Journal, 9:129–138.

Sand, W., Gerke, T., Hallmann, R., and Schippers, A., 1995. Sulfur chemistry, biofilm, and the (in)direct attack mechanisms. A critical evaluation of bacterial leaching. Applied Microbiology and Biotechnology, 43:961–966.

Santana-Casiano, J.M., Gonzalez-Davila, M., Perez-Peña, J., and Milleron, F.J., 1995. Pb^{2+} interactions with the marine phytoplankton *Dunaliella tertiolecta*. Marine Chemistry, 48:115–129.

Santelli, C.M., Welch, S.A., Westrich, H.R., and Banfield, J.F., 2001. The effect of Fe-oxidizing bacteria on Fe-silicate mineral dissolution. Chemical Geology, 180:99–115.

Santschi, P.H., Bower, P., Nyffeler, U.P., Azevedo, A., and Broeker, W.S., 1983. Estimates of the resistance to chemical transport posed by the deep-sea boundary layer. Limnology and Oceanography, 28:899–912.

Sasaki, K., Tsunekawa, M., Ohtsuka, T., and Konno, H., 1995. Confirmation of a sulfur-rich layer on pyrite after oxidative dissolution by Fe(III) ions around pH 2. Geochimica et Cosmochimica Acta, 59:3155–3158.

Sassen, R., Roberts, H.H., Carney, R., et al. 2004. Free hydrocarbon gas, gas hydrate, and authigenic minerals in chemosynthetic communities of the northern Gulf of Mexico continental slope: relation to microbial processes. Chemical Geology, 205:195–217.

Sayer, J.A. and Gadd, G.M., 1997. Solubilization and transformation of insoluble inorganic metal compounds to insoluble metal oxalates by *Aspergillus niger*. Mycology Research, 101:653–661.

Schelske, C.L., 1985. Biogeochemical silica mass balances in Lake Michigan and Lake Superior. Biogeochemistry, 1:197–218.

Schelske, C.L. and Stoermer, E.F., 1971. Eutrophication, silica depletion, and predicted changes in algal quality in Lake Michigan. Science, 173:423–424.

Schidlowski, M., 1988. A 3,800-million-year isotopic record of life from carbon in sedimentary rocks. Nature, 333:313–318.

Schidlowski, M., 1989. Evolution of the sulphur cycle. In: Brimblecombe, P. and Lein, A.Y. (eds), Evolution of the Global Biogeochemical Sulphur Cycle, SCOPE 39. John Wiley, Chichester, pp. 3–20.

Schidlowski, M., 1993. The initiation of biological processes on Earth. In: Engel, M.H. and Macko, S.A. (eds), Organic Geochemistry. Principles and applications. Plenum Press, New York, pp. 639–655.

Schidlowski, M., 2000. Carbon isotopes and microbial sediments. In: Riding, R.E. and Awramik, S.M. (eds), Microbial Sediments. Springer-Verlag, Berlin, pp. 84–95.

Schidlowski, M., Hayes, J.M., and Kaplan, I.R., 1983. Isotopic inferences of ancient biochemistries: carbon, sulfur, hydrogen, and nitrogen. In: Schopf, J.W. (ed.), The Earth's Earliest Biosphere: its origin and evolution. Princeton University Press, Princeton, pp. 149–186.

Schieber, J., 1999. Microbial mats in terrigenous clastics: The challenge of identification in the rock record. Palaios, 14:3–12.

Schiewer, S. and Volesky, B., 2000. Biosorption processes for heavy metal removal. In: Lovley, D.R. (ed.), Environmental Microbe–Metal Interactions. ASM Press, Washington, DC, pp. 329–362.

Schink, B., 1997. Energetics of syntrophic cooperation in methanogenic degradation. Microbiology and Molecular Biology Reviews, 61:262–280.

Schinner, F. and Burgstaller, W., 1989. Extraction of zinc from an industrial waste by *Penicillium* sp. Applied and Environmental Microbiology, 55:1153–1156.

Schippers, A. and Sand, W., 1999. Bacterial leaching of metal sulfides proceeds by two indirect mechanisms via thiosulfate or via polysulfides and sulfur. Applied and Environmental Microbiology, 65:319–321.

Schleifer, K.H. and Kandler, O., 1972. Peptidoglycan types of bacterial cell walls and their taxonomic implications. Bacteriological Reviews, 36: 407–477.

Schleifer, K.-H. and Ludwig, W., 1994. Molecular taxonomy: classification and identification. In: Priest, F.G., Ramos-Cormenzana, A., and Tindall, B.J. (eds), Bacterial Diversity and Systematics. Plenum Press, New York, pp. 1–15.

Schleper, C., Puehler, G., Holz, I., et al. 1995. *Picrophilus* gen nov, fam nov – a novel aerobic, heterotrophic, thermoacidophilic genus and family comprising Archaea capable of growth around pH 0. Journal of Bacteriology, 177:7050–7059.

Schneider, J. and Le Campion-Alsumard, T., 1999. Construction and destruction of carbonates by marine freshwater cyanobacteria. European Journal of Phycology, 34:417–426.

Schnell, H.A., 1997. Bioleaching of copper. In: Rawlings, D.E. (ed.), Biomining. Theory, microbes and industrial processes. Springer-Verlag, Berlin, pp. 21–44.

Schnitzer, M. and Khan, S.U., 1972. Humic Substances in the Environment. Marcel Dekker, New York.

Scholl, M.A. and Harvey, R.W., 1992. Laboratory investigations on the role of sediment surface and groundwater chemistry in transport of bacteria through a contaminated sandy aquifer. Environmental Science and Technology, 26:1410–1417.

Schoonen, M.A.A. and Barnes, H.L., 1991a. Reactions forming pyrite and marcasite from solution: I. Nucleation of FeS_2 below 100°C. Geochimica et Cosmochimica Acta, 55:1495–1504.

Schoonen, M.A.A. and Barnes, H.L., 1991b. Reactions forming pyrite and marcasite from solution: II. Via FeS precursors below 100°C. Geochimica et Cosmochimica Acta, 55:1505–1514.

Schoonen, M.A.A., Xu, Y., and Bebié, J., 1999. Energetics and kinetics of the prebiotic synthesis of simple organic acids and amino acids with the $FeS-H_2S/FeS_2$ redox couple as reductant. Origins of Life and Evolution of the Biosphere, 29:5–32.

Schopf, J.W., 1993. Microfossils of the early Archean Apex chert: new evidence of the antiquity of life. Science, 260:640–646.

Schopf, J.W., 1994. The oldest known records of life: early Archean stromatolites, microfossils, and organic matter. In: Bengtson, S. (ed.), Early Life on Earth. Columbia University Press, New York, pp. 270–286.

Schopf, J.W. and Packer, B.M., 1987. Early Archean (3.3-billion- to 3.5-billion-year-old) microfossils from Warrawoona Group, Australia. Science, 237:70–72.

Schopf, J.W., Kudryavtsev, A.B., Agresti, D.G., Wdowiak, T.J., and Czaja, A.D., 2002. Laser-Raman imagery of Earth's earliest fossils. Nature, 416:73–76.

Schrenk, M.O., Edwards, K.J., Goodman, R.M., Hamers, R.J., and Banfield, J.F., 1998. Distribution of *Thiobacillus ferrooxidans* and *Leptospirillum ferrooxidans*: implications for generation of acid mine drainage. Science, 279:1519–1522.

Schulte, M. and Shock, E., 1995. Thermodynamics of Strecker synthesis in hydrothermal systems. Origins of Life and Evolution of the Biosphere, 25:161–173.

Schultze-Lam, S. and Beveridge, T.J., 1994. Nucleation of celestite and strontionite on a cyanobacterial S-layer. Applied and Environmental Microbiology, 60:447–453.

Schultze-Lam, S., Harauz, G., and Beveridge, T.J., 1992. Participation of a cyanobacterial S layer in fine-grain mineral formation. Journal of Bacteriology, 174:7971–7981.

Schulz, H.N., Brinkhoff, T., Ferdelman, T.G., Hernández Mariné, N., Teske, A., and Jørgensen, B.B., 1999. Dense populations of a giant sulfur bacterium in Namibian shelf sediments. Science, 284:493–495.

Schwartzman, D.W. and Volk, T., 1989. Biotic enhancement of weathering and the habitability of Earth. Nature, 340:457–460.

Schwartzman, D.W. and Volk, T., 1991. Biotic enhancement of weathering and surface temperatures on earth since the origin of life. Palaeogeography, Palaeoclimatology, Palaeoecology, 90:357–371.

Schwertmann, U. and Fechter, H., 1982. The point of zero charge of natural and synthetic ferrihydrites and its relation to adsorbed silicate. Clay Minerals, 17:471–476.

Schwertmann, U. and Fitzpatrick, R.W., 1992. Iron minerals in surface environments. In: Skinner, H.C.W. and Fitzpatrick, R.W. (eds), Biomineralization. Processes of iron and manganese. Catena Verlag, Cremlingen, Germany, pp. 7–30.

Segerer, A.H., Neuner, A., Kristjansson, A., and Stetter, K.O., 1986. *Acidianus azoricus* gen. nov., sp. nov. represents a novel genus of anaerobic, extremely thermoacidophilic Archaebacteria of the Order Sulfolobales. International Journal of Systematic Bacteriology, 41:495–501.

Seitz, H.-J. and Cypionka, H., 1986. Chemolithotrophic growth of *Desulfovibrio desulfuricans* with hydrogen coupled to ammonification of nitrate or nitrite. Archives of Microbiology, 146:63–67.

Seitzinger, S.P., 1988. Denitrification in freshwater and coastal marine ecosystems: ecological and geochemical significance. Limnology and Oceanography, 33:702–724.

Senior, E., Talaat, M., and Balba, T.M., 1990. Refuse decomposition. In: Senior, E. (ed.), Microbiology of Landfill Sites. CRC Press, Boca Raton, Florida, pp. 18–57.

Seong-Joo, L., Browne, K.M., and Golubic, S., 2000. On stromatolite lamination. In: Riding, R.E. and Awramik, S.M. (eds), Microbial Sediments. Springer-Verlag, Heidelberg, pp. 16–24.

Shapiro, R., 1999. Prebiotic cytosine synthesis: a critical analysis and implications for the origin of life. Proceedings of the National Academy of Sciences, USA, 96:4396–4401.

Sharma, A., Scott, J.H., Cody, G.D., et al. 2002. Microbial activity at gigapascal pressures. Science, 295:1514–1516.

Sharma, M.M., Chang, Y.I., and Yen, T.F., 1985. Reversible and irreversible surface charge modification of bacteria for facilitating transport through porous media. Colloids and Surfaces, 16:193–206.

Sharma, P.K. and McInerney, M.J., 1994. Effect of grain size on bacterial penetration, reproduction, and metabolic activity in porous glass bead chambers. Applied and Environmental Microbiology, 60:1481–1486.

Sharp, M., Parkes, J., Cragg, B., Fairchild, I.J., Lamb, H., and Tranter, M., 1999. Widespread bacterial populations at glacial beds and their relationship to rock weathering and carbon cycling. Geology, 27:107–110.

Shaw, D.J., 1966. Introduction to Colloid and Surface Chemistry. Butterworth, London.

Shen, Y., Buick, R., and Canfield, D.E., 2001. Isotopic evidence for microbial sulfate reduction in the early Archaean era. Nature, 410:77–81.

Shen, Y., Knoll, A.H., and Walter, M.R., 2003. Evidence for low sulfate and anoxia in a mid-Proterozoic marine basin. Nature, 423:632–635.

Shinn, E.A., Steinen, R.P., Lidz, B.H., and Swart, P.K., 1989. Perspectives: whitings, a sedimentologic dilemma. Journal of Sedimentary Petrology, 59:147–161.

Shively, J.M., van Keulen, G., and Meijer, W.G., 1998. Something from almost nothing: carbon dioxide fixation in chemoautotrophs. Annual Reviews of Microbiology, 52:191–230.

Shock, E.L., 1992. Chemical environments of submarine hydrothermal systems. Origins of Life and Evolution of the Biosphere, 22:67–108.

Shock, E.L. and Schulte, M.D., 1998. Organic synthesis during fluid mixing in hydrothermal systems. Journal of Geophysical Research, 103:28513–28527.

Shock, E.L., McCollom, T., and Schulte, M.D., 1998. The emergence of metabolism from within hydrothermal systems. In: Wiegel, J. and Adams, M.W.W. (eds), Thermophiles: the keys to molecular evolution and the origin of life? Taylor & Francis, London, pp. 59–76.

Shuttleworth, K.L. and Unz, R.F., 1993. Sorption of heavy metals to the filamentous bacterium *Thiothrix* strain A1. Applied and Environmental Microbiology, 59:1274–1282.

Silver, M. and Kelly, D.P., 1976. Rhodanese from *Thiobacillus* A2: catalysis of reactions of thiosulfate with dihydrolipoate and dihydrolipoamide. Journal of General Microbiology, 97:277–284.

Silver, M. and Torma, A.E., 1974. Oxidation of metal sulfides by *Thiobacillus ferrooxidans* grown on different substrates. Canadian Journal of Microbiology, 20:141–147.

Silver, S., 1996. The bacterial view of the periodic table: specific functions for all elements. In: Banfield, J.F. and Nealson, K.H. (eds), Geomicrobiology: interactions between microbes and minerals, vol. 35. Mineralogical Society of America, Washington, DC, pp. 345–360.

Silverman, M.P. and Lundgren, D.G., 1959. Studies on the chemoautotrophic iron bacterium *Ferrobacillus ferrooxidans*. II. Manometric studies. Journal of Bacteriology, 78:326–331.

Simoni, S.F., Bosma, T.N.P., Harms, H., and Zehnder, A.J.B., 2000. Bivalent cations increase both the subpopulation of adhering bacteria and their adhesion efficiency in sand columns. Environmental Science and Technology, 34:1011–1017.

Singer, P.C. and Stumm, W., 1969. Oxygenation of ferrous iron. FWQA Report 14010–06/69.

Singer, P.C. and Stumm, W., 1970. Acidic mine drainage: the rate-determining step. Science, 167:1121–1123.

Skidmore, M.L., Foght, J.M., and Sharp, M.J., 2000. Microbial life beneath a high Arctic glacier. Applied and Environmental Microbiology, 66:3214–3220.

Sleep, N.H. and Zahnle, K., 2001. Carbon dioxide cycling and implications for climate on ancient Earth. Journal of Geophysical Research, 106:1373–1399.

Sleep, N.H., Zahnle, K.J., Kasting, J.F., and Morowitz, H.J., 1989. Annihilation of ecosystems by large asteroid impacts on the early Earth. Nature, 342:139–142.

Sleigh, M.A., 1992. Protozoa and Other Protists. Cambridge University Press, New York.

Small, T.D., Warren, L.A., Roden, E.E., and Ferris, F.G., 1999. Sorption of strontium by bacteria, Fe(III) oxide, and bacteria–Fe(III) oxide composites. Environmental Science and Technology, 33:4465–4470.

Small, T.D., Warren, L.A., and Ferris, F.G., 2001. Influence of ionic strength on strontium sorption to bacteria, Fe(III) oxide, and composite bacteria-Fe(III) oxide surfaces. Applied Geochemistry, 16:939–946.

Smit, J., 1987. Protein surface layers of bacteria. In: Inouye, M. (ed.), Bacterial Outer Membranes as Model Systems. John Wiley, Chichester, UK, pp. 343–376.

Søgaard, E.G., Medenwaldt, R., and Abraham-Peskir, J.V., 2000. Conditions and rates of biotic and abiotic iron precipitation in selected Danish freshwater plants and microscopic analysis of precipitate morphology. Water Research, 34:2675–2682.

Sogin, M.L., 1991. Early evolution and the origins of eukaryotes. Current Opinion in Genetics and Development, 1:457–463.

Sogin, M.L., 1994. The origin of eukaryotes and evolution into major kingdoms. In: Bengtson, S. (ed.), Early Life on Earth. Columbia University Press, New York, pp. 181–192.

Sogin, M.L., Gunderson, J.H., Elwood, H.J., Alonso, R.A., and Peattie, D.A., 1989. Phylogenetic meaning of the kingdom concept: an unusual ribosomal RNA from *Giardia lamblia*. Science, 243:75–77.

Sokolov, I., Smith, D.S., Henderson, G.S., Gorby, Y.A., and Ferris, F.G., 2001. Cell surface electrochemical heterogeneity of the Fe(III)-reducing bacteria *Shewanella putrefaciens*. Environmental Science and Technology, 35:341–347.

Sonnenfeld, E.M., Beveridge, T.J., Koch, A.L., and Doyle, R.J., 1985a. Asymmetric distribution of charge on the cell wall of *Bacillus subtilis*. Journal of Bacteriology, 163:1167–1171.

Sonnenfeld, E.M., Beveridge, T.J., and Doyle, R.J., 1985b. Discontinuity of charge on the cell wall poles of *Bacillus subtilis*. Canadian Journal of Microbiology, 31:875–877.

Sørensen, J., 1982. Reduction of ferric iron in anaerobic, marine sediment and interaction with reduction of nitrate and sulfate. Applied and Environmental Microbiology, 43:319–324.

Sørensen, J., 1987. Nitrate reduction in marine sediment: pathways and interactions with iron and sulfur cycling. Geomicrobiology Journal, 5:401–421.

Sørensen, J., Jørgensen, K.S., Colley, S., Hydes, D.J., Thompson, J., and Wilson, T.R.S., 1987. Depth localization of denitrification in a deep-sea sediment from the Madeira Abyssal Plain. Limnology and Oceanography, 32:758–762.

Soudry, D., 2000. Microbial phosphate sediment. In: Riding, R.E. and Awramik, S.M. (eds), Microbial Sediments. Springer-Verlag, Berlin, pp. 127–136.

Soudry, D. and Champetier, Y., 1983. Microbial processes in Negev phosphorites (southern Israel). Sedimentology, 30:411–423.

Sousa, C., Kotrba, P., Ruml, T., Cebolla, A., and De Lorenzo, V., 1998. Metalloadsorption by *Escherichia coli* cells displaying yeast and mammalian metallothioneins anchored to the outer membrane protein LamB. Journal of Bacteriology, 180:2280–2284.

Southam, G. and Beveridge, T.J., 1992. Enumeration of *Thiobacilli* within pH-neutral and acidic mine tailings and their role in the development of secondary mineral soil. Applied and Environmental Microbiology, 58:1904–1912.

Southam, G. and Beveridge, T.J., 1993. Examination of lipopolysaccharide (O-antigen) populations of *Thiobacillus ferrooxidans* from two mine tailings. Applied and Environmental Microbiology, 59:1904–1912.

Southam, G. and Beveridge, T.J., 1994. The in vitro formation of placer gold by bacteria. Geochimica et Cosmochimica Acta, 58:4527–4530.

Stal, L.J., van Gemerden, H., and Krumbein, W.E., 1985. Structure and development of a benthic microbial mat. FEMS Microbial Ecology, 31:111–125.

Staudigel, H., Chastain, R.A., Yayanos, A., and Bourcier, W., 1995. Biologically mediated dissolution of glass. Chemical Geology, 126:147–154.

Staudigel, H., Yayanos, A., Chastain, R., et al. 1998. Biologically mediated dissolution of volcanic glass in seawater. Earth and Planetary Science Letters, 164:233–244.

Steefel, C.I. and Van Cappellen, P., 1990. A new kinetic approach to modeling water–rock interaction: the role of nucleation, precursors, and Ostwald ripening. Geochimica et Cosmochimica Acta, 54:2657–2677.

Steele, A., Goddard, D.T., Stapleton, D., et al. 2000. Investigations into an unknown organism on the Martian meteorite Allan Hills 84001. Meteoritics and Planetary Sciences, 35:237–241.

Stetter, K.O., 1994. The lesson of Archaebacteria. In: Bengtson, S. (ed.), Early Life on Earth. Columbia University Press, New York, pp. 143–151.

Stevens, T.O. and McKinley, J.P., 1995. Lithoautotrophic microbial ecosystems in deep basalt aquifers. Science, 270:450–454.

Stevenson, D.J., 1983. The nature of the Earth prior to the oldest known rock record: the Hadean Earth. In: Schopf, J.W. (ed.), Earth's Earliest Biosphere, Its Origin and Evolution. Princeton University Press, Princeton, New Jersey, pp. 14–29.

Stewart, M. and Beveridge, T.J., 1980. Structure of the regular surface layer of *Sporosarcina ureae*. Journal of Bacteriology, 142:302–309.

Stillings, L.L., Drever, J.I., Brantley, S.L., Sun, Y., and Oxburgh, R., 1996. Rates of feldspar dissolution at pH 3–7 with 0–8 mmol oxalic acid. Chemical Geology, 132:79–89.

Stone, A.T., 1987. Microbial metabolites and the reductive dissolution of manganese oxides: oxalate and pyruvate. Geochimica et Cosmochimica Acta, 51:919–925.

Stone, A.T., 1997. Reactions of extracellular organic ligands with dissolved metal ions and mineral surfaces. In: Banfield, J.F. and Nealson, K.H. (eds), Geomicrobiology: interactions between microbes and minerals, vol. 35. Mineralogical Society of America, Washington, DC, pp. 309–344.

Stone, A.T. and Morgan, J.J., 1984. Reduction and dissolution of manganese (III) and manganese (IV) oxides by organics. 2. Survey of the reactivity of organics. Environmental Science and Technology, 18:617–624.

Straub, K.L., Benz, M., Schink, B., and Widdel, F., 1996. Anaerobic, nitrate-dependent microbial oxidation of ferrous iron. Applied and Environmental Microbiology, 62:1458–1460.

Strauss, H., 2003. Sulfur isotopes and the early Archaean sulfur cycle. Precambrian Research, 126:349–361.

Stribling, R. and Miller, S.L., 1987. Energy yields for hydrogen cyanide and formaldehyde syntheses: the HCN and amino acid concentrations in the primitive ocean. Origins of Life and Evolution of the Biosphere, 17:261–273.

Strohl, W.R. and Schmidt, T.M., 1984. Mixotrophy of the colorless, sulfide-oxidizing, gliding bacteria *Beggiatoa* and *Thiothrix*. In: Sprott, G.D. and Jarrell, K.F. (eds), Electrochemical Potential and Membrane Properties of Methanogenic Bacteria. Ohio State Press, Colombus, Ohio, pp. 79–95.

Strous, M., Fuerst, J.A., Kramer, E.H.M., et al. 1999. Missing lithotroph identified as new planctomycete. Nature, 400:446–449.

Stumm, W. and Morgan, J.J., 1996. Aquatic Chemistry, 3rd edn. John Wiley, New York.

Stumm, W. and Sulzberger, B., 1992. The cycling of iron in natural environments: considerations based on laboratory studies of heterogeneous redox processes. Geochimica et Cosmochimica Acta, 56:3233–3257.

Suess, E., 1979. Mineral phases formed in anoxic sediments by microbial decomposition of organic matter. Geochimica et Cosmochimica Acta, 43:339–352.

Suess, E., 1980. Particulate organic carbon flux in the oceans – surface productivity and oxygen utilization. Nature, 288:260–262.

Sugio, T., Tsujita, Y., Katagiri, T., Inagaki, K., and Tano, T., 1988. Reduction of Mo^{6+} with elemental sulfur by *Thiobacillus ferrooxidans*. Journal of Bacteriology, 170:5956–5959.

Sugio, T., Tsujita, Y., Inagaki, K., and Tano, T., 1990. Reduction of cupric ions with elemental sulfur by *Thiobacillus ferrooxidans*. Applied and Environmental Microbiology, 56:693–696.

Summers, D.P. and Chang, S., 1993. Prebiotic ammonia from reduction of nitrite by iron (II) on the early Earth. Nature, 365:630–633.

Summons, R.E., Jahnke, L.L., Hope, J.M., and Logan, G.A., 1999. 2-methylhopanoids as biomarkers for cyanobacterial oxygenic photosynthesis. Nature, 400:554–557.

Sumner, D.Y., 1997. Carbonate precipitation and oxygen stratification in late Archean seawater as deduced from facies and stratigraphy of the Gamohaan and Frisco Formations, Transvaal Supergroup, South Africa. American Journal of Science, 297:455–487.

Sun, H.J. and Friedmann, E.I., 1999. Growth on geological time scales in the Antarctic cryptoendolithic microbial community. Geomicrobiology Journal, 16:193–202.

Sunda, W.G., 2000. Trace metal–phytoplankton interactions in aquatic systems. In: Lovley, D.R. (ed.), Environmental Microorganism–Metal Interactions. ASM Press, Washington, DC, pp. 79–107.

Sutherland, I.W., 1972. Biofilm exopolysaccharides. In: Wingender, J., Neu, T.R., and Flemming, H.C. (eds), Microbial Extracellular Polymeric Substances: characterization, structure, and function. Springer-Verlag, Berlin, pp. 73–92.

Sweeney, R.E. and Kaplan, I.R., 1973. Pyrite framboid formation: laboratory synthesis and marine sediments. Economic Geology, 68:618–634.

Swift, M.J., Heal, O.W., and Anderson, J.M., 1979. Decomposition in Terrestrial Ecosystems. University of California Press, Berkley.

Sylvester, P.J., Campbell, I.H., and Bowyer, 1997. Niobium/uranium evidence for early formation of the continental crust. Science, 275:521–523.

Szostak, J.W., Bartel, D.P., and Luisi, P.L., 2001. Synthesizing life. Nature, 409:387–390.

Takai, K., Moser, D.P., DeFlaun, M., Onstott, T.C., and Fredrickson, J.K., 2001. Archaeal diversity in waters from deep South African gold mines. Applied and Environmental Microbiology, 67:5750–5760.

Tappan, H. and Loeblich, A.R., Jr, 1988. Foraminiferal evolution, diversification, and extinction. Journal of Paleontology, 62:695–714.

Taylor, P.D., Jugdaohsingh, R., and Powell, J.J., 1997. Soluble silica and high affinity for aluminum under physiological and natural conditions. Journal of the American Chemical Society, 119:8852–8856.

Tebo, B.M., Ghiorse, W.C., van Waasbergen, L.G., Siering, P.L., and Caspi, R., 1997. Bacterially mediated mineral formation: insights into manganese (II) oxidation from molecular, genetic and biochemical studies. In: Banfield, J.F. and Nealson, K.H. (eds), Geomicrobiology: interactions between microbes and minerals, vol. 35. Mineralogical Society of America, Washington, DC, pp. 225–266.

Thamdrup, B. and Dalsgaard, T., 2002. Production of N_2 through anaerobic ammonium oxidation coupled to nitrate reduction in marine sediments. Applied and Environmental Microbiology, 68:1312–1318.

Thamdrup, B., Finster, K., Hansen, J.W., and Bak, F., 1993. Bacterial disproportionation of elemental sulfur coupled to chemical reduction of iron or manganese. Applied and Environmental Microbiology, 59:101–108.

Thamdrup, B., Fossing, H., and Jørgensen, B.B., 1994. Manganese, iron, and sulfur cycling in a coastal marine sediment, Aarhus Bay, Denmark. Geochimica et Cosmochimica Acta, 58:5115–5129.

Thauer, R.K., Jungermann, K., and Decker, K., 1977. Energy conservation in chemotrophic anaerobic bacteria. Bacteriological Reviews, 41:100–180.

Theron, J. and Cloete, T.E., 2000. Molecular techniques for determining microbial diversity and community structure in natural environments. Critical Reviews in Microbiology, 26:37–57.

Thiele, J.H. and Zeikus, J.G., 1988. Control of interspecies electron transfer during anaerobic digestion: significance of formate transfer versus hydrogen transfer during syntrophic methanogenesis in flocs. Applied and Environmental Microbiology, 54:20–29.

Thomas, D.J., Zachos, J.C., Bralower, T.J., Thomas, E., and Bohaty, S., 2002. Warming the fuel for the fire: evidence for the thermal dissociation of methane hydrate during the Paleocene-Eocene thermal maximum. Geology, 30:1067–1070.

Thomas-Keprta, K.L., Bazylinski, D.A., Kirschvink, J.L., et al. 2000. Elongated prismatic magnetite crystals in ALH84001 carbonate globules: potential Martian magnetofossils. Geochimica et Cosmochimica Acta, 64:4049–4081.

Thompson, J.B., Ferris, F.G., and Smith, D.A., 1990. Geomicrobiology and sedimentology of the mixolimnion and chemocline in Fayetteville Green Lake, New York. Palaios, 5:52–75.

Thompson, J.B., Schultze-Lam, S., Beveridge, T.J., and Des Marais, D.J., 1997. Whiting events: biogenic origin due to the photosynthetic activity of cyanobacterial picoplankton. Limnology and Oceanography, 42:133–141.

Thompson, J.G. and Ferris, F.G., 1990. Cyanobacterial precipitation of gypsum, calcite, and magnesite from natural alkaline lake water. Geology, 18:995–998.

Thorseth, I.H., Furnes, H., and Heldal, M., 1992. The importance of microbiological activity in the alteration of natural basaltic glass. Geochimica et Cosmochimica Acta, 56:845–850.

Thorseth, I.H., Furnes, H., and Tumyr, O., 1995. Textural and chemical effects of bacterial activity on basaltic glass: an experimental approach. Chemical Geology, 119:139–160.

Thorseth, I.H., Torsvik, T., Daae, F.L., Pedersen, R.B., and Keldysh-98 Scientific Party, 2001. Diversity of life in ocean floor basalt. Earth and Planetary Science Letters, 194:31–37.

Tice, M.M. and Lowe, D.R., 2004. Photosynthetic microbial mats in the 3,416-Myr-old ocean. Nature, 431:549–552.

Tiedje, J.M., 1988. Ecology of denitrification and dissimilatory nitrate reduction to ammonium. In: Zehnder, A.J.B. (ed.), Biology of Anaerobic Microorganisms. John Wiley, New York, pp. 179–244.

Tiedje, J.M, Sexstone, A.J., Myrold, D.D., and Robinson, J.A., 1982. Denitrification: ecological niches, competition and survival. Antonie van Leeuwenhoek, 48:569–583.

Tipping, E., 1981. The adsorption of aquatic humic substances by iron oxides. Geochimica et Cosmochimica Acta, 45:191–199.

Tobin, J.M., Cooper, D.G., and Neufeld, R.J., 1990. Investigation of the mechanism of metal uptake by denatured *Rhizopus arrhizus* biomass. Enzyme and Microbial Technology, 12:591–595.

Toporski, J.K.W., Steele, A., Westall, F., Thomas-Keprta, K.L., and McKay, D.S., 2002. The simulated silicification of bacteria – new clues to the modes and timing of bacterial preservation and implications for the search for extra-terrestrial microfossils. Astrobiology, 2:1–26.

Torres de Araujo, F.F., Pires, M.A., Frankel, R.B., and Bicudo, C.E.M., 1986. Magnetite and magnetotaxis in algae. Biophysical Journal, 50:375–378.

Towe, K.M., 1990. Aerobic respiration in the Archaean? Nature, 348:54–56.

Tozzi, S., Schofield, O., and Falkowski, P.G., 2004. Historical climate change and ocean turbulence as selective agents for two key phytoplankton functional groups. Marine Ecology Progress Series, 274:123–132.

Tréguer, P., Nelson, D.M., van Bennekom, A.J., DeMaster, D.J., Leynaert, A., and Quéguiner, B., 1995. The silica balance in the world ocean: a re-estimate. Science, 268:375–379.

Trendall, A.F., 2002. The significance of iron-formation in the Precambrian stratigraphic record. Special Publication from the International Association of Sedimentology, 33:33–66.

Trevors, J.T., 2003. Origin of the first cells on Earth: a possible scenario. Geomicrobiology Journal, 20:175–183.

Trevors, J.T., van Elsas, J.D., van Overbeek, L.S., and Starodub, M-E., 1990. Transport of genetically engineered *Pseudomonas fluorescens* strain through a soil microcosm. Applied and Environmental Microbiology, 56:401–408.

Trüper, H.G., 1984. Microorganisms and the sulfur cycle. In: Müller, A. and Krebs, B. (eds), Sulfur. Its significance for chemistry, for the geo-, bio- and cosmosphere and technology. Elsevier, Amsterdam, pp. 367–382.

Trüper, H.G., 1989. Physiology and biochemistry of phototrophic bacteria. In: Schlegel, H.G. and Bowien, B. (eds). Autotrophic Bacteria. Springer-Verlag, Berlin, pp. 267–281.

Tsezos, M., 1986. Adsorption by microbial biomass as a process for removal of ions from process waste solutions. In: Eccles, H. and Hunt, S. (eds), Immobilization of Ions by Bio-Sorption. Ellis Harwood, Chichester, UK, pp. 201–218.

Tugel, J.B., Hines, M.E., and Jones, G.E., 1986. Microbial iron reduction by enrichment cultures isolated from estuarine sediments. Applied and Environmental Microbiology, 52:1167–1172.

Turick, C.E., Tisa, L.S., and Caccavo, F., Jr, 2002. Melanin production and use as a soluble electron shuttle for Fe(III) oxide reduction and as a terminal electron acceptor by *Shewanella algae*. Applied and Environmental Microbiology, 68:2436–2444.

Tyagi, R.D., Couillard, D., and Tran, F.T., 1990. Studies on microbial leaching of heavy metals from municipal sludge. Water Science and Technology, 22:229–238.

Tyrrell, T., 1999. The relative influences of nitrogen and phosphorus on oceanic primary production. Nature, 400:525–530.

Tzeferis, P.G., 1994. Leaching of low grade hematitic laterite ore using fungi and biologically produced acid metabolites. International Journal of Mineral Processing, 42:267–283.

Ullman, W.J. and Aller, R.C., 1982. Diffusion coefficients in nearshore marine sediments. Limnology and Oceanography, 27:552–556.

Umbreit, W.W., 1976. Oxidation of metallic iron by *Escherichia coli* and other common heterotrophs. Developments in Industrial Microbiology, 17:265–268.

Urrutia, M.M. and Beveridge, T.J., 1993. Mechanism of silicate binding to the bacterial cell wall in *Bacillus subtilis*. Journal of Bacteriology, 175:1936–1945.

Urrutia, M.M. and Beveridge, T.J., 1994. Formation of fine-grained metal and silicate precipitates on a bacterial surface (*Bacillus subtilis*). Chemical Geology, 116:261–280.

Urrutia, M.M., Kemper, M., Doyle, R., and Beveridge, T.J., 1992. The membrane-induced proton motive force influences the metal binding ability of *Bacillus subtilis* cell walls. Applied and Environmental Microbiology, 58:3837–3844.

Urrutia, M.M., Roden, E.E., Fredrickson, J.K., and Zachara, J.M., 1998. Microbial and surface chemistry controls on reduction of synthetic Fe(III) oxide minerals by the dissimilatory iron-reducing bacterium *Shewanella alga*. Geomicrobiology Journal, 15:269–291.

Valentine, D.L., Blanton, D.C., Reeburgh, W.S., and Kastner, M., 2001. Water column methane oxidation adjacent to an area of active hydrate dissociation, Eel River Basin. Geochimica et Cosmochimica Acta, 65:2633–2640.

Vali, H., Förster, O., Amarantidis, G., and Petersen, N., 1987. Magnetotactic bacteria and their magnetofossils in sediments. Earth and Planetary Science Letters, 86:389–400.

Valley, J.W., Peck, W.H., King, E.M., and Wilde, S.A., 2002. A cool early Earth. Geology, 30:351–354.

Van Cappellen, P., 1996. Reactive surface area control of the dissolution kinetics of biogenic silica in deep-sea sediments. Chemical Geology, 132:125–130.

Van Cappellen, P. and Berner, R.A., 1988. A mathematical model for the early diagenesis of phosphorus and fluorine in marine sediments: apatite precipitation. American Journal of Science, 288:289–333.

Van Cappellen, P. and Berner, R.A., 1991. Fluorapatite crystal growth from modified seawater solutions. Geochimica et Cosmochimica Acta, 55:1219–1234.

Van Cappellen, P. and Gaillard, J-F., 1996. Biogeochemical dynamics in aquatic sediments. In Lichtner, P.C., Steefel, C.I., and Oelkers, E.H. (eds), Reactive Transport in Porous Media. Reviews in Mineralogy, vol. 34. Mineralogical Society of America, Washington, DC, pp. 335–376.

van de Vossenberg, J.L.C.M., Driessen, A.J.M., and Konings, W.N., 1998. The essence of being extremophilic: the role of the unique archaeal membrane lipids. Extremophiles, 2:163–170.

van den Hoek, C.C., Mann, C.D., and Jahns, H.M., 1996. Algae – an introduction to phycology. Cambridge University Press, New York.

van Gemerden, H., 1986. Production of elemental sulfur by green and purple sulfur bacteria. Archives of Microbiology, 146:52–56.

van Gemerden, H., 1993. Microbial mats: a joint venture. Marine Geology, 113:3–25.

Van Kranendonk, M.J., Webb, G.E., and Kamber, B.S., 2003. Geological and trace element evidence for a marine sedimentary environment of deposition and biogenicity of 3.45 Ga stromatolitic carbonates in the Pilbara Craton, and support for a reducing Archaean ocean. Geobiology, 1:91–108.

van Lith, Y., Warthmann, R., Vasconcelos, C., and McKenzie, J.A., 2003a. Sulfate-reducing bacteria induce low-temperature Ca-dolomite and high Mg-calcite formation. Geobiology, 1:71–80.

van Lith, Y., Warthmann, R., Vasconcelos, C., and McKenzie, J.A., 2003b. Microbial fossilization in carbonate sediments: a result of the bacterial surface involvement in dolomite precipitation. Sedimentology, 50:237–245.

van Loosdrecht, M.C.M., Lyklema, J., Norde, W., and Zehnder, A.J.B., 1989. Bacterial adhesion: a physiochemical approach. Microbial Ecology, 17:1–15.

van Loosdrecht, M.C.M., Lyklema, J., Norde, W., and Zehnder, A.J.B., 1990. Influence of interfaces on microbial activity. Microbiological Reviews, 54:75–87.

van Veen, W.L., Mulder, E.G., and Deinema, M.H., 1978. The *Sphaerotilus–Leptothrix* group of bacteria. Microbiology Reviews, 42:329–356.

van Zuilen, M.A., Lepland, A., and Arrhenius, G., 2002. Reassessing the evidence for the earliest traces of life. Nature, 418:627–630.

Vargas, M., Kashefi, K., Blunt-Harris, E.L., and Lovley, D.R., 1998. Microbiological evidence for Fe(III) reduction on early Earth. Nature, 395:65–67.

Vasconcelos, C., McKenzie, J.A., Bernasconi, S., Grujic, D., and Tien, A.J., 1995. Microbial mediation as a possible mechanism for natural dolomite formation at low temperatures. Nature, 377:220–222.

Veizer, J., Hoefs, J., Lowe, D.R., and Thurston, P.C., 1989. Geochemistry of Precambrian carbonates: II. Archean greenstone belts and Archean sea water. Geochimica et Cosmochimica Acta, 53:859–872.

Vellai, T. and Vida, G., 1999. The origin of eukaryotes: the difference between prokaryotic and eukaryotic cells. Proceedings of the Royal Society of London, B, 266:1571–1577.

Vellai, T., Takács, K., and Vida, G., 1998. A new aspect to the origin and evolution of eukaryotes. Journal of Molecular Evolution, 46:499–507.

Verrecchia, E.P., 2000. Fungi and sediments. In: Riding, R.E. and Awramik, S.M. (eds), Microbial Sediments. Springer-Verlag, Berlin, pp. 68–75.

Verrecchia, E.P., Freytet, P., Verrecchia, K.E., and Dumont, J.L., 1995. Spherulites in calcrete laminar crusts – biogenic $CaCO_3$ precipitation as a major contributor to crust formation. Journal of Sedimentary Research, 65:690–700.

Vestal, J.R., 1988. Biomass of the cryptoendolithic microbiota from the Antarctic Desert. Applied and Environmental Microbiology, 54:957–959.

Vidal, G. and Knoll, A.H., 1982. Radiations and extinctions of plankton in the late Proterozoic and early Cambrian. Nature, 297:57–60.

Vincent, W.F. and Roy, S., 1993. Solar ultraviolet-B radiation and aquatic primary production: damage, protection, and recovery. Environmental Reviews, 1:1–12.

Visscher, P.T., Reid, R.P., Bebout, B.M., Hoeft, S.E., Macintyre, I.G., and Thompson, J.A., 1998. Formation of lithified micritic laminae in modern marine stromatolites (Bahamas): the role of sulfur cycling. American Mineralogist, 83:1482–1493.

Volcani, B.E., 1983. Aspects of silicification in biological systems. In: Westbroek, P. and de Jong, E.W. (eds), Biomineralization and Biological Metal Accumulation. D. Reidel, Dordecht, Holland, pp. 389–405.

Volesky, B., 1990. Removal and recovery of heavy metals by biosorption. In: Volesky, B. (ed.), Biosorption of Heavy Metals. CRC Press, Boca Raton, pp. 7–44.

Volesky, B. and Holan, Z.R., 1995. Biosorption of heavy metals. Biotechnology Progress, 11:235–250.

Volesky, B. and May-Phillips, H.A., 1995. Biosorption of heavy metals by Saccharomyces cerevisiae. Applied Microbiology and Biotechnology, 42:797–806.

von Wolzogen Kuehr, C.A.H. and van der Vlugt, I.S., 1934. The graphitization of cast iron as an electrochemical process in anaerobic soils. Water, 18:147–165.

Vreeland, R.H., Rosenzweig, W.D., and Powers, D.W., 2000. Isolation of a 250 million-year-old halotolerant bacterium from a primary salt crystal. Nature, 407:897–900.

Wächtershäuser, G., 1988a. Pyrite formation, the first energy source for life: a hypothesis. Systematic and Applied Microbiology, 10:207–210.

Wächtershäuser, G., 1988b. Before enzymes and templates: theory of surface metabolism. Microbiological Reviews, 52:452–484.

Wächtershäuser, G., 1994. Vitalysts and virulysts: a theory of self-expanding reproduction. In: Bengtson, S. (ed.), Early Life on Earth, Nobel Symposium no. 84. Columbia University Press, New York, pp. 124–132.

Wada, E. and Hattori, A., 1978. Nitrogen isotope effects in the assimilation of inorganic nitrogenous compounds by marine diatoms. Geomicrobiology Journal, 1:85–101.

Wada, E., Kadonaga, T., and Matsuo, S., 1975. ^{15}N in nitrogen of naturally occurring substances and global assessment of denitrification from isotopic viewpoint. Geochemistry Journal, 9:139–148.

Wada, S.-I. and Wada, K., 1980. Formation, composition and structure of hydroxy-aluminosilicate ions. Journal of Soil Science, 31:457–467.

Waelbroeck, C., Labeyrie, L., Michel, E., et al. 2002. Sea-level and deep-water temperature changes derived from benthic foraminifera isotopic records. Quaternary Science Reviews, 21:295–305.

Wagner, M., Roger, A.J., Flax, J.L., Brusseau, G.A., and Stahl, D.A., 1998. Phylogeny of dissimilatory sulfite reductases supports an early origin of sulfate reduction. Journal of Bacteriology, 180:2975–2982.

Wakeham, S.G., Lewis, C.M., Hopmans, E.C., Schouten, S., and Damsté, J.S.S., 2003. Archaea mediate anaerobic oxidation of methane in deep euxinic waters of the Black Sea. Geochimica et Cosmochimica Acta, 67:1359–1374.

Wakeman, S.G. and Lee, C., 1993. Production, transport, and alteration of particulate organic matter in the marine water column. In: Engel, M.H. and Macko, S.A. (eds), Organic Geochemistry. Plenum Press, New York, pp. 145–169.

Walker, J.C.G., 1984. Suboxic diagenesis in banded iron formations. Nature, 309:340–342.

Walker, J.C.G., 1990. Precambrian evolution of the climate system. Palaeogeography, Palaeoclimatology and Palaeoecology, 82:261–289.

Walker, J.C.G. and Brimblecombe, P., 1985. Iron and sulfur in the pre-biologic ocean. Precambrian Research, 28:205–222.

Walker, J.C.G., Hays, P.B., and Kasting, J.F., 1981. A negative feedback mechanism for the long-term stabilization of Earth's surface temperature. Journal of Geophysical Research, 86:9776–9782.

Walker, J.C.G., Klein, C., Schidlowski, M., Schopf, J.W., Stevenson, D.J., and Walter, M.R., 1983. Environmental evolution of the Archean–early Proterozoic Earth. In: Schopf, J.W. (ed.), Earth's Earliest Biosphere, Its Origin and Evolution. Princeton University Press, Princeton, pp. 260–290.

Walker, S.G., Flemming, C.A., Ferris, F.G., Beveridge, T.J., and Bailey, G.W., 1989. Physiochemical interaction of Escherichia coli cell envelopes and Bacillus subtilis cell walls with two clays and ability of the composite to immobilize heavy metals from solution. Applied and Environmental Microbiology, 55:2976–2984.

Walsh, M.M. and Lowe, D.R., 1985. Filamentous microfossils from 3,500-Myr-old Onverwacht Group, Barberton Mountain Land, South Africa. Nature, 314:530–532.

Walter, M.R. 1976a. Hot-spring sediments in Yellowstone National Park. In: Walter, M.R. (ed.), Stromatolites. Elsevier, Amsterdam, pp. 489–498.

Walter, M.R. 1976b. Geyserites of Yellowstone National Park: an example of abiogenic "stromatolites." In: Walter, M.R. (ed.), Stromatolites. Elsevier, Amsterdam, pp. 88–112.

Walter, M.R., 1994. Stromatolites: the main geological source of information on the evolution of the early benthos. In: Bengtson, S. (ed.), Early Life on Earth. Columbia University Press, New York, pp. 270–286.

Walter, M.R., Bauld, J., and Brock, T.D., 1972. Siliceous algal and bacterial stromatolites in hot spring and geyser effluents of Yellowstone National Park. Science, 178:402–405.

Walter, M.R., Buick, R., and Dunlop, J.S.R., 1980. Stromatolites 3,400–3,500 Myr old from the North Pole area, Western Australia. Nature, 284:443–445.

Walter, M.R., Bauld, J., Des Marais, D., and Schopf, J.W., 1992. A general comparison of microbial mats and microbial stromatolites: bridging the gap between the modern and the fossil. In: Schopf, J.W. and Klein, C. (ed.), The Proterozoic Biosphere. Cambridge University Press, Cambridge, pp. 335–338.

Wan, J., Wilson, J.L., and Kieft, T.L., 1994. Influence of the gas–water interface on transport of microorganisms through unsaturated porous media. Applied and Environmental Microbiology, 60:509–516.

Ward, D.M., Weller, R., Shiea, J., Castenholz, R.W., and Cohen, Y., 1989. Hot spring microbial mats: anoxygenic and oxygenic mats of possible evolutionary significance. In: Cohen, Y. and Rosenberg, E. (eds), Microbial Mats: physiological ecology of benthic microbial communities. American Society for Microbiology, Washington, DC, pp. 3–13.

Ward, J.B., 1981. Teichoic and teichuronic acids: biosynthesis, assembly and location. Microbiological Reviews, 45:211–243.

Warren, L.A. and Haack, E.A., 2001. Biogeochemical controls on metal behavior in freshwater environments. Earth Science Reviews, 54:261–320.

Warren, L.A. and Ferris, F.G., 1998. Continuum between sorption and precipitation of Fe(III) on microbial surfaces. Environmental Science and Technology, 32:2331–2337.

Warren, L.A., Maurice, P.A., Parmar, N., and Ferris, F.G., 2001. Microbially mediated calcium carbonate precipitation: implications for interpreting calcite precipitation and for solid phase capture of inorganic contaminants. Geomicrobiology Journal, 18:93–115.

Warthmann, R., van Lith, Y., Vasconcelos, C., McKenzie, J.A., and Karpoff, A.M., 2000. Bacterially induced dolomite precipitation in anoxic culture experiments. Geology, 28:1091–1094.

Watanabe, Y., Martini, J.E.J., and Ohmoto, H., 2000. Geochemical evidence for terrestrial ecosystems 2.6 billion years ago. Nature, 408:574–578.

Waterbury, J.B., Watson, S.W., Guillard, R.R.L., and Brand, L.E., 1979. Widespread occurrence of a unicellular, marine, planktonic cyanobacterium. Nature, 277:293–294.

Weckesser, J., Hofmann, K., Jürgens, U.J., Whitton, B.A., and Raffelsberger, B., 1988. Isolation and chemical analysis of the sheaths of the filamentous cyanobacteria Calothrix parienta and C. scopulorum. Journal of General Microbiology, 134:629–634.

Wefer, G., Suess, E., Balzer, W., et al. 1982. Fluxes of biogenic components from sediment trap deployment in circumpolar waters of the Drake Passage. Nature, 299:145–147.

Weiss, B.P., Kirschvink, J.L., Baudenbacher, F.J., et al. 2000. A low temperature transfer of ALH84001 from Mars to Earth. Science, 290:791–795.

Weiss, T.H., Mills, A.L., Hornberger, G.M., and Herman, J.S., 1995. Effect of bacterial cell shape on transport of bacteria in porous media. Environmental Science and Technology, 29:1737–1740.

Welch, S.A. and Banfield, J.F., 2002. Modification of olivine surface morphology and reactivity by microbial activity during chemical weathering. Geochimica et Cosmochimica Acta, 66:213–221.

Welch, S.A. and Ullman, W.J., 1993. The effect of organic acids on plagioclase dissolution rates and stoichiometry. Geochimica et Cosmochimica Acta, 57:2725–2736.

Welch, S.A. and Ullman, W.J., 1996. Feldspar dissolution in acidic and organic solutions: compositional and pH dependence of dissolution rates. Geochimica et Cosmochimica Acta, 60:2939–2948.

Welch, S.A., Barker, W.W., and Banfield, J.F., 1999. Microbial extracellular polymers and plagioclase dissolution. Geochimica et Cosmochimica Acta, 63:1405–1419.

Wen, J-S., Pinto, J.P., and Yung, Y.L., 1989. Photochemistry of CO and H_2O: analysis of laboratory experiments and applications to the prebiotic Earth's atmosphere. Journal of Geophysical Research, 94:14957–14970.

Westall, F., 1997. The influence of cell wall composition on the fossilization of bacteria and the implications for the search for early life forms. In: Cosmvici, C., Bowyer, S., and Werthimer, D. (eds), Astronomical and Biochemical Origins and the Search for Life in the Universe. Editori Compositrici, Bologna, pp. 491–504.

Westall, F., Steele, A., Toporski, J., et al. 2000. Polymeric substances and biofilms as biomarkers in terrestrial materials: implications for extraterrestrial samples. Journal of Geophysical Research, 105:24511–24527.

Westbroek, P., Brown, C.W., van Bleijswijk, J., et al. 1993. A model system approach to biological climate forcing. The example of *Emiliania huxleyi*. Global and Planetary Change, 8:27–46.

Westrich, J.T. and Berner, R.A., 1984. The role of sedimentary organic matter in bacterial sulfate reduction: the G model tested. Limnology and Oceanography, 29:236–249.

White, D.C., 1993. *In situ* measurement of microbial biomass, community structure and nutritional status. Philosophical Transactions of the Royal Society of London. 344:59–67.

Whitfield, C., 1988. Bacterial extracellular polymers. Canadian Journal of Microbiology, 34:415–420.

Whitman, W.B., Coleman, D.C., and Wiebe, W.J., 1998. Prokaryotes: the unseen majority. Proceedings of the National Academy of Sciences, USA, 95:6578–6583.

Whittaker, R.H., 1969. New concepts of kingdoms of organisms. Science, 163:150–160.

Widdel, F., 1988. Microbiology and ecology of sulfate- and sulfur reducing bacteria. In: Zehnder, A.J.B. (ed.), Biology of Anaerobic Microorganisms. John Wiley, New York, pp. 469–585.

Widdel, F., Schnell, S., Heising, S., Ehrenreich, A., Assmus, B., and Schink, B., 1993. Ferrous iron oxidation by anoxygenic phototrophic bacteria. Nature, 362:834–836.

Wieder, R.K., 1993. Ion input-output budgets for five wetlands constructed for acid coal mine drainage treatment. Water, Air and Soil Pollution, 71:231–270.

Wiegert, R.G. and Fraleigh, P.C., 1972. Ecology of Yellowstone thermal effluent systems: net primary production and species diversity of a successional blue-green algal mat. Limnology and Oceanography, 17:215–228.

Wikjord, A.G., Rummery, T.E., Doern, F.E., and Owen, D.G., 1980. Corrosion and deposition during the exposure of carbon steel to hydrogen sulfide-water solutions. Corrosion Science, 20:651–671.

Wilde, S.A., Valley, J.W., Peck, W.H., and Graham, C.M., 2001. Evidence from detrital zircons for the existence of continental crust and oceans on the Earth 4.4 Gyr ago. Nature, 409:175–178.

Wilhelm, S. and Trick, C.G., 1994. Iron-limited growth of cyanobacteria: multiple siderophore production is a common response. Limnology and Oceanography, 39:1979–1984.

Wilkin, R.T. and Barnes, H.L., 1997. Pyrite formation in an anoxic estuarine basin. American Journal of Science, 297:620–650.

Wilkin, R.T., Barnes, H.L., and Brantley, S.L., 1996. The size distribution of framboidal pyrite in modern sediments: an indicator of redox conditions. Geochimica et Cosmochimica Acta, 60:3897–3912.

Williams, L.A., Parks, G.A., and Crerar, D.A., 1985. Silica diagenesis, 1. Solubility controls. Journal of Sedimentary Petrology, 55:301–311.

Williams, R.J.P., 1953. Metal ions in biological systems. Biology Reviews, 28:381–415.

Williams, R.J.P., 1981. Physico-chemical aspects of inorganic element transfer through membranes. Philosophical Transactions of the Royal Society of London, B294:57–74.

Wilson, W.W., Wade, M.M., Holman, S.C., and Champlin, F.R., 2001. Status of methods for assessing bacterial cell surface charge properties based on zeta potential measurements. Journal of Microbiological Methods, 43:153–164.

Windley, B.F., 1984. The Evolving Continents. Wiley, New York.

Winklemann, G., 1991. Specificity of iron transport in bacteria and fungi. In: Winklemann, G. (ed.), CRC Handbook of Microbial Iron Chelates. CRC Press, Boca Raton, pp. 65–105.

Woese, C.R., 1987. Bacterial evolution. Microbiological Reviews, 51:221–271.

Woese, C.R., 1998. The universal ancestor. Proceedings of the National Academy of Sciences, USA, 95:6854–6859.

Woese, C.R. and Fox, G.E., 1977. Phylogenetic structure of the prokaryotic domain: the primary kingdoms. Proceedings of the National Academy of Sciences, USA, 74:5088–5090.

Woese, C.R., Kandler, O., and Wheelis, M.L., 1990. Towards a natural system of organisms: proposal for the domains Archaea, Bacteria and Eukarya. Proceedings of the National Academy of Sciences, USA, 87:4576–4579.

Wogelius, R.A. and Walther, J.V., 1991. Olivine dissolution at 25°C. Effects of pH, CO_2, and organic acids. Geochimica et Cosmochimica Acta, 55:943–954.

Wolery, T.J. and Sleep, N.H., 1976. Hydrothermal circulation and geochemical flux at mid-ocean ridges. Journal of Geology, 84:249–275.

Wolfaardt, G.M., Lawrence, J.R., and Korber, D.R., 1999. Function of EPS. In: Wingender, J., Neu, T.R., and Flemming, H.C. (eds), Microbial Extracellular Polymeric Substances: characterization, structure, and function. Springer-Verlag, Berlin, pp. 171–200.

Wolfe, R.S., 1972. Microbial formation of methane. Advances in Microbial Physiology, 6:107–146.

Wollast, R., 1993. Interactions of carbon and nitrogen cycles in the coastal zone. In: Wollast, R., Chou, L., and Mackenzie, F. (eds), Interactions of C, N, P and S Biogeochemical Cycles and Global Change. Springer-Verlag, Berlin, pp. 198–210.

Wood, P.M., 1988. Chemolithotrophy. In: Anthony, C. (ed.), Bacterial Energy Transduction. Academic Press, London, pp. 183–230.

Wright, D.T., 1999. The role of sulfate-reducing bacteria and cyanobacteria in dolomite formation in distal ephemeral lakes of the Coorong region, South Australia. Sedimentary Geology, 126:147–157.

Xiong, J., Fischer, W.M., Inoue, K., Nakahara, M., and Bauer, C.E., 2000. Molecular evidence for the early evolution of photosynthesis. Science, 289:1724–1730.

Xu, Y. and Schoonen, M.A.A., 1995. The stability of thiosulfate in the presence of pyrite in low-temperature aqueous solutions. Geochimica et Cosmochimica Acta, 59:4605–4622.

Xue, H., Oestreich, A., Kistler, D., and Sigg, L., 1996. Free cupric ion concentrations and Cu complexation in selected Swiss Lakes. Aquatic Sciences, 58:69–87.

Yamaguchi, K.E., Johnson, C.M., Beard, B.L., and Ohmoto, H., 2005. Biogeochemical cycling of iron in the Archean-Paleoproterozoic Earth: constraints from iron isotope variations in sedimentary rocks from the Kaapvaal and Pilbara Cratons. Chemical Geology, 218:135–169.

Yang, J. and Volesky, B., 1999. Biosorption of uranium on Sargassum biomass. Water Research, 33:3357–3363.

Yee, N. and Fein, J.B., 2001. Cd adsorption onto bacterial surfaces: a universal adsorption edge? Geochimica et Cosmochimica Acta, 65:2037–2042.

Yee, N. and Fein, J.B., 2002. Does metal adsorption onto bacterial surfaces inhibit or enhance aqueous metal transport? Column and batch reactor experiments on Cd–Bacillus subtilis–quartz systems. Chemical Geology, 185:303–319.

Yee, N., Fein, J.B., and Daughney, C.J., 2000. Experimental study of the pH, ionic strength, and reversibility behavior of bacteria-mineral adsorption. Geochimica et Cosmochimica Acta, 64:609–617.

Yee, N., Phoenix, V.R., Konhauser, K.O., and Benning, L.G., 2003. The effect of bacteria on SiO_2 precipitation at neutral pH: implications for bacterial silicification in geothermal hot springs. Chemical Geology, 199:83–90.

Yee, N., Fowle, D.A., and Ferris, F.G., 2004. A Donnan potential model for metal sorption onto Bacillus subtilis. Geochimica et Cosmochimica Acta, 68:3657–3664.

Young, G.M., von Brunn, V., Gold, D.J.C., and Minter, W.E.L., 1998. Earth's oldest reported glaciation: physical and chemical evidence from the Archean Mozaan Group (~2.9 Ga) of South Africa. Journal of Geology, 106:523–538.

Young, J.P.W., 1992. Phylogenetic classification of nitrogen-fixing organisms. In: Stacey, G., Evans, H.J., and Burns, R.H. (eds), Biological Nitrogen Fixation. Chapman and Hall, New York, pp. 43–86.

Zachara, J.M., Kukkadapu, R.K., Fredrickson, J.K., Gorby, Y.A., and Smith, S.C., 2002. Biomineralization of poorly crystalline Fe(III) oxides by dissimilatory metal reducing bacteria (DMRB). Geomicrobiology Journal, 19:179–207.

Zahnle, K.J., 1986. Photochemistry of methane and the formation of hydrocyanic acid (HCN) in the Earth's early atmosphere. Journal of Geophysical Research, 91:2819–2834.

Zamaraev, K.I., Romannikov, V.N., Salganik, R.I., Wlassoff, W.A., and Khramtsov, V.V., 1997. Modelling of the prebiotic synthesis of oligopeptides: silicate catalysts help to overcome the critical stage. Origins of Life and Evolution of the Biosphere, 27:325–337.

Zavarin, G.A., 1993. Epicontinental soda lakes as probable relict biotopes of terrestrial biota formation. Microbiology, 62:473–479.

Zettler, L.A.A., Gómez, F., Zettler, E., et al., 2002. Eukaryotic diversity in Spain's River of Fire. Nature, 417:137.

Zhang, J., Lion, L.W., Nelson, Y.M., Shuler, M.L., and Ghiorse, W.C., 2002. Kinetics of Mn(II) oxidation by Leptothrix discophora SS1. Geochimica et Cosmochimica Acta, 65:773–781.

Zierenberg, R.A. and Schiffman, P., 1990. Microbial control of silver mineralization at a sea-floor hydrothermal site on the northern Gorda Ridge. Nature, 348:155–157.

ZoBell, C.E., 1952. Part played by bacteria in petroleum formation. Journal of Sedimentary Petrology, 22:42–49.

Zuckerkandl, E. and Pauling, L., 1965. Molecules as documents of evolutionary history. Journal of Theoretical Biology, 8:357–366.

Index

Page references in *italics* refer to figures.

Printed and bound by CPI Group (UK) Ltd, Croydon, CR0 4YY

Printed and bound by CPI Group (UK) Ltd, Croydon, CR0 4YY

27/10/2024

14580393-0003